Liberal Dreams and Nature's Limits
Great Cities of North America Since 1600

James T. Lemon

D1141815

Oxford University Press
Toronto New York Oxford
1996

Oxford University Press
70 Wynford Drive, Don Mills, Ontario M3C 1J9

Oxford   New York
Athens   Auckland   Bangkok   Bombay
Calcutta   Cape Town   Dar es Salaam   Delhi
Florence   Hong Kong   Istanbul   Karachi
Kuala Lumpur   Madras   Madrid   Melbourne
Mexico City   Nairobi   Paris   Singapore
Taipei   Tokyo   Toronto

and associated companies in
Berlin   Ibadan

Oxford is a trade mark of Oxford University Press

Canadian Cataloguing in Publication Data

Lemon, James T.
  Liberal dreams and nature's limits : great cities of North America since 1600

Includes bibliographical references and index. ISBN 0-19-540793-8

1. Cities and towns—North America—History. I. Title.

HT122.L45 1996   307.76'4'097   C96-931733-6

Page xii constitutes an extension of the copyright page.

Design: Brett Miller
Cover illustration: François Chartier
Formatting: Linda Mackey
Cartography: free&creative

To Margaret, Janet, and Catherine

CONTENTS

∎

LIST OF FIGURES

■

PREFACE

■

After 1600 Europeans sought to fulfil their dreams of opportunity in North America. For 300 years, the agricultural frontier beckoned them. Increasingly after 1800 they sought improvement in cities that grew up with the technological frontier, the vast outpouring of ingenuity that unravelled the mysteries of nature and applied them to create material progress. The liberal dream is the American dream: people could pursue material riches if they have the will to succeed, and they should because it is virtuous to do so.[1] Although obviously not every American followed Benjamin Franklin's admonitions, many leaders and many ordinary people, probably a majority, have deferred to these tenets. Canadians, meanwhile, have not been quite as able to enjoy this luxurious belief, though they have been close to embracing it. The term liberal, I realize, is confusing: the most fervent liberals in America call themselves conservatives. They defend most tenaciously the liberal ideology. They call those who are less committed, those more inclined to defend the collective public interest, liberals. Yet, conversely and historically, one could argue that those with the most legitimate claim on the word conservative are those concerned about conserving community and environment.

Today the modern world of historically new means of production seems now to have reached its limits. Nature (as it were) has imposed a brake to further advancement in the qualities that defined progress—goods and services that provided speed, convenience, and comfort to a wide range of people. The limits of urbanization were reached about 1975. Thus, today North Americans and others in the Western world face a new frontier. That frontier is social: how to maintain civility and equity in an economically stagnant world.

There are, of course, strong voices of the expansionist growth culture that has dominated our past, voices that promise a marvellous future through the information highway. I am sceptical of this claim of futurists. What is being promised is insubstantial, a kind of ersatz spiritual experience on the Internet or watching hundreds of channels. Since, like everyone else, I am a poor predictor of the future, I could be wrong, but, lest we forget, we are still earthbound creatures with feet of clay.

In the past cities have been the centralizing locus of organization for society. The early colonial cities expedited settlement, communication, trade, and finance. Then later they became the major locales for the rising tide of manufacturing. Recently, however, manufacturing began to recede from cities, and employment has become divided between the privileged few and part-time service work for a rising tide of the masses. One is tempted to take the view that big cities are no longer generative but parasitic like ancient Rome. Financial activities seem to be the only action of importance to the powerful in these huge cities, which the media insist are engines of growth.

Even in the past, cities were never engines of growth in themselves. They could not be generative without something to generate. To be useful to society, their residents have always needed the resources of hinterlands, if not in their own nation state then elsewhere in the world. Nature has been the source of wealth from those hinterlands.

This book looks at the present condition of life within large cities and then how North Americans, through business, government, and other organizations, brought us to the current situation. I focus on five large cities at various times in the past in my attempt to understand how American and Canadian societies work. Although Sam Bass Warner, Jr, has dealt with four of the five, and I agree with much of what he says, I present a basically different interpretation, namely that the future cannot lie in liberal democracy as we have known it.[2] While preserving personal rights, North Americans have to move towards social democracy in the workplace and in other aspects of our common life. The 'creative destruction' of the economic system we have lived under has become just too enervating for people and too despoiling of the environment.[3] The Great Depression came close to galvanizing North Americans to face the new social frontier. Then war and new technologies put off the inevitable. Now citizens and their leaders have to act to maintain a sustainable future for all people.

As a Canadian, I have long been interested in the workings of American society, first through my interpretation of early Pennsylvania. I have taught courses on urban development and American society generally. In one sense as a non-American, but I believe not as an anti-American, I have attempted to respond to historian Warren Susman's lament about America that 'we have contributed to the making of mass culture without developing any position from which to evaluate it'.[4] From a position on the margins in a state with only one person to every nine Americans, it is perhaps possible to cast a critical eye on American society. What the giant United States does profoundly influences public policy and personal behaviour in this smaller country.

This study is also a critique of Canada. Certainly until 1989, Canadians, and especially I and my fellow Torontonians, could be pleased, perhaps we were smug, that life in our inner city was far different from that in most large American cities, not least those nearby on the Great Lakes.

One reason is that Canada did better economically than the United States. Over the three decades after 1960, comparative gross domestic product data indicate that Canada enjoyed nearly 50 per cent more growth per person than the US, which grew at the second slowest pace among twenty-two Western countries. I will argue too that Canadians managed their cities better partly because they were willing to pay higher taxes and partly because they held lower than utopian expectations. If the lower pitch of patriotic fervour in Canada is any indication, Canadians are less preoccupied with their nationhood, notwithstanding the issue of many Quebecois' desire for independence. Americans' more flamboyant patriotic displays suggest that they seem less confident about their place in the world. But the 1990s has revealed to unbelieving Canadians that the party is over, that the limits to growth are upon us.

In the five central chapters on particular cities, the organization is similar. I describe the national and regional demographic and economic patterns preceding each of the dates specified. Then I consider the local economy and society, the management of public resources, land development, and finally the local geography—central area, manufacturing, circulation, and residential districts. This study is based primarily on secondary sources, except for the period since 1945, in which I have obviously used many more contemporary sources.

I must thank a host of people from among my own colleagues and local friends in academe, politics, and local life, with whom I have had many intense debates; also those with whom I have engaged at seminars in a variety of places, and with whom I have had brief but insightful encounters. I hope I have struck an Aristotelian balance between structure and agency. Many are cited in the bibliography. Specifically I am grateful to those who undertook research for this study: William Holowacz, John Bacher, Norah Johnson, and especially Douglas Browne. I am indebted to those who read various chapters: Larry Bourne, Mark Egnal, Joseph Ernst, Frances Frisken, Richard Harris, Gary Nash, Peter Rees, and Kenneth Scherzer. I recall with pleasure a splendid outing one Sunday in Chicago with Florence and the late Harold Mayer. None of the above are responsible for the outcome. Special thanks go to James Simmons, Carolyn Lemon, two anonymous readers, and, not least, to Valerie Ahwee at Oxford University Press. Besides others at the Press, I would also like to thank two who were formerly there: Richard Teleky, who long ago signed me up to write this book, and Brian Henderson, who encouraged me.

Finally, I acknowledge the support of the Guggenheim Foundation in 1974–5, the Canada Council, the Social Sciences and Humanities Research Council, and those in American Studies at Flinders University in Adelaide and the Geography Department in Wellington for my visits in 1988–9. I regret that virtually all the wonderful ideas about cities I absorbed in Britain in 1975, and Down Under more recently, have not found a place in this book. But they helped me think about Canada and the United States, two other offspring of Britain. One needs to add that they were also progeny of France, west Africa, Spain, and other regions and countries too. The colonizers were, of course, indebted to First Nations people.

J.T.L.
March 1996

ACKNOWLEDGEMENTS

This is an extension of the copyright page. The following previously published materials have been reproduced with permission:

James R. Beniger. Extract reprinted by permission of the publisher from *The Control Revolution: Technological and Economic Origins of the Information Society* by James R. Beniger, Cambridge, Mass.: Harvard University Press, Copyright © 1986 by the President and Fellows of Harvard College.

Winston Crouch and Beatrice Dinerman. Figures 6.1 and 6.2 are reprinted from *Southern California Metropolis: A Study in Development of Government for a Metropolitan Area* by Winston Crouch and Beatrice Dinerman. Copyright © 1963 The Regents of the University of California. Reprinted by permission of University of California Press.

Herman E. Daly and John B. Cobb, Jr. Extract from *For the Common Good* by Herman E. Daly and John B. Cobb, Jr. Copyright 1989, 1994 by Herman E. Daly and John B. Cobb, Jr. Reprinted by permission of Beacon Press.

Eshref Shevky and Marilyn Williams. Figure 6.6 is reprinted from *Social Areas of Los Angeles: Analysis and Typology* by Eshref Shevky and Marilyn Williams. Copyright © 1949 The Regents of the University of California. Reprinted by permission of University of California Press.

Dylan Thomas. Quote from 'Do Not Go Gentle into That Good Night' from *The Poems of Dylan Thomas* by Dylan Thomas (J.M. Dent). Reprinted by permission of David Higham Associates. Copyright © 1952 by Dylan Thomas. Reprinted by permission of New Directions Publishing Corp.

CHAPTER 1

■

Today's Stagnating Metropolises: Unravelling in Cyperspace?

Late in 1994 *Fortune* proclaimed its list of 'the world's best cities for business'.[1] Soon after, Corporate Resources Group (CRG) offered its list, ranking cities by 'quality of life'.[2] *Fortune*'s consultants, Moran Stahl & Boyer (MSB), surveyed 500 corporate executives in thirty-two countries. From a final list of sixty cities, MSB analysed business-support programs, market-growth potential, and cultural support of business to pick the top ten cities. CRG measures of 118 cities were more clearly focused on social than business issues as its categories included level of security, public services, medical and health structures, and political and social stability. In MSB's top ten are five American cities, one Canadian, two western European, and two Asian. On CRG's listing of the leading ten are five western European, three Canadian, and two Pacific cities. Of *Fortune*'s sixty cities, half are American; of CRG's, the leading American city, Boston, appears only in the thirtieth spot. Obviously, one reason for the divergence is that *Fortune* is American and CRG is western European, in fact, it is a consultancy based in its number-one city, Geneva. Boosterism is at work in both cases.

Not surprisingly in today's world, business and social assessments mostly fail to match; the only two in the top ten on both lists are Toronto and Singapore because, in spite of itself, *Fortune* does, on closer reading, consider quality of life as important for including those two, but not so much for its other leading eight cities. Needless to say, Toronto and Singapore are two quite different places, though a careful observer might note that

they both help world business. Are these rankings worth anything? Canadians and western Europeans, not only Americans who would question the Geneva rankings, should pause to consider their value. That Boston ranks only thirtieth points to widely held judgement that American cities are perpetually in crisis. However, most Americans no longer pursue their dreams in central cities but on the suburban frontier, so for most of them the urban crisis is largely out of mind. They seldom visit cities anyway. Vancouver's second position in the CRG listing points to its liveable inner city. Still, Canadian cities too are weakened, caught in the throes of economic stagnation that, despite appearances, has persisted for two decades. They are also caught in an electronic upheaval that conceivably could end their traditional organizing role in society. Not only that, since 1975 the gap between rich and poor in North America has widened. This is in sharp contrast to greater social equity over the previous three decades. As we will see in later chapters, American cities, if not Canadian, have been troubled places, 'boiling pots' rather than melting-pots over the last century.[3] Urban realities have overcome the dream that has brought migrants from Europe since 1600.

Stagnation and Dissipation

Do rankings like these denote world-class status? The media in many cities perpetuated this notion in the 1980s when finance was on a speculative binge. During the depression of the early 1990s, such talk became muted as cities seemed to drift with a less clear purpose. Few North American city publicists are now as boastful as they were a few years ago. Atlanta, fourth in the *Fortune* league, claimed status by hosting the 1996 Olympics and serving as the southeastern hub, but it is widely known as among the worst in terms of income distribution between central city and suburbs. Vancouver can boast about its trade connections to the Pacific realm and the massive influx of Hong Kong and other Asian money, yet its great regional resource base is weakening. Miami (tenth in the MSB ratings) is the 'Hong Kong of Latin America', but Latin America seems very shaky.[4] Certainly San Francisco, eighth in the MSB, still has great drawing power, yet nearby San Jose's Silicon Valley's economy is slowing down. Utah's Software Valley has been up and coming, as does Silicon Valley North in Kanata near Ottawa, but these are minor players. The energy cities—Dallas, Houston, Calgary, and Edmonton—looked like winners before the mid-1980s' collapse in oil prices. Recent rising prices have again put them on the upswing. Washington and Ottawa and all of the state and provincial capitals, which have been so important in holding together the economy through spending to speed up consumption since the 1930s, are hamstrung financially and nearly gridlocked politically. Governments have joined the downsizing parade.

Consumption-based cities like those catering to entertainment, seem to shine, like Orlando with its Disney World, or even old more and more versatile Montreal (seventh on the CRG list) and New Orleans. A new star, Branson, Missouri, has nosed out Washington and Toronto and even that other country music mecca, Nashville, as the 'Best All-Around City' for motor coach operators.[5] Then too, not to be forgotten, are the many retirement communities in the sun. Maybe one should take a step further to argue that the most important places are the ones of total dissipation, those with casinos. 'Las Vegas exists because it is a perfect

> 'Las Vegas exists because it is a perfect reflection of America.'

reflection of America,' claims one of its dominant hucksters.[6] Does he have a point? That city has doubled its population in a decade; two-thirds of its workers are in the gambling 'industry' that was underground not so long ago. Las Vegas's success is being emulated in many states and provinces, from cities to Mississippi showboats to Indian reservations, often under the guise of local economic development, thus receiving government help. Americans, mostly those with modest incomes, bet $500 billion a year, while the benefits flow to a few.

The five great cities of the past in this book are in trouble today. In order of current metropolitan population, they are New York, Los Angeles, Chicago, Philadelphia, and Toronto. The dominant investment and therefore organizing centre of the continent since the early nineteenth century, New York, is less effective than it was earlier. Its metropolitan regional population has grown to a gargantuan 18 million. After sliding for several decades, its share of national income actually rose from 1985 to 1991, but the 'Capital of the American Century' has lost a great deal of its influence, even if it is still billed with London and Tokyo as one of the global cities, and second to Hong Kong on *Fortune*'s tally.[7] No doubt over the past two decades, Wall Street has recaptured from Washington some of the economic power it held before 1929; then, in the wake of the Great Crash, the government had to step in as never before, so 'the formalization of financial policy shifted control to the Federal Reserve'.[8] The return of economic power to Wall Street is evidenced by government debts rising more rapidly than private corporate debts, which are in turn the result of lower taxes (that began as early as the mid-1960s) on the affluent and on corporations. Today, bond and currency dealers dominate federal reserve and White House strategy, but they are located in Japan and Europe as well as on Wall Street. Permanent power seems shaky.

Beyond finance, the fabric of New York society has become frayed. In the three years after 1989, the city has lost 342,000 jobs and only recently regained a small proportion, clearly an indication of decay. Manhattan lost half of these jobs. The massive lay-offs of manufacturing workers are well known. Significant too has been the reduction of Manhattan's role as America's front office. Of 750 company headquarters, New York's region in 1969 hosted one in four, in 1979 one in six, in 1989 only one in nine—less than half that of a generation ago. Although it still has the largest share of multinational, entertainment, and media headquarters, and still attracts tourists to shop and play, much of its economic power has been dispersed elsewhere to Texas and Japan. On CRG's ranking by quality of life, New York is only forty-fourth. If 'communities of money and capital are communities without propinquity', the centre cannot hold.[9]

Los Angeles, the model city of the mid-twentieth century, has, to some, become the 'capital of the third world'.[10] It does not even make *Fortune's* top ten. Still, it is the place to go to seek success in screenwriting and acting; the land of make-believe has been attracting the ambitious since 1910. Entertainment is the focus of its apparent modest recovery, but in a recent year actors made the grand average sum of $12,500 a year; few are lucky to make it into the assembly of stars. Aerospace jobs were cut in half after 1988. Between 1985 and 1991 in this metropolitan area of nearly 15 million, the per capita personal income fell from 116 per cent of the American average to 108 per cent as high-income earners became swamped by the flows of poor Hispanics, who comprise well over a third of the 14 million people in the region. In competition with the poor of the Third World, they provide cheap labour to the booming clothing industry. The Asian share of the population (now one in ten after San Francisco, sharing second place with Vancouver and Toronto among the big cities) has not overcome this influx either. The Rodney King affair in 1992 showed that the problems the Watts riot made conspicuous are still boiling. In a world with little of interest, the O.J. Simpson affair riveted the attention of the idle and dissolute. These now, it seems, compose a majority, rather than a minority vilified in nineteenth-century lexicons of immoral behaviour.

The great shock city of the nineteenth century, Chicago, remains formidable and was fifth on the *Fortune* scale. Still, although in 1993 *The Economist* claimed that Chicago made a 'great rebound', soon after the city worried about a 'bleak future' for its Board of Trade and Mercantile Exchange, the central financial institutions.[11] Ten years ago, one of these large futures-marketing firms did all of its business in the United States; now only half of that number do so. The incomes of futures traders have fallen sharply. Even a new electronic trading system has yet to pay off, if it ever will. One should also note London's failure in a similar attempt to computerize its stock market system after the big bang. Office rents there and everywhere fell dramatically after 1989, reflecting vacancies of up to 25 per cent. The top end of the economy was overextended during the speculative 1980s. Besides, Chicago has lost enormous numbers of manufacturing plants and jobs. Although the region has slowly grown to over 8 million, the central-city population continued to fall in the 1980s, dropping a quarter since 1950.

Philadelphia's metropolitan area population is now estimated at nearly 6 million people, 300 times the number in 1760. Like Chicago and New York, Philadelphia's regional share of national per capita income rose in the six years after 1985. In 1994 it hosted nearly 500 foreign firms, but this cannot hide the enormous decline in manufacturing employment in a city that depended on industry more than other large cities. Nor can these hide the disparities in income and less taxes between the central city and the suburbs that have grown substantially since 1950. No doubt the still-revered Ben Franklin would encourage hard work as the way to wealth, and many today would still believe him. Like the other cities and Camden, its small neighbour across the Delaware, Philadelphia attracts tourists to its waterfront and heritage sites, but it has some of the worst slums. The opening of its suburbs to low-income housing has been difficult. Philadelphia lost a quarter of its population after 1950, if not as dramatic a fall as some others.

Even Toronto, which has been fiscally and socially the healthiest of our five great cities, has finally felt the bite of a stagnating economy. This late twentieth-century alternative to Los Angeles still remains liveable, but is less ebullient. Its central role as the financial power in Canada helped to increase its status until 1989 to make it a very affluent metropolitan area, one of the richest on the continent on average, at the top in equity. So immigrants came in abundance, and it is second only to Miami in its proportion of foreign born, nearly two in five in 1994, who helped to increase its population to nearly 5 million (on the American metropolitan measurement scale). However, unemployment jumped from 4 to 11 per cent after 1989, a level that was unthinkable during the financially driven decade just before. In the modest recovery after 1993, while industrial cities beyond its immediate region regained employment, largely because of auto production, Toronto lagged behind. This suggests a major structural shift: its diversified economy cannot cushion the blows of the world economy. Most visible are the vacant industrial lands. To be sure, unlike American cities, Toronto does not have slums and abandoned housing. Its inner-city population began to increase again after 1980, in contrast to nearly all large American cities. The gloom has deepened. The government of Ontario has cut welfare payments and other social programs. Many ordinary people believe that Toronto's glory days are over.

Decentralization and Social Polarization within Cities

Some voices celebrate the *new* American city with its 'galactic' cities—an array of new stars shining around the edges of the dying giants.[12] They have been called 'stealth' cities because they lack local governments.[13] The more banal term 'edge cities' has, however, captured the imagination of the business and academic world, probably because a journalist invented and publicized it.[14] These 'centres' are new downtowns of sorts, with postmodern offices to house corporate headquarters and insurance companies, spacious industrial research laboratories in some cases, huge shopping malls, hotels, entertainment attractions, and upscale condominium apartments, surrounded by low-density

housing for the upper third of the income scale. Corporate executives living on their nearby estates often move their head offices to the edge city. A short journey to work is still important to them, as it was for owners of old perched in their grand houses on the hill overlooking their mills and workers' villages. As a result, lower-paid workers are inconvenienced; public transit is usually absent. Cleaning people are compelled to drive long distances. Public spaces are usually nil. Tyson's Corner, south of Washington and a rural crossroads hamlet not long ago, is the most publicized edge city. Between Dulles Airport and the Pentagon, and of course the Capitol, it seems well located for the legions of corporate lobbyists, or at least the supporting cast for the K Street 'political influence brokers', who practise the highest art in this stagnating information age.[15] Tyson's Corner rose in the 1970s at a time of 'intensifying corporate activism' in seeking government largesse.[16]

In a sense these edge cities are not new: they have been a long time coming. In fact, some activities have been suburban or exurban from the beginning of settlement. Satellite industrial communities appeared in early times. Retailing sprang up in new suburbs by 1800. Before 1900 commercial strips emerged along tram lines, then later with branch banks at main intersections. The trend accelerated when department stores and insurance companies moved outward from downtowns in the 1920s, followed after 1945 by shopping malls and outlying office clusters, some adjacent to airports. The process of decentralization culminated in concentrated new centres. By the 1970s, as will be noted in Chapter 7, developers, planners, and politicians saw the virtue of concentrated decentralization. In Houston developers put up a dozen major ones; much more prudent Toronto established six. Without question, these complexes have, to many commentators, become such formidable competitors to the old downtowns that the latter floundered, especially in the United States.

However, the original centres have not been entirely lost. If downtown retailing is only a shadow of its former fame, in most big cities with financial strength, the commanding heights of finance remain downtown. Indeed, developers like

Olympia and York were busy putting up new downtown office buildings in the 1980s. Those dealing with liquid money on a moment's notice and on a face-to-face basis remain there, even with thinned ranks, after the 1987 crash. What the new clusters provide is space for those companies that do not need to be downtown, or for the back offices of financial institutions themselves, though some of the routine data processing operations are now scattered thither and yon to South Dakota, New Brunswick, Ireland, or even India. Telecommuting has picked up, so not everyone has to go to the office.

This latter change could be an ominous portent for those who still advocate metropolitan concentration. Even more startling is computer trading of stocks on screens. If the stock market is no longer a place of intense personal action, then what is left for a financial core? That leaves the big banks. If they give up face-to-face action, the financial system as we know it could be finished. If the key players are scattered, the energy will be lost. One commentator recently predicted that 'New York's once-bustling financial district will become a litter-strewn ghost canyon'.[17] During the 1980s, the overbuilding of offices both in the dominant centre and on the edges led to vacancy rates of over 20 per cent, reminiscent of the 1930s. Although the stagnant economy has shown no assured signs of renewal, the great world banks remain downtown, at least in some major financial cities.

The downtown has kept some other dominant functions that will not go away. Local government power, even if greatly weakened, remains in city halls that still try to provide services. The downtown remains the premier high-culture locus, as in new structures along Broad Street in Philadelphia. In Los Angeles, culture is a legacy of urban renewal decisions in the 1950s by the downtown élite. The Los Angeles Music Center hosts the Oscar ceremonies. Even those with the weakest financial centres (Detroit, St Louis, and others in the Middle

> **If the stock market is no longer a place of intense personal action, then what is left for a financial core?**

West) retain their cultural establishments, though Tiger Stadium, Cobo Hall, Joe Louis Arena, and Greektown alongside the Renaissance Center provide a pretty narrow base for any kind of serious recovery. Cleveland has attempted to encourage retail in a new development at the Terminal Tower. In Pittsburgh, the Golden Triangle hosts most of its metropolitan office space. Then too, retail and cultural developments on waterfronts that cater to tourists (though overhyped in the 1980s) have created some stability in central areas. The downtown élites, a continually changing group of players, have not given up trying to maintain development. As discussed in Chapter 7, the costs in American cities have, however, been borne by the poor. The core has not been stripped of other functions as dramatically as office work. Although overbuilt with offices in the 1980s as elsewhere, Canadian downtowns are less beleaguered than American ones because far more people of diverse incomes live close by.

Even so, income disparities have sharpened between Canadian cities and their suburbs, if less so than in American metropolitan areas. Central-city household incomes relative to those in the metropolitan areas have not changed dramatically; in some cases since 1970, this indicates stability; the City of Toronto's share as of 1990 was still almost dead on 90 per cent, even though it has the greatest range of incomes in the region and a very high proportion of single people (most counted as households) with very modest incomes. In fact, Toronto's share dropped only slightly over the two decades before 1970. It retained high- and middle-income households, yet its two old working-class suburban municipalities have fallen relative to the newer and richer suburbs. Among the other largest Canadian urban areas, Montreal and Quebec City are below 85 per cent of their metropolitan incomes. Unlike Toronto and others, they have fallen since 1960.

However, this is nothing compared to Detroit and Newark, the worst American cases. On a per capita measure, city incomes have dropped to under 50 per cent of the metropolitan areas; in 1950 the average city family's income in Detroit actually exceeded those in suburbia! Other large northeastern and Midwest cities—such as Philadelphia, Baltimore, Milwaukee, and Chicago—run at about three-fifths the per capita incomes of their metro areas, New York and Boston about two-thirds. Western and southern cities come closer to Canadian central cities in part because of the many annexed middle-income areas. But even in the expansive Los Angeles region, the huge central city is only four-fifths of its metropolitan area income, despite many extremely poor Latino and Black communities outside city boundaries.

The relative difference between the two countries is the immediate result of greater losses from American central-city populations after 1950, the decline of the middle-class households thus leaving the poor behind, and Americans' dislike of cities. As we will see, public policies, legislatures, and the courts encouraged that sentiment. In Canada in the 1970s, several large central cities lost population, with Montreal and Toronto losing the most. This reflected in large measure rapidly dropping household sizes, which affected central cities more than other urbanized areas. A suburban flight of the middle class did not happen. In fact, during the 1980s central cities expanded again, if not back to their earlier peak. Most important, in contrast to the US, household numbers did not fall, so residents continued to occupy virtually all housing units, but with more space per person.

In American central cities, the 1970s were the most rapid decade of decline, accelerating after 1950, a trend that started as early as the 1920s. Then they lost more in the 1980s, if less rapidly. This occurred most markedly in the Midwest where, in contrast to the South and West, few cities could cover their losses through annexation. St Louis led the fall by losing over half of its residents after 1950, but was followed closely by six others. In fact, virtually all old cities in the East lost at least a quarter of their population. Columbus and Indianapolis were saved by annexations. Even after annexing, Atlanta lost a fifth after 1970. The slowdown of losses in the 1980s probably signalled a

kind of stabilization, but hardly with the social diversity and income variation of earlier times. Washington still loses people.

Income disparities between the central city and suburbs are growing in both countries. Consider the enormous growing social distance between the richest and poorest census tracts representing neighbourhoods, for example, in Atlanta and Toronto. In Atlanta between 1959 and 1989, average *family* incomes grew by six times to $35,000 in current dollars. The richest tract increased in value by eight times, the poorest by two, so the differential rose from eight to thirty times—$5,000 versus $150,000! Assuming that all the poor are on welfare, and if goods (like food stamps) and services (notably Medicaid) are added in, then perhaps we should add another (generous) 50 per cent, cutting the differential to twenty times.

In Toronto the differential was sharpened greatly too, if not as much: in 1990 the difference in average *household* incomes between the highest and lowest tract was twelve times—$231,000 compared to $19,000. In 1970 the difference was only seven times. Taking families only, the incomes would have been higher for most tracts because single people living alone—and therefore on only one income—count as households. Both the richest and the poorest benefit from public health insurance, thus at least cancelling out the Medicaid increase attributed to Atlanta. At least some of the poor resort to food banks, a sure sign that the safety net has not been working as well as it did earlier. Yet Toronto's poorest census tract had an average income of twice or more that of its equivalent in Atlanta, pointing to the still then more generous welfare support in Canada. That tract is Toronto's original public housing project near downtown. Other public housing tracts, mostly in the suburbs, have 50 per cent higher incomes. Even so, poor people and deprivation are more likely to be found in Canadian central cities than in suburbs, and both are becoming more polarized.

What is striking also is that Canada's central-city tracts have remained more mixed by income. In fact, the difference between the median and average household incomes has grown over the past twenty years. This means that *within* neighbourhoods, disparities have widened. For example, in one central-city census tract that now has an average household income four-fifths that of the metropolitan area, the median over the average has fallen sharply from over 80 to 70 per cent. The affluent households (most of them home-owners and those with double incomes) are beating out the far more numerous single and often younger renters in the district, some of whom lease space from these owners. The Atlanta differential *between* rich and poor tracts has grown wider than those in Toronto, where the differential has become greater *within* tracts.

Gentrification, a misleading term, is often invoked to explain the improvement of some inner-city areas. It supposedly denotes the return of professionals to renovated erstwhile working-class housing. The term also includes new housing or the upgrading of former middle-income housing. New housing, built on waterfronts as in Buffalo, seems to count for some revitalization. In New York and Toronto, at least some older office buildings and warehouses have been converted to housing. Yet in American cities, despite the conspicuous presence of St Louis's Lafayette Village, Philadelphia's Rittenhouse Square, and Columbus's German Village, the influx of the affluent professionals has fallen far short of earlier expectations. The hope that affluent baby boomers 'might serve to reinvigorate the nation's largest central cities seems to have been misplaced'.[18] New York, San Francisco, and a few other cosmopolitan cities continue to attract the marginal gentrifiers, like artists, who may have led the way in minor upgrading in a few neighbourhoods. However, for many other cities, the potential was not so apparent; the Midwest's metropolitan areas have not been growing rapidly anyway. Even in relatively healthy San Jose, the new attractions around downtown have not persuaded suburbanites to move nearby. Hence, retailing for local populations cannot revive seriously, even though Vietnamese and Mexican restaurants and hairdressers have brought back some vitality. Of course, some central cities have retained earlier White neighbourhoods, as in Washington, San Francisco, New York, Los Angeles, and even Detroit.

In Canada so-called gentrification has been much less conspicuous because housing has not been abandoned, the turnover has been more piecemeal, and, as already noted, social mixing is so widespread as to obscure the process. Certainly there have been some returnees, but most middle-class people grew up in the city, many the offspring of working-class parents.[19] In fast-growing cities, many people moved in from elsewhere. Renovations in many neighbourhoods reflect postwar Canadian affluence that rose faster than it did in the United States. Over the past two decades, though, many well-educated but poor young people have occupied tiny apartments, as discussed in Chapter 7 (Figure 1.1).

What is more apparent in the United States than in Canada are the social pressures that have profoundly pulled apart the residential patterns: city versus suburbs and even rich versus poor suburbs. Inner-city neighbourhoods have decayed further over the last twenty years, adding to earlier woes. On a hot summer day in 1991, a New York editor drove into Bushwick, passing 'burned-out stores and shattered side streets where only a few rotted and abandoned houses stood', a legacy in part of a 1977 riot, a consequence of the loss of jobs and abandonment. It was no longer what it was thirty years ago, then 'a self-satisfied Brooklyn neighborhood, warm … with three-story wooden houses along tree-lined streets' where children rode bicycles.[20] In 1993 over a million New Yorkers were on welfare (Figure 1.2). Elsewhere, large tracts lay abandoned (Figure 1.3).

This scene is repeated elsewhere. Some of the poorest outlying suburbs are just the same, like Ford Heights (formerly Chicago Heights) south of Chicago, with only a tenth of the income of that metropolitan area's richest suburb, Kenilworth. The mayor hoped the name change would 'urge people to try harder and not lose hope. It's easy to lose hope here.'[21] Poor exurban communities survive in flood plains and in remote hill lands, not just in Appalachia. In fact, three-quarters of America's poor live outside central cities in suburbs and rural areas.

1.1   Inner-city houses in Toronto. Built in the late nineteenth century near a brickworks, they are within 2 mi. (3.2 km) of city hall and now occupied by ethnically mixed professional, clerical, and manual workers. (Janet Lemon)

oft-told formula for American success—that their interest and the public interest 'are one'. So, 'driven by insatiable fears ... to achieve at least the illusion of safety, residents must buy in to an enormous measure of corporate domination'.[26] In these new subdivisions, lots are walled as if neighbours are not neighbours. In the 1880s George Pullman discovered that laying down rules in his company town did not work. Zoning of residential districts, supposedly a panacea to protect property values and stop change in the 1920s, hardly prevented the undesirables from taking over many of them after 1950. This superprotectionism will not necessarily prevent people from 'bringing problems with them', as one resident rued.[27]

The forces of decay have run their course, it seems, the senses dulled by long accommodation. The massive outpouring of outrage and rebuttals of the 1960s, then the popular press's sorrowful acceptance of the decline (paralleled by the fawning over Toronto as we note in Chapter 7), was followed in the 1980s by some cheerful notes of revival, but now the 'Voices of Decline' have reasserted themselves as Americans see the enormity of the problems.[22] On Chicago's south side, the slums are among the 'most horrible places on earth' or worse. To federal Housing and Urban Development secretary, Henry Cisneros, they are 'what the approaches to hell must be like'.[23] He has no money to do anything about it. Among poor people, local community action has little chance.

Deeply worrying to some Americans is the rise, by contrast, of 'fortress domesticity' in new suburbs.[24] Orange county has become 'obsessed with fear' about the 'advancing tide of crime'.[25] Walled and gated, protected by guards, the new 'master-planned communities' near Las Vegas and elsewhere try to escape what they perceive as mounting violence. Their developers repeat the

In Canadian cities, the level of apprehended fear is considerably lower because violent crime is far lower, less concentrated, and not associated with one minority. When visiting American planners ask to see slums, they are taken to the lowest-income census tract, one with public housing. They come away astonished; it does not look like a slum. Despite management problems in housing authorities, financial shortfalls, and residents' complaints about crime and drug dealing, Canadian projects are still tolerable if not pleasant. Gated suburbs are still rare, though they are appearing. In the summer of 1994, Toronto experienced the first fatal shooting of a police constable in a dozen years, which panicked many residents, so calls for more police echoed American fears. Home-protection devices are now much more widely advertised than a generation ago. If urban Canadians seem to be moving towards more protection (at least in older mixed-income districts), they are more clearly shielded by neighbourhood associations whose presence reduces the danger. Many women still ride the subway trains alone at night. The flight or even rout of the middle class, the devastation of urban renewal and freeway building, leading to abandonment (as discussed in chapters 6 and 7), have not afflicted Canadian inner cities even though they have become relatively poorer.

Excessive development and social exclusion have marked the American urban experience, especially in the twentieth century, and even more so over the past two decades.[28] Vast numbers in the middle class and the blue-collar working class with money answered the siren call of developers, both helped by government financial support, to move out to the new suburbs. Many perceived the deterioration of their city housing

1.2   Inner-city houses, graffiti, and a citizen in Philadelphia. (Christopher Pillitz/Network Photographers)

as a result of Black invasion. American class differences have, as ever, been covered over by race. Neither the left nor the right can leave the race issue alone. The left seeks integration and equity, while the right, it almost seems, deny that people of colour are truly American. To many of the latter, inner cities are a Black problem, not an American one. Many Blacks still pursue the goal of integration, but many middle-income families have escaped to segregated suburbs. The more circumspect reject the 1970s' hope for 'the declining significance of race' or attempts to rescue 'the truly disadvantaged' of the 'underclass' who are mostly stranded in the central city.[29] The more radical embrace a longer-term view that racial discrimination is permanent, that Blacks are and will be 'designated scapegoats' in the most liberal and open of societies.[30] To the more radical, celebrity gains are only token, and those of most others between 1965 and 1975 were temporary. Still, the 'million men march' may be a sign of positive change, a strike against apartheid.[31] Besides, violent crime rates are falling.

In a more racially homogeneous society, which Canada still is, class issues have been more apparent while in America they are constantly hidden, especially in the North over the past half century. This is not to say that Canada can proclaim purity; Blacks and Native peoples would strongly object.

However, the cumulative historic experiences of the two countries have been markedly different. In the next chapter, I will lay out what I believe are the salient differences and similarities between the two countries and how they arose. Given that the consciousness of race in the inner cities and the largely unknown marginal suburbs that are overwhelmingly Black or Hispanic is so high, I can hardly avoid this stark social question.

The population of several big American central cities today are less than half White. Most conspicuous is Detroit, which now has only half of its 1950 population and an average income that is less than half that of the metro area. Four out of five Detroit residents are Black and less than one in six is White, many of whom are the Appalachian poor. The Detroit metropolitan area has the highest share of Blacks outside the South—one in five—though most are in the city. Segregation of Blacks persists in most cities and even in middle-class suburbs. Two-thirds of Blacks live in neighbourhoods that are Black, virtually the same as in the Chicago of 1930. In many cities the degree of segregation actually increased in the 1980s. Inner-city Blacks have incomes half that of Whites. In relatively affluent San Jose, Blacks' incomes are three-quarters that of Whites, but in Washington, their incomes reached only two-fifths that of Whites in 1989, despite non-discrimination in government hiring. In Washington, Black infants die at twice the American rate, itself 40 per cent above Canada's and even more compared to Japan. In Baltimore, despite the slogan 'the city that reads', a fifth of residents are functionally illiterate and half of school pupils live below the poverty line.[32] Drugs and crime haunt the scene. In 1992 among twelve- to seventeen-year-old males, relatively far more Blacks than Whites were the victims of violence. Many would rather work as drug dealers than live on welfare with their mothers and sisters. In Baltimore, over half of Black men between the ages of eighteen and thirty-five are in prison, on parole, or on probation. At the same time, Republicans have taken to 'demonizing the welfare moms'.[33]

In Toronto, for comparison, Blacks are far less segregated, partly because there are far fewer of them, and partly because Whites have less fear of losing property values. Their Index of Dissimilarity is far below that of Blacks in the United States and even considerably below that of some European ethnic groups in Toronto. Fewer than 40 per cent would have to move to be evenly spread out across the city. They compose only one-quarter of public housing residents, compared to near totality in US projects. Even so, their incomes are lower than those of Whites without question, and drug dealing is an issue. The Rodney King affair in 1992 in Los Angeles was echoed in Toronto, but at least no one was killed.

A prominent story in the last twenty years has been an influx of Hispanics into certain American cities. The numbers in Los Angeles, San Diego, and Miami have shot up. Los Angeles is now over one-third Hispanic (mostly Mexican), and they are well on the way to forming a majority. Miami has the largest immigrant population on the continent.

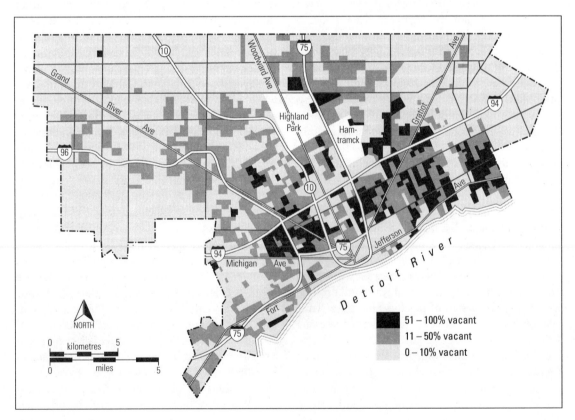

1.3   Detroit thinned out. Suburbanization without compensating inflows to the city weakened the historic fabric. *Source*: Detroit City Planning Commission.

Houston is another city with a population that is less than half White, with one-quarter Hispanic and one-quarter Black. The same holds for New York; Hispanics are now a majority in Bushwick. In San Diego, authorities pulled down shack towns like Rancho Los Diablos (apparently not considered an edge city), wiping out the shelter of citizens in the futile hope, like referenda and fences, of rooting or keeping out illegals. (One ray of hope: in the mid-1800s officials cleared out the Irish shack towns in Manhattan too; the Irish made it anyway.) Although they are less segregated than Blacks, Hispanics in most cities and metropolitan areas earn less than Blacks, except, interestingly, in Washington and Baltimore. However, Black poverty in the United States remains higher than that for Hispanics. In 1992 one-third of Blacks were under the official poverty line, compared to 28 per cent for Hispanics, and this poverty rose faster during the 1980s.

Asians have made their presence felt in West Coast cities from Vancouver, Seattle, Portland, the Bay area (with the most at 15 per cent) to southern California. In the New York region, they make up 5 per cent, mostly in the city. Toronto has a higher share, nearly one in ten, most of them Hong Kong Chinese. Many are suburbanized. Income-wise Asians do better in most cities than Blacks or Latinos, as in Washington, but at the level of only half of the richest concentration of central-city Whites in the country. Despite many poor southeast Asians, most Asians will no doubt vault ahead of Blacks and Hispanics, and Hispanics over Blacks. This is obvious in Toronto and San Francisco. Residential segregation is not an issue with a lot of Asians. Money counts in their case, as they replace White engineers, for example, in upscale Cupertino near Apple Computer.

In sum, population shifts have led to 'trends toward balkanization' within American cities and from city to city, though less so in Canada. Movements have been 'uneven … rewarding corporate nodes, information centers, and other tie-ins to the global economy'.[34] Places with high-technology activities, recreation, and retirement have done well in the northwest states and British Columbia, and in other states east of California and the south-

east. Blacks drifted to Washington and Atlanta. Minority immigrants are concentrated in ports of entry. Much of the vast interior of the continent is thinning out and populated mostly by Whites.

Stagnation or Crisis in Running Cities?

Governing cities with their mixed residential districts is relatively easier in Canada than in America. Few neighbourhoods are seedy. Even the poorest have some high-income households. However, in Canada the fiscal pressures are mounting as transfer payments from higher levels of government are strained. Transit ridership, which is higher than in American cities except New York and less subsidized, is declining as more commute to work in cars. Arguments between suburbs and cities have escalated. Local councils vie with one another more strenuously for what industry there is, their economic development departments desperate for investment. There are few office and public developments on the horizon to squabble over. In some areas, taxpayers appeal their assessments because no one has thoroughly reassessed properties in many years. Some say there is a crisis in planning; the only serious crisis is that planning graduates have trouble finding jobs, as little is going on in these now slow-moving cities. Heritage preservation projects, prominent in the 1980s, have dried up.

'Crisis' has been a more appropriate word to describe American rather than Canadian cities, but the word seems hackneyed after its frequent incantation over the past two or more generations. However, no one doubts that American central cities are even more worse off financially today than they were in 1975; the 'disparities … between rich and poor local and state jurisdictions got larger'.[35] Because neighbourhoods are far more homogeneous by race and income than those in Canadian cities and suburbs, the local housing market is more fragmented as a result, like the municipal jurisdictions in most places. When public choice is a guiding philosophy for planners and politicians, who expect people and firms to vote with their feet to find lower municipal taxes but higher status, then this could hardly be otherwise. Today the urban renewal federal bulldozer or highway engineers are not bullying low-income households out of the way as they were between 1950 and the early 1970s. Instead, they are largely left to rot. The new federalism of Reagan and Bush 'succeeded in reversing fifty years of American domestic policy by cutting back the constellation of federal grants ... used to help poor people and needy city jurisdictions'.[36]

Nonetheless, some people still think they can govern American cities. Recently 500 mayors have been Black. After the first Black mayor was elected in Atlanta in 1973, it seemed for a moment that Blacks might actually possess power. The excesses of urban renewal and freeway building displaced one-seventh of the city's population; there would be less of that. However, the Black leaders accommodated the White financiers and their wishes. Also, the 'Black élite extended benefits to themselves', leaving the poor Blacks even worse off. Three-quarters of the council members who are Black live in the suburbs and promote progress—that is, offices and the Olympics—putting the cost on a 'quiescent Black citizenry'.[37] The imposition of land-use controls in 1989 favours money, not people. Black mayors' inability to do little more than go along with big projects while presiding over residential decay, continuing crime, and relentless drug trafficking has resulted in a return of fiscally tight, right-wing mayors. Rich White men have taken the mayoralty reins in New York and Los Angeles; their austerity ideas will just make matters worse for the poor: more police, more prisons, and more homeless. They claim that improved policing has reduced violent crimes recently. If so, fewer teenagers and fewer hard drugs are thought to have contributed.

The Economist, ever fascinated with the foibles

> **Today the urban renewal federal bulldozer or highway engineers are not bullying low-income households out of the way as they were between 1950 and the early 1970s. Instead, they are largely left to rot.**

of American society, has believed that the 'decline of inner cities is far from irreversible'. A great deal of the problem lies in bad government and 'too few politicians with the guts to do anything about it'.[38] If 'hell is a dying city', then what can be done? It has proposed many solutions for New York with its 'overloaded' political system.[39] One would be to cut taxes, then cultural programs; another would be to cut civic and school board jobs. The city has, they believe, too many employees who are members of too many powerful unions. The New York City school district pays out much more per pupil than Chicago and Los Angeles. However, in 1992 Toronto paid out 16 per cent more than New York, and a whopping 40 per cent more than Chicago.[40]

New York unions respond by saying the council and school board employed people when the private sector abandoned the city, and union members spend at least some of their money in the city. *The Economist* admits, and thus seems to undercut its advocacy, that 'the idea of nibbling at the perks of city employees earning (on average) under $50,000 a year, while millions are made on Wall Street, plays poorly on the hustings'.[41] The vast inequities within the region are much less the responsibility of explicit unions than they are of the implicit union of executives and brokers.

In 1989 the New York metropolitan area, now defined as including rich Fairfield County, Connecticut, where many Wall Street brokers live, was still the richest in the country, despite the Bedford-Stuyvesants and Newarks. *The Economist* does suggest regional government and an equalization of taxes as a means of sharing wealth. A former mayor of a smaller city who, in his book, *Cities without Suburbs*, presents data showing that cities that have annexed large areas—'elastic' cities—have fewer fiscal and social problems than those that have not.[42] Planners optimistically suggest that politicians have 'recently begun to explore new regional planning approaches to urbanization'[43] Although the Twin Cities have had a somewhat effective regional body implementing tax-based revenue sharing, other metropolitan areas can only claim voluntary and largely ineffective advisory groups. The 'shadow' governments set up in corporate-run suburbs are resistant, probably even more so than the older municipal fragments that are so widespread in this century.[44] Besides, regional planning has been on the table since 1909 when Daniel Burnham proposed his solutions for Chicago. None of the prescriptions are likely to work as long as income and job discrimination continue to the detriment of central-city residents.

Do city power élites know what they are doing?

> In 1900 no one could yet articulate a clear vision of how the city worked, much less a direction in which to steer. [Today,] no class or interest group can yet attain the high ground in directing urban policy, because no one knows where the high ground is. [O]ur form of city is *too new* to sustain an analysis of any clear set of policy decisions ...[45]

Possibly no group can rise above its own interest. Given the public choice model of the city as simply a marketplace of goods and services operating like the private sector, élites are not likely to act decisively. Strong political leadership and the cohesive business élites of the 1950s who brought urban renewal and downtown superprojects (to the detriment of the poor) have given way to weak and 'dispersed' power.[46] However, to say that the form of the city is new is clearly illusory; how many more years are needed for a city to be mature or old when presumably one could act? Ultimately this view sounds like another version of American exceptionalism, a belief in eternal youth, so urban management bounces from crisis to crisis. Better-managed Canadian cities have at times implemented clear policies. Today rational policies in Canada are also hard to formulate, let alone implement; stagnation, if not crisis, is the worrying reality.

Since the Great Depression, higher levels of government have had the primary financial responsibility of solving or at least ameliorating big-city problems. Their financial contributions through conditional and unconditional grants have been enormous. However, in the 1980s it could be said that 'Washington Abandons the Cities'.[47] In the 1980s Reagan and Bush cut federal payments to cities nearly in half in constant dollars down to $31 billion, minus 46 per cent. In 1980 the federal and state governments contributed 50 to 70 per cent of six large central cities' general expenditures, and 40 to 50 per cent of another eleven. In 1989 only five got over 40 per cent. Cutting New York City's funding from over a half to one-third clearly contributed to yet another fiscal crisis and cutbacks there in 1990. Poorer local governments and poorer states lost out to richer ones. In all, federal aid to large cities fell by over two-thirds.

Under the 'new federalism' of the 1980s, states failed to increase their share, which stagnated at just over 15 per cent. Already gradually undercut during Nixon's second term and then again near the end of Carter's, Lyndon Johnson's Great Society programs unravelled and the cities were hung out to twist in the wind. Because many cities have also had to contribute up to a quarter of rapidly rising Medicaid costs, that meant a decrease in other services, higher taxes, and thus led to more middle-class residents leaving the cities. Many vital community programs such as legal aid evaporated. On the other hand, direct federal benefits to poor citizens rose by just over 40 per cent, mostly in Medicaid and housing assistance, not far off from the cuts to cities. However, this individualistic approach could hardly replace the services that cities have collectively provided. Then the depression beginning in 1989 made it even worse: there were more homeless with no possessions, many more with a roof over their heads but with no discretionary incomes, contrasted with top-income earners and their still-rising incomes. Public housing apartments stand empty and unrepaired because cities lack the funds to fix them. To add to the crisis, suburbs like Orange county are financially beleaguered. Cramped for revenues, the treasurer unwisely speculated in financial markets, so in the end, residents 'never got the tax-break they naively thought they had voted for' in 1979 to limit property taxes.[48]

In Canada the federal government has transferred money to individuals and to provinces to create a degree of equity, but little to city governments directly. Until recently, it supported social housing programs. Stable financial relationships between provinces and municipalities emerged by the 1960s. From about 1914 when local politicians first requested financial help and then asked for more in the 1930s, to the working out of the complex transfer system in the mid-1960s, provinces raised their share of local funding to just over 40 per cent. In 1988 this was about the same ratio as both higher levels provided in the United States, though as we just saw, the American federal share fell. In contrast to the United States, the pattern of grants and taxation in Canada has been far more consistent, with grants carefully worked out on the basis of need. All schools, at least in theory, are similar in quality. The power of the provinces has maintained stability; grants have often been conditional for specific purposes, though municipalities always chafe on this leash. Local borrowing has been strictly regulated; standards are imposed, just as the federal government has imposed tight rules on provincial management of their health insurance systems.

Although higher level grants may have been similar in both countries, the ability of central cities (many as partners in regional governments) to raise taxes remained (until now) far higher in Canada. The urban condition and incomes have not deteriorated as far. Although property taxes as a share of revenues have dropped markedly since early in the century, they remain significant for most property holders. As a percentage of all local taxes, those on real property and businesses in Canada reached 85 per cent, 10 per cent more than in the United States. Local governments are responsible for over one-third of total government expenditure, of which nearly $2 in every $5 is for education, though the amount varies considerably depending on provincial arrangements. Local fiscal imbalance, that is

expenditures over revenues, is slight. Provinces have contributed more to revenue, in Toronto up by 10 per cent more between 1992 and 1994, even though rich Toronto receives no education subsidy.

As the depression after 1989 worsened, the provincial debt levels rose to such a peak that in 1993 the province cut the salaries of all public workers who earned over $30,000. Those affected protested, of course. Although, as observed earlier, crisis has been hardly a notion to apply to Canadian cities, the bond rating agencies are downgrading their credit worthiness. Municipal efforts to transfer school funding to the provinces would mean even more provincial taxes. Although this might create a more equitable system, it will likely lead to more complaining about tax hikes. Canadians seem to be moving towards American ways.

Voluntary action has provided a supplement, even an alternative to formal government, especially in the United States where government is always suspect. In the 1830s French aristocrat Alexis de Tocqueville was astonished at the plethora of voluntary groups that Americans set up. Today American commentators extol this virtue. Instead of governments providing housing, ex-president Jimmy Carter signs up volunteers to build 'habitats for humanity'. Unfortunately, though, such a program will supply few houses. Recent new inner-city housing schemes are the result of government subsidies to financial institutions that reap big rewards for few results. Charity alone will not do the job; only taxes will, as Europeans have long since known. Local initiatives for community services need outside money; few inner-city businesses can afford to support swimming-pools.

With so much empty space in big cities, now would be the time for American governments to build on that cheap land instead of continually trying to engage financiers in a fruitless task. That would help to revive central cities, but I would not bank on this. Cities have been straitjacketed by hostile states that hold the constitutional card. The cities are also at the mercy of the liberal dream, which atomizes society.

> The cities are ... at the mercy of the liberal dream, which atomizes society.

Social Forces Fraying the Urban Fabric

Immediate and long-term pressures in society are pulling down American cities and threatening Canadian cities. Most obviously, American inner cities and their poor residents lack money and jobs, which seem to become ever more remote possibilities. Poor people are unable to finance poor cities. Income distribution generally has become more skewed. The public and private debts piled up in recent times are pinching urban life. Behind these is the slow economic growth over the past two decades, but, as I will argue in subsequent chapters, there has actually been less certainty about growth for several decades, despite a revival after 1940 through military and consumer spending. Only recently has this become apparent. More fundamentally, hostility towards cities, and by that I mean towards the public collective life, underlies the American predicament. I hope to make that clear in assessing Jane Jacobs's philosophy shortly.

The decline of the quality of life in cities reflects the winding down of a resolve to distribute wealth and income more equitably. In the past, gains in material well-being spread erratically down the income scale. After 1850, the great expanding white-collar middle class clearly enjoyed some improvement in quality of life. The social welfare support system established after the Great Depression improved the lot of blue-collar workers and even the poor, but that support has been weakening over the past two decades, especially for the most vulnerable. Restructuring the economy has rendered more and more blue-collar and now even white-collar workers obsolete. Two incomes have sustained many families, especially in the upper half of the income scale. The great gains of the liberal era of four centuries, of fulfilling European dreams if you will, seem to be turning to dross. If the word 'postmodern' has any social meaning, it surely lies in this turnaround from progress to regress.

Weakening wealth distribution, unemployment, and underemployment are immediate factors contributing to urban malaise. Not since the 1920s have wealth disparities been so sharp. Redneck sentiment bubbles up in the froth of frustration owing to a weakening job certainty. Wealth distribution, to put it mildly, has become enormously skewed in favour of the top, more clearly so in the United States than in Canada. Only the few élites in the Third World stand out as clearly. The top half of 1 per cent of American families hold over a third of national wealth, which actually exceed 1929 levels. At the bottom, four in ten minority families have nothing and are largely invisible in their isolation. The ultraprivileged hold half the value of stocks, over three-quarters of the bonds, and over three-fifths of business assets. In 1949 the richest held only 21 per cent of the wealth after losing mightily in the stock market crash through New Deal correctives, and especially because of wartime price controls, heavy taxation, and virtual full employment. At that time, money as such became less of an issue; military hardware and housing were higher on the agenda. In recent times, money has become a preoccupation. The acquisition of wealth, as in the flapper age of the 1920s and the gilded age of the 1880s, bodes ill, not just for the poor but for those who are risking too much.

Higher incomes for the top end have led to this wealth distortion. In 1989 the top 1 per cent's share from 'market' sources was 13 per cent of the total. Their take had increased since 1977 from 8.7 per cent, an enormous shift upward, partly through investments, though élite earnings too have shot up. In 1992 the incomes of these 600,000 families averaged $675,000. The high salaries of executives have had wide press coverage. Interestingly, the big incomes and wealth accumulators are not only in finance, as usual, but also today in entertainment. Entertainment and communications have vaulted to the top as darlings of finance. Everybody else seems hooked on the wonders they provide. Perhaps that is why it is possible for the top managers to run away with a disproportionate share of the largesse. 'Chief executives, it seems, are determined to reap all the rewards of swashbuckling entrepreneurs, while living the risk-free lives of cosseted bureaucrats'.[49]

Even discounting the very wealthy, after social transfers and taxes, the top fifth of American households in 1992 received *eleven* times as much as the bottom fifth, who 'suffered an 11% drop in real income' over twenty years.[50] By contrast, in Canada, the difference was *seven* times, considerably more equal, and a major reason why Canadian cities are better off. As in many countries, social support has been needed to compensate for a widening gap in earned and investment income. Recent Canadian budgets have reduced social supports: Canada ranks only about twelfth among Western countries in its level of social benefits, though it is the best among English-speaking countries. Australia and New Zealand have regressed from earlier equity. The poorest 10 to 15 per cent of Americans benefited slightly in the 1980s from tax cuts and transfers of cash, food stamps, and housing allowances, yet overall suffered a fall of over 10 per cent in real income. The lower middle ranks did worse, as they paid higher taxes, while the top fifth saw its effective rate of tax fall in the Reagan years while their incomes rose. Indeed, about 15 per cent of the working poor above the poverty line—over 30 million people—have been without health insurance. At least in Canada that is less of an issue, though its health system has never become totally universal; only the poor, handicapped, and elderly get public support for drugs and dental care, and these services are threatened. Besides, controlling health costs has now become a major political issue. Over the two decades to 1990, the elderly were better off than previously in both countries, but single mothers were worse off.

The United States has been less generous to the bottom quintile than have western European countries or Canada, so it has the higher number in poverty, nearly a fifth compared to less than half that in western Europe. When single parents and

their children are counted, Canada is not far behind the United States, both with over 50 per cent of all living below the poverty line, though Canada has a more generous cut-off line. Scandinavian countries have only about 10 per cent, though the business press likes to point out that they cannot afford this luxury. Prices may be lower in the US, but so are wages. Besides, the constant hectoring of those on welfare grinds down everyone. The lower middle ranks of the working poor have been pushed down into the poor category; the poor, with social supports, fall more quickly. The income scale, which was shaped like a pyramid earlier, then briefly more like a diamond in the postwar era, now resembles an hourglass, whose bottom globe is much larger and growing faster than the top globe.

Business has finally shown a concern about the effect of inequality on the economy. In 1994 *Business Week* opined that 'the growing gap ... is hurting the economy'.[51] *The Economist* mused on a report that 'inequality may even be harmful to growth'.[52] Probably so, though I see little evidence that the business sector would seriously grapple directly with incomes. They trot out the old nostrums of better education, apparently believing that the money for it grows on trees, but even if everybody were better educated, there would still be a fifth at the bottom of the income scale. Besides, I am sceptical that a massive upgrading of education would translate into a great spurt of growth. Quite probably the greatest gains to human capital came through nineteenth-century literacy and numeracy training in elementary schools. Participants in retraining programs wonder if jobs await them. The last great spurt of growth needed, it seems, the hot war of the early 1940s and then a cold war.

Unemployment and underemployment are a large part of the social 'impasse'.[53] Over the past

> When single parents and their children are counted, Canada is not far behind the United States, both with over 50 per cent of all living below the poverty line, though Canada has a more generous cut-off line.

two decades, social supports have increasingly compensated for declining earnings in the lower ranks, together with second and third earners in families. On the surface, according to unemployment figures, conditions are worse in Canada and Europe than in the US, but the American method of recording undercounts discouraged workers, and there are more part-time workers.

'The rise of the losing class' has generated anxiety.[54] Those who manage to get a job do well, but fewer have. As one recent American graduate puts it about his part-time job, 'The Silver Fox types own everything and the boomers run it … and I'm living in a basement apartment working at Copy Cat.'[55] Cutting the taxes of the rich in the 1980s and deregulating the economy was supposed to release entrepreneurial energy to lift the economy and thus create employment. The recovery after the sharp recession of 1981 and 1982 when many factory workers lost jobs did improve employment levels briefly. However, after 1989 the losses became more severe, only rising again slowly during the so-called 'jobless recovery'. Restructuring of companies has continued. Restructuring is a euphemism for downsizing, itself a euphemism for laying off workers. Governments, once the certain haven of job security and economic stabilization because of the large share of total employment they provided, are downsizing too. Then too *The Economist* has pondered, 'Will a less secure, more mobile workforce be a less skilled and hence less productive one?'[56] A promoter of re-engineering, now disillusioned, rues all the 'mindless corporate bloodshed'.[57]

Polarizing Societies Reflect Weakening Slow-Growth Economies

The great cities have not collapsed, reflecting an economy that still has some strength, and some sectors seem to boom. Yet to what effect? In recent times, computers, telecommunications, and the information they carry have probably kept the economies from falling further than they have. The speed of digital movement has been remarkable. Certainly quantum jumps have increased flows so that messages are nearly instantaneous. E-mail, subsidized by the American military, is advantageous for academics; they waste less time, at least in theory, especially if they can discipline themselves to ignore most of the messages. Cellular phones keep some people in touch more easily. Financial companies have upgraded their computer power to steal the march on competitors by nanoseconds. The annihilation of space and time continues.

But there are underlying doubts. The enthusiasm may be misplaced. Ingenuity has already virtually overcome time in the movement of money, though people have to sleep. Ingenuity will never overcome space. In terms of contributing to economic growth, the recent upgrading of speed probably has had less power in quickening investment and the economy than the telegraph, a point I will argue in the next chapter. True, each new computer manufactured adds to the gross domestic product, but it is interesting that over the past half decade, the enormous expenditure on new machines in the home, schools, and workplaces has not inspired very much economic growth. Office productivity did not rise, that is, until employers put more work on fewer staff. It appears that equity has not been served, as the information age has witnessed wider divisions in society. The expectations that everyone would do better have not been met.

New advances in biotechnology and the health sciences have been stunning. Each cow will now produce more milk, supposedly cutting costs, though critics point to possible serious side effects and fewer farmers. The big drug companies have insisted on more years for patent protection so they can increase prices to pay for research. Ultrasound scanning devices provide a service to decaying bodies, postponing the inevitable. Making the machines contributes marginally to growth, but does prolonging lives do the same? The apostles of the 'new' brainy economy—telecommunications, computers, and biotechnology—have hectored us to buy stocks as the road to riches.

Entertainment fields boom. No longer a diversion, they now occupy nearly centre stage in the economy, if the salaries of the top entertainers and executives are a measure. Huge major league sport salaries for the few suggest a major preoccupation. Advertising costs for major events continue to rise. The Canadian Broadcasting Corporation, for example, put up nearly $30 million for the 1996 Atlanta Olympics compared to a paltry $5 million for the one in Seoul, six times more in eight years.

What other fields are growing? Lawyers, like the poor, we have always had with us, and, like the poor, their presence as players in the economy has increased in recent times. The mergers, deals, and protection of information have led to prolonged court cases, with the top lawyers as the winners and the consumers and tax-paying citizens as the losers. The fact that American companies seem more inclined to fight one another in the courts to protect their market share than attend to business may in fact reflect slow growth. The law may be a growth area, but to what end? George Bernard Shaw was surely right when he saw professionals as a conspiracy against the laity.

One growth field is certainly visible: prisons. In America, where violent crime rates remain 'freakishly high', though now falling compared to elsewhere, more are being incarcerated under the policy of imprisonment after three offences.[58] Unfortunately, tending prisoners does not persuade us that economic growth is served very much. Building new prisons, as does infrastructure spending generally, whether on roads, subways, or dams, does pay off a bit. But the relative impact on growth of *new* roads and dams was far greater early in the century than it is today. Replacements add less to growth. New household gadgets too, though they proliferated in the 1980s, seem to have less influence on economic growth than earlier. So contributions to growth in any of the above fields are largely illusory.

The workings of the financial world over the past two decades do not inspire confidence that growth has been boosted. Those who commanded money in the past have been the citizens most responsible for what has happened to society and the environment, but money worked best when it was not such a preoccupation as it is now. Recently, *rentier* incomes through investments have risen sharply as a share of the national income. Even the crash of 1987 did not stop the sellers of mutual funds because the top income earners still had plenty of money. In the early 1990s pension funds were saviours of commercial real estate like offices and shopping centres, which were overbuilt in the limitless 1980s. However, can the pension funds sustain themselves? The continent still has too many offices and shopping centres. Telecommuting may obviate the need for more offices, so maybe pension funds have just postponed an even greater shakedown than the Great Depression out of which they sprang.

The giant financial firms have created even greater instability with their instantaneous movement of trillions of dollars over the world. Instability in the derivatives market has led to spectacular defaults in both private banking and government. The International Bank of Settlements in Geneva has been very worried that one big failure to settle accounts by one firm could set off a serious chain reaction so that 'the entire financial system will collapse'.[59] Much of the flow through the electronic global network has, apparently, no connection with productivity. In fact, 'the slowdown in capital investment … has been caused more by the lack of opportunity to profit in a slow-growing economy than by a dearth of savings.'[60] Savings through huge mutual and pension funds do not necessarily result in productive investment. Futures markets that formerly dealt with commodities now deal much more in money—a sign of lower productivity and speculation.

The structuring and restructuring of companies, the take-overs, have been praised for improving efficiency and raising productivity. Mergers that created oligopolies at the turn of the century probably improved productivity in goods, that is, until 1929. From 1945 to the 1960s mass production brought material satisfaction to far more people than earlier. Yet since the 1960s, flexible production in some fields has not increased productivity to past levels. Marketing has become more intense, taking a larger proportion of corporate spreading. Fads like empowering workers and total quality management come and go. The worse the conditions, the more enthusiastically they are embraced. Holding market share seems to be more the purpose of these moves than increased productivity. We live in a zero-sum world; in 1993 Lester Thurow reckoned that there is 'overcapacity' of 30 per cent in the world in many sectors.[61]

North Americans can no longer count on arms and space-race spending. Today's American military spending is half that of Reagan's Star Wars appropriations, far below the 1950s' cold war spending as a share of national production. Southern California employment in these activities has fallen by half. The glory days of the early cold war triggered far more production, and it was the Second World War that rescued North American production from the Great Depression. Policy makers no longer believe that extravagant cold war spending contributes to economic growth, but in the meantime, the United States and Canada, like other peace-loving big nations, supply arms to constantly warring Third World countries and then wonder why peacekeeping operations are necessary. Forgotten are the peace dividends promised at the demise of the cold war.

Neither technology, organizational changes in business, nor finance seem to be contributing to growth. More fundamentally, long-term problems with population and the environment could well be at the bottom of this. (In the next chapter, we will look at these more fully.) Demographically, as the sperm counts fall, the continent cannot reproduce itself, but politicians respond to populist worries by trying to reduce the number of poor immigrants and births to single mothers. Thus the population ages and pension fund sustainability is being questioned. Older people do not buy as much as younger ones do. If population growth in the past was a large contributor to economic growth, as I will argue, the pace today bodes ill.

The resource base of the continent is less secure than earlier, the cheap minerals are gone, and what remains is less accessible. Oil imports rise. East Coast and West Coast fisheries have lost much of their resource base. Countries fight over the few remaining fish in the seas while planning more airports, as oblivious now to the limits of petroleum as they have been about the limits of the bounty of the sea. Environmentalists have signalled the loss of forests. Worries about ozone holes and global warning keep resurfacing. Even if economic and population growth is much slower, our way of life conflicts with the environment. Nature is surely warning us of limits. More profoundly, it is actually slowing growth.

Over the past two decades, debt levels have become serious, doubly so since foreign investors began buying bonds. With only one Canadian for every nine Americans, Canada is more vulnerable. Its size makes it potentially less stable, and Canadians feel this: the image of the mouse living with the elephant is often invoked. From 1972 to 1992 in Canada, total private and public debt rose from about 110 per to 200 per cent of gross domestic product. Of this total, government debts increased the most, more than doubling to 90 per cent. In both countries, it hovers near 100 per cent. Private consumer debt went up by about 50 per cent. Interestingly, business debt increased the least from about 35 to 50 per cent.[62] The trend has been similar in the United States, though public debts are at a lower level. This pattern has clearly put pressure on the public sector, and so on the services that everyone depends on, not least the most vulnerable. The main clue to the rise of government deficits would appear to be in lobbying. Almost 100,000 lobbyists and think-tanks with charitable status in Washington lend credence to the view that 'the weight of influence in the capital comes from interest groups, not voters'.[63] Those representing business, such as the sugar-cane growers, have more power than underfinanced social advocates and the strapped city of Washington itself.

The fact that foreigners invested in earlier times is no comfort. Before 1914 the British pumped capital into North America when this continent was growing. Then that well ran dry and it was America's turn to be the supplier of capital to the world. By the 1970s this gusher too had slowed to a trickle. Although plenty is still invested overseas, the net flow has been inward again, but is subject to the whim of Japanese and German investors. Weakening dollars are evidence of effete economies. The free trade deal with Mexico has drained away wealth and jobs. As long as the United States and Canada remain affluent, though, the money will come, if encouraged by higher interest rates. Can the debts keep mounting? Despite disclaimers, it seems that the bond merchants will try to make sure they do and encourage higher interest rates.

Why has public debt risen faster than business debt? The business world blames government profligacy, though it is obvious that subsidies to business have been a major reason for the problem, often in the name of spurring economic activity. The Generation X (or the thirteenth American generation since 1776) certainly has some sense of the real reason. A 1994 graduate from a well-endowed California college puts his finger on another obvious reason:

> The federal debt has been created because Boomers and their elders are unwilling to pay higher taxes, though they continue to demand higher spending to pay for programs that benefit them. It has fallen to the Thirteeners to be realistic about our fiscal, environmental, and social predicaments, and to tighten our belts accordingly.[64]

Compared to the war crisis of the 1940s, taxes have fallen dramatically for the affluent and corporations. The American tax rate for the top 1 per cent dropped from 36 to 29 per cent between 1977 and 1992. Although taxes rose again recently for those in the upper ranks, the tax revolt of the late 1970s that brought Reagan to the presidency still lingers to the detriment of public services, to the bottom half of the income scale, and, ironically, to economic growth. The Republican victories in November 1994 suggest that the revolt more than lingers. The graduate goes on to a vague threat of action, but hardly makes a clarion call for politicizing to an atomized generation. The quagmire of stagnation has, it seemed, dulled the spirit. Many from Generation X are pursuing the almighty dollar, but for many more, there is a growing chasm between them and the 'young rich'. 'Boomer teens who got in trouble heard political leaders call for social services; 13rs who get in trouble mainly hear calls for

boot-camp prisons—or swift execution.'[65] The solution to the public debt is not in further impoverishing the poor. Their total government entitlements only equals the support of Medicare for the elderly. The largest entitlements are deductions for retirement plans, deduction for home mortgage interest (far more than in any other country), and exemption of health-insurance premiums. If 'you rule out [higher] taxes and Social Security and most Medicare, you're not serious' about cutting the debt, says one commentator.[66]

Why Jane Jacobs Failed to Win the American City Debate

When Jane Jacobs captivated many city watchers with *The Death and Life of Great American Cities* in 1961, it seemed to her and to others that large cities could be revived in their centres. Seven years later, she was living in Toronto while American central cities continued to deteriorate. Although American urban commentators still often quote her, usually without noting her present place of residence, her prescriptions have not been followed. The slums were not 'unslummed', except in the most

> **The solution to the public debt is not in further impoverishing the poor.**

nihilistic fashion of abandonment.[67] In Toronto, city advocates listened to her, though what happened would likely have occurred anyway, as we will see in Chapter 7. She settled in a place that was stable and vibrant at the time. I will briefly assess Jacobs's views on cities, the economy, and politics. I believe her stance is more typically American than she might admit. In fact, her liberal philosophy is fundamentally detrimental to city life.

Jacobs set forth solutions to the enormous problems that had by 1960 beset inner-city areas, especially in New York. She was preoccupied with design, finance, and the need to retain small-scale businesses and medium-density neighbourhoods. Cities should be designed to ensure social and economic diversity and intensity and exclude cars. Development should be very small scale through 'gradual' financing rather than the 'cataclysmic' application of money for large projects, which destroyed the fabric of many neighbourhoods.[68] One can hardly disagree, but how do we keep excess in check?

In later books Jacobs extended the argument to economies in general. Spontaneous bottom-up, small-scale enterprises were healthy in inducing growth; big monopolies were not. In her recent *Systems of Survival*, she continues to advocate small-scale enterprises as the salvation of economies. Her chief argument is that governing by 'guardians' should not be allowed near creative business entrepreneurs, and that they should 'shun trading'.[69] Guardians are to protect and regulate business, though she does not systematically address the latter. With this set of mind, she condemns state monopolies, it would seem, as excessive regulation. Corporate monopolies are let off more easily, except for the odd one that threatens to corrupt the rest like the unproductive conglomerates of the past few decades. This has been consistent with her work at Energy Probe, where she and her fellow American expatriates have condemned the Ontario Hydro-Electric Power Commission. Ontario Hydro has been one of Canada's most revered public enterprises, set up in 1906 to deny power to the monopolistic electricity rings and provide cheap power widely. Even though Ontario Hydro has overexpanded in recent times, Jacobs's advocacy faces an uphill battle in Ontario where utilities are seen as services, not profit-making businesses.

American cities got worse after her pronouncements in 1961, abandonment quickened, and the slums became more hellish. The middle class continued to leave, and the hoped-for return of gentrification was very sparse. Cataclysmic central-city urban renewal projects, which were funded in large part by Washington, planned locally, and encouraged by the corporate élite, continued to eat away at poor neighbourhoods for a few years, then left large tracts abandoned when the money ran out. The new public housing projects in inner cities deteriorated. Why did Jacobs's ideas have so little influence in the US? Well-known urbanologist Lewis Mumford, who had

already declared that American cities were a lost cause, dismissed her prescriptions as 'home remedies for urban cancer'.[70] Although Jacobs and other advocates did contribute to saving a few inner-city White neighbourhoods and won a battle in stopping the Lower Manhattan Expressway (though public officials also saw rising costs as an impediment), the war was over. Maybe it was over before her book came out: the damage had been done. The horse of development had escaped before the barn door was closed.

Jacobs's position is seriously flawed because she picked the wrong targets. First, she blamed planners for ruining the city. She had a point with Corbusier, the advocate of vertical living, though less so, I believe, than with early English garden city advocate, Ebenezer Howard. Robert Moses, the forceful public builder of New York, was an ideal target. Like other dominant city bureaucrats, his power was, as another has said, 'entrenched by law, supported by tradition, the slavish loyalty of the newspapers, the educated masses, the dedicated civic groups, and, most of all, by the legitimized clientele groups enjoying access under existing arrangements.' They have so much power that the 'mayor stands impoverished'.[71] To blame planners—and Jacobs is still attacking them today in Toronto—is to miss the mark: behind the planners have been the developers, their bankers, the chambers of commerce, and politicians who were desperate to overcome decay. Bankers who blacklisted inner-city properties were, she said, 'not villainous'.[72] Probably not, but their money (and the money and power of eminent domain coaxed out of Washington) backed these projects. They could have said no. Jacobs never mentioned the powerful business leaders who promoted and supported the great plans earlier in this century—Chicago in 1909 and New York in 1929. While I share her condemnation of the City Beautiful *cum* 'Monumental', even 'Centre Monumental', planners Daniel Burnham and Thomas Adams were only the acolytes to serious money. She shoots the messengers. Planners and politicians in American cities have always accommodated people with money and power, more so, I would submit, than those in her adopted city. Money and power also created the suburbs that attracted the middle class. Jacobs correctly condemned money employed for suburbanization 'at the expense of starving city districts',[73] but then blamed 'high-minded social thinkers', not the big developers and bankers, for this.

More fundamentally, my second criticism is that the persistent ideologies of America made Jacobs's case largely untenable. Three dimensions need consideration: expansion, newness and freedom, and the contradictions arising from these. The United States has been in the thrall of expansion since the beginning. After all, the colonies were England overseas, spreading over Native lands to make money or to provide escape from the corruption of London. Ben Franklin advocated expansion; American land speculators, including Franklin, bridled at the Proclamation of 1763 and the Quebec Act limiting westward movement. Andrew Jackson was the Indian exterminator; manifest destiny was the slogan. Eventually the limits of agricultural land were reached. Meanwhile, the workings of the economy thinned out the ranks of farmers. Their children sought the cities, as did European immigrants, and later Black sharecroppers. Cities thus grew and expanded outward through suburbanization. In the process, urban land became increasingly more expensive than farmland, so suburbanization became the natural successor to the agricultural industry. Like the pied piper of Hamelin, suppliers persuaded ordinary folk that life was better closer to nature. Expansion is built into the American way.

Closely conjoined is newness. English settlers sought a new start, to escape the corruption of England. Americans have led the world in seeking new consumer goods. Thus, other than the shrines of patriotism, Americans rejected the old. By 1960 only Jacobs and a relatively small band loved the old city. Greenwich Village and even Manhattan were exceptional in drawing the literati. Few other American cities were attractive to most people; in fact, many were downright seedy. Landlords who controlled residential property failed to provide services, so many families responded to the expansionist ethic. Up to 1913, immigrants temporarily occupied the old, on their way to fulfilling their dreams. Then European immigration virtually

ground to a halt, and was virtually proscribed in 1924. After 1945 too few Blacks filled the widening void in central cities. Too few renewal projects could fill them up either, so abandonment ensued. After all, the economy rides on newness. Horace Greeley told the young man to go west; latter day publicists advocated the suburbs.

The third ideological strand is freedom, which is obviously closely allied to expansion and newness. Prominent throughout Jacobs's writing is business spontaneity. Freedom has meant control over property. Freedom to do what one wants with one's property was the basis for household independence, which was the great attraction of America for immigrants. But corporations arose, blessed in legislation and by the courts as persons like the farmers, so they are no different than the corner storekeepers. Since they're not, Americans and Canadians cannot muster sufficient will to combat their power.

Finally, deep in American lore is the theme of populist voluntarism. Historian Frederick Jackson Turner, who expressed the powerful frontier ideology, stressed voluntary collective action where governments were weak, but then too, England has produced all kinds of voluntary societies, and Canada has also. Citizens in both countries, however, rely on government more to achieve social, even economic ends, despite the Margaret Thatchers and Brian Mulroneys of recent times. Certainly Torontonians could show volunteer-organized resistance to expressways as vigorously (perhaps even more so) than Manhattanites. American private health management organizations are voluntary and are thus considered good, while a government system is not, the dominant message that overturned the recent Clinton initiative. Never mind that the Canadian or British systems are actually less bureaucratic, less costly, and more universal.

Jacobs has ignored these problems in her philosophy and therefore ignored the contradictions that arise from her liberal stance. First, American society, ironically, is more bureaucratic. Excessive devotion to freedom, and thus to insecurity and instability, has produced the opposite. Excessive protection of freedom results in more prisons.

Another contradiction is that small entrepreneurs can become big ones. Today, if successful, owners of small firms find it hard to resist the dollars dangled before them by the large corporations, or they serve the large firms. All seek concession from government. When big companies downsize, they send out 'work they used to perform in-house' to small ones who will continue to be vulnerable to the vagaries of the great powers.[74]

A further contradiction is that Americans have limited choices. Health management organizations limit choices of doctors, while the Canadian system does not. Jacobs would agree that suburbs limit choice because they are not diversified. Her adopted city has a good public transit system that facilitates commuting in suburbs for children, older people, and the poor more so than in other cities where only cars can be used. However, in a recent letter, Jacobs and her Energy Probe colleagues want to privatize this system to allow competition with the vague promise that service will improve through flexibility. They have chosen the example of Thatcher's Britain where companies outside London service commuters during rush hours, but stop running after these are over! This solution is disastrous for sanity and equity. As European experience shows, transit can only work seriously if the use of cars is severely limited. Of course, to most North Americans that would deny choice.

Her philosophy of liberalism results in social exclusion; some people have more choices than others, but this never comes up for discussion. In 1835 Tocqueville observed that Blacks had not voted in a Pennsylvania election. He then asked an informant why this was so. The informant said they had the legal right, but did not exercise it. Why not, Tocqueville persisted. They did not because 'they are afraid of being mistreated'.[75] The authorities were unable to protect them because 'the majority entertains very strong prejudices against the Blacks'. Tocqueville queried mischievously: 'Then the majority claims the right not only of making the laws, but of breaking the laws it has made?' He noted that debate on a particular issue will ensue among the majority until a decision is reached, then silence. Minority voices are shunned. In this absolutist regime of the people, 'there is so

little independence of mind and real freedom of discussion …,' and too few checks on majority power.[76]

The majority, that is, the media power abetting the Republican view, resolved the debate on health insurance into either health management organizations or the unworkable 'bureaucratic' Clinton model. As one commentator asserted on public television, neither a more moderate Republican model nor the Canadian 'single payer' had a chance to be heard. Since American publishers have reissued Tocqueville's work in translation a number of times, perhaps it shows that not all Americans are classical liberals, but then Blacks are still excluded, for the most part, from full participation.

The central problem with sovereignty of the people is that there are no leaders with authority. As Tocqueville regretted, 'public opinion constitutes the majority'[77] and the legislature and executive obey it. Gun control and capital punishment are current matters that spring to mind. As I write, Amnesty International has condemned the United States for executing minors, putting it in a league with Iraq and Iran. The people want revenge, so the authorities give in to them. The powerful lobby government for more support while proclaiming that government is bad. The United States is headless. It is in the hands of a small number of an often changing élite who pursue the bottom line rather than the public interest, or they subsume the public interest under theirs when they lobby collectively. Corporate power rides higher than elsewhere in the Western world because, ironically, Americans defer to their authority more so than others. The fact that fewer than half bother to vote suggests a deep-seated despair that government cannot help. In the November 1994 elections, only 20 per cent of the electorate created the Republican majority in Congress. America may be one nation under God, it seems, but God needs an intermediary in government, not profit-seeking managers or entrepreneurs to whom the courts have bowed.

Jacobs notes 'apathy on the part of the rest of us' had 'something to do with the situation', with the tearing down of cities in the 1950s for redevelopment.[78] It would seem the majority are antiurban, according to the media. Thus the federal government aided and abetted the banks and developers, private and public, in overrunning the cities with supposedly creative destruction. The Great Depression had greatly worried many of the efficacy of the American way. Its leaders had to renew the dream with a vengence. Jacobs lost the war over cities, in large measure because of the power of property with which her own views actually coincide. She might say that rebuilding cities is not productive, but then who decides this in a society where money overwhelms social concern? One wonders why Jacobs believes that old cities should be privileged. Liberalism, in its guise of American conservatism, has privileged private property, and therefore the private corporation, through legislation and especially the courts. It has hindered the public corporation, and so the sense of commitment to the common public life, just as it has subordinated workers and unions. It has turned citizens into consumers.

Many have considered the city as a place to live morally inferior to rural life, its older districts where the poor lived worthy only to be torn down. Little wonder that this was done in the 1950s. Why did Americans with power reject the cities? The poor have always been conspicuous in cities, more so when Blacks arrived from the South. In America, the frequent but fatal efforts at purification and eliminating crime and violence (that is, Blacks and the poor), and moving to new edge cities to forget the past, killed many central cities. The author who coined the term 'edge cities' seduces readers with his book's subtitle, 'Life on the New Frontier'.[79] Like those she criticizes, Jacobs apparently believes that the world has limitless frontiers for entrepreneurs if only governments would let them get to work. Unfortunately, there are social and environmental limits to liberal dreams. Sustainable cities and a sustainable world depend on collective management.

CHAPTER 2

■

Urban Influences:
Persistent and Changing Forces Since 1600

Cities have played the central role in organizing societies over four or five millennia. So it has been in North America. The most clever and ambitious, and those controlling money, skills, and machines, have based themselves in cities to tap the riches of the world and organize the lives of ordinary mortals. Publicists have encouraged these entrepreneurs, arguing that their action facilitates growth through speed, convenience, comfort, and safety. The unspoken message is a promise of status to those who move to cities. All this worked in the past. Largely spared the devastation that many inventions of war have wrought, and despite severe financial crises, over the past 400 years North Americans have increasingly shared an ever more abundant life, which is certainly an absolute ingredient in buying status.

Yet cities, like Rome, have risen and fallen, suggesting that they are ultimately subject to uncontrollable forces. Less ultimately, cities and their populations are also subject to the sometimes rational but often willy-nilly decisions of people with power. When conditions are good, actions seem rational; when not, politicians are hauled over the fires for their incompetence. In short, cities and even their most powerful residents live within the limits of the human condition.

These limits seem more apparent today than at any time over the past 400 years. Rapid economic growth in earlier centuries slowed in the more volatile twentieth century, and has faded to a virtual no-growth state. In the past, we have banked on ingenuity to sustain growth and provide pros-

perity. The power of cities to organize the economy may well be petering out. Recent proclamations that cities are the 'engines of growth' are mere bravado. The more strident the claim, the more one should be wary. Every upswing is cheered and the instability is momentarily forgotten until another crisis, so frequent in recent times.

In a slow-growth state, maintaining healthy public environments and social equity should be the priority of those with power—not just politicians but the great leaders of finance. In these essays on five great cities, I am concerned primarily with how leaders have responded to growth and slowing growth, and how the vast majority earned a living and enjoyed or endured living. The question for our common future, to which I will return in the last chapter, is whether a permanent state of slow or no-growth can be a steady state of equitable well-being and environmental responsibility for future generations. The last prolonged modern no-growth, even negative growth, era—the 1930s—was rescued in large measure by reliance on military spending and the creation of the 'welfare state'. Those solutions have, it seems, lost their punch.

Like Alice in Wonderland, we do not understand why we grew and shrank, yet we try. To attempt to understand cities today, we can start by looking at six major groupings of influences: (1) land and space, (2) people (numbers and kind), (3) business organizations, (4) government and nonprofit bodies, (5) technologies, and (6) ideologies or culture. For this study, they are the main independent variables for understanding cities and the fulfilment of their residents. Taking what nature provided, people organized themselves and their businesses and governments, and shaped their lives using various technologies to create and perpetuate societies. Cities, the locus of power, have been a central outcome. Geography and history can be combined in historical geography. On the geographical side, the generosity and limitations of environment and space influenced action; on the historical side, the flow of human action altered the environment, including the people themselves.

Land: Resources, Space, and the Bounds

Land provided the basis for agriculture without which cities would not have been possible—at least for long. Nowadays, urbanites take food and fabrics for granted. For shelter, people have relied on the plentiful trees, rock, clay, and other materials for construction. For heat, they have used wood, coal, oil, gas, and electricity. Water power and minerals became the basis of industrialization. As raw materials, these goods provided the staples for interregional and international trade. In contrast to Aboriginal ownership, these have been the property of persons, corporate 'persons', or governments.

Space too has been a dominant concern. Overcoming the constraints of space to increase the speed of movement of information, goods, and people, has been a major preoccupation of urbanites since the beginning. At the end of this chapter, we will consider some key spatial principles dealing with cities on the national, regional, and local scales.

Nature not only provided but constrained. By the mid-nineteenth century, North Americans began to think not only about water provision but water pollution. Later in the century, forest conservation, often associated with recreation, became an issue. Later added to the conservation agenda were agricultural land degradation, flood control, waste disposal, loss of biodiversity, and (only recently) the dramatic depletion of fish stocks. Only now

have we begun to recognize resource shortages: not all minerals were not used up, but the cheap and accessible ones in North America, such as oil and base metals, have been. This appears to be a major reason for slowing growth. There have been unintended negative qualitative results of what seemed good, such as the excessive use of fertilizers in agriculture. Nature has also imposed limits by creating new diseases, as well as seasonal climates, sporadic weather, and tectonic action.

Less apparent—but at the exact centre of my underlying argument—is nature's subtle ability (as it were) to slow the growth of population and even the economy in a finite world. As curves drawn on logarithmic graphs show, growth can never continue at the same pace and eventually slows down. People are time-bound natural creatures. All attempts to annihilate space and time, to simplify tasks, like Sisyphus, reach limits. The harder we try, the greater the cost. Cities may have been organizing centres for exploitation, but exploitation may well have gone beyond the bounds of material sustainability. Put another way, cities are probably too big and complex for the good of everyone. More obvious in many Third World countries, cities seem to absorb redundant people, like writers of books. Subsequent sections discuss how technology has pushed the limits back and how liberalism, which fostered growth, led to the current debate over whether an economic goal should be pursued.

People

Demographic variables have also been crucial in creating society. First and foremost, our chief measure of economic health has been growing populations. Sheer numbers are the simplest way to measure growth (even if powerfully modified by technological improvements). However, the rate of population growth inevitably declines. In any new open area, early growth is rapid, but gradually the pace slows. The bigger the population, the harder it is to match further increments through births or immigration. To double the current population of Canada and the United States would mean adding about another 300 million people, an impossible outcome over a decade (or even several decades), unless a dramatic rise in birth rates occurs. A birth today has a less marked effect on change than one in an earlier age.

In early years of settlement in many regions, the population doubled over one or several decades as, for example, Ontario in each decade from 1790 to 1860. (From 1780 to 1790, it was far faster starting from a tiny base.) Malthus was thus wrong in assuming continued geometric growth, though probably not in our ability to strip certain resources. This growth curve will be a fundamental base for understanding changes in economy and society since, given the open North American environment, the pace at which people arrived was the major determinant of early economic growth. The act of settling, of establishing farms, may have had the greatest impact ever.

On the other hand, we know that immigration and internal migration have been a response to perceived opportunity. Population movements today remain a reaction to potential job opportunities, though obviously the proportional magnitudes are smaller than earlier. Central to the story is how leaders and citizens in cities responded to growth and its subsequent slowing down of growth.

The long-term trends of population growth were as follows (Figure 2.1). Until 1670, the colonies that later formed the United States and Quebec experienced very rapid expansion. Then until 1860, the population increases per decade ranged from 30 to 40 per cent. By that time, the humid prairies were largely occupied. From 1860 to 1910, the rate was above 20 per cent but falling, so that subsequently the pace dropped below 20 per cent, and far below during the 1930s. After a rise during the baby boom (slighter than during earlier booms), the rate of population growth was down to 9 per cent during the 1980s. Canada's

> ... Canada with its 30 million will never match the United States with its 265 million in population, though Canada has become large and rich enough to join the Group of Seven nations.

path was less smooth, in part because settlement was not evenly spread out and its population was smaller. In 1790 the Canadian White population was just under 5 per cent of the American one, rose to 10 per cent in 1860, fell to 7 per cent in 1900, then climbed back to the current 11 per cent or so. After 1945, immigration to Canada was proportionately far greater than it was to the United States, largely because the Canadian economy was stronger. Of course, Canada with its 30 million will never match the United States with its 265 million in population, though Canada has become large and rich enough to join the Group of Seven nations. Since the onset of settlement, the indigenous First Nations population fell dramatically as disease and wars took their toll, and has began to increase again only recently. It is remarkable that most nations or tribes actually survived the European onslaught.

Other demographic variables will be used in this discussion, though they have an influence secondary to my main thrust of sheer numbers. First, vital statistics. Over the long term, falling death rates, followed later by falling birth rates, created a demographic transition that altered the age profile of the population, thus raising the average age. Aging today has become an increasingly central preoccupation in social policy. Immigrant ethnic and religious characteristics have influenced politics and life within cities, including the response of descendants of earlier immigrant groups to later ones. Roman Catholic Irish, who were vilified in the mid-nineteenth century, respond no differently than the descendants of earlier Protestant charter groups to new immigrants or to Blacks. Quite possibly too the most ambitious Canadian migrated to the United States.

Then too the interplay of race and social class is central to understanding cities. How Whites and Blacks have dealt with one another is obviously crucial. A salient difference between Canadian and

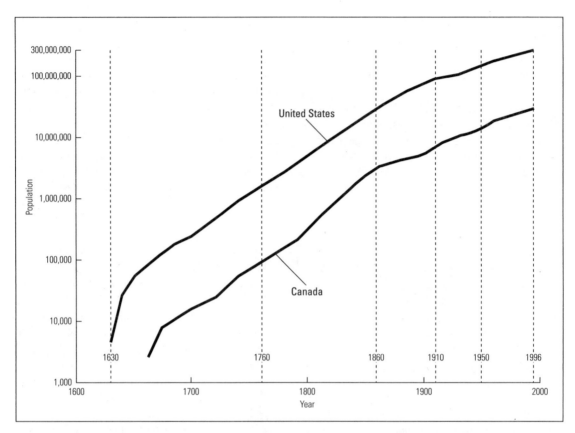

2.1   Population growth in the United States and Canada, semilogarithmic scale. Early, very rapid growth gave way to a still-fast pace to about 1860, followed by decelerating growth to the 1930s. Growth picked up again, though immigration has recently compensated for the lower reproduction rate. Because Canada has a smaller and patchier agricultural land area, its path has been more erratic, dropping from about 10 per cent of the United States in 1850 to 7 per cent around 1900 before rising to over 11 per cent today. *Sources:* Pre-1790 US from US Bureau of Census, *Historical Statistics of the United States: Colonial Times to 1970* (Washington: US Government Printing Office, 1975):1156. Pre-1867 Canada from Canada Bureau of Census, *Census of Canada 1870–1871*, vol. 4 (Ottawa: Canada Bureau of Census, 1876, 4):xv–lii. Others, decennial censuses.

American twentieth-century cities has been the relatively larger numbers of Blacks in the latter. As suggested in Chapter 1, race is not, however, the fundamental determinant of the sicker American central cities compared to healthier Canadian ones. Blacks commit more violent crimes, but they are just as American as Whites. Blacks have served as scapegoats, as a necessary social glue in the most liberal of societies. To paraphrase a title dealing with early America, American freedom depended on slavery. After emancipation, they served as the convenient bottom rung of class. Even the poorest of Whites could claim higher status. Put another way, without Blacks, this most liberal of societies,

espousing freedom above all virtues, would have had to invent a category of people by caste, say green, to maintain the myth of classlessness. This long-standing mechanism of exclusion still continues, though many would like to believe it will eventually disappear. Some Blacks have risen economically, some to high celebrity status, which might suggest the beginning of a dissolution of race as an issue. Miscegenation is, however, likely the only final cure and, in fact, intermarriages are becoming more frequent. If race were to disappear, then debates and action based on class would emerge more strongly, given that power over others has been a cardinal reality.

Still, social classes defined by income, organizational participation, and subjectively by people themselves are in reality important even in American society. Public reactions to policy by class continue to recur. Where people reside is a reflection of class as well as race. Social standing has been the most important input into establishing leadership. In the following chapters, we will try to understand how leaders responded to growth and its decline in shaping the public social and material environment.

Organization of Business

By business I mean all those activities that can be classed as the pursuit of material well-being and status through profit, ranging from the corner store to huge entertainment corporations. It includes farming, forestry, mining, trading, leisure provision, media, finance, and the property industry in cities. Governments, which we discuss separately, have also had a hand in promoting growth by running some businesses, as well as by regulating the private sector and maintaining order. The line between business and government has been by no means clear. Although it apparently became more distinct in the nineteenth century, the two sectors were never separated: the post office, for example, became crucially important in the middle years of that century and remains a conspicuous example of government enterprise. For better or worse, local and higher levels of government pass rules to regulate the economy. Since the Great Depression, governments have taken a more active role in the economy, though that has been weakening in recent years. In a sense, mercantilism has never disappeared. The wealth of nations is still a collective pursuit.

Corporations of varying scale and spatial reach today employ most people; the self-employed comprise just over 10 per cent. Historically, many more ran their own operations, like the independent farm household and the corner store. Yet people have organized others from the beginning. Under the aegis of the Massachusetts Bay Company, groups of investors established the first towns in Massachusetts. Southern plantations and ironworks employed many slave workers, while many waged workers worked in shipbuilding. Most wholesaling, retailing, and manufacturing operations were proprietary run. Central finance operations were in the hands of unspecialized merchants or retired successful merchants and, of course, government. Over the first two centuries, the population actually became increasingly rural and those in the countryside engaged in a range of activities. Prominent farmers and millers were financiers too, for example. A graph of increasingly specialized employment obscures the range of activity wrapped up in one person or household in earlier times: a majority were farmers, processors, traders, and communicators (Figure 2.2).

By the end of the eighteenth century, specialization accelerated and so urbanization began to rise, most dramatically in the United States between 1845 and 1855, and later in Canada. The number of farm operations continued to rise until about 1913, then fell dramatically. Independent shopkeepers began to lose out to corporate operations. About then urban populations caught up with rural ones in most regions, with the exception of the American South, the Canadian prairies, and American central plains. Some regions were urbanized much earlier, most obviously New England, because of the rise of factory operations there through the so-called Industrial Revolution. (I refer to it as so-called because the notion obscures the immense degree of industry before then.)

After 1750, factories increasingly replaced small shops and household manufactures in England. At the beginning of *The Wealth of Nations*, Adam Smith in 1775 approvingly noted eighteen different steps needed to make a common pin and that someone integrated them by organizing workers to carry out the tasks. Eighteen workers who specialized could produce far more in a day than one person undertaking all the steps. This was the basis for mass assembly-line operations later. America and the world followed Britain and Smithian virtue, factories integrating operations under one roof. Helping America along, Britain provided enormous amounts of indirect investment in North America from the beginning and,

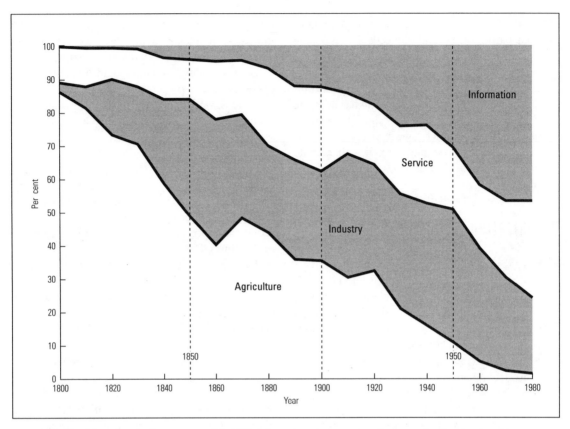

2.2 US civilian labour force by four sectors, 1800–1980. Information refers mostly to bureaucracies that developed in spurts after 1830 to organize flows within the economy. Throughout history, people have used 'information' to order economic life. Diversification gave way to increased specialization. James Beniger, *The Control Revolution: Technological and Economic Origins of the Information Society* (Cambridge: Harvard University Press, 1989):23.

although later declining relatively as more money was generated internally, remained a prominent investor until 1914.

From the 1780s onward, paralleling and pushing urbanization, corporate organization was far more prevalent in the United States than in Britain, where partnerships prevailed. Apparently the freer and more liberal, the greater the urge to incorporate. By the 1820s, corporations were frequent. After 1850, the truly great impersonal corporations ran extensive operations, overshadowing smaller operations. Smith's invisible hand of a myriad of small units gave way to the 'visible hand' of managers of these hierarchically structured enterprises.[1] The now legendary moguls created railway and telegraph systems (which were then giants of

production and distribution), media empires, and all financial and investment institutions. The 'robber barons' integrated horizontally by grabbing market share through buying out competitors. John D. Rockefeller and a small number of others used cost cutting to drive out their competitors. They integrated vertically to control the material and informational resources and the marketing of commodities. This dominance of scale and reach should not be overestimated, however; many smaller companies arose to fill niches and supply bigger ones. Production and productivity rose to such an extent that prices fell through most of the century (Figure 2.3). Businessmen who managed large operations persuaded legislators to limit their personal liability and to define their companies as legal persons, impersonal as they increasingly became.

Around 1900, mergers created even greater enterprises that dominated several fields of manufacturing and distribution. By then, corporate hierarchical structures and lines of authority in complex systems were established and have largely remained unchanged since then. Gone were the great entrepreneurs of the previous generation. Stockholders were owners who were more interested in earning profits than in watching the quality of goods as managers took the reins. Financial giants like J. Pierpont Morgan controlled boards of directors. The enterprises possessed enough oligopolistic power, it seems, to control prices, so prices began to rise. The biggest firms became not

only binational (that is, mainly American companies taking over many Canadian plants) but multinational.

The nineteenth century had witnessed the 'corporatizing' era; the early twentieth century up to 1929 was the 'corporate' regime, that is, when corporations rode high, defining not only production but welfare. They smoothed out the rough, chaotic, *laissez-faire* of the nineteenth century. It has been referred to as 'welfare capitalism' because some owners took a paternal interest in their workers to maintain their loyalty rather than overworking them as earlier.[2] Companies were beginning to discover that to sell, they had to sell to workers too. Marketing superseded the drive to increase productivity. Although a high corporate auto exec-

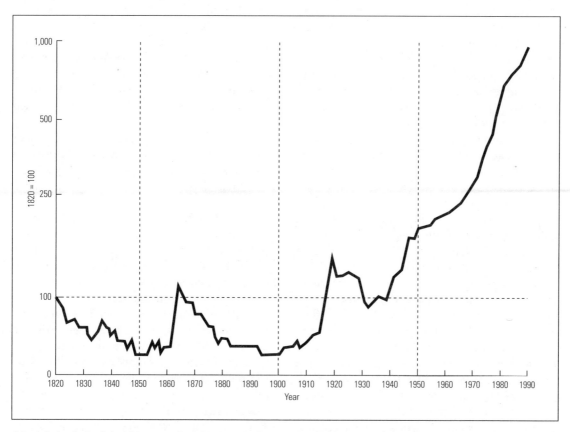

2.3   Inflation in the United States, semilogarithmic scale. Except for the Civil War era, nineteenth-century prices dropped, reflecting more productivity than consumption. In the 'corporate' era after 1896, prices rose again. After the Great Depression, among other factors, government spending on military hardware to keep the economy moving contributed to the sustained inflation until the recent slowdown. The Canadian path was no doubt similar. *Source: The Economist*, 22 February 1992.

utive later uttered the slogan, 'What is good for General Motors is good for the country,' the term was more appropriate for the 1920s when Alfred Sloan perfected management flows, just as Henry Ford had on the shop floor.

In the wake of the crash that devastated production, corporations revived primarily through war. Federal governments managed the economy through the purchase of military goods for hot and then cold wars, spurred consumer spending through welfare and tax measures. One result was the rise of the corporate urban land industry to build houses, offices, and malls. The ensuing brief near 'corporatist' era, when government, large business, and (to a degree) labour cooperated, lasted until after 1970, though less completely and less explicitly than in western Europe. Even before then, the most powerful created conglomerates combining many different kinds of enterprises within holding companies. Unfortunately, these combinations reflected not so much an interest in improving productivity as increasing power to those who executed the deals. In fact, after 1965 productivity per worker fell even as large manufacturing firms downsized because of 'flexible' methods.[3] An expanded money supply, largely the result of profligate government spending on military goods, led to dramatic inflation. A major outcome of the welfare state has been the rise of large public and private pension funds. Their managers, not surprisingly, abhor entrepreneurial risk, yet want quick returns to maintain their funds' viability.

Since 1975 North America has been in a 'silent depression', noted in Chapter 1. Let us call it a 'chaotic silent depression' because the great holding firms take over companies, reject those deemed unprofitable, so firms change ownership and their executives frequently. Foreign control over large parts of the economy has contributed to this chaos in an economically weakening continent. The American dollar floated after 1972, loosening stability. The preoccupation with quick gains undercut order by contributing to ill-conceived faddish experimental measures designed willy-nilly to raise productivity. Executives search for solutions through re-engineering.

A major social result of corporate concentra-tion was the rise of unions. Big companies resulted in big unions. A long-held but erroneous view is that labour was scarce in the past and therefore expensive compared to Europe. The abundance of land, it has been argued, drew off workers, thus raising the price of labour. This worked to a degree, but as land became increasingly expensive in the nineteenth century, ordinary people had to seek jobs in the expanding factories. Industrializa-tion was facilitated because unskilled 'labor was cheap', especially when immigrants arrived in droves.[4] Immigrants came because they perceived greater opportunity, or at least less social harsh-ness, than in Europe beyond the first difficult years. Later waves of immigration helped to push earlier ones upward. The descendants of the poverty-striken Irish in the 1840s improved far beyond their parents' material lot. This push upward has hardly been easy for Blacks, most of whom remain at the bottom of the scale. Poverty levels have remained higher in America than else-where in the Western world.

From associations of craftsmen before 1800, transport, media, and factory operatives combined increasingly if erratically in the nineteenth century to enhance their status. Wages, workplace safety, and job security became battlegrounds between unions and bosses. Unions in new crafts were cre-ated, while those in older ones lost out. Skilled workers generally did better, though, as in the 1880s (a period of rapid technological change), certain skills became obsolete, so workers orga-nized to hold back the tide of change. But even for new crafts workers, organizing was not easy. Denied bargaining rights by law, industrial workers did not gain independent recognition until the Great Depression. Then unskilled unionists enjoyed income security for a time, though (unlike Europe) few had much stake in management.

In the United States, legislation and the courts historically disadvantaged unions while privileging private corporations. So after 1945, reflecting on management's position, the courts and legislators whittled away at unions' powers, while in Canada they made it easier to organize. For Americans, the gains of the 1930s turned out to be 'a counterfeit liberty' that promised to lift the subordinate label off workers.[5] Thus, since 1965 union membership has fallen dramatically to 15 per cent in the United States, while in Canada it has remained more than twice as high, suggesting that Canada moved closer to European ways. The loss of membership in the Canadian manufacturing sector has been compensated for by higher rates in services and government, though it was weakened recently.

The American system of mass production has also tended to segment its workforce more rigidly than in other Western countries. Operatives in corporations in Japan especially are not as remote from their bosses, managers, or the technical experts between them. In the United States flexible work team arrangements of recent times have hardly altered the most extreme hierarchical system in the Western world, ironically in the most liberal of countries. How can it when the bottom line looms larger than elsewhere? This is indeed one of the ironies of American history: freedom begets rigidity. Supporters of the American way respond by arguing that workers get cheaper goods than elsewhere through mass production. Much mass production has moved offshore. To be sure, unions with pension funds have some financial clout, but in the American system, government regulation also has fostered 'strong managers' and fragmented 'weak owners', so that their power is limited, the result of 'a deeply ingrained popular mistrust of concentrated financial power'.[6] The executives who run corporations have, as we know, great power over property.

More flexible labour markets and lower unemployment than in other Western countries are the results of the dramatic fall in the American rate of unionization. It has also meant more low-paid, part-time jobs. Taking into account a greater number of discouraged workers who are not included in the statistics when they stop looking, and part-timers who would prefer full-time work, the recent American rate is actually closer to those of Canada and European countries. One American business columnist asserts that 'U.S. job statistics are dangerously wrong', based on sloppy counting.[7] Peo-

ple who are not working in Canada 'are more likely to be classified as unemployed'.[8] Although the 9 or 10 per cent who are officially unemployed today in Canada does not reach the level of the Great Depression, the plethora of baby-boom skilled workers has shut out the unskilled and the young, even from lower-status jobs.

The workings of business were a major reason for concentrating populations in towns and cities after 1790, then later drawing more people to the largest metropolises with huge factories and also offices where a new middle class of white-collar workers toiled. Then too private real estate developers created towns, subdivided land within cities, after which others built on their lots. Behind them and other entrepreneurs were the investors. In the last analysis, a handful of financiers have been the chief actors controlling the flow of investment and thus, for better or worse, the lives of most (if not all) people in cities.

> ... one of the ironies of American history: freedom begets rigidity.

Government: Formal and Informal

Governments have exerted great influence on urban patterns at the macroregional and local levels. Less easy to assess in city building are the nonprofit organizations or, if you will, informal governments. They have directly acted and influenced legislation and government action. Public expenditure by all levels in the United States rose from 12 per cent of the gross national product to 45 per cent between 1929 and 1945, the federal share of that from one-quarter to over half. Since 1945, it has remained higher than before 1930. The pattern has been similar in Canada, but even before 1929, government employment was growing dramatically. Over the fifty years after 1870, the American federal government expanded at over twice the rate of population growth, and even faster between 1940 to 1970. Even in the 1980s, reputedly a decade of privatization, Reagan added federal employees at a faster rate than any of his postwar predecessors, though in the 1990s cutbacks have hit public sector employment. This reflected the federal government's increasing power since the 1930s compared to states, provinces, and municipalities. The latter are constitutionally 'creatures' of the states and provinces, but cities have, as we will see, exerted autonomy, so relationships have not been smooth, especially in America.

Governments started and shaped cities in five ways: to maintain order, regulate the economy, create municipalities, provide environmental infrastructure and social support, and promote economic growth. It is not easy to separate these overlapping functions. Government and business have always been intimately connected, one way or another. First, executive authority needs places from which to keep order through military and police coercion. Colonial administrators built structures to house these operations. Subsequently, imposing buildings with their flags, other symbols, and at times uniformed soldiers made the point about coercive power and defined boundaries. Second, officials set up legislatures, courts, and administrative bodies, not only for public order but to regulate the economy. They congregated in cities, thus acting as multipliers for other activities, as did private business. Colonial officials cosily shared power with the prominent merchants and lawyers, as is often the case today. Thus capital cities were the locales of many, probably a majority, of the highest-status men in early days. In fact, in those less specialized days, the officials themselves engaged in trade and real estate. The early capitals—the seats of authority—were the nuclei of several of today's large cities.

Higher authorities established local municipal governments and delegated responsibility to regulate the immediate urban (as distinct from region-wide) questions. In the nineteenth century, local control over businesses waned, but as corporations grew, legislators themselves laid down more and more rules for business, so today they are massive in number. As legislators anxious to reduce government spending and bolster productivity discover, eliminating regulations is not easy, and cutting them back can result in unintended consequences. By and large, rules have helped the cause of the large firms. A well-known American economic historian concluded a generation ago that the 'regulated regulate the regulators, in the interest of the regulated—rather than that of the public'.[9] Yet on the other hand, consider the recent multitude of environmental laws that have emerged from the legislative chambers, and the countless court cases that have attempted to strengthen the public good. Given the power of property in the more litigious United States, the American courts have been more involved in creating law than the courts in Canada.

A fourth role has been to build public works and systems that have improved the quality of living, helped business, ensured social support, connected regions, and created links within cities. During the late nineteenth and early twentieth century, for example, the heyday of local assertiveness, city governments taxed property to pay for expanding material and social services, including education, which in many jurisdictions was extended to the compulsory age of sixteen. When the limits of local ability to tax were reached, municipal politicians pressured states and provinces for more funding.

The stage was being set for federal governments to intrude into local planning and the lives of urbanites beyond the constitutional power that states and provinces had over municipalities. The Great Depression provided the occasion for massive increases in government action on social supports and material infrastructure. Once the federal governments committed themselves to full employment just after 1945, social supports and urban planning no longer remained local or even exclusively state concerns, though many financial transfers passed through state, provincial, and local governments, as well as directly to individuals. Through social transfer payments, governments succeeded in stabilizing incomes and thus security (Figure 2.4). Education was expanded in the name of human capital, though its greatest importance was job creation. Postsecondary students got unpaid work, teachers paid positions. The great expansion of social spending weakened in the late 1970s. Today a sense of financial crisis hangs over infrastructural maintenance, needed to connect various regions of the countries, and especially

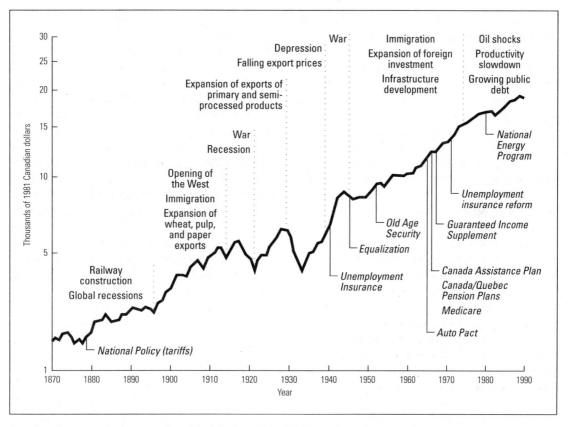

2.4 Canadian per capita income, adjusted for inflation, 1870–1990. The path was bumpier before 1945 than it was later. Government welfare transfers and the larger number of workers in services (especially in the public service) smoothed out the trend. Income gains were substantial in the decade after 1880. Compare to Figure 2.3. In the 1990s incomes fell or stagnated for most people. *Source:* Economic Council of Canada, *A Joint Venture: The Economics of Constitutional Options, Twenty-Eighth Annual Review* (Ottawa: Minister of Supply and Services Canada, 1991):2.

social support. Tax revolts, successful in the United States, but less so in Canada at least until 1994, have hurt the bottom half of the income scale, eliminating some of the gains after 1940 and threatening cuts.

Promoting economic growth has been a fifth role, hardly separable from the fourth. As has been argued vigorously, authorities 'politicized the economy in favor of special interests since colonial times: it is the All American way'.[10] Mercantilism never disappeared. Social welfare payments have had as much an economic as a social purpose, which is to maintain consumption. Earlier governments subsidized rural settlement and land speculators by massively subdividing the formerly Native lands into rural lots, townships, and counties. They financed canals, the first telegraph lines, and railways, the last in part through land grants, thus contributing spatially, for example, to distinctive linear urban patterns in areas not previously occupied. Municipal governments handed out grants, bonuses, and tax abatements to manufacturing businesses. This local promotion of economic development was most prominent after 1840 to the early twentieth century. Some municipal, state, and provincial governments, responding to populist concerns early in the twentieth century (especially in Canada), set up their own utility enterprises to promote growth and establish a stronger method of regulation. The recent urge to privatize has weakened but not completely eliminated the early twentieth-century resolve for collective ownership, most clearly in Canada, which has a stronger tradition of direct public involvement. Rhetoric to the contrary, privatization has not weakened the resolve of business to pursue government subsidies.

Historically, wars have led to increased permanent government employment. After 1940 federal governments became far more directly involved in promoting economic growth, especially after 1950 as a result of the cold war. In the United States, the federal government's presence in defence was so powerful that it altered the regional urban pattern dramatically. Through its own military and space bases and indirectly through contracts, much of the urban system in the North shifted to the South and West, but in recent years, the impulse to spend for martial reasons has been thwarted by the end of the cold war. It is not easy to reverse the decades of government growth that rose in response to the Great Depression and to the growth of cities and corporations that began before 1800. In the 1980s the public sector employed nearly one-quarter of the Canadian workforce and one-fifth of the American one. As with regulation, governments expanded. 'The American people believed axiomatically in sturdy individualistic values, yet freely legislated against those beliefs.'[11] Ask any (candid) lobbyist.

Voluntary groups continue to govern certain aspects of city life, functionally and symbolically maintaining the fabric of society. Americans have been more inclined than Canadians to see groups as a substitute for government. Churches have been the most conspicuous, and still draw by far the largest share of so-called charitable contributions. More addicted to religion than Canadians, Americans remain the greatest bastion of Christian churches. Two-thirds of Americans claim attendance at religious services least once a month and believe in the efficacy of prayer. Education remains next in the non-profit (even profit) line for donations, more so in the United States than in Canada because in the latter, state support of all levels is near universal. Cultural institutions too attract funding, though government contributions are needed, whether for sports or operas. Voluntary action founded many health organizations, and donations still support them, though obviously taxes have provided the bulk of funding since 1930 in Canada, but less so in the more corporate-run United States, as we saw in Chapter 1. As for relief and social welfare, local volunteer bodies have supplemented government action since the beginning, but despite heart-rending stories of deprivation, only a trickle of private money flows to the down and out. Despite the enormous publicity in each city, the local chapters of the United Way deliver only a small share of social spending, relatively more in the United states than in Canada. In sum, volunteer action for social welfare purposes results in less than 10 per cent of funding. Although advocates claim that 'yeast like' United Ways spur further action, voluntary governing has to be put in the context of big government.[12]

Technology

Over the past few centuries, coinciding with growth and the unprecedented rise of urban life, technological advances have been spectacular by long-run historical standards. Harnessing technology has been crucial to the urbanization process. With the help of government, business people and the technically skilled engineered a 'control revolution' that innovated the inventions of the ingenious and reorganized work.[13] From the Renaissance onward, human ingenuity has been persistent. In the eighteenth century the British led the way with machines. The clocks in factory towers forcefully reminded workers of their place in the production system. Luddites have not stopped the march of machines. Adaptations of inventions and innovations contributed to economic development and therefore to growth beyond what was possible simply with agricultural settlement and more primitive technologies. The power to move and make things has been central to these. Human energy and animal energy have been displaced to a substantial degree, though the latter obviously much more so than the former.

Wave after wave of innovations have swept over the land (Figure 2.5).

Without technical adaptations of energy sources, business adaptations of many other inventions could not have happened, and, of course, vice versa. Wind for sails on ships and mills had been around for centuries before 1800. Use of water to turn stones in mills also has a long history, especially during the medieval era, and continued to be used by settlers in North America. The abundance of wood led investors to cut hardwoods for charcoal, which was burned to reduce ores to pig-iron and other metals. Eventually coal became the great energy source during the nineteenth century for ships, factories, and railways, following its use in Britain from the sixteenth century onward. Compared to Britain, the exploitation of coal in

North America progressed much more slowly because of the abundance of cordwood for heating and stoking early steam engines and hardwood for charcoaling. However, coal became king after the mid-nineteenth century. By 1870 steam engines were rapidly displacing water-power, allowing the concentration of factories in cities. Americans could not resist the call of British industrialist, Matthew Boulton, who, in 1776, boasted: 'I sell here, Sir, what all the world desires to have— POWER.'[14]

People began using two other fossil fuels, oil and gas. From the late 1850s onward, petroleum supplied lubricants for gears in machines and kerosene for lighting, replacing whale oil and candles, then later fuelled the internal combustion engine. In turn, policy makers used the car to

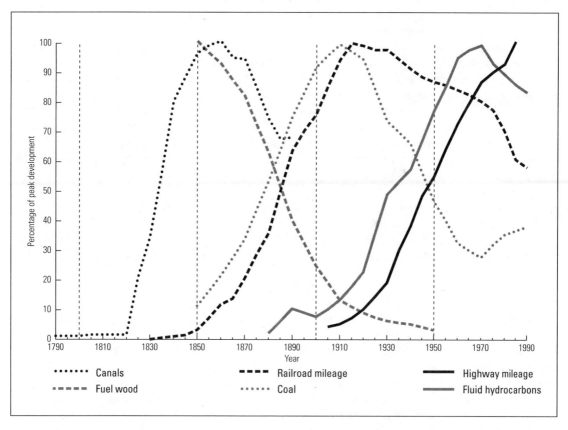

2.5   Successive technological change in the US: waves of 'creative destruction' and fuel. Note the revival of coal. *Source:* Brian J.L. Berry, E.J. Harpham, and Euel Elliott, 'Long Swings in American Inequality: The Kuznets Conjecture Revisited', *Papers in Regional Science* 74 (1995):157.

reshape cities in a dramatic fashion. Foreign sources of oil gradually became more prominent, making it clear that the heartland could not operate without hinterlands. America is relatively weaker than previously, mostly as a result of this dependence on foreign sources. Exploitation of petroleum in more inaccessible offshore sites has become increasingly costly. Less-glamorous gas is steadier, but shortages are inevitable in the long run.

While steam engines replaced water-power, electricity burst on the scene. Electromagnetic waves were first harnessed to power the telegraph. Then direct and alternating currents were generated first from coal, supplemented around the turn of the century by hydro in certain places such as Niagara Falls, later by the other fossil fuels, and finally uranium, so electricity shared with oil the status of the wonder energy of the twentieth century. Harnessing water-power sites has seemingly reached the limits, as testified by the resistance of environmentalists and Native peoples.

Electricity enabled people to transmit messages. Recent writings would have us think that only now do we live in an 'information' age, but people have always communicated. In 1778 Samuel Johnson asserted that 'society is held together by communication and information'.[15] The key invention for the modern era was the telegraph, first adapted for use in 1844: 'computers notwithstanding', *The Economist* asserts, 'the most important innovation in international financial markets ... was probably the ... telegraph ...'[16] Just think of the quantum jump from the pony express, carrier pigeon, and semaphores to the nearly instantaneous clicks of the Morse code from one place to another initiated in 1844. Then undersea intercontinental cables transmitted messages that were previously carried by the fastest clippers and steamships. In 1876, at the time of the Philadelphia World's Fair, the *Enquirer* confidently proclaimed that the telegraph and locomotive 'annihilate time and space'.[17] Then came the telephone in the 1870s, though long distance was not commercially feasible until the 1920s. Since then the messages and techniques have piled up. In recent years, computers became obsolete quickly

through waves of 'creative destruction'. The speed of electronic messages has reached the limits, yet compared to earlier take-offs, recent gains contribute less to business and government.

Electricity alters chemicals and metals, but electricity itself depends on materials. The telegraph was possible through wires (and the wires imposed a limit on possibilities). The wires had to be fabricated from copper and alloys of metals, which in turn originated as minerals mined from the earth. Long before the telegraph, experts fashioned iron, copper, zinc, and lead and their alloys into useful tools, including wire. The village smith fashioned horseshoes, the wheelwright iron tires for carriage wheels. The great expansion of mining and metal manufacturing occurred over the last two and a half centuries. Textile machinery in English factories built in the late eighteenth century (around the time industrial cities began to grow) was another trigger for mining and metal manufacturing. Arkwright and others invented spinning and weaving machines that increased output enormously. Although much of the early machinery was shaped from wood and leather, crucial parts were increasingly made of iron and then its derivative, steel. In the nineteenth century, machine tools to make machines appeared.

The British Industrial Revolution was initially associated with canals and later turnpikes. In North America workers dug canals, only to have most of them replaced by railways; 'machines in the garden', the iron horses, were the most visible of the great inventions.[18] With government help, entrepreneurs extended their rails across the landscape. To the ordinary person, the trains seemed far more important than the wires strung along their right of ways. However, while the rails carried the mail, passengers, and goods, the telegraph carried urgent messages to speed up the flow of money and commodities. Since then, chemicals have wrought miracles, but with serious side effects. New ceramic materials expand the limits of productivity further, but how far is uncertain.

Business people, their investors, and even governments have adapted the discoveries for convenience and efficiency. That is, they have built on past inventions and recognize new possibilities for utilizing various sources of energy and materials. Ingenuity continues because people just cannot resist. If science opens the doors to technical inventions, business people have been the chief innovators. While independent and unheralded individuals sometimes successfully introduce new ideas to commerce, either they grow to become corporations or are gobbled up by corporations. These processes have been occurring for well over a century. Organizational innovations encouraged production through research and development. General Electric and Bell Telephone took the inventive process inside the corporation. In the 1870s Thomas Edison expected to create 'a minor invention every ten days, and a big one every six months'.[19] In retrospect, corporations have been effective instruments in creating workable technologies.

What Technology Has Delivered

What all these technologies delivered through business organization with government help was not simply consumer goods themselves but speed, convenience, comfort, and safety. Speed in moving ideas, goods, and people has been on the agenda for 400 years. Using machines to overcome limits has been a central quest, so if new gadgets deliver speed, people want them. Factories delivered goods faster and more of them than shops or craftsmen. The mid-nineteenth century marked the key turning-point in speed. The telegraph greatly boosted the velocity of money and therefore of business. Trains dramatically increased the speed of moving people and goods. Later refinements raised the ceiling, but not as much. Computers have added marginally to the speed in moving money. Financial firms hardly need more power; they now seek small advantages in organizing and analysing data. Trains probably reached their peak speeds by the 1880s. Because of the competition from autos and aircraft, only in a few dense corridors in the United States would fast trains of the type in Japan and France be financially possible. Cars reached their speed limit in the 1930s, though paved roads and freeways reduced time and made just-in-time delivery possible. With jet engines, airplanes reached their maximum feasible speed about 1960, the Concorde the exception to the rule. Thus it appears that 'speed has plateaued'.[20]

For all these improvements, convenience was a twin of speed. The computer station can facilitate work more rather than having to visit the telegraph office. Unlike public transit or trains, the car can deliver one to the driveway. Airlines strive to provide ease for checking in and the like, though for many people, train stations are still more easily accessible than airports. In the 1960s containerization boosted convenience for shipping, and with it productivity per worker. Will more fiddling achieve further improvements to convenience?

Comfort followed speed and convenience. The first trains had unpadded, convenient benches; later ones had more comfortable seats. A trip on the subway where one can read may be more comfortable and even more convenient in some cases than driving a car. Listening to soothing stereo music while sitting on a gridlocked freeway can bring comfort too, unless the cellular phone announces an urgent matter that one does not want to deal with. Comfort goods have been mass produced, though the élites and even in recent times many others have insisted on custom-made goods to satisfy their presumed needs, as we will see shortly, to maintain status. One person's comfort can be to the discomfort of others; one thinks of the daily grind of typists at computers and low-paid Third World workers producing sports shoes in sweatshops.

Safety is more ambiguous in contributing to growth and satisfaction. Communication devices obviously contributed to aircraft speed, operating convenience, and passenger comfort, but inspec-

> What all these technologies delivered through business organization with government help was not simply consumer goods themselves but speed, convenience, comfort, and safety.

tion at airports has undercut these. Today property safety devices, like house alarms, can create inconvenience to owners and discomfort to neighbours. Regrettable necessities can become onerous. In our discussion on cities, we will note that preoccupation with safety has weakened many qualities that make for comfortable and convenient living, even of expeditious travel. The rising concern with safety may in fact signal the limits of growth—the limits nature imposes. Information, people, and goods cannot, it seems, travel any faster, conveniently, or more comfortably.

There is no doubt that applied technological developments have contributed to the gross domestic product (GDP), which has mostly risen by conventional standards (Figure 2.6). No doubt whatever added to speed and convenience has improved our material condition far beyond that of our ancestors. It would be folly to suggest that inventiveness will cease, yet each increment of speed added is less than the last one. Vehicle production too has reached its limits of growth, it seems, to satisfy a craving to keep up with the Joneses; production is now largely driven by design (Figure 2.7). The world financial system now has twenty-four-hour global share trading and is as instantaneous as one can imagine. The effects of ingenuity, as in all earlier improvements, suffer

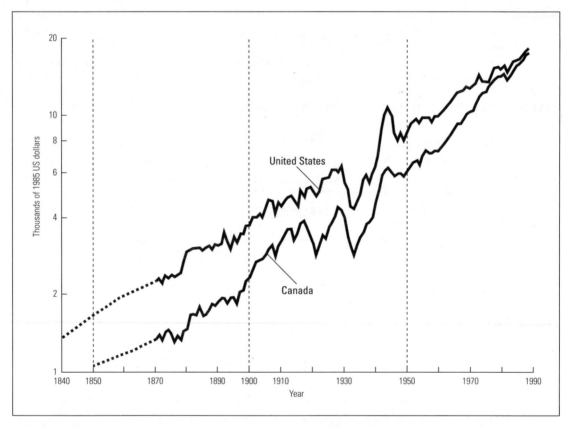

2.6   Gross domestic product per capita, 1840–1989, semilogarithmic scale. This pattern is similar to Figure 2.4. Pre-1870 data are few, so long waves are not easily recognized. From 1945 to 1989, GDP growth was less erratic than the violent swings in the early twentieth century, and even less so in Canada, largely owing to government action encouraging consumption. Quarterly data, not shown, exhibit the swings even more dramatically. Canada had virtually caught up to the US by 1989, and has since fallen back somewhat. Note how the 1973 oil (price) crisis affected the US more so than Canada. *Source:* Angus Maddison, *Dynamic Forces in Capitalist Development: A Long-Run Comparative View* (New York: Oxford University Press, 1991):Appendix A.

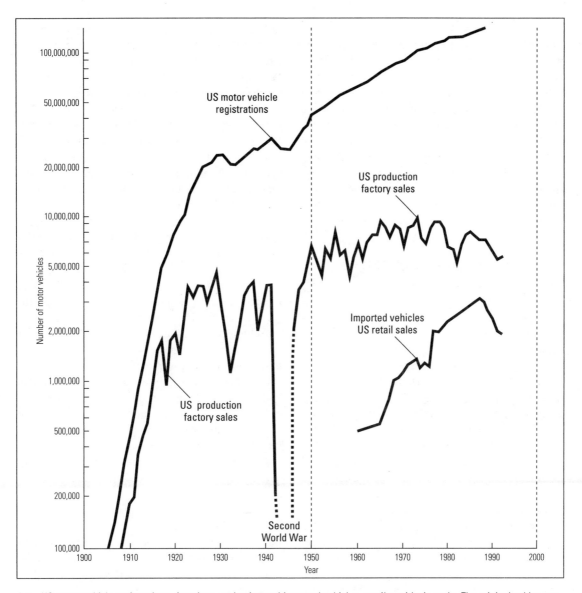

2.7 US motor vehicle registrations, American production and imported vehicles, semilogarithmic scale. The original golden age take-off was over by 1915, after which used cars contributed more to registrations. Thus marketing and organization became more crucial. The post-1945 revival was completed by 1957. Over three decades, overseas imports plus Canadian-made domestics became more important. More 'imports' are now included in US production. The Autopact of 1965 integrated Canada's production with that of the US. Trucks are more prominent than earlier. A stagnant pattern of domestics has been apparent since 1978 and recently imports have shown a similar pattern. *Sources:* US Bureau of Census, *Historical Statistics of the United States* (Washington, DC: US Government Printing Office, 1975); US Bureau of Census, *Statistical Abstracts* (Washington, DC: US Department of Commerce, 1975).

from diminishing returns. Production and consumption have reached their limits.

Perhaps too the limits are imposed even by the content of the messages themselves. Perhaps human beings, especially those with power in finance and government, cannot absorb any more speed or any more convenience and comfort, for that matter. Let us speculate that part of the enormous difficulty politicians face today is that they have no time to take in and sort out the bombardment of messages. Instant gratification is hard to satisfy. In 1994, halfway through William Clinton's four-year-term, a citizen said she would not vote for him again because he has been in the White House too long! Perhaps we should have taken Mark Twain seriously when, on the inauguration of the long-distance telephone, he supposedly said something like, 'Do the people of Maine have anything to say to the people in Boston?' Perhaps people did, probably to more effect for growth than the webs for today.

Ideology: The Liberal Culture

The two dominant ideologies of this continent have gone hand in hand: liberalism (stressing individualism, choice, and private property) and growth (stressing expansion and development). The occupation of what seemed to be limitless space reinforced and deepened the liberal commitment. The origins of these ideologies have their roots in Europe before 1600. Even though earlier customary and natural constraints hindered rapid transformation of the landscape to full-blown liberalism, frontier expansion had occurred there following the decrease in population as a result of the Black Death. In Britain, the enclosure movement of medieval manorial strips elevated the notion of individual control and operation, if not always ownership of parcels of land. In America from colonial times onward, settlers of New France, New England, Virginia, and subsequent colonies, and later of states and provinces, sought to set up independent households on their own holdings. Although some became tenants, the cultural norm was ownership, even in New France where ostensibly seigneurs controlled most of the land. Before John Locke said it, the settlers were already asserting the right to enjoy the fruits of their labour. Before Adam Smith advocated freehold as the backbone of the wealth of nations, the independent farm operated by the nuclear family predominated in America. Europeans of every national stripe and virtually all religious groups, no matter

how conservative, sought freehold property. A late eighteenth-century French immigrant, Hector Crèvecoeur, moved from Connecticut to frontier Pennsylvania to fulfil 'the desire of contributing to the success of my family, and of assuring its independence after my death'.[21] The European landlord was anathema, though American ownership of Blacks was not. Since farmers and ranchers took three centuries, even into the twentieth century, to occupy land for agriculture and ranching, the veneration of property was sustained, reinforced through every generation.

However, the line between providing for one's offspring and speculation in land was not clear in the slightest. In fact, speculators or developers expedited settlement (or sometimes held it back for greater gain), most obviously during the early nineteenth-century era of massive spatial expansion. Successful independence was contingent too on selling commodities in the market; few sought self-sufficiency and isolation, and even fewer could find it. Inevitably, people were caught up in social and commercial networks. Independent farmers had no defence against them because they too traded in land and commodities. The era of the 'common man' in the 1820s and 1830s in Jacksonian America may have been the apogee of freehold sentiment, as creating the 'fee simple empire' on 'virgin land' was in full flood, helped along by large speculators.[22]

High home-ownership in mid-twentieth-century cities may have emerged from the drive for freehold in farming. Just as the rural freehold empire began its decline after 1900—that is, as the number of farmers fell dramatically when they were driven off or attracted to the city—home-ownership began to increase in most cities. Poor immigrants often led the way. The biggest cities, most notably Manhattan and Chicago, lagged, however, as it seems status was equated with high-value, time-consuming jobs rather more than property. But ownership was held back until government financial support began in the 1930s. Home-ownership as the post-1945 mainstay of suburban expansion has exerted a powerful influence on social patterns within cities. The expectation of home-ownership seems to be held more deeply in America (and other English-speaking countries), though in recent times Europe has followed the trend to an extent. Where it has been weaker (in urban Quebec and Britain), the urge has recently quickened. It seems that Thatcher's chief project was as much to prevent a return of Tory landlordism as a dismantling of Labour council housing by promoting freehold. Status, pursued through the illusion that households are independent of the social fabric and that life in the suburbs is not really urban at all, satisfies a strong antiurban bias that has built up in American minds since the late nineteenth century.

Liberal Culture in the United States and Canada Contrasted

I will pair the following tendencies: excess/restraint, exclusion/inclusion, polarization/compromise, antigovernment/progovernment, populist/élitist, corporate/small scale, consumer/citizen. The United States tends more to the former, while Canada leans more to the latter and is thus more like northwestern Europe. America is exceptional, but mainly in its extreme liberal ideology, which persists despite its advanced age. Contradictions arise, as we saw briefly in the first chapter when discussing problems with Jane Jacobs's views.

In the chapters that follow, cases of American excess will appear: Ben Franklin's expansionism, mid-nineteenth-century Broadway consumerism, overbuilding of canals and railways, utopian planning of Chicago before and after 1900, Los Angeles captivated by the car by 1925, extreme suburbanization, urban renewal, and urban freeways after 1945 that ruined perfectly good cities. The literature of America is replete with titles invoking the frontier (as if any new thing is a frontier of choice), the American dream (or often the California dream), or covenants, as Clinton used it in his 1995 state-of-the union 'message', as in a sermon. One can hark back to the earliest settlements, particularly to early Massachusetts, where Winthrop proclaimed that his covenanted Puritans would establish a city on a hill, a beacon unto the world, and an errand into the wilderness, to tame the natives (one way or another) and to conquer nature. The utopian strain has persisted. Consider persistent Protestant 'primitivism' that gave rise to 'illusions of innocence'.[23] While nostalgically 'restoring' the New Testament church and more, quintessentially frontier denominations, among others the Disciples of Christ and the Mormons, sought to deny the intervening European years of the Church, replicating the early Puritans. They emphasize individual salvation. 'Given the chronic insecurity Americans feel about their place and status in society', they are fair game for the charismatic preachers bent on 'Selling God'.[24] In a secular vein, a civil religion of expansion by individuals arose. The French and then the British, through the Proclamation of 1763 and the Quebec Act of 1774, sought to constrain settlement from western Indian lands, but independence rendered these inoperative. The pursuit of property has dominated.

Canadians never had the luxury of visions of freedom largely because in a much smaller arable area, every action, implicitly if not explicitly, has been measured against American practice, against the giant next door. Canadians indulge in patriotic pep rallys to boost national self-confidence, but are usually self-deprecating or more often restrained. Many original Canadians were conquered, the Acadians and Québécois victims of British expansion. Loyalists and later settlers from the United States and from Britain and Europe, if they remained, accepted the limitations and British institutions and, in some respects, remained more British than the British. Early nineteenth-century British free-enterprise liberalism hardly displaced tory protectionism in Canada. Not all settlers stayed, of course. Even now, ambitious Canadians go off to seek their fortunes in America. It almost seems that a different genetic disposition lies under both countries: ambition and excitement compared to contentment and quiet. At least at certain times, not only socialists but Tories could be 'red'.[25] Canada too had trouble creating a west: the vast, rocky Canadian Shield held back settlement until late in the nineteenth century. Frontier-type religion failed to gain a strong foothold. In Canada today, four in five Christians belong to only three denominations; in the United States, to reach that share, one needs to count people in 300 groups. Restraint and stability within cities has been more obvious in Canada, as we will see.

Paradoxically, open America is more exclusive. Blacks especially have not achieved full participation in citizenship, a reality many Americans have agonized over. The liberal ideology leads to exclusion. To clamber ahead means to climb over others: as already noted (it can hardly be avoided), Blacks have been a convenient bottom layer, but even without Blacks, would America ever accept universal social programs? The gated and walled communities and the history of zoning are attempts to fend off anyone, not just Blacks, who does not fit in. When the media constantly parade the dream of success in what is a nearly zero-sum world, the failure of many to reach higher ground haunts the land of opportunity. Fatalism about the collective good sets in; few complain about excessive corporate salaries. Alienated Americans, suffering from *anomie*, 'search' for communities that suffered 'eclipse' sometime in the past.[26] Exclusion rules out the impossibility of neighbours.

Canadians have tended to be more socially inclusive. The pursuit of order is a firmer ground for community than liberty in the pursuit of property. Treatment of Native peoples, while hardly a model to follow, was gentler than that in the United States. Not surprisingly, after 1776 many of the Iroquois crossed the Niagara frontier, though admittedly full status is only now being achieved. Acceptance of post-1945 settlers from a wide variety of countries has probably been easier in Canada than it is south of the border, though economic growth helped. Ironically, Canada may be more of a melting-pot and America more of a mosaic, despite mythologies to the contrary. In fact, English-speaking Canadians have been more accepting of Québécois and less deprecating than they were earlier. Universal Medicare became the pride and joy of most Canadians. There are fewer homeless people. Canadians are more conscious of an ancient notion (there but for the grace of God—or

luck—go I), so they have done more about the condition of the poor. Americans, it seems, lean more to a view that God's grace falls on those who look after themselves. Still, Canadian virtue should not be overstated: austerity programs are falling harder on the poor than on anyone else.

Closely related concepts are intellectual polarization and compromise. Barry Goldwater's 1964 admonition that extremism in the defence of liberty is not vice, considered even too extreme for a majority of American voters at the time, was aimed at godless communism. Liberty here seems to have meant a super-organic America, cocooned against the evil empire. Within the cocoon, people are free; within the Soviet world, everyone is oppressed. But American socialism foundered because most people could not distinguish between totalitarian and democratic collectivities. Democratic western European and even Canadian socialism could not be tolerated. The band of political discourse has been narrower, the acceptance of corporations greater. Intellectual and moral polarization has shown up in prohibition, with the unhappy outcome of Al Capone gangland shootings, the war on drugs, carnage on the streets, antiabortion violence, and the gun control lobby. Handgun murders in the United States are ten times the rate in Canada, though that is high by Japanese standards. It seems no middle ground can be found on many issues. Strident moralistic prudery runs up against strident excessive permissiveness. American society is easily the most litigious of societies, reflecting rigidity of positions. Many of these issues do, of course, come up in Canada too, but are less ardently debated.

American distrust of government lands the citizens into deeper contradictions. Business and the media deny government legitimacy and therefore foster and tolerate Dickensian conditions among the poor while hypocritically pursuing public hand-outs. Canadian businesses do the same, but with less vigour. The key issue is whether politicians only represent the most vocal and well-

> ## The liberal ideology leads to exclusion. To clamber ahead means to climb over others . . .

heeled constituents, or whether they possess some kind of autonomy of commitment, following British parliamentarian, Edmund Burke. Public health insurance was created in spite of the powerful medical lobby. Capital punishment was outlawed in Canada when politicians bucked public opinion polls suggesting that Canadians were nearly as bloody-minded as Americans. American politicians give in so easily to the worst populist instincts, egged on by the Rush Lumbaughs or their successors, who in turn are financed by advertisements of corporations. Leaders in Canada at least sometimes rise above the people.

Ironically, a major outcome of the antigovernment sentiment is that Americans are more in the thrall of corporations who fill the power vacuum. Despite their professed populist view that corporations are too big, ordinary people defer to the advertisers who define their status as consumers in a kind of 'new feudalism' or 'homage' to power.[27] Consumership overwhelms citizenship. Canadians 'defer' to government because they expect legislatures to act for the common good.[28] Obviously Canadians are consumers too, but, though strained, retain a bit of the notion of citizenship. They are not quite addicted to the principle of I am what I buy. To many Canadians, 'If [Canada is] the kinder, gentler welfare state, then America is the shining city on the hill, surrounded by a walled moat outside of which roams a marauding and dangerous underclass.'[29]

The Clash of Growth/No-Growth Advocates

Canadians are as committed to growth as Americans, or nearly so. Built on liberal freedom and expansion has been another dominant ideology: growth is good and necessary for prosperity. An old saying is 'grow or die'. It fits with the progressive view of history: now is better than ever before. Hope springs eternal. As a conscious ideology, the Great Depression was the inducement as never before. Although earlier business people and politi-

cians pushed for growth, the 1930s' funk demanded more deliberate pursuit of the holy grail of growth.

To growth advocates, record keeping and measurement became fine arts. As a result of the Great Depression, the gross national and domestic product became central measures of collective health in toting up national accounts. Economists measure the value-added per worker in the transformation of raw materials to finished products. Even as late as 1960, economic historians averred that value-added measures were 'available for manufacturing only'.[30] Nonetheless, services have been measured. Because service employment rose so markedly compared to goods processing, the productivity of stockbrokers, office workers, and hamburger flippers had to be calculated. In October 1987, with so much business after the meltdown of the stock market, brokers accounted for a 'remarkable' 40 per cent increase in the Canadian GDP (and presumably something similar in the US).[31] Was that real growth? Does real work deal with money, the most insubstantial of commodities, which can evaporate in the twinkling of an eye during a crash?

I am sceptical that the productivity of services can be measured meaningfully. Recent conventional studies conclude that office productivity (that is, per worker) has not risen, despite all the fancy equipment. Firms 'slim themselves down in search of higher productivity.'[32] Yet service employment, whether upscale or downscale, is about all that is left. Waiters who stand around on slow days are in one obvious sense not productive, but they are paid. Academics who write books are obviously productive in one sense, I suppose, but how can the writing itself be measured as contributing to the GDP? Governments, the business press, and the public are preoccupied with the concept, so it is not just growth at any cost, it is measurement by any means to make us look good. Downsizing offices supposedly increases productivity per worker, so this is calculated to make the national accounts look better than they are. The outcome is that some people are overworked to the point of stress, while others are underworked to the point of stress.

Publicists of the information society argue that 'copper and oil come out of our minds. That's really where they are.'[33] It is as if land were no longer a factor in production. Slightly less blatant, but reeking of hubris, is a comment that the oil and gas industry, having recently shed too many 'knowledge' workers, has acted in a way that is 'not a healthy sign in an economy where wealth is created more from what's in people's heads, rather than what's beneath their feet'.[34] The notion of human capital has run rampant, the importance of resources and manufacturing downgraded. A more reasoned approach to 'our common future' is 'sustainable development', trying to combine growth with environmental concern.[35] Even then, some critics see sustainable development as an oxymoron. Sustainable survival depends on a degree of management hitherto unimagined.

More cautious environmental economists have attempted to measure 'net' domestic product seeking a more realistic appraisal of growth by subtracting the exploitation of non-renewable resources, regrettable necessities such as military spending, other wasteful practices, and indeed waste management from the gross domestic/national product. Cleaning up oil spills is put on the other side of the ledger and not counted as production. In contrast to the conventional measure of gross domestic product, the Index of Sustainable Economic Welfare shows slowing growth for several decades (figures 2.6, 2.8). In this index, the distribution of income is the strongest indicator of welfare, presumably because it affects people directly and activities indirectly. The referent is social well-being in pursuing 'The Green National Product'.[36] Stands taken on behalf of the environment have to be balanced by a concern for human work. Politicians in Washington state and British Columbia are acutely aware of the direct clash as they attempt to balance cutting forests and preserving them, as well as with social demands.

Yet another view focuses on the 'social limits to growth'.[37] In affluent societies, status is harder to accumulate as material satisfaction reaches its limits. This was the case in the 1960s when mass production put two cars in every garage, TVs in every room, and barbecues in every back yard of new suburban houses. In earlier, more spartan times, those without reached for more to improve their health and standing. By the 1960s, nearly everyone was satiated. Even so, many became dissatisfied with the egalitarian mass-produced world that resulted in rising competition for positional goods and services.

The demand for custom-made quality goods led to flexible production that the Japanese recognized before American mass producers. The latter switched gradually to flexible production, but this undercut productivity, as we noted earlier. So consumers came to have a wider choice, but since wages have fallen because of lower productivity, fewer consumers can buy. Paradoxically, affluence brought 'the end of affluence'.[38]

Another outcome of flexibility can be seen in universities. In the 1960s many students rebelled against standardized programs. So picking and choosing courses cafeteria-style became the norm. Students revolted against the governing hierarchy, so the universities were democratized. But the result in the long run was more competition for less and less status. Grades inflate and quality deflates. At the same time, because productivity slowed down, calls for improved skill training became more strident. This cost government money, but ran up against the demand to cut taxes. The latter demand came from those who were losing out to lower productivity, and from those who were already ahead who saw an opportunity to distance themselves further from the crowd. The direct losers in this downward spiral of competition have been those in the bottom half, even the bottom four-fifths, of the income scale. Adding to the woe, each restructuring of business organization, each increment of new technology, and each innovation in education brings less and less growth and less and less satisfaction. The social limits combine with the natural limits to growth. The most liberal of societies seems to be concluding its run as a stagnating neo-feudal regime—the few at the top surrounded by well-paid courtiers, while the mass of common folk are outside the fold.

This view is consistent with the 'rise and fall of great powers'.[39] Power runs out like a product life cycle. Britain peaked in the mid-1800s, and the

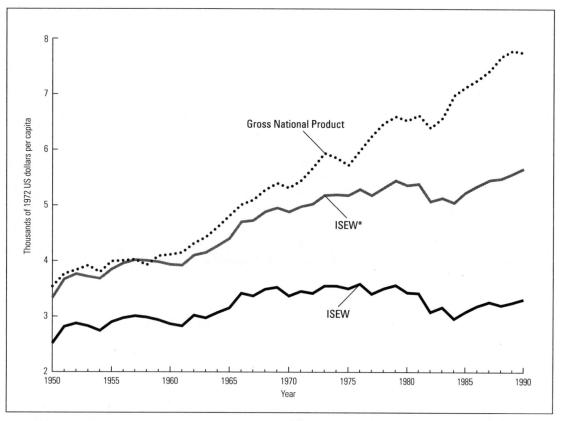

2.8   Alternative measures of economic welfare. ISEW refers to Index of Sustainable Welfare. ISEW* excludes depletion of non-renewable resources and long-term environmental damage. GNP or GDP are often used as measures of well-being. This graph attempts to measure social and environmental well-being, accounting for items not often included in the market measures and subtracting those that are detrimental. *Source:* Herman E. Daly and John B. Cobb, Jr, *For the Common Good: Redirecting the Economy toward Community, the Environment, and a Sustainable Future*, 2nd ed. (Boston: Beacon Press, 1994):464.

United States in the mid-1900s. Some would say that Japan has already peaked, so it too will likely decline, at least relatively. Now that the world is technologically and financially tied together, whichever country rises to the top perch will not stay there for long.

How to Periodize the Past?

As I am telling a story covering a long period of nearly 400 years, a shape—a historical periodization—has to be imposed on the shifts in economic and military power. One approach would be to identify processes leading to 'progress' in which everything before was preparatory for the glorious present. In such a scenario, setbacks are seen as temporary. This will hardly do. Defining periods by great men, such as the reigns of kings, prime ministers, and presidents, will not do either, since so-called great men had to be in the right place at the right time with the right credentials. I insist, however, that even in the worst of times, leaders can act for the public good, though they will rarely earn credit for doing so. Given the liberal and growth culture, it is at down times that equity rather than economic matters deserve the most attention. As Samuel Johnson averred, 'a decent provision for the poor is the true test of civilization'.[40]

The long-term pattern of growth is rapid at first and then slows down on the growth curve. Within this framework we can use the notion of long waves of rises and falls. In the 1920s a Russian economist, Nikolai Kondratieff, proposed that in the capitalist West, economies gathered steam and then stagnated or even deflated through forty- to sixty-year waves. Thus since 1790 four Kondratieffs, as they are popularly referred to by the many aficionados, have been suggested, though the precise dates of major take-offs out of depression remain debatable. In my schema, 1793, 1845, 1897, and 1940 were take-off points in North America, with a slow-down about half way between them. Even earlier dates can be added: 1713 and 1745. Each long wave has two phases of growth, a secondary one following a moment of stagnation about halfway through.[41] Severe depressions mark the end, as after the 1780s, 1837, 1890 or so, and, of course, 1929. Regional variations alter the dates somewhat. The current prolonged trough, beginning with the 1987 stock market meltdown, seems to fit (though less decisively than in 1929), and is popularly referred to only as a recession, not a depression. Recently the scholars and journalists who believe that progress is ever upward have speculated that we are on the verge of the fifth upswing based on the 'carrier wave' of communications and information—after the debt, which was piled up through speculative binges of the 1980s by business and sustained by governments, is wiped out.[42] Social and political waves have been connected to these basically economic waves. In the secondary growth stage after stagnation, it has been argued that because new technologies demand increased skilled labour, those with the right training pull ahead of the unskilled.[43] This seems to have happened with computerization. Whether the wage imbalance will soon return to equity is obviously uncertain.

Measurement of such long-term and sweeping generalizations has tantalized many, but there are sceptics, mainly because they are virtually impossible to measure. Kondratieff tracked wholesale prices. Productivity measured in various ways has been tried. Wars are also fitted into the sequence,

usually halfway through the cycle, the 1939-45 conflict notwithstanding. To the unconvinced, the various measures amount to illusory grand changes. Except for the Great Depression, all the other slow times are recessions or financial panics that happened too often to fit into a long wave pattern. Scholars have objected that long waves impose a deterministic schedule for human events. Assuredly, one cannot predict precisely enough what will happen in such a pattern. Today no one can say when downturns begin or end because we are all inside the events and processes. Certainly one has to grant some autonomy to political action and social processes, overlapped with economic ones. Yet aggregate financial problems like debt have a significant effect on politics and social well-being. Basically, material nature imposes limits such as scarcities, expressed through economics. Politicians' task is to ensure that the pain is fairly spread. Interestingly, long waves as an issue fell out of favour in the booming 1950s, only to resurface when conditions deteriorated.

In our discussion of North American cities, I will not attempt expert measurements, but I partially accept the notion of long waves. Following some current writing, I will hypothesize that relative stability marked upswings, increasing instability marked downswings, leading to strong restructuring of economies (urban and otherwise), some for the better, some for the worse. A progressive view can absorb long waves—each take-off was at a higher plane than the next. My hunch is that the 1870s marked the long-term turnaround, the beginning of the slow-down. Prior to that, the processes of circular and cumulative causation had taken the economy upward through 'self-perpetuating momentum'.[44] In cities, new enlarged industries triggered multipliers; that is, a host of other firms and new inventions, and also induced local urbanization which fostered building and infrastructure. After the 1870s the momentum slowed. In an attempt to maintain the pace, large bureaucratic corporations took over from the smaller scale, but developed more productive methods. These large corporations energized the economy until the 1920s. Locally, infrastructures were largely completed. Around the 1870s too, consumption to spur production began to replace production for 'production's sake'. The limits of possibilities for greater consumption seem to have been reached. It is no longer easy to deliver any more speed, convenience, comfort, safety, or now status. The revival after 1940 has run its course.

In my essays on various cities, the dates on which I focus are not fitted exactly to the long waves as such. Philadelphia in 1760 does come at the end of a major upswing noted earlier (though the peak may have been reached by 1756), as does New York in 1860 (the peak in 1857), Chicago in 1910 (just before the end of the upswing in 1913), and Los Angeles in 1950 in the upswing. Toronto comes at the end of the long primary spurt after 1940 that was spent in part by the oil crisis of 1973. The discussion in Chapter 1 focuses on the short secondary growth spurt and the present depression.

To generalize periods in which to situate cities, in part using long waves, we can use the following labels. The first period, encompassing nearly two centuries, was an open and extensive era of rural expansion from until 1760 or so, when populations in the country grew more rapidly than in the cities. As farmers spread out, they used the land extensively. Operations were mostly small scale. Economic growth, according to convention, was slow, though I am sceptical that it was as slow as suggested because people were radically modifying Native peoples' space. Also, terms of trade became better for the colonies; prices for their products rose relatively higher than those for textiles, the chief import from Britain. I prefer not to use the term 'preindustrial'. Before factories, there was industry—hands were busy processing as well as planting seeds. Because the population was predominantly rural, income and wealth were more equitably distributed than at

any time until after 1940, but the small cities showed marked disparities.

The year 1793 marks a take-off after several depressed years. From then until 1860, the continent remained open for settlement. While overall population growth was slower than earlier, it was rapid west of the Appalachians in newly settled areas. By 1860 most of the best agricultural land was occupied, and agricultural practices intensified and specialized regionally more distinctly. With the introduction of factories, industrial productivity sped up. So did transport and communications, especially after 1845. The most rapidly growing decades of urbanization in the United States occurred then too. Even this era was punctuated by a severe depression in the wake of the panic of 1837. Partly because of urbanization, and partly because the lawmakers privileged business over labour, inequities in wealth and income grew sharply.

The third period from 1860 to 1896 was the 'corporatizing' era in full flood until it ended with the 1890s' depression. Corporatizing kept the ball rolling when friction against growth was intensifying, even while great inventions burst on the scene. Little of the continent was left for settlement, except in the Canadian West. Despite the rise of the new service middle classes, social inequities remained marked, though fluctuating.

There was a brief fourth intensive corporate era to 1929. The great corporations were in charge, and the courts and the new middle class considered them legitimate. Perhaps we can even speak of a corporate ideology as never before. Prices rose again after falling through most of the nineteenth century (except during the wars), suggesting that the corporations after 1896 managed structures that could control prices. Canadian growth (particularly until 1913) was far more rapid than American growth, catching up through western settle-

> **The limits of possibilities for greater consumption seem to have been reached. It is no longer easy to deliver any more speed, convenience, comfort, safety, or now status.**

ment and mineral exploitations. As in the previous period, social inequality was strong.

Then came the great break of the 1930s. What followed was a corporatist period when government and large corporations worked together, and the large industrial unions negotiated generous pay for their workers. Prices rose even more dramatically as governments handed out money to businesses and consumers. Military spending, justified by the cold war, worked for a while as America led the world. An unheard of degree of income equity was achieved, but finally the US came up against the limits of its power in tiny Vietnam and devalued the dollar.

A chaotic, silent depression set in about 1975. Income and wealth equity deteriorated. Even so, historian and futurist Newt Gingrich teaches us that 'America is exceptional, moral, and just (though also innovative, technologically sophisticated, and geared to success for all its "real" inhabitants, those who make themselves into pioneering businessmen like Benjamin Franklin).'[45] North Americans live not in an open environment but in a closed space that has to be managed with care.

As America faltered, Canada had its finest brief hour as a nation. In the mid-1960s, it established the social safety net that Americans could only envy. If the great world's fairs on the continent—1876, 1893, and the others that followed—had been American celebrations of good times, then Expo in 1967 far overshadowed the anaemic 1964 fair in New York. Now Toronto itself has plunged into a trough. The next century may not be as kind to Canada as the twentieth century has been.

Cities for Making a Living and for Living?

Places persist because nature persists: all the speed generated in communications and trade has not overcome the limits of environment and space. The natural human condition is to congregate, though huge cities are probably not necessary to satisfy gregariousness. In each chapter that follows, we will discuss cities on two scales: wide and local. The wide scale looks at cities globally and within regions of varying dimensions. Hierarchies of cities have existed for millennia. A vast literature describes the array of cities in space. Two key notions are, first, major network cities, gateways, or entrepôts connecting the region with other countries.

The early settlements had network anchor points on the Atlantic facing Europe. One of those, New York, became a global finance centre and, together with Washington, oversees the dominant institutions of the United States (and financial institutions of Canada, for that matter). It has been the major continental cultural focus, and also serves its immediate region. New York joined London early in this century as a centre of world finance, though the latter still remains the European financial (if obviously not political) gateway, and was followed more recently on the global scene by Tokyo. Messages may be instantaneous today, but the important ones dealing with money seem inevitably to concentrate. Subsidiary to New York and Washington on the continental scale are the major regional gateways, some of which we will discuss at various times: Philadelphia, Chicago, Los Angeles, and Toronto. These and several others all have global and continental connections too, though less so than New York and in varying degrees. Below these are a larger number of smaller regional centres.

The second notion is that of central place. Network cities are at the top of national and regional commercial hierarchies, sometimes referred to as nesting hierarchies: there are few large ones and more small ones. Central places serve a definite population within regions. In the concept, at the bottom of the hierarchy, retailers serve local rural populations, though modified by other aspects of trade, manufacturing, mining, transportation, culture, and certainly government. In the theoretical formulation, there are layers of circular (more precisely, hexagonal) regions of varying sizes, each with an urban focus. Again, there are more smaller regions than larger ones. The threshold of a good or service is a crucial concept: the higher the city is in the hierarchy, the higher are its main services for wider areas and larger populations (besides its own low-order, local-serving occupations), whereas the smaller urban places provide few and only lower-level activities. For example, bank headquarters and stock markets are found in only a few cities; the smallest hamlets may each have one crossroads convenience store, maybe a gas station and maybe a church. In the twentieth century, the hamlets at the bottom of the hierarchy thinned out greatly.

Manufacturing was a great modifier of the neat commercial agricultural pattern; specialized industrial towns sprang up, often near large cities, first at ironworks and major mill sites. Larger commercial places always hosted processing, though their share increased in the late nineteenth century and then dwindled in the latter part of the twentieth century. Large commercial cities had the attractions of large pools of labour and access to transportation. Some places emphasized transportation. Mining centres were unique and often ephemeral. Hardly ephemeral, capitals have been unique decision-making centres, making clear that politics are as important as economics to social life.

Once established, inertia characterized the urban systems of the continent and the regions. The larger the city (usually, if not always), the greater its stability. Many small places became ghost towns. Over the past half century, the hierarchical pattern became less stable, more clearly so in the United States. Much of the economic power of the dominant heartland created by 1860 shifted to the South and West. The largest cities with the weakest financial and corporate bases have been hit hard, most obviously Detroit. At the beginning of each of the next five chapters, I will consider the urban pattern of the era based on overall population and economic changes.

Within cities, similar notions of hierarchy and centrality can be used. In the core have been the region-serving economic activities in finance, wholesaling, higher-status retail, high culture, and local government. As we will see, the central business district reached its greatest relative prominence about 1910, as indicated by the highest land values, noted in Chapter 5 on Chicago. Since then, many enterprises have spread to outlying centres, now often clustered in edge cities as we noted briefly in Chapter 1, but finance and government—the dominant powers—remain in the centre. Downtown retailing and other work have remained relatively more important in large Canadian cities than in American ones. Well before 1910, much manufacturing moved outward from the city centres, even as activities clustered increasingly in the larger cities. In fact, some manufacturing had been on the edge of the cities since the beginning. The very large, space-consuming operations were located on the outskirts, often in satellite towns. Today many are located in rural areas and indicate a return to an earlier time when most processing took place in the countryside.

Home is as important as work, even for the powerful and restless. Cities grew outward. Residences became increasingly separated from work compared to early cities where work and home were often in the same buildings. That trend has recently reversed somewhat. In the affluent twentieth century, the amount of space per person increased. Residential patterns became increasingly spatially segregated by class and race in neighbourhoods; more precisely, the segregation widened from micro to macro scales, though mixed areas remained. The more powerful appropriated the more attractive sites. As we will see, developers and builders constructed (and reconstructed) the cities, often at the behest of moneyed power, including government. The right to develop one's property has been one dominant and constant principle, and it has often overcome equity and the public good.

The most startling difference between major Canadian and many large American cities, as we have already seen in Chapter 1, is the near, yet violent, vacuum in former inner-city residential areas of the latter. Those who were able to leave evacuated them as if they were war zones. The heady mix of neglected housing and racial tensions was combined with an excessive penchant for newness on the suburban frontier in the pursuit of protection and status, while governments and business were anxious for city growth. One prominent urban planner in 1950 warned that the failure to plan for several decades placed a 'tremendous burden of social and physical regeneration upon urban government'.[46] As a result, instead of regeneration came hell. No one, it seemed, could maintain a balance between city and suburbs. American suburbs themselves became more sharply divided by income and race.

Suburbanization can be attributed in part also to the automobile, or rather to people who were hooked on them as a mode of transportation. Circulation, including the journey to work, has to command our attention in order for us to understand the public order of cities beginning in an era when most people walked, through the development of public transportation beginning in the 1820s, and through to its heyday about 1910 when auto commuting began and presaged the decline of the trolley (if not the subways) in a few cities. Canadian cities have maintained more vigorous public transport systems. In large measure because of these changing means of circulation, over the long run the form of the city shifted. The walking city had a triangular or elliptical shape, the public transport city took on a star shape, and the automobile city produced a circular pattern and was obviously of far greater spatial magnitude than earlier.

A way to sum up this discussion is to consider whether cities have been 'generative' or 'parasitic'.[47] North Americans are not likely to entertain the idea that those with power in North American cities have turned more and more to exploiting the wealth of nature as an end in itself rather than helping society, including city dwellers, create new growth. There are signs that cities have less and less generative energy. Most people have had no choice but to go to the few big cities in the same way as they drifted to an increasingly parasitic Rome. Urbanization, once a force for economic growth, has, it seems, reached the limits like so many technologies we depend on. Many people seem to be permanently (or nearly so) on welfare. Most jobs are low-paying and more are part-time. Top managers appropriate more of the income pie. The bankers chastise politicians as spendthrifts, but lend money to governments to pay the interest on ever-increasing debts. Indeed, financiers have had their hands in the taxpayers' pockets for some time. They occupy ostentatious offices and promote entertainment. Like the circuses of a decaying Rome, diversions command attention. Perhaps I am too pessimistic: this could be only short-term parasitism, but we have to recall the spending for a world war that got us out of the mess left behind by the Wall Street debacle of 1929. Nature, it seems, has triumphed. Can we settle down to a steady state? We can explore that possibility only by first looking at our common past as Americans and Canadians on what has been a bountiful continent and the wonders and tragedies of great cities.

CHAPTER 3

∎

Franklin's Philadelphia in 1760: Fulfilling European Peasant Dreams

When the ship *Bountiful* sailed up the Delaware River on a fine May morning in 1760, John Brown, the business agent of a British merchant, and his fellow passengers stepped out on deck to survey the scene before them. After passing lush green fields, woods, farmsteads, and tiny villages, Philadelphia came into view. The agent remarked to his fellow passengers about the similarity of Philadelphia's skyline to that of British cities. As in the larger English ports, there were a few score of masts along the shoreline and beyond them the spires of several churches. Lining the waterfront were rows of houses and warehouses. Coming closer, he could see the men and wagons hauling barrels and sacks to and from the ships. The bustle on the waterfront not only fit his customary images of waterfronts but was also one indication of a land with a promising future. This was a welcome scene to the agent from London who had sailed on the Atlantic for two months. To this Englishman, the picture heightened his sense of pride. To another traveller at the time, it 'must certainly be the object of everyone's wonder and admiration', carved out of the wilderness only a scant eight decades earlier.[1]

The Seven Years War in Pennsylvania's back country had been over for some time; Quebec had fallen and Montreal was about to. The British empire embraced the Atlantic and the French presence verged on the collapse. Here was a city of nearly 20,000, slightly larger than New York. Apart from gigantic London, in 1760 few cities were larger than Philadelphia in the British Isles where

Bristol, Liverpool, Glasgow, Dublin, and Cork owed their rise to the great Atlantic network of power. Philadelphia's ascent less than eighty years since its founding had been a remarkable achievement. As the Swedish naturalist, Peter Kalm, exclaimed in 1749, 'such grandeur and perfection ... [is] by no means inferior to ... any, even the most ancient towns in Europe'.[2]

The moment the ship docked at Pemberton's wharf near the foot of Chestnut Street at nine, prominent merchants or their agents and government officials met men of like ilk on the ship. While some passengers went off to the offices of merchants in nearby warehouses or to inns where they would be accommodated, our businessman headed to the London Coffee House, the exchange, the central spot for trading information and making deals. There, greetings were quickly followed by a demand for news (and rumours and gossip) of war, politics, prices, fashions, and new technologies in England. Because the *Bountiful* had carried important men, William Bradford, the founder of the London Coffee House and the publisher of the weekly *Pennsylvania Journal*, and David Hall of the *Pennsylvania Gazette* made the effort to be on hand. The famous Benjamin Franklin, Hall's now silent partner and formerly the postmaster, had recently become the province's agent in London, but his views frequently came up in conversation. Once merchants knew the prices of English goods then flooding into Philadelphia, they would submit advertisements to the newspapers. At the same time, the visitor listened intently to reports about military actions in America, the extent of the grain harvest in Pennsylvania, and other business news from elsewhere in the colonies, but he had his word first: Philadelphia was mainly on the receiving end of 'information' from Europe and the empire. The first British empire embraced the northern part of the New World, much of the Caribbean, and even much of India, and was at its zenith with the signing of the Treaty of Paris in 1763.

Some passengers could take time to leave the ship. A few families had enough money and connections to buy land. They were luckier than the German families on a ship docked nearby, who waited for someone to pay for their passage so they could land. These 'redemptioners', as they were called, depended on city and back-country coreligionists, mostly Lutherans and Reformed Calvinists, to hire them for a term of several years. More abject were the indigent indentured single men and women from the city streets or from farming regions in northern Ireland and Britain. They too had to stay on board until their passage was paid by artisans who put them to work in their shops or by merchants, some of whom employed them in their homes. Some found jobs in ironworks in the countryside.

The indentured sought escape from a class-ridden and mean mother country to seek independence, the independence of the New World farmer or craftsman. If they were healthy (and many were not, in part because of poor food provided on the Atlantic passage), these young people could harbour positive expectations; their terms of servitude, unlike slaves, would last only four to seven years. But, unlike servants in the early decades of settlement, only the lucky ones would earn enough to own property. Those slaves who were freed were even less likely to advance. The streams of redemptioners, servants, and slaves had reached its peak in Philadelphia and the province. Future employers would rely more and more on the wage labour of native Pennsylvanians, for labour had become plentiful both in the city and surrounding counties, as evidenced by increasing population densities. The frontier was an option for only a minority, albeit a large minority, with the means to head

west. A major role of Philadelphia was to expedite their occupation of the land.

Besides attending to business, Mr Brown, would observe the social realities within Philadelphia's changing environment regionally and locally. The British agent's admiration for Philadelphia, its rise to become the pre-eminent city in North America, was predicated on the widespread belief in Europe that America was the land of opportunity, that Pennsylvania was 'the best poor man's country'.[3] Nonetheless, he would be sobered to discover over the following winter that not all of Philadelphia's poor men or women shared in the bounty.

We will look closely at the society, that urban space that John Brown encountered in 1760, but first we must consider the broader sweep of urbanization and regionalization on the continent in this era when it was open to extensive settlement. A vast majority of people then lived in the countryside. Cities were small but vital in promoting the well-being of Europeans who sought to fulfil their dreams on the land.

The Cities That Linked the New World to the Old

The century and a half leading up to our Philadelphia story was momentous in Western world history. The British and the French had contended for the occupation of the eastern half of the continent, both gradually overcoming the formidable presence of Native peoples. Until 1664, the Dutch had been a rival to the English, and would leave a cultural legacy especially in and around New York. Spain's role was marginal, restricted to Florida, the Gulf of Mexico, and the southwest. By 1760, after periodic struggles, the British prevailed, but not only because of their superior military strength. The much larger and more or less united British population prevented France from remaining a serious competitor in North America. In 1760 1.6 million people occupied what would soon become the United States. By contrast, the French along the St Lawrence numbered only 60,000, with much smaller populations on the Mississippi and in what became the Maritime provinces of Canada. In 1760 it seemed that the world belonged to English-speakers and the British empire.

European expansion across the Atlantic had been led by fishermen from European countries facing the Atlantic pursuing cod on the Grand Banks off Newfoundland. The cod's protein provided a fundamental basis for European growth and expansion. Explorers followed in the 1500s. The Spanish were lured by gold and silver into what became Latin America. Those metals contributed to western Europe's wealth too, thanks in part to British privateering.

More important for our story were the colonies on the mainland with farming populations beginning about 1607: Jamestown in Virginia and Quebec City in New France. Soon after came New Amsterdam (New York after 1664); Salem and Boston on Massachusetts Bay, among other fishing ports mostly to the north of Cape Cod; then Montreal in 1642. Subsequently, companies and proprietors organized other colonies and towns along the coast, notably Providence and Newport in Rhode Island, New Haven, Annapolis (replacing shortlived St Mary's in Maryland), Charleston, and Philadelphia in 1682. By 1763 when the peace treaty was signed with France, the Atlantic Coast urban pattern was largely established with the addition of Savannah, Baltimore, Norfolk, Halifax, and Louisbourg (the latter two mainly fortified towns), plus many smaller places. These 'network' or entrepôt cities and towns connected the colonies and their farmers, traders, fishermen, and soldiers to London and its outports (such as Bristol, Liverpool, and Glasgow) in Britain, which had developed extraordinarily faster than any other European country. America's great success was built on that enormous political and economic power.

These connections were political, military, commercial, and cultural. Organizers of colonies, who were mostly commercially minded investors and the home government, established capitals where governors, bureaucrats, and legislatures met to order and manage rules of conduct, some carried over from Europe, others developed for local circumstances, such as the regulations for distributing land to settlers. It was also necessary in times of war to provide military protection against other European powers and, internally, to keep Aboriginal peoples in check.

Commerce and trade on the Atlantic to and from these colonies was by and large concentrated in the capitals, except in Chesapeake Bay. Along with top officials, the better-off merchants dominated the life of these cities. The merchants exported agricultural and other goods to other colonies, including the West Indian colonies, southern Europe, and to Britain itself. Some, in the North especially, developed beaver fur and deerskin trading.

Acting as wholesalers, merchants imported raw materials, especially sugar from the West Indies, and an increasingly vast array of manufactured goods from England. Domestic manufactures could not easily compete with these relatively cheaper British imports. The merchants controlled credit in the colonies, though they in turn were dependent on merchants in Britain and indirectly on government spending. Bills of exchange and similar devices—more or less like cheques traded for goods and services and toted up in bookkeeping ledgers—were the means of making payments and shifting money. Paper money helped some colonies to foster economic growth. By 1760 the scale of trade had expanded so much that merchants began to specialize, some in exports, others in imports. In the biggest places, retailing was starting to separate from wholesaling.

The superior financial position of merchants allows one to speak of 'commercial' cities. Another term, 'mercantile', means much the same, though it also carried the connotation of an era when Parliament attempted to control trade through mercantile acts. Parliament designed these acts to enhance Britain's wealth, for example, by limiting Atlantic trading to the British, including colonial ships. Despite complaints by American revolutionaries later, the British mercantile system probably helped rather than hindered their economic condition. Although craft manufacturing added to the bustle in the cities, the great era of factories lay ahead.

While commerce and public order became the functional cores of colonies and cities, cultural concerns laden with symbolism clearly provided justification for establishing some places, and even their basic form on the land. In the early seventeenth century, the English experienced an economically difficult time with its last mass starvation. That period was thus politically and religiously volatile. The Pilgrims, who established Plymouth in 1621, and the Puritans soon after expressed their desire to build God-fearing colonies. Massachusetts became a beacon of Calvinist rectitude unto the world. Boston was founded as the site of the figurative lighthouse for a sinful England, a 'Citty upon a Hill'. However, John Winthrop's utopian communitarian vision quickly evaporated: he 'had rejected the vision of modern man he had seen in England' only to see his commonwealth become 'distinctly modern' as 'acquisitive instincts' rushed 'to the surface'.[4] On arrival, some of the more wealthy and aggressive 'hurried out to engross large tracts', so that within a decade thirty gentry families owned half the land in and near Boston.[5] The settlements still carried moral imperatives on behaviour as defined by Calvinist theology. Just as moral was the 'errand into the wilderness', a justification for conquering Native lands and setting in motion what would later be called 'manifest destiny', the American expansionist ideology. From this seed-bed of moralistic Protestantism, more stringent Calvinists went to found New Haven, its urban plan following Ezekiel's vision. Less extreme dissidents established Providence and Newport. In a less fervently religious Chesapeake, colonial planners laid out Annapolis and in Williamsburg about 1700 in a baroque fashion, the latter symbolizing the plantation élite's desire to parade their status.

By 1682, as Calvinism faded, conditions in Britain improved, and the earlier colonies struggled successfully, William Penn and his coreligionist Quakers and other sects settled Pennsylvania with less stress. With less religious certainty than John Winthrop, Penn promoted his plan as an experiment, though indeed still holy. Penn intended Philadelphia as a model for others to follow, a 'greene countrie towne' on a grid pattern of streets and lots reflecting his goal of amity, a more secular sense of landscape, but also a desire to perpetuate baroque aristocratic rectitude. The actual plan, however, hardly allowed for greenery, nor was it destined to be anywhere near the low density he envisaged. Besides, as in Boston, purchasers and settlers quickly sought land nearest the central point at Market and the Delaware River. Many were disgruntled when Penn changed the rules arbitrarily for his own material benefit. The city of brotherly love was destined to be as property-minded as any city. German-speaking Moravians later established communitarian towns in rural Pennsylvania, though they did not stay communal for long because settlers moved onto their own private lots. Some communitarian towns eventually became cities: the best-known were Bethlehem, Pennsylvania, and Salem, North Carolina. In 1730 Savannah was yet another carefully designed but failed experiment; the communally motivated founders sought to bring over the riff-raff from the streets of London to create a closed society.

> **Perhaps Americans' well-known antagonism to government was rooted in the very process of founding.**

Unlike New France and New Netherlands, the religious and moral peculiarities of these various English strands, some of which had an antiestablishment bias, muted the royal governing and military presence. Perhaps Americans' well-known antagonism to government was rooted in the very process of founding. Yet Puritans, Quakers, and others shed some of their English ways, reinforced others (not least property relations), and kept the most important basis for culture, which was a common and thus unifying language. While religious and communitarian impulses revived from time to time, these places concentrated on what America came to be renowned for—business.

Coastal cities served as anchors for inland settlement and town development. First came the fur trading and military posts. French settlers founded

Montreal in 1642, which was later a major ocean port in its own right. The French established other posts on the upper Great Lakes, in Detroit, and later St Louis, in what became Pittsburgh, New Orleans, and other places on the Mississippi River system, and in Mobile on the Gulf of Mexico. Meanwhile, very early English colonial entrepreneurs founded Springfield and Hartford on the Connecticut River, and the Dutch established Albany on the Hudson. After 1670 the Hudson's Bay Company established posts in the far North. The French presence seemed to have the British hemmed in to the east by the early eighteenth century, but the British colonies had established far more populous colonies, and so prevailed.

The survival of English colonies depended upon their own agricultural hinterlands. Even though farms produced for their own or local consumption, they also traded in world markets through coastal merchants and within their colonies. As commercial farming production expanded, especially in the North, inland central places or service centres provided the facilities for the transfer of information on markets and for the movement of goods. County seats became important because they drew not only retailers and agents of wholesale merchants from coastal cities dealing in country produce and imported goods but also government, courts, registry offices, and lawyers. As in the coastal cities, government and commerce provided mutually reinforcing attractions.

Urban populations nonetheless remained small by subsequent measures. On the eve of conquest on the St Lawrence in the 1750s, one in five or more people lived in the three towns, but in the British colonies only one in twenty, a smaller share than earlier. The strong agricultural colonies actually experienced a *fall* in the proportion of urban dwellers over several decades, as expanding rural settlement outran urban. By comparison, urbanization in the British Isles continued to increase, and not only in London, while the highly urbanized Low Countries experienced a declining urban share after 1700. That did not mean that North American cities declined in importance: from the beginning, cities were crucial to the health of the

colonies and would remain so. Once established, the government and commercial apparatuses could handle great inflows of immigrants and the expansion of those born in the colonies while they themselves grew more modestly. After 1760, as their regional economies intensified, Philadelphia, New York, and Boston grew into metropolises by European standards.

Agricultural outran urban populations because most settlers sought their own land, to seek 'comfortable independence through competency', and they were assisted by public policies and speculators.[6] Obviously it was in the interest of home and colonial governments to establish stable rural communities, in the North through family settlements. Most people settled in these communities on their own land, creating relatively equitable societies, the first great middle class. By and large, people sought fee simple, which was the least restraining tenure, and rejected feudal vestiges, which were finally virtually wiped out in 1776, except in Quebec. Those who were smart enough and affluent enough to rent good quality land instead of buying poorer land operated as independent producers preparing to become property owners (often farther inland where prices were cheaper), though not all of them did. Only in the Hudson Valley did farmers settle on large estates that would persist. The will to possess property on open land—or rather on the conquered or deeded land of Natives—was undoubtedly powerful, though that determination was not possible in cities, and would not be for the majority until about 1950. Until the late eighteenth century at least, settlement expanded at a rate equal to the amount of land, that is, at roughly the same density. As recorded on deeds in registry offices, property was widely sold and traded. Early America was in this sense 'liberal': private property (with only a few attempts at collective ownership) prevailed. The 'freehold empire' renowned in the next century started at the beginning.

Affirming this, by and large the colonists settled in dispersed communities rather than in clustered agricultural villages, even within most New England towns, the latter contrary to what most scholars have assumed. In Quebec, the communities were strung out along the St Lawrence. By the time of the earliest American settlements, the enclosure movement had been underway in England, a process that dispersed agricultural villagers. Without doubt, though, cooperation among neighbours was life-saving at the beginning of a settlement. In occupying new lands, groups of households settled close to one another for protection and interaction. Few sought isolation. They cooperated in barn raisings, in processing livestock into meat, and in churches, though not always without conflict and stress. In English colonies, religion could be almost as divisive as unifying.

The drive for independence and ownership of rural property by households initially overrode commercial interests. Self-sufficiency had precedence over making money through cash crops and livestock, yet obviously at the beginning of settlement, an independent household had to acquire seed and livestock unless content with hunting and gathering, which few were. Almost universally, settlers attempted commercial production as soon as possible. In fact, to be comfortably independent, they had to trade. Subsistence farming for self-sufficiency could hardly be the desired goal because it did not build up value. To establish sons on their own land, and to have daughters marry well, farmers needed to increase capital. Land was not free. Obviously some farmers were more aggressive than others, worked harder, acquired better tools and horses, occupied better land, or were lucky at timing. Those who were more marginal and less successful might have resented this and resisted deeper commitments to the market, but they fell behind. After 1830, as indicated by population decline in communities, those on poor land who failed to sell abandoned land to the forest and sought opportunity elsewhere.

Who would buy farm products? If every farmer produced the same thing, then local trading would be minimal. Specialization in farming and processing emerged with comparative environmental and access advantages, so local and regional trading developed. Early farmers, through merchants, provisioned ships and sold the products of fields, mines, and forests to the West Indies, southern Europe, and Britain.[7] By 1760, imports to

these places from America had increased greatly. Labour markets developed side by side with unfree slaves and servants, most clearly so in the best farming areas. After 1760, older areas did not experience shortages of labour. Each farmer calculated his need on how much family labour was available. If he had several teenage sons, they might have worked seasonally on other farms. Even urban dwellers were hired to harvest grain or cut trees in woods in the winter, while some labourers went to the city. Local and regional pay scales were well known. Although not as commodified as in the factory era later, free labourers were obviously under a time discipline and in the thrall of property owners. Bound servants and slaves often ran away, but owners hunted them down, or tried to, through classifieds in the newspapers.

Regions in Early America

In the early days of occupation, city dwellers helped to establish regional economies. These became entrenched and so persist to the present time. Despite changes and intensification through specialization, subsequent settlers operated within the constraints and opportunities provided by access and resource quality. To set the context more precisely for life inside Philadelphia and for later development, we will focus on its economic region and then briefly on seven others to the north and south. Businessmen crossed colony boundaries, just as transnational corporations operate in many countries today, though obviously at a far more expansive scale now than then. Still, as today, provincial governments had some influence on business decisions through regulation, and it is simply impossible to discuss regions without referring to the names of colonies, for they defined residency.

Pennsylvania
After 1680, Pennsylvania rose quickly, reaching a population of about 185,000 in 1760, one-third behind Virginia and Massachusetts, which were founded much earlier. The wider economic region focused on Philadelphia included west New Jersey

and Delaware, increasing the population of its region to over 250,000. Then too Philadelphia's entrepôt merchants reached into western Maryland and even farther into the back country of the South, though these areas would be shared increasingly with Baltimore's merchants after 1760. The Quaker city and its hinterland had been the success story of the 1700s, thanks to Penn's ability to find investors and settlers from England, Wales, Ireland, and the continent. Indeed, within a decade of its founding in 1682, Philadelphia's population had reached 2,000, slowing down early in the next century, catching up with New York by the 1730s and Boston in the 1750s.

Philadelphia's early and meteoric rise can be attributed to the 'vigorous spirit of enterprise'[8] of merchants but also to the luck of timing: the rapid expansion of West Indian sugar production and the lower South's rice production opened new opportunities for farmers and merchants to supply food and other commodities in Pennsylvania with its temperate climate. The province thus attracted the largest number of immigrants in the eighteenth century. The more affluent farmers rented out small parcels of land to inmates or cottagers and hired their labour periodically. During harvest time, farmers mobilized day labourers. Together with a modest number of tenant farmers, the cottagers brought a taste of English rural society, though widely diffused ownership prevailed and there were no large permanent estates as in Britain.

By 1760, the region sold a wide range of staple products, most prominently wheat and flour, as well as lumber and potash from forests, and iron from its furnaces and forges to new destinations in New England, southern Europe, and even Britain. Like fish farther north, the region's food and forest products contributed to western European development. The terms of trade were good. Prices generally were good. Reflecting regional affluence and the ability of its merchants, Philadelphia imported more British-manufactured goods than other cities, thus the provisioning of ships too came to be of major importance. Philadelphia's merchants had created strong commercial linkages backward to the land and forward to the Atlantic world, and helped to constitute the most affluent equitable population anywhere on the globe.

Pennsylvania's success led to the most distinct pattern of market towns, though they were not as well developed in 1775 as the English system, which dated back to medieval times. The largest were county seats where government, courts, lawyers, retailing, and even wholesaling emerged, the last subordinate to Philadelphia. In response to the growing material well-being, a clearly defined policy for establishing counties and their seats emerged. Initially in the early 1680s, when the proprietor William Penn and his agents organized the settlers, they set up three counties with townships that had modest responsibilities.

Subsequently, additional counties and seats (Lancaster, York, Carlisle, Reading, and Easton) appeared. The Penns and others used three clearly defined principles: adequate distance from Philadelphia so as not to be overwhelmed commercially, adequate distance from one another so that they did not compete, and central locations within their counties. These inland county seats and second-tier central places were described as miniature replicas of Philadelphia: gridded, but also with central squares. During the boom to 1760 (and even up to the mid-1760s), and then again in the late-century take-off, urban speculators emulated the Penns in laying out towns. A few, probably with more wealth and confidence, chose their locations and timing wisely or luckily and turned out to be successful. Thomas Harris's town not only

> Philadelphia's merchants had created strong commercial linkages backward to the land and forward to the Atlantic world, and helped to constitute the most affluent equitable population anywhere on the globe.

became a county seat but the state capital in 1812. A similar urban pattern developed in western Maryland, up the Shenandoah Valley, in New Jersey, and to the west in the next century.

In those days, most manufacturing occurred in the countryside on farms and shops and in small towns of Pennsylvania (and other colonies). A few places specialized in certain kinds of manufacturing and mining. Widely available raw materials—such as crops and livestock, wood, metals, and clay—spawned a great variety of industrial categories. Processing and fabricating added value to raw materials. Most prominent was the milling of winter wheat and spring grain into flour or feed. Many mills, more or less evenly spaced along streams, served local populations. Concentrated at sites with the strongest waterfalls, a few became merchant mills, grinding flour for export. By 1760, a major cluster of mills was emerging on Brandywine Creek and adjacent creeks near Wilmington, and on the Wissahickon (now inside Philadelphia) where some mills also turned out higher-value paper. Mill owners built workers' cottages, creating rudimentary mill towns—the first 'edge cities'. Carters and farmers from adjacent counties hauled wheat there in Conestoga wagons. Mill races also powered sawmills, fulling mills (for cleaning wool), and the occasional oil mill (flax) that dotted the margins of streams. Exporters sent fine-quality flax-seed to northern Ireland.

Farms themselves were complex industrial sites. The men slaughtered livestock in the fall and preserved meat with smoke and salt. They girdled trees to kill them, burned them for potash, and logged them. They fashioned the plentiful supply of wood into many tools. Women baked, made butter, and retted flax. Some farmers brewed beers, while others distilled whiskey, mostly from rye. By 1760 each township seemed affluent enough, for example, to encourage a clock maker. A fine grandfather clock gracing the front hall of the now substantial farmhouses of the affluent was a sign of status. Some of their produce was sold locally and regionally, though the magnitude is unclear as records are scant.

Two other major sets of activities presaged industrial expansion and urbanization in the next century. Textile production, mainly from wool and linen, became an important commercial activity. Women spun at home. Men wove for commercial enterprises, organized by jobbers in small shops or put out yarn for work through the winter. These male weavers had emigrated from Ireland and the German states. The gender division of labour would carry over into the factory era later.

Iron production provided the second means of industrial expansion. Several iron villages operated in rural Pennsylvania and other colonies. Furnaces fired by charcoal reduced limonite bog ore. Forges, located not far from Philadelphia, shaped bar iron. Slitting mills altered the iron again, and steel furnaces further refined it. Finally, metalworkers fashioned ever more sophisticated products in the colonies. The vast hardwood forests of Pennsylvania provided charcoal and thus slowed, by English standards, the exploitation of its extensive beds of bituminous coal in the western part of the state.

Since urban populations added up to less than 10 per cent of the total, it is safe to say that most manufacturing developed in the countryside, not in the towns. However, as we will see later, the manufacturing sector expanded in the city where high value-added production gave it some advantages over the countryside. Increased affluence fostered wider and higher expectations, which imports alone could not satisfy. Import substitution worked its way slowly into Pennsylvania's economic structure, and was certainly helped by the non-importation agreements a few years before the revolution and then by separation from Britain. Generally, as we noted in Chapter 2, the quantity of rural processing belies efforts to make a distinction between the preindustrial or even protoindustrial character of the era and the industrial. This era was very industrious.

Pennsylvania had a negative trade balance with Britain because many English goods were in demand. More so than elsewhere, returns on trade with the West Indies and southern Europe, and the coastwise carrying trade redressed the balance. During wars, especially just before 1760, British military payments helped too, far outweighing any taxes collected. London merchants extended long terms for credit; despite interruptions by war and subsequent gluts, as between 1760 to 1763, London merchants could be assured that the growing population and abundance of the region would return more and more to them. Paper money, prudently issued by the provincial government beginning in 1722 through a land bank, helped economic growth and diversification. A likely sign of this was that inflation did not result, unlike the case of Massachusetts, which had less vigorous economic possibilities and exercised less restraint in issuing the money.

New York City Region

The second of the eight regions focused on New York City, which, in 1760, was nearly as large as Philadelphia. Its merchants controlled the Hudson valley, Long Island, and the eastern half of New Jersey. This region had a population of about 175,000, and exercised influence over western Connecticut, Massachusetts, and even Quebec. New York merchants, like Philadelphians, built their own ships and so strengthened their role in the carrying trade. Like Philadelphia, New York found that being capital of the province gave it advantages, and its economic strength suppressed the growth of other towns on Long Island Sound, such as New Haven, and across the Hudson. A metropolis in the making, New York would later move to the front of the urban league. Its exports were similar to Philadelphia's and it handled nearly as many imports. In 1774, on the average, the wealth of each free person in the Middle Colonies of New York, New Jersey, and Pennsylvania was about £51, which was more than that of New Englanders and three-fifths that of White Southerners.

Boston

Boston, the entrepôt of the third region, was less central to New England than Philadelphia and New York were to their regions. The third region's population of nearly 450,000 was spread out along the coast to the north, south, and inland in its then four provinces. It had contributed more militarily than any other British place to the defeat of the French through taxes, goods, and the sacrifice of sons. The prolonged sporadic struggle with the French drained Boston's strength. Little wonder that it was the earliest focus of agitation for independence.

The largest North American city in 1700 with about 8,000 people, Boston stagnated after 1740 at about 16,000 for three decades. For a century prior to 1740, Boston's chief economic role had been carrying trade on the Atlantic, connecting the colonies with Britain, the West Indies, and Africa. This earned revenue to pay for imports. Boston's post-1740 stagnation also reflected rising competition in the carrying trade from Newport (a growing place of 7,500 in 1760) and other New England ports. Together they provided the home base for three-quarters of all ships owned by Americans. New York and Philadelphia were catching up too; the latter was building more ships than any other port. Both cities hosted far more British vessels than Boston, many of which were built in the colonies with their surfeit of wood. About half the total of New England's dominant exports were composed of lower-value fish and whale oil. Livestock exports exceeded those of the Middle Colonies, but very little grain and flour was shipped out. By 1760, these had to be imported from the Delaware and Hudson valleys to make up a regional shortfall. Despite some fertile areas, Boston's hinterland was less productive than lands to the west and south. On the other hand, Yankees showed much ingenuity in crafts that laid the basis for rapid industrialization a few decades later. By 1770, Lynn became the first major shoemaking centre, a harbinger of factory towns.

The net worth of each free White person in New England in the early 1770s has been estimated at only £33, about two-thirds that of people in the Middle Colonies, which has led some historians to claim that a 'moral economy' predominated over a commercial orientation among New England farmers.[9] While it is no doubt true that most farmers consumed more at home than they sold, and that some produce was traded in local networks, ambitious cosmopolitans in most communities sought commercial markets. From the beginning, Springfield was a commercial fur trading centre dominated by the Pynchon family. Religion may have impelled Winthrop and his coreligionists to cross the Atlantic, but they saw no contradiction in pursuing gain where possible. In fact, when establishing the New England towns, which were fabled as communitarian theological utopias, the more affluent Puritan investors often took the best land for their settlers, leaving the less affluent with the poorer land.

Atlantic Canada

The fourth region, Atlantic Canada, had still more limited agricultural opportunity, being even more dependent on the sea. Before 1760 the Newfoundland Grand Banks fishery had expanded greatly and was of utmost importance to European growth. It established St John's as a commercial centre. Around 10,000 lived scattered in outports. By 1760 in Nova Scotia, only about 13,000 farmed and fished, the latter providing its chief export. As settlers from New England and Britain were then arriving in larger numbers, the region was increasingly drawn into 'Greater New England' under Boston's merchant sway.[10] Although many were expelled, some French-speaking Acadians clung to isolated settlements.

Quebec

The last region to the north of Philadelphia, that of Quebec, is conspicuously different from the others to the present day because the majority speak French. Its 60,000 people (a low number because its early immigration was small, short-lived compared to other colonies, and many returned to France) established themselves along the St Lawrence River, the main highway for the province. Quebec City, with 8,000 people in 1754, ranked among the five largest places on the continent. Montreal (4,000) and Trois Rivières also stood out. Because of church, military, and other government functionaries, fur merchants and traders, and those who served them, the urban population in Quebec exceeded that found elsewhere in America and even Europe—over 20 per cent of the population. Quebec's urban share would fall after 1760. Farmers were relegated to a lower status than the freeholders to the south, though seigneurs who leased land had only modest power over them in this open environment.

Quebec farmers occupied land with a shorter growing season compared to that of English colonies to the south. They could sustain urban populations and themselves, and even exported modestly, but could hardly match the Middle Colonies. Besides, the *habitants* were a conquered people after 1760. The British merchants took over the fur trade, especially in beaver pelts which were so sought by the western European élite. Although, unlike Philadelphia, it did not promote settler expansion to the West, Montreal would eventually become one of the continent's great metropolises.

The South

South of Philadelphia, the region encompassed a total population of about 700,000 in 1760. The South differed from the other regions in its massive dependence on slave labour, especially after 1670; by 1760, slaves made up two-fifths of the population (though far less a share than in the West Indies). Because of this, the wealth of free White people stood at £87 in the early 1770s, nearly three times that of New Englanders. Distribution of this wealth was extremely skewed towards the top, to the élite planters, who held the greatest number of slaves and also produced a disproportionate share of its high-value exports. We can divide the region into three parts: the upper South around tidewater Chesapeake Bay, where the largest share of people resided; the lower South focused on Charleston and secondarily on Savannah; and the back country. North Carolina came under all three spheres. All three depended on metropolises outside the colonies for organizing their economies—a long-term limitation on southern development.

Settlement around the Chesapeake lacked a clearly articulated urban pattern, at least until after 1750. In fact, English commentators, visitors, and officials fussed about the lack of market towns they saw as necessary for the civilizing process; directives from London and attempts by governors had little effect over the first century and a half. Courthouses stood alone more often in the countryside, as did stores of the factors of Glasgow merchants, than in towns. Large plantations took on the semblance of urban places, since planters housed slaves near their houses, barns, and shops, where local trade was often focused. For Virginia planters, Williamsburg served mainly as a focus for periodic legislative decisions and social affairs, and secondarily and only seasonally as a merchants' town. To handle its main product, tobacco, the Chesapeake's metropolis lay across the Atlantic. This situation contrasted strongly with the North where merchants in Philadelphia, New York, and New England cities exercised considerable autonomy as middlemen in seeking out trade channels on the Atlantic within the rules of not easily enforced mercantilism.

In the lower South lived about 150,000 people; the Blacks made up 60 per cent of South Carolina's population, making it demographically most like the Caribbean colonies. Charleston, founded in 1670 and complemented by Savannah six decades later along with smaller ports, focused its activity in the region, similar to a northern city and unlike the Chesapeake. But its large plantations, producing rice and indigo for export, distinguished it from the North. By 1760, Charleston was among the largest cities, housing 8,000 people. Pine forests provided naval stores to Britain. Like ports in the North, ship provisioning occupied some merchants and stevedores.

Finally, the southern back country, on the Piedmont as far south as South Carolina and the Great Valley (including the Shenandoah), differed from the Tidewater. Settlers from the North (mostly Pennsylvania) and from the coast, came together in a mixture of people, which included more slaves than in the North, but far fewer than in the Tidewater. Large planters there, like George Washington, speculated in back country land and became the developers who were basically responsible for settling people. Rural land, like urban land, was carved out of Indian space.

By 1760, much of the area, like the Middle Colonies, came to depend on growing small grains and raising livestock, so a similar urban pattern developed along the Great Philadelphia Wagon Road. On this extended base, Baltimore emerged a great northern city after 1760. As the back country developed (and Chesapeake planters diversified), Norfolk and Richmond emerged as substantial towns too, yet urban development would remain weaker in the South than in the North. Until well into the twentieth century, while the North forged ahead, the South would remain dependent on outside organization of its trade and on slaves for much of its production.

The Bases for Future Development

These eight regions were tied together by water and increasingly by better roads. Packet ships on regular schedules called at ports. On the roads, traffic flows intensified. Small streams were bridged. Stagecoaches became more frequent and faster in the corridor from Boston to Philadelphia and the Potomac—the later Megalopolis. In the 1760s the 'Flying Machine', now running daily, cut the time from Philadelphia to New York (the most arduous stretch) from three days to two. By then fast horses delivered mail in a day. Mail deliveries increased after post offices were established in 1709 under the direction of the British government. Even in 1676 the Duke of York called for legislation to ensure 'Speedy and true Information of Publique Affairs' through the hands of a constable. In 1697 Pennsylvania sought 'publick post' for the 'maintaining of mutual and Speedy Correspondences' for the king and the 'encouragement of trade'. Apparently, establishment of the public post came about because 'great Prejudice hath accrued to affairs of merchants and others' owing to many 'Loose hands'. After 1709, 'certain, safe and speedy Dispatch' was more likely.[11] By 1760, there were eighteen post offices; by 1775, sixty-five post offices connected cities, towns, and well-off farmers, carrying public and business correspondence and newspapers.[12] These links strengthened the internal economy, solidified public opinion, and increased Americans' resolve to break away from Britain. Those today who believe we live in *the* age of information should take note.

For our discussion on Philadelphia, we can draw several important points from this overview of the array of cities and regions. First, by 1760, despite the strong commercial agriculture of the South and Charleston, urban growth was far stronger in the North. The North, especially Pennsylvania, had the most strongly articulated urban hierarchy. This reflected not simply the economic potential of these regions but the ability and skill of their merchants to operate successfully and largely autonomously within the British mercantile system. The large ones would develop as metropolises.

Second, although all colonists shared in the economic expansion of England in North America, the powerful merchants of northern cities (especially in Philadelphia and New York) and the great planters of the South gained the most in wealth, although many artisans, mostly in the North, also did well. (In turn, North American prosperity helped English expansion.)

Third, the presence of provincial government was a key factor in the growth of most major cities. Philadelphia and New York would later lose their claim as the site of the national capital to Washington and their role as state capitals to back country, smaller places, Harrisburg and Albany. But that they grew mightily suggests that, once established, their organized institutional commercial power was sufficient to propel them upward. Still, these moves would symbolize a split between the interests of the metropolises and the country, although obviously the centre and periphery were complementary. Political battles of representation in assemblies pitted the back country not only against the city but also against the older settled rural areas.

Fourth, the relative decline in the urban share of total population, even in Pennsylvania, pointed to an extensive and open (if at the same time very productive) set of societies. Landed property was the goal of most immigrants, a pattern that would continue, though at a decreasing rate after the middle of the next century. Most farmers did not intensively cultivate the land, other than their gardens and orchards. Market gardeners near towns and cities were the exception. Most industries were located in the countryside, and most were not concentrated. Nineteenth-century intensification of industrialization would increasingly encourage the establishment of more towns in these regions and in the West, and increasingly favour population shifts to cities and then even more to large cities in the North. In the long run, Jefferson would continue to win the hearts of Americans through the belief in the prosperous independent yeoman, but Alexander Hamilton, an advocate of industry after 1776, would win (along with the few big farmers remaining) the economic battle. Cities did grow, but a strong antiurban sentiment would develop in the US. In Canada, this division would be far less marked, and hence in the long run, cities were less alien places to rural folk.

Fifth, by 1760 staple exports provided a diminishing share of wealth. Gradually, locally manufactured goods replaced imports from Britain, although obviously many British goods remained cheaper, at least until the 1820s. By 1760, many regions in America had lots of cheap labour, but British labour was still cheaper and British manufacturers operated the new machines before Americans did.

Finally, mercantilism notwithstanding, free enterprise flourished about as strongly in this era as at any time in our past. As always regulations helped some actors and hindered others. Governments had only modest direct influences on trade flows and little on privateering during wartime. Certainly, as would always be the case, business was subsidized in the name of what we now call economic development. The postal service subsidized business, as did the modest improvements of roads. The biggest subsidy to America had been military spending at various times, though to many poorer Bostonians, it had brought dislocation. Taxes were generally light, hardly onerous. Philadelphia's most renowned man, Benjamin Franklin, may have asserted that taxes were as certain as death, but Americans found it difficult to warm to the idea, as the response to the Stamp Act would soon show. Maybe he should have added that evasion of taxes by the rich, one way or another, was a dead certainty too.

Philadelphia's Open but Unequal Society

Since its founding in 1682, Philadelphia had been bustling, though it had some years of stagnation, especially over the two decades or so after 1700. Free enterprise in this 'private city' was balanced by a need for public order.[13] Despite rapid growth, economic enthusiasm, and frequent political clashes that were increasing by 1760, the city remained more socially stable compared to other cities and was less fractious than New York. However, considerably more so than in rural communities, income and wealth in Philadelphia were unequal and became increasingly so. That polarization contributed to radicalization and the revolt of 1775.

The founding Quaker élite (composed of merchants, higher-status craftsmen, and proprietary officials) set the tone. Following Penn's injunction to admit western Europeans regardless of religion and ethnicity, they had no trouble ideologically in absorbing the newcomers. They were prepared to receive a wide variety of groups in the city and colony: German Lutherans and Reformed, Mennonites and other denominations, Scotch-Irish Presbyterians, French Hugenots, Welsh Anglicans, and even deists like the perceptive Ben Franklin, who moved from Boston to Philadelphia precisely when growth picked up in the 1720s. They less enthusiastically accepted Roman Catholics and Jews. Compared to the Puritans who arrived in New England a half century earlier, for several decades after 1682, the Friends (though tightly knit themselves) were extremely tolerant by world standards. Religious controversy did not wrack Pennsylvania or Philadelphia like it did in early New England.

The Quakers' principles would, however, eventually present an obstacle to their continued domination in politics and social life. In the early 1740s they lost some influence in town affairs. More seriously, in the mid-1750s, when the pressure for military spending during the Seven Years' War proved too great for most others to resist, several principled Friends resigned from the provincial assembly that they controlled. By then too their share of the city's population had diminished to about one-quarter or less, even though half of the grandees were still Quaker. Some prominent Quakers, including Penn's sons, who were pursuing higher status, became Anglican. Some Presby-

terian merchants also became similarly influential in society.

Although many Quakers retreated into exclusive sectarianism, their early permissiveness became the bedrock for the most multicultural region in early America, and perhaps the one with the least tension. People were related in overlapping kin and organizational networks. Although slavery marred the openness, many Quakers were uneasy that Blacks were debased. The fact that the French were an outside enemy probably also helped to maintain stability and held the people to the British cause.

Penn hoped that the city of brotherly love would result in a community more cooperative than any in England, but, as in Boston, his early investors expressed a stronger interest in returns from a new society than in a cooperative society. Thus, openness to religious and ethnic freedom did not translate into an egalitarian distribution of wealth; those with the right financial connections increasingly took advantage of them. Some, though not many, with ambition and skill and Franklin as the model, could move ahead during the frequent spells of growth. The British government's spending and privateering during wartime greatly helped those who controlled resources. Luck was another factor in success: the right cargo arriving at the right time in Jamaica could harvest a windfall. Ill-timed investments, just on the verge of a weakening in the economy, could result in failure for those who overreached and had kin to support. However, over the long run, the successes outweighed the failures among the able. As an indication of a rise in wealth, goods, and services for the affluent, newspapers increasing advertised items such as four-wheel carriages, expensive furniture, dancing schools, theatre, country homes replicating those of the English aristocracy, and the like. Still, only a minority rose through the ranks.

As Philadelphia's economy grew, wealth shares increasingly showed 'great disparities', skewed towards the top.[14] The top 10 per cent's share of personal estate and taxable wealth rose, according to one estimate, from about 40 per cent in 1700 to 65 per cent or more in the 1760s, roughly double the top decile in nearby rural counties, which probably still had a more equitable division among residents than elsewhere in the colonies. The 'fabulously wealthy' 5 per cent controlled a half of the wealth, a few dying with personal estates of £50,000, even £100,000.[15] If real estate had been calculated in estate settlements, their share would have been considerably higher. In poorer Boston, where real estate was included in calculations, the top decile's share was nearly two-thirds of the total wealth.

Incomes were, as always, less skewed; incomes are calculated annually, wealth cumulatively. The top 10 per cent took in 90 per cent of rent paid, and about another 10 per cent took the remaining share. More than four in five households rented. The élite lessors gained much of their rent from higher income, even very high-income, tenants, not poor ones. Unfortunately, the data are inadequate for a complete understanding. Those with property of insufficient value to be assessed for taxes, not just poor boarders but even some frugal and modestly successful persons, also paid rents. Although wealth was skewed greatly towards the top, incomes, as indicated by rent paid, were considerably less so. One calculation suggests that, though those in the top decile held two-thirds of the wealth, their earnings amounted to just over a third of the total income.[16] If so, this is a bit higher than at the present time.

One-fifth of the taxpayers held all the landed property, and only one-tenth held the most valuable land. The property ownership level was low only by recent standards, although it would be even lower in the next century. However, it was a sharp contrast to the countryside where two-thirds owned their property on the best-quality and most market-accessible land. In poorer locations, ownership was even higher. As has generally been the case, ownership will be lower where opportunities and hence values are higher. The fact that many high-income people rented in the city probably meant that they kept their assets more liquid; the fact that the less affluent rented reflected their hope for improvement.

The 'better sort', the wealthiest and generally the most powerful, were the high public officials, successful wholesale merchants, a few top craftsmen, a handful of clergy, and, often one and the same, land developers. Penn and élite 'First Purchasers', who were mostly Quaker, held properties in the city. Penn created 'manors' in the countryside, though these long-term holdings differed from permanent English estates with farming tenants because eventually the holders expected to sell parcels. Owners put renters on these lands to clear the forests, then appropriated the appreciation of value through improvements. Merchants who retired to become 'gentlemen' financed and managed properties. As in Britain, the élite too had to work at maintaining status through wealth accumulation. Some combined public with their private activities, their public affairs helping to sustain private economic power.

Next to land, the chief way to wealth, if riskier, was through trade. Prominent Quaker James Logan, Penn's first provincial secretary and chief agent, not only held land but traded in furs. His administrative connections with Indians no doubt helped. Around 1760 the city's merchants made up only just over one in twenty taxed people (fewer than 200), but some of these were successful and thus powerful. Quaker and Anglican merchants like the Shippens, the Pembertons, and the Willings, started with money brought from England and elsewhere and with the continued trust of fellow merchants overseas. Some organized partnerships, such as Baynton, Wharton, and Morgan. Their sons and grandsons followed—indeed, Philadelphia's élite today boasts many colonial names. Some of the most successful invested in rural ironworks.

Frugality and austerity marked the early Quakers, though Logan led the way to expressions of status with the first elegant country house near the city. By 1760 most at the top followed his lead. The ostentation of silver buckles, elegant gowns, and fine furniture among the grandees offended the more pious and less affluent of the Friends in city and country. Rich sons took the grand tour of Europe, as wealth permitted leisure pursuits, though on their return their families expected them to start working in the family firm. For some Quakers, shifting to the Anglican fold solved the moral problem of too much finery. Others rose in status too, such as Presbyterian William Allen, though would-be merchants without connections within old-boy networks or without inherited wealth could hardly be counted among the merchant aristocracy unless they were lucky in the Atlantic. The top congregations and clubs, such as the Schuylkill Fishing Club and the city council, sustained the old-boy networks. In 1766 some élite organized the first hunt club to terrorize foxes.

Between 1756 and 1775, the affluent held more than £400 at death. This bracket of about 30 per cent possessed over 90 per cent of the wealth. Nearly one-third of probated widows were in this rank. Three-quarters of merchants died with a personal worth of £400 or more, as did over a half of the professionals, most of whom were lawyers. Ten per cent of mariners, presumably ship captains, had reached that level. Manufacturers who reached near the top included one in seven shipbuilding artisans (probably the entrepreneurs and organizers); over one in five house builders (probably those who built houses for the élite and buildings for the military during the war); and surprisingly between one quarter-and one-third of the shoemakers and tailors, as well as a few who made carriages, fashioned silver and gold, and built fine furniture. By 1760, the latter competed effectively against still higher-status English imports. These leading leather-aproned craftsmen outstripped others in their industries by producing higher value-added goods (put another way, goods the affluent wanted). Making buttons and glass brought Richard Wistar into this class.

Some printers and publishers, notably William Bradford and Benjamin Franklin, achieved high status among the forty or so who established businesses between 1720 and 1775. By the latter date, weeklies became tri-weeklies and magazines appeared, which was a sign of cultural intensification. These people sold their own and other books, and ran circulating libraries. Media authorities, using their wordsmithing both in reporting and diagnosing society's problems, profoundly influenced the course of events. Even though Franklin retired in 1748 after twenty years of practice in printing books and the *Pennsylvania Gazette*, he remained the symbol of the successful, manual, leather-aproned artisan who organized the work of others.

Below the high-status craftsmen, other self-employed master artisans and many journeymen produced lower-valued goods. Thirty per cent in these groups possessed far less wealth, between 5 and 8 per cent of the total wealth, according to tax rolls and probate inventories, and their share fell over the decades. In shops they made and retailed goods such as pewterware, pottery, harnesses, and leather pants, largely for those at the same income level or below, though women dressmakers catered to those with money. Those with modest incomes dressed mostly in cheap imported clothing. Others built, outfitted, and provisioned them. Some women who ran taverns and ordinaries (restaurants), and probably the most reliable draymen and carters belonged in this category as well, as would the small number of public employees who did manual work. Carters nonetheless (if the New York experience fits Philadelphia) possessed a 'status far above their skills and financial worth'. At least in New York they were 'anything but deferential', believing in the equality of all men.[17] Licensed by the city, they organized sufficiently to limit their numbers and so maintained control over the movement of goods.

Measured by the worth of their estates on death, those known collectively as the 'poorer sorts', who formed the bottom 30 per cent, held only 2 per cent of the total wealth. Over the decades they had lost ground proportionally from perhaps double or more than that share. Wages at the bottom, even in good times, did not keep up with others, partly because rural young people without means, who sought out opportunity in the city, pushed them down. Even in stagnant Boston, the poor weakened compared to the rich. So no matter whether the economy rose or fell, it favoured those who had more. Yet the fact that the absolute amount of personal goods of those poor Philadelphians dropped from £39 to only £22 is hard to fathom.[18] Even the rise in wages in good times after 1745 did not allow half of the deceased poor (the few whose estates were actually recorded between 1756 and 1775) to leave behind more than £50. This situation applied to four in five in poorer Boston. These sharp differentiations continued, except when many of the rich Loyalists lost their wealth after 1776.

One in five or six was very poor—the 'underclass'—was similar to today's figure, though today most have access to food and obviously other amenities like television. During the severe downturn after 1760 when the British military largesse was cut off, one in six of the 'deserving' poor received relief from charitable organizations and the overseers of the poor. Undoubtedly, more needed help in their struggle to make ends meet; many did not have enough money to pay taxes. The number of taxpayers without any taxable property rose from just over one-quarter to over two in every five people—an astonishing rise in this city of riches. The number would climb even further in the 1770s as the rich controlled more property and gained disproportionately from trade and finance.

The working poor were the marginal craftsmen, many of them journeymen and most of the labourers. A few labourers who had steady jobs (the one in ten taxed above the minimum), or the jacks of all trades were in higher income levels. The marginal labourers often worked only during the warmer months. Most of the ordinary mariners made modest incomes; women who washed, cooked, and baked earned little; and one suspects that prostitutes in brothels for the pleasure of the élite likely earned much more than streetwalkers. For most, 'economic mobility could be no more than a dream'.[19] Most of their earnings covered food and rent. Fortunately for low-income people, food was cheap most of the time. In America, generally the total cost of food was 40 to 50 per cent of income, somewhat below that of England. That was largely a function of abundance, though the municipal corporation's increasingly shaky control of bread prices and quality eventually threatened the survival of low-income people.

Below the working poor, but hard to distinguish separately from them, were what are today

> **Franklin may have been gentle to deserving elderly indigents, but he was more hard-nosed about the able-bodied.**

called the underclass, who earned little. Most were probably not recorded. They can be separated into what were later called the deserving and the undeserving as seen in the eyes of the better-off. The first were objects of pity and charity: the old who were indigent, or the disabled. Some found a home in poor houses, called bettering houses or workhouses, which suggests that officials thought they might undertake productive work, like sheltered workshops today. A frequent worry was the poor and sick arriving from Europe, especially from the Rhineland, who received some help from immigrant aid societies. Rural counties set up almshouses, which (at least those near New York) auctioned off paupers to buyers who needed workers.

The unemployed able-bodied were undeserving of public or philanthropic aid. Many of the servants lived in Philadelphia. After serving three to four years and then turned out in the world, they did not receive much pity, and had little chance of advancement. The undeserving included older orphans, rejected children, and adolescents from the countryside who were attracted to the city, hoping to find work or at least to evade attention in the anonymity of the city. Even more equitable rural Pennsylvania had its casualties; increasingly, landless married cottagers and single men from poorer families lived and worked sporadically on the farms of the more affluent. With some exceptions, free Blacks, though hardly as numerous as those in New York, were probably worse off materially than those who remained slaves. Starvation among the homeless tramps, who were euphemistically called the 'strolling poor', and even the imprisoned was not unknown in the city of brotherly love. Those imprisoned included debtors who fell from grace.

Franklin may have been gentle to deserving elderly indigents, but he was more hard-nosed about the able-bodied. He no doubt believed in Poor Richard's admonitions of frugality and industry; with regard to London's poor, who in 1765 mobbed grain wagons, Franklin commented, 'the more public provisions were made for the poor, the less they provided for themselves, and of course they became poorer'.[20] He actually advocated abolishing the poor law; hard work and sobriety would then banish the St Mondays that followed weekends of debauchery. Franklin's opinion of the poor did not quite prevail among the overseers of the poor in Philadelphia at the time, who clung to the English notion that the public had a responsibility, minimal though it was, to the truly disadvantaged. Religious voluntary charity could not do the job. However, Franklin's view has continued to the present among the privileged, more so in the United States than elsewhere in the Western world.

Franklin was the quintessential American, who, because of his varied guises, many of today's so-called conservatives and liberals can claim as their forerunner. Little wonder he has been admired. Having learned the printing trade in the most literate of colonial British cities, Boston, Franklin arrived in Philadelphia with ambition. With little money, though likely more than he claimed, he managed to advance in society. At first he identified with the craftsmen, or at least the better-off among them. By 1748, he retired after reaching an adequate level of material well-being. He then made money through investments in land and others' businesses, and pursued politics and his scientific experiments. Although his was not quite the rags-to-riches story, his rise was to be a model for others to follow.

He achieved his elevated status through a combination of qualities, not the least his uncanny ability to read the signals of those with power. On certain occasions, he deferred to the élite, cleverly ingratiating himself while lobbying behind the scenes to become postmaster and later clerk of the assembly. He became city councillor and then a higher-status alderman, a position that also carried the prestige of justice of the peace. On other occasions, he was aggressive. When the leaders of the relatively high-status Masonic lodge rejected his application, in the *Gazette* he cleverly ridiculed those responsible and was soon admitted. The élite no doubt agreed with his admonitions through Poor Richard to work hard and be frugal; most believed their success was the result of these traits.

Not least, this remarkable chameleon could play off political factions—sometimes standing with the Quakers in the assembly, sometimes with the proprietary powers, though these blocs were often antagonistic to one another. His 'quiet but assiduous efforts for … self-advancement, and his willingness to bend a principle when it was to his personal advantage' explains his shifts of allegiance, and his agility paid off.[21]

Although he failed to win over British officials to his view that Pennsylvania should become a royal colony, or to his secretive pursuit of the governorship, he managed to ensconce his son as governor of New Jersey. He was successful in advancing his own status while promoting the cause of spatial, economic, and military expansion against those who were content to stay within already established bounds, such as rural and many urban Quakers and Mennonites, or to advance only with the consent of the Native peoples. Despite his close association with Quakers, he promoted military defences. He thought it was wonderful that the population was doubling every generation at an exponential rate that would fill the back country, yet at the same time he could cool frontiersmen's tempers riled by Aboriginal resistance to expansion. He kept slaves, but eventually when it was expedient, he joined the abolitionists of slavery. In 1775 he supported separation; though uncertain of the outcome, three years later he talked clandestinely to British agents just in case the empire prevailed.

In the religious sphere he was nominally a Presbyterian, but was actually a deist, who complained about sermons 'seeming to be rather to make us Presbyterians than good citizens'.[22] Yet he could befriend Great Awakening evangelical George Whitefield, admire his expository skills (to the point of emptying his pockets into the collection bag), and print his sermons, even though he disagreed with Whitefield's decidedly otherworldly message. What they had in common was selling God, setting a time-honoured precedent for America. He was the first notable media mogul who extolled free enterprise, hard work, expansion, and optimism to fulfil the American dream, and who had much to do in influencing Americans to defer to these principles and to those who extolled this civic religion.

Local Government Fragmentation and Voluntary Initiative

Franklin's leadership in shaping Philadelphia's social and material public services can hardly be denied. He was a powerhouse of ideas in what was a confusing and even chaotic government. The city's Common Council often seemed anaemic. To compensate for its inaction, the provincial assembly set up other public bodies to provide services, though it could be equally argued that it also thwarted council action. The limited extant record suggests both. Voluntary agencies often initiated programs. To manage public and quasi-public matters, officials and others provided social support, sought to maintain order and safety, regulated parts of the economy, promoted public health, and served material needs through infrastructure, all of which can be seen as encouraging economic development to a modest degree.

Philadelphia's local council is often seen as the clearest American example of the medieval 'closed corporation', self-perpetuated through appointments drawn from freemen of the city. In contrast to New York's elected body and the town meetings of Boston and even of Lancaster in Pennsylvania, the corporation has been described as 'an impotent anachronism', a remnant of earlier English governing bodies.[23] It was not quite true: the corporation acted in a number of fields, especially in earlier years.

After a failed attempt in 1691 to organize a local government, using his executive power, William Penn granted Philadelphians a charter for the closed system in 1701 for the 'better ordering' of public life. Apparently conditions in England brought this about. Before the Glorious Revolution of 1688, the Stuarts tried to centralize decision making and finance in the Crown and so weaken traditional city and borough rights. But William and Mary reinstituted charters of corporations, however, putting power back in the hands of urban bourgeois oligarchies. These members of the mer-

cantile élite were hardly paragons of democracy, though status required a degree of action and *noblesse oblige*, which Philadelphia's Quaker élite gladly conceded.

Although the charter gave certain powers to this corporation, Penn added another and formidable check: through the Charter of Privileges in 1701, he designated the assembly as the only provincial legislative body with power to regulate cities. This was close to the modern reading that municipalities are creatures of the state. Led by David Lloyd, a rural Quaker, the assembly forced Penn to put this provision in the Charter. Despite their religion, many rural Quakers sought to curtail proprietary power, and with it the city's power, mostly in the hands of 'weighty' Friends. The assembly sometimes refused to increase the corporation's ability to act, though it occasionally imposed obligations. In contrast to other places, it also created special elective commissions to act on particular issues. Only elected people could, it seems, raise taxes.

Yet Penn was not without more direct power over the city and province. He and then his sons appointed the governors, who signed or rejected legislative bills, though final authority lay in London and also with members of the provincial executive council. In fact, William Penn's son Thomas reinvigorated the proprietary urban interest in the 1740s, using his power to create county seats, though curiously not as boroughs with some powers independent of the counties. Philadelphia County Quarter Sessions Court and then later, after 1722, the elected county commissioners, as it related to major roads, also influenced local governance.

The assembly also sought to take control of the scales of justice from the proprietor, his officials and the corporation. When Lloyd, the powerful speaker of the assembly, became chief justice in 1719, this shift seemed accomplished. Rural Quakers seemed to ride high, presaging later times when hostile states controlled cities, but this was not quite the outcome. Near gridlock persisted. Until 1776 citizens lived with the contradiction between the closed corporation and proprietary power on the one hand and that of the assembly on the other.

They had to endure the confusion when prominent people, including Franklin, were at times members of both bodies.

In 1701 Penn defined the corporation's composition and appointed the first mayor (Edward Shippen, a prominent Quaker merchant and landowner), recorder, clerk, councilmen, and, from among them, a smaller group of aldermen who were also justices of the peace. Subsequently, councilmen appointed the mayor from among themselves and their replacements from among the freemen of the town. As in England, freemen had to hold real and personal property amounting to £50 and be permitted to practise trades. The freemen were not, however, the only ones plying trades in this growing city. Besides, oddly, on occasion corporate members appointed some men to council prior to their admission to freedom, so in this sense, the corporation was not entirely closed. It also appointed a commissioner of property to manage private lands (an unknown amount) held by the corporation, which were distinct from land held by the proprietor. Through the years, as in city councils elsewhere, the corporation was composed mostly of top merchants and artisans and, more likely than not, people appointed relatives. If Philadelphians did not elect councillors, elsewhere the franchise was restricted to the few with property.

Adding to the complexity, other official bodies operated within Philadelphia. Following English practice, the overseers of the poor were a separate body appointed by the corporation but dependent on the corporation for financial help raised through a poor tax. After 1749 that body was able to raise funds independently from the corporation. Also, in response to petitions from grand juries appointed by the county sheriff, the assembly set up separate boards elected by freemen: assessors of property in each ward for taxation; wardens over policing, street lighting, and cleaning; eventually commissioners for street paving; and regulators of party-walls. All were enjoined to cooperate with the corporation, though divided authority hardly favoured efficiency. This multiplicity of structures allowed for passing the buck on occasion. The corporation's lack of standing committees probably inhibited consistent action.

Confusion thus reigned in financing and delivering public services. Unfortunately for the council, Penn had not stipulated the corporation's power to raise money. As early as 1706, it unsuccessfully asked the assembly to legislate that power. It continued to find ways to raise revenue through fees, fines, purchase of freedom, rents, and even lotteries. Elected boards could raise tax money, but all were limited by a widespread reluctance to pay taxes; resistance to British taxes during the decade before independence obviously had a precedent. Sometimes the bodies did not pay their public employees, most of whom were part-time but numbered two or three score around 1760. While other levels of Pennsylvania government worked reasonably well, Philadelphia's did not seem to.

Ceremonial parades and the corporation's posh annual members' banquet gave the impression that it was a club for the élite. The mayor paid for this banquet, at least until Mayor Andrew Hamilton in 1745 decided to donate the money instead for a public building. Some others followed this more sensible path. Some who were appointed mayor sometimes refused to serve, suggesting that what had been an unpaid position until about 1745 was a burden costly in time and responsibility, outweighing the status accorded. The rules stipulated a stiff fine for refusing the position. Thus it seems the corporation was not quite a moribund club. Perhaps in a more open economic environment after 1740, its élite members inwardly chafed at the obligations, but resistance to the assembly, Quaker rectitude, and the medieval legacy did not allow a full break to modern American-style city government until after independence.

Riding the wave of democratic enthusiasm in 1776, revolutionaries abolished the closed corporation. The chief reasons seem to have been its perceived aloofness from the people and the already noted confused relations between it and the other elected bodies. Overlapping jurisdictions had hampered clear action. For thirteen years after 1776, the assembly ran the city. Then it created an elective corporate body in 1789, though voters still had to own a certain amount of property. Even so, suggesting that the closed corporation had not quite been a 'kind of tyranny on paper', the first council selected the same person who had been the last mayor in 1776![24] Eventually around 1840, during the era of the 'common man', all male citizens there and in other cities were considered freemen and could vote, so mayors were directly elected. As we will see in subsequent chapters, democratization would not necessarily bring good government.

Voluntary non-profit organizations, many of British precedent, had an impact on the social fabric and public environment. In fact, in a society where government was viewed suspiciously, as Penn's appointed governors and the corporation often were, public action depended on these groups. Franklin has often been praised as the initiator of many projects, though he also took public offices too, even for a while on the council. The volunteer fire company was the model of the effective non-profit urban organization. Beginning with the Union, started by Franklin in 1736, these companies, some ethnic based, provided not only protection but mutual support among their members. They convivially met in taverns.

> **Voluntary non-profit organizations, many of British precedent, had an impact on the social fabric and public environment.**

Some craft-oriented groups can be classed as quasi-unions; the Leather Apron Club of the 1720s led a popular movement, mooted as 'the beginning of America's first working class'.[25] One suspects, though, that, like Franklin, more of its members were drawn from the high-income mechanics than from the lower ranks, and they can thus be considered, along with farmers, as part of the first middle class. The former became prominent leaders in the revolutionary movement. Increasingly after 1760, the affluent masters acted more like bosses and journeymen more like employees, as indicated by the first full-blown strike in 1786. Shipwrights of the White Oak Club paraded their barge in celebrations and generally supported the political status quo, at least until the non-importation agreement in 1769 when they became radicalized. There

was, however, some internal labour stress: in the 1750s, the Cordwainers Fire Company attempted to stop the monopolistic practices of the leather dealers who supplied them. Other voluntary groups—such as the diverse religious organizations, immigrant aid societies, scientific and literary societies (such as the Junto, the forerunner of the American Philosophical Society, founded by Franklin soon after he arrived) also brought together at least the 'better' and upper 'middling sorts' of the community. These voluntary groups and public bodies delivered social and material services to Philadelphians.

Social Services

Improved education was on the minds of some, not least the energetic Franklin, who supported many proposals such as schools to Anglicize German youth. In his widely read Germantown newspaper, Christopher Sauer, who supported the independence of quietist sects, criticized this idea. Although Franklin failed in a first attempt to start a German-speaking newspaper, local Lutherans who favoured integration and wider learning countered Sauer, beginning in 1762, with the weekly *Philadelphische Staatsbote*.

Others pursued pedagogical initiatives. The Presbyterian and Anglican merchants hired the Reverend William Smith in 1754, in the midst of that rapid-growth era, to implement higher education through the College of Philadelphia, to which was added the Academy (a related secondary school) and later a medical school. Churches and private masters sought to balance children's religious, classical, and pragmatic educational needs. Lutherans started a seminary. In 1753 William Dawson started a night school for apprentice and working girls to raise their literacy and numeracy and inculcate morality. In 1767 a boarding school for élite young ladies began. A charity school served some poor children to raise their opportunities and control their behaviour. Although all these efforts hardly added up to an education system, they presaged a much more vigorous effort in the next century.

Libraries sprang up; the Library Company was the first. On Franklin's initiative and with the financial support of James Logan, the proprietors granted a charter and a site to its members in 1742. At first, its membership was limited though diverse, but then became caught in the struggles between Quakers and others, so in 1747 a group of professionals, tradesmen, and high-status artisans opened the Union Library, which, by 1759, had 100 members who subscribed. Then others were formed, including one for mechanics. In 1769 they all merged into the Library Company, which was now more open to the public with lower fees. It was linked to the philosophical societies. At a time when Americans were chafing under British rules, the consolidation of libraries promoted colonial unity. All these libraries, like the schools, emphasized practical learning, though they were still balanced somewhat with classical and theological writings. English novels were opening the door for the acceptance of popular writing.

In addition to these semipublic libraries were specialized collections, such as that of the Carpenter's Company and those of the churches. Some of the very rich also assembled their own libraries. Writing, painting, music, and theatre also began to flourish in mid-century, as did scientific enquiry, obviously triggered by Franklin. David Rittenhouse became an inventor of sophisticated instruments. Better-off intellectual Philadelphians (and those in other cities and towns), through their networks of interest, were in tune with the Enlightenment. Philadelphians partook of the richness of enquiry that flowed across the Atlantic. The ideas of some travelled back across.

Following English custom and the Elizabethan poor law, the corporation, through the overseers, provided 'outdoor' relief (food and firewood) to the poor, usually in winter when employment was hard to find. As would be the case in the subsequent two centuries, rural poor who were 'reduced to great Straits … flocked' into the city.[26] Over the tough winter of 1761–2, charity groups (such as the St Andrew's Society and other national and religious associations) and the city council provided material relief to perhaps one in six families. Given the number of poor we noted earlier, it would seem that the task became more and more difficult when conditions deteriorated. The church bodies set up pension funds for the widows and orphans of their mostly poor clergymen.

Supplementing an almshouse begun by Quakers for their own disabled poor, the overseers themselves eventually built one that was run by supervisors. Then, in an attempt to reform at least those capable of some work, in 1766 they started a bettering house, a sheltered workshop to keep 500 or more disabled people busy and thus disciplined. It was financed by both tax and charity funds. How many were bettered is uncertain. In 1751 Presbyterian cleric Frances Allison and Benjamin Franklin successfully promoted the financing (largely from Quakers) of the Pennsylvania Hospital, the first in the colonies that served the indigent sick. Even so, 'the punishment of dependence' forced them to behave deferentially if they hoped for aid. They even had to wear a badge to indicate public assistance.[27] Other humanitarian groups sought to relieve prison inmates, including debtors, who, unable to pay for their keep, were left to shiver and rot in wretched conditions.

Public Order and Environment

As for protection and order, ward-based constables and the watch, together with the county sheriff and coroner, attempted to maintain order. In 1750 wardens increased protection by taxing citizens for a paid watch and for street lighting. They fostered public safety by ordering citizens to put up whale-oil lamps on the streets and installed public lamps about 1760. Worries about vice and immorality led to severe punishments; as in Britain, the courts meted out capital punishment for a wide range of crimes. The corporation's members, including the mayor, acted as a court. The mayor and recorder registered sales and resales of bound servants to maintain a measure of orderly relations.

To fight fires, the corporation bought buckets and then pumps to draw water from wells and hired mechanics to look after them. Before Franklin organized the first voluntary fire company in 1736, apparently the corporation expected everyone to pitch in. Fire companies played a role in controlling the behaviour of their own members. Also, when the corporation apparently did not act, the assembly appointed regulators of party-walls to ensure safety; prohibited wooden buildings in the city proper about 1760; and required clean chimneys, which created miserable and unhealthy work for poor boys. The Contributionship Fire Insurance Company, initiated by Franklin in 1752, insured only brick houses. Although Philadelphia had not experienced a large fire, the memory of London's massive conflagration in 1666, as did some fires in Boston, remained.

The most obvious of the corporation's concerns was regulating public markets and semiannual fairs, a time-honoured practice. As in cities and boroughs elsewhere, the corporation attempted to prevent merchants from cornering the market. Cartmen sometimes tried to do that. The price for bread especially was controlled, if increasingly ineffectively. Although the clerk of the market assigned stalls and collected rents, no one other than the mayor was enjoined to inspect bread, a medieval vestige destined to fade away. Other officials inspected the quality of grain shipped in sacks and meat in barrels destined for export. This was absolutely essential to ensure the province's good name in the Atlantic world. Apparently to control quality, the corporation set up a public abattoir outside the built-up area. This move may also have helped to banish slaughtering from the public market stalls and other inner-city places. Firewood was virtually the only source of fuel other than used tanbark and a little coal from England and Cape Breton Island. Because it was sometimes depleted late in long winters and was in short supply for the poor, resale was prohibited from September to March to prevent cornering the market. Public wood corders were responsible for orderly supplies of correct size and sufficiently dry wood.

As for public health, wardens were responsible for inspecting pumps and wells, which came increasingly under public rather than private purview. Early in the century, the annual deaths exceeded annual births, but the death rate fell around 1760, which suggested a healthier city.[28] The wardens' inspections may have improved the quality of water. Drinking water in the guise of small beer and whiskey was one way to avoid infection. Also by 1760, epidemics became less frequent as populations became naturally immunized against some diseases. Perhaps moving the quarantine hospital—the 'pesthouse'—from the city to an island warded off some diseases infecting immigrants, who were making up relatively fewer of the population as it grew. Isolation of the ill in the hospital may have helped. Smallpox vaccinations began too, and soon nearly banished that dreaded disease. Epidemic fevers returned sporadically, however; the yellow fever outbreaks of the 1790s lowered Philadelphia's reputation. Infant mortality remained high by modern standards, in part because cow's milk was carelessly processed. Another century and a half would pass before this changed. Around 1760 the corporation paid scavengers to clean the streets weekly, probably improving public health too, though pigs roamed the streets, eating refuse. As for privies, cleaners undoubtedly came to remove human waste to dumps somewhere on the edges of the city, maybe even on the waterfront where lessees of waterlots slowly extended the shore outward. Lime temporarily kept privy pits somewhat tolerable.

Public control over waterways and wharves was essential for health and orderly growth. The corporation appointed an inspector of watercourses, presumably to improve drainage and answer protests about pollution. Dock Creek was a continued point of contention. The creek drained a major part of the town. In early days, besides its natural run-off, the slight tides helped to cleanse it daily, but persistent dumping of garbage, butchers' offal, dead animals, as well as run-off from privy pits, several tanners' and skinners' lime pits, and breweries fouled the stream. Clean-ups were sporadic from 1739 to the early 1760s. In 1748, for example, the assembly would not permit the corporation to tax for improvements recommended by a select committee, though the corporation may have had enough revenue on hand for the job. Tanners held considerable clout and resisted paying for any improvements. Through the 1760s a series of acts, codified in 1769, showed that leaders could respond 'in a comprehensive fashion to the long environmental crisis'.[29] In the 1780s the city began to create an underground sewer. Elsewhere on waterways, the county sheriff was responsible for policing the rivers. Ferries across the Delaware and Schuylkill were essential for movement and therefore for economic growth, though the assembly did not allow the council complete control of franchising them, which was another sign of conflict.

The most important facilities were the wharves. A public wharfinger supervised their use. The council leased out water lots to individuals, mostly merchants, who built their own docks. It also retained the river ends of streets for public access, and maintained two wharves. In 1733 the members of the elected bodies had to thwart the corporation, which was intent on raising cash, from privatizing one of them. Had this happened, poor boatmen would have had to pay a charge. The corporation also held an unknown share of private property that it rented out, and held land for development.

Public roadways were a concern for the council and eventually for elected road commissioners. The failure to raise sufficient funds led to ordinances requiring abutting owners to pave the streets with pebbles, which no doubt led to patchy quality. Around 1760, lotteries provided money for more public paving with cobblestones in better areas. To protect roadbeds, the corporation and commissioners sought help from the assembly to regulate tire widths and the weight of wagon and cart loads. They also tried to have the streets sloped inward to provide a modicum of proper drainage. They were only modestly successful; heavy rainstorms could fill cellars. No doubt, however, the streets were neater and less muddy in 1760 than they were earlier. However, the overlapping jurisdictions hampered cooperation, for example, in rebuilding the drawbridge over the mouth of Dock Creek at Water Street in 1770.

In sum, the general impression gleaned from the few analysts who have written on local government (using the few extant records) is one of confusion, yet of halting environmental and even social improvements. The corporation was active to a degree, but was gradually superseded in many areas by assembly-mandated elected bodies. Obviously the city worked with enough efficiency for those who were well-off. There was modest concern for the poor. While few expected to wipe out poverty, the reformation of the poor (very apparent in the next century) was coming on the agenda.

Like other cities, the civic bodies loosened their regulatory control over the growing economy. The city could not stop pedlars from competing with market-stall owners, for example, nor were the latter necessarily freemen, yet public markets continued for many decades. In fact in some cities, they never disappeared and have recently gone through a revival. Even Boston, where earlier opposition had prevented market-places, finally allowed Faneuil Hall around 1760. Despite conflict, confusion, and inequities, Philadelphia was growing about as rapidly as any city in the Western world at the time and was probably about as well serviced. To overseas visitors, it was a great success story. The patterns of economy and society on the landscape revealed a less optimistic evaluation.

Spatial Order: The Modern Commercial Walking City

In 1682 the surveyor-general Thomas Holme followed Penn's wishes in laying out a grid pattern of streets, creating 176 rectangular, though not totally symmetrical, blocks within the official city of 1 mi. (1.6 km) wide and 2 mi. (3.2 km) long between the Delaware and Schuylkill rivers. Other blocks followed in Southwark beyond South Street and Northern Liberties north of Vine Street (Figure 3.1). The idea of a loose, open, green country town was abandoned when Holme laid out much smaller and narrow (but still quite deep) lots than originally intended; Penn came to realize that First Purchasers (as they were appropriately called), who had rights to a share of city property by buying rural land on which to settle people, wanted not to spread out but to be as near the centre of action as possible. Not surprisingly, those to whom he rented urban properties felt the same way. It

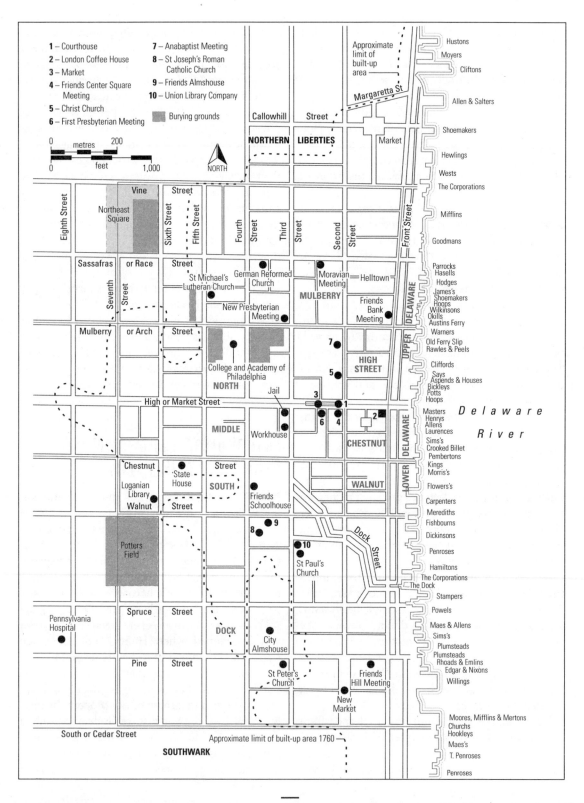

1 – Courthouse
2 – London Coffee House
3 – Market
4 – Friends Center Square Meeting
5 – Christ Church
6 – First Presbyterian Meeting
7 – Anabaptist Meeting
8 – St Joseph's Roman Catholic Church
9 – Friends Almshouse
10 – Union Library Company

Burying grounds

0   metres   200
0   feet   1,000

NORTH

Approximate limit of built-up area

Margaretta St

Hustons
Moyers
Cliftons
Allen & Salters
Shoemakers
Hewlings
Wests
The Corporations
Mifflins
Goodmans
Parrocks
Hasells
Hodges
James's
Shoemakers
Hoops
Wilkinsons
Okills
Austins Ferry
Warners
Old Ferry Slip
Rawles & Peels
Cliffords
Says
Aspends & Houses
Bickleys
Potts
Hoops
Masters
Henrys
Allens
Laurences
Sims's
Crooked Billet
Pembertons
Kings
Morris's
Flowers's
Carpenters
Merediths
Fishbourns
Dickinsons
Penroses
Hamiltons
The Corporations
The Dock
Stampers
Powels
Maes & Allens
Sims's
Plumsteads
Plumsteads
Rhoads & Emlins
Edgar & Nixons
Willings
Moores, Mifflins & Mertons
Churchs
Hookleys
Maes's
T. Penroses
Penroses

Callowhill   Street

NORTHERN   LIBERTIES

Market

Vine   Street

Eighth Street
Northeast Square
Sixth Street
Fifth Street
Fourth   Street
Third   Street
Second   Street
Front Street

Sassafras   or Race   Street

Seventh   Street

St Michael's Lutheran Church
German Reformed Church
Moravian Meeting
Helltown
MULBERRY
Friends Bank Meeting

New Presbyterian Meeting

Mulberry   or Arch   Street

7
HIGH STREET
5

College and Academy of Philadelphia

NORTH

Jail

High or Market Street

3
1
6   4   2

MIDDLE
Workhouse

CHESTNUT

UPPER   DELAWARE

LOWER   DELAWARE

Delaware   River

Chestnut   Street

State House
Loganian Library

SOUTH

Friends Schoolhouse

WALNUT

Walnut   Street

Potters Field

8   9

10

St Paul's Church

Dock   Street

Pennsylvania Hospital

Spruce   Street

DOCK

City Almshouse

Maes & Allens

Pine   Street

St Peter's Church

Friends Hill Meeting

New Market

South or Cedar Street

SOUTHWARK

Approximate limit of built-up area 1760

may also have occurred to him that revenues would be higher too than they might have been had he stayed with his original bucolic vision. He kept only a hint of the latter by naming most east-west streets after trees. In keeping with the rising secular baroque sensibilities, and following English or Irish city plans, Holme laid out five public squares, with either the notion of playing fields or private parks for the rich in mind. The central square (much later the site of city hall), where 100-ft (30-m) wide High (later Market) and Broad streets met, was equidistant between the rivers. Penn and Holme hoped that growth would grow eastward from the Schuylkill and not only west from the Delaware, but it did not because people sought centrality.

Philadelphia, of course, was not the first gridded city; Spanish-American cities were rectangular, as were new towns in Britain and Ireland, of which Holme was aware. Philadelphia is, however, often seen as the rectangular norm for most subsequent urban development on the continent. Neither Quebec, Boston, nor early New York with their irregular streets were spatial models. The earliest rural surveys of lots in Pennsylvania were regular too (that is, defined prior to settlement), though the land office responsible abandoned its control over subsequent occupation for several decades. By the end of the 1700s, the United States and the British colonies in what became Canada embarked on dividing the wilderness—rather, what had been Native communal lands—into rectangular orthogonal blocks. Many new towns on the frontier were gridded, and New York City followed soon after. The rectangular pattern of streets and lots within blocks was then the most rational pattern for the modern world in which individual property ownership was the goal of so many. Not only was it the easiest way to survey, subdivide, and provide effi-

cient movement on streets, but it allowed individual properties to be easily differentiated and recorded.

Subsequent to settlement, owners of lots split up many of them for denser occupation to accommodate buildings 15 to 20 ft (5 to 6 m) wide and often in rows. Owners also gradually created alleys fostering subdivision and denser building, since urban space was valued at a factor of ten or more than rural land. Although Philadelphia clearly had periods of boom and bust, the market for urban land went upward. In boom times, speculators could cash in as they sold to speculative builders or to rich individuals who engaged builders. When conditions weakened and population growth slowed down, it was the workers in the construction trades who suffered, plus a few guileless speculators. The city boomed after 1725, most especially during the wars. In the decade before 1760, rapidly growing Philadelphia's real estate values reached 'dizzy heights'.[30] During the good times, developers who had assembled large parcels of land cashed in. In 1739 the Shippens opened Society Hill for housing development, and the family continued to be the largest landowner. In the 1740s, beyond the city boundaries in the Northern Liberties, the city's recorder, Ralph Assheton, put 80 acres (32 ha) up for division into lots. Samuel Powel, 'the rich carpenter', was the first large builder. By 1756 he owned over ninety houses, which he rented out, as well as land for further building.[31] As we saw earlier, urban property was in the hands of a few.

By 1760, as in other commercial walking cities replicating London, Philadelphians occupied a roughly triangular, or at least a semielliptical area, that covered less than one-quarter of the 2 sq. mi. (5 km²) rectangle from the Delaware.[32] The western built-up apex of the triangle was then about

3.1 *(Opposite)* Nicholas Scull's Philadelphia, 1762, with a population of about 20,000. Most of Southwark strung out along the river has been cut off from the original. High-level functions—the coffee-house, courthouse, market, jail, and churches—were centralized. Other public and quasi-public buildings and graveyards were more dispersed. Surveyor Holme's southeast square, intended as a park, was a potter's field used for grazing. Dock Creek was filled in later. Scull did not show ship-building yards and ropewalks depicted on later maps, nor the quality of housing. *Source*: Modified from Lester J. Cappon et al., eds., *Atlas of Early American History: The Revolutionary Era* (Princeton: Princeton University Press for The Newberry Library and Institute of Early American History and Culture, 1976):10.

Eighth Street at Market. The triangle was splayed out along the Delaware; there activities spread beyond the boundaries of the city proper, up the river into the Northern Liberties and farther down into Southwark. Beyond the contiguous built-up area, there were small concentrations of urban populations in nearby crossroad hamlets. The most developed was Germantown, 7 mi. (11 km) north-west of the city centre, and definitely an urban village with a modest range of services and manufacturing. The Wissahickon water-powered grain and paper mills and housing were close at hand. In 1854 the city would annex these and other villages like Moyamensing and Kensington, suburban extensions of the city, and farms in the whole of what was then Philadelphia county.

The triangular shape provided the greatest access for all to the centre of the town at High and First streets. Residents and activities initially tended to concentrate at the mouth of Dock Creek, later at High and Front, then up to Third. As the city expanded along the river, it also extended deeper inland along or at least near the axis of High Street. The centre's attractive power was apparent; affluent residents, not surprisingly, occupied this axis as it allowed them the quickest trips. As for Holme's squares, by 1760 surveyors laid out three. The southeast at Walnut and Sixth, which was then used as the potters' field for burying the indigents, Blacks, the unbaptized, and strangers, also doubled as a pasture. The one to the north provided space for a church and cemetery. Only later would these two squares become parks. A racetrack and probably the site of the semiannual fairs occupied the central square.

Residential density and rent gradients declined roughly outward from the centre, a pattern consistent with modern cities, though later residential uses would decrease in the central area when land values became prohibitive in the face of higher and presumed better uses. Modifying the symmetry somewhat, property values were higher south of Market Street more so than to the north, but den-

sities were higher to the north, assuming the pattern of 1790 was similar in 1760. Only about ten of the 176 original blocks in the city were surrounded by buildings. No doubt the blank spaces within these usually large blocks were occupied by stables, workshops, and possibly even housing. There were a few alleys by then. Laetitia Court, named after Penn's daughter, was the first. Others followed, including Benjamin Franklin's Elbow Lane, which joined Market to Third, and Elfreth's Alley, today a tourist attraction. People spread onto twenty-five or so other blocks, some beyond South and Vine streets. By 1760 brick buildings of two to three storeys, mostly in rows with common walls, predominated over the clapboard-clad. On the edges of the city, where the line between urban and rural was not distinct, wooden shacks of the poor were common.

The central district was the linchpin in connecting the colony with the Atlantic world.

Within the template of survey, urban space was structured politically, socially, economically, and hence materially. People were not quite 'jumbled together' in chaos or even rough equality.[33] Without question, though, activities and people of various stripes were only microsegregated. Sleeping, eating, and working in the same building was the norm for many people. Points and areas deemed by those with money and power to be more valuable were assessed higher than others. Access to the highest-quality information—that is, social power in politics, business, and religion—defined the most valuable properties, thus creating a hierarchy of land.

The central district was the linchpin in connecting the colony with the Atlantic world. To call it the central business district is to define it too narrowly, as government and culture were central too, serving the whole city and the region. Within this area the key public ideological and functional points were situated mostly along or near the busiest and widest street: High between Front and Fifth streets on the central axis at right angles to the river. The axis delineated the central square, similar to those found in medieval and many modern cities: the courthouse in front of the market shambles at Second, the London Coffee House at the southwest corner of Front, the post office further along at Hall and Franklin's newspaper shop, and the main Quaker meeting house at Second. Long since, though, Quaker power had been shared with Anglicans who worshipped in Christ Church. The market house itself was city serving.

The courthouse, a harbinger of those found in central squares in later towns, was the central focus. Perhaps this should not be surprising considering that the courts have held a higher status in the United States than in other English-speaking jurisdictions. But it also indicates that government has not been as irrelevant as many Americans would like to believe. Its porch, reached by a pair of flanking stairs up to the second floor meeting room, was the podium for gubernatorial and mayoral proclamations, as well as for speeches, celebrations, demonstrations, sermons, and gatherings of people below. Its meeting room hosted the corporation as a municipal body and as a local court. The provincial assembly met there before 1739. Then, breaking the symbolic symmetry of the Market Street axis, the assembly moved into the new imposing State House (later Independence Hall) on the edge of the city on Chestnut between Fifth and Sixth. (It was on the side of westward expansion to the south, where the élite were beginning to concentrate). The proprietary functions, on the other hand, were not as centrally located. The provincial council met at the homes of Penn's sons or their agents from time to time. The latter provided offices, one of which dispensed land. The British government, since it had an intermediary in the proprietor, was not very visible; a custom-house of modest proportions was on the docks.

The London Coffee House was no ordinary tavern or inn. In 1754 William Bradford, a printer and publisher, organized a group of merchant investors to turn it into an exchange, a place where men could make deals, buy insurance for their ships, and the like. It was the central point of business and political information from Europe. Newspapers from Britain and other colonies, bills before the assembly, descriptions of public auctions (then called vendues), notices of runaway slaves and ser-

vants were posted there, 'a public space of record'.[34] Little wonder that the most prominent visited it frequently to meet with agents from New York or Boston, from London or Kingston, Jamaica, or with back-country merchants in Lancaster, Reading, and York. The City Tavern at Chestnut and Second, 'the largest and most elegant in America', later took over the exchange (signalling the bias of commerce slightly more to the south and west, it seems, towards an emerging high-income residential sector); later, in the 1780s, when finance became more specialized, the exchange built its own classical structure on Dock Street that had replaced the once-smelly creek.[35]

Publishers and printers who were granted monopoly franchises for the post office were central figures for the city and region; their presses (the type all set by hand) turned out books, business papers, proclamations, pamphlets, and broadsides. They published newpapers; the first reached the streets in Boston before 1700. Andrew Bradford's *American Weekly Mercury* was Philadelphia's first in 1719. Franklin took over the *Pennsylvania Gazette* in Philadelphia soon after he arrived from Boston in 1722. In 1737 he replaced Bradford as postmaster. Although he retired from active work in 1748 to pursue politics, he remained a silent financial partner to David Hall. In 1742 William Bradford, the organizer of the exchange, began circulating the *Journal*. From Germantown, Christopher Sauer's *Zeitung* catered to German settlers, as did the *Staatsbote* from the city later. A sign of an intensifying economy and society, after 1760 several other papers, gradually becoming biweekly and even triweekly, appeared, such as the *Chronicle* and *Packet*. They provided news, government notices, merchants' advertising, real estate and other classifieds, and a forum for political comment for a wide spectrum of society.

While scant by the standards of the mass circulation dailies of the 1840s onward, these papers were powerful instruments for influencing views. The application of Gutenberg's invention increased the flows of information, especially about money, by whetting avaricious appetites. Writers were less reticent than they are today; at the height of debate in 1756, after the French and Indian attacks on the back country of the region ignited passions, the politically active and prominent Anglican clergyman, William Smith, attacked the pacifist Quakers as traitors. A daring critic of Smith went further, saying 'the Vomitings of this infamous Hireling ... betoken that Redundancy of Rancour, and Rottiness of Heart, which renders him the most despicable of his Species'. He was a 'Frantick Incendiary, A Minister of the Internal Prince of Darkness, the Father of LIES'. However, Smith won and many hard-line Quakers quit the assembly.[36] Maybe this excessive language helps to explain the incendiary environment of the next two decades. Rhetoric was crucial to political if not economic change and, one might add, to religious sensibilities.

In Philadelphia the main Quaker meeting house, next to the courthouse, was the spiritual centre for the provincial Friends, where the local Monthly and regional Yearly meetings that brought delegates from a wide region to regenerate spiritual energy and define religiously correct language, were held. It was also the site of the 1736 treaty with the Delaware Indians and many other secular gatherings. This diminutive structure could hardly be considered a cathedral. Any sense of religious symmetry was altered by the great Christ Church and its spire just to the north on Second, and the First Presbyterian church was just west on Market. In the long run, none of the four central structures—the courthouse, coffee house, market house or Quaker meeting house—would survive on their sites, though the State House and Christ Church would endure. Representing a central contradiction of America to the present, the Church of England (renamed Episcopalian) has, ironically, remained the highest-status denomination.

Behind the courthouse, the market house was set out in the median of wide Market Street between Second and Third (later extended to Fourth) to serve the city's population. Butchers of domesticated and wild meat, farmers, bakers, and others, perhaps a hundred in all, rented space from the corporation. A hay market was located just to the west. Twice weekly the town crier announced its opening. Additional space was allocated east of Second for New Jersey gardeners who ferried fish, meat, eggs, vegetables, and fruit across the Dela-

ware. Market gardeners practised the most intensive agriculture on the edges of all cities (and some are found there even today). Dances and drinking after the Saturday market closed attracted lower-class males and females, including prostitutes.

Public and quasi-public buildings, other than those in the centre, were less clearly symbolically placed. In 1745 public demand for access to food led the corporation to put up a new south market house on Second between Pine and South streets, attracting customers in Society Hill and Southwark. Later, Callowhill market operated in the Northern Liberties, its square indicated on Scull's map. The jail and workhouse on Third near Market were soon to be rebuilt on the edge of the city. Structures serving the poor were on the margins, at least when they were built. The public almshouse south of Spruce between Third and Fourth supplemented, if inadequately, the Quaker house of refuge just to the north on Walnut. The Pennsylvania hospital was built on Eighth Street south of Spruce.

The college and academy were at Fourth and Mulberry (Arch) on land shared with the Anglican cemetery. The Quaker and Presbyterian burying grounds were nearby. Note that these were north of Market, which suggests that early expectations had been for less intense development in that direction and more to the south. Churches, other than the central ones noted earlier, were already neighbourhood oriented. The three German-speaking congregations were to the north in an area that was tending towards concentration of that ethnic group. All congregations ran schools; the Moravians for example, had theirs on Race Street. The other two Anglican churches were established in the incipient higher-income area south of Walnut. Even the Roman Catholic chapel was there too. A theatre was built on the south side of South Street, thus outside the city proper. Thomas Mullen operated the Vauxhall theatre for summer parties on the banks of Schuylkill River, emulating London's pleasure garden in a modest way. The precise siting of these buildings was probably a function of land availability at the time.

Turning to more strictly economic spatial patterns, the waterfront obviously stood out, as in all ports of entry. Penn had legally specified the Port of Philadelphia. The most prominent buildings were the warehouses of the great merchants, with the counting houses on the first floor. The warehouses extended along busy Water and Front streets and out onto their wharves on lots leased from the corporation. Especially near Market and Chestnut, merchants stored high-quality imported goods of the world on the upper floors for distribution not only to Pennsylvania but also for re-export to other colonies. Counting houses and warehouses pushed west beyond Front Street in the centre, but as the city grew, many more had been added along the river, over a mile to the south, though not as far to the north.

By 1760 too, affluent merchants like Thomas Willing and Israel Pemberton, the King of the Quakers, in Philadelphia, Thomas Hancock in Boston, and James DeLancey in New York, no longer had to share these buildings as residences as the earliest merchants did. In fact, some residences of an earlier generation of successful merchants were totally converted to commercial use, as was the case in London, but redevelopment had already led to new purpose-built warehouses with their small offices.

By 1760 an incredibly wide range of imports found their way to the merchants' warehouses. Luxuries arrived—such as spices and pepper from the East Indies, brass household ornaments from India, silks from China, wine from Madeira, brass clock fixtures for the clock makers of the city and country—all for the elegant lifestyle of the affluent merchants, lawyers, public officials, and even the better-off farmers. More mundane goods, now widely considered necessities—such as iron kettles, tin pots, some woollen and cotton cloth, and clothing—arrived from England. The list was almost endless, though not quite: most of what Pennsylvanians used was produced by Pennsylvanians. One other import commodity handled extensively by some Pennsylvania merchants was slaves, but by 1760 auctions on Market Street were becoming rare. Quaker sensitivities and a more fluid free market in labour reduced demand. The waterfronts of Charleston and the Chesapeake wharves were more likely to witness this activity.

The wharves farther from the centre handled bulky goods. Merchants imported new breeds of livestock from England. Among others, John Bartram, the naturalist, imported seeds for experiments. Salt arrived from Portugal. Great piles of cordwood and lumber came from Pennsylvania and New Jersey down the Delaware and Schuylkill, and from inexhaustible but increasingly remote forests. Some merchants organized the export of wheat, flour, and maize. Established soon after 1760, Levi Hollingworth, for example, brought flour in from the Christiana River and its tributaries at Wilmington, and then exported it. Dealers shipped livestock elsewhere, though most of the beef and pork left the waterfront already slaughtered, salted down in barrels for provisioning ships or for slaves on the sugar islands. Chesapeake tobacco was re-exported. Merchants handled a few other agricultural products, such as flax-seed for Ireland and hemp grown in the back country. As yet, few metal goods passed through their hands. Britain was anxious to buy pig-iron and bar iron from America's inland furnaces and forges, but in 1750 legislated that the production of higher-value added iron and steel goods would be prohibited in the colonies—needless to say, to no avail—though exports may have been slowed.

The waterfront was a locale for much processing. Although manufacturing also occurred elsewhere in the city, processing was a major set of activities, if usually often not separable from service and selling. Shipbuilding was prominent early; in 1692 twelve ships were constructed. By 1760, the city led all places on the continent in shipbuilding, though New England ports collectively built more. This activity, financed by merchants who sent vessels to sea, was highly visible on the northern margins. Shipbuilding gave rise to a whole host of ancillary processing activities nearby: the making of rope (walks were 900 or more feet in length in the open air and so were located inland a few blocks), sails, nails and pegs, and guns, along with mast trimming, plank and lumber milling, and brass and iron founding. When the ships were completed, coopers, chandlers, distillers, breweries, and food provisioners supplied goods to consume. Not all of these activities, it should be noted, were dependent on shipbuilding and servicing, or some not even primarily, but all contributed to Philadelphia's prowess in shipbuilding. Sugar from the West Indies was refined or converted to rum, though New England provided much of this product. Sugar refining has occupied waterfronts ever since.

Polluting industries created a problem for politicians on Dock Creek, as we have seen. Citizens preferred to have distilleries and breweries located on creeks on the edge of the city. Spent mash from these industries were used to fatten livestock. Drovers located stockyards in Spring Garden, a western part of Northern Liberties, which received livestock fattened in nearby farms or distillers' pens after being driven in from back-country ranges. Calls to banish slaughtering from the central market led to abattoirs.

Brickyards with access to clay, as shown in later maps, dotted the peripheral landscape. Most buildings were clad in brick, which indicates the good quality of the area's clay. Market gardening, orchards, and increasingly nurseries (to supply flora for gentlemen's gardens) intensified land uses on the margins. As we have seen, most of Pennsylvania's manufacturing still occurred in the countryside. Intermediate goods, like barrel staves to supply city-based coopers, were made in the country at sawmills.

Craftsmen processed most of the final consumer goods. They were also retailers for the most part and thus the most conspicuous business people in the mercantile city. Although several analyses of crafts have been done for the city as a whole or of particular wards, none have established a spatial hierarchy for the time. An obvious way to distinguish among crafts is to establish whether they were citywide, therefore central and generally more valuable (if they were clean), or neighbourhood oriented. Except for printing and publishing on Market Street, the spatial picture seems undifferentiated for consumer goods. Most occupations were apparently ubiquitous: more than two-thirds of the 150 occupations noted in tax lists and estate inventories during that era were manual, and over one-third of the people created goods with their hands. They sometimes worked alone or with a few

helpers, journeymen, and apprentices (including wives and children in some cases), though some furniture makers and carriage makers may have had larger shops. In 1774 10 per cent of Middle Ward's population consisted of slaves and indentured and hired servants, about the same as the city as a whole.

Often, if their materials were clean, they operated in the front rooms of their residences, just as most merchants kept a counting room in their warehouses. Makers of turned goods, boots and shoes, harnesses, perukes, hats, and clothing—the list is long—were among the leather-aproned artisans. They were supplied by merchants who brought to Philadelphia cowhides, deerskins, furs, horsehair and human hair, lumber, and linen and woollen cloth. Rural ironworks supplied pig-iron and bar iron for making pots and pans, horseshoes, and the like. These works required hot charcoal fires and were often located in the backyards, the larger ones on the margins of the city. Obviously blacksmiths needed sheds to stable horses waiting to be shod. Three steel furnaces were enumerated in 1760, all in the western suburbs. Bakers who operated ovens in rear yards were neighbourhood oriented, though perhaps there was a concentration of them near market-places.

Many people, including the craftsmen just mentioned, retailed goods. The divisions between occupations in those days of less specialization were not distinct. As for imported goods, many wholesale merchants still doubled as retailers, though shopkeepers, (who were mostly concentrated north of Market), tobacconists, and druggists began to appear as such on tax returns. Despite the complaints of shopkeepers and market-stall holders, pedlars were paragons of free enterprise, never giving up to regulators. Dairymen retailed milk. During the day they turned their cows over to herders for pasturing in fields on the edge of the city (such as the potter's field). Maids milked them morning and evening in stables behind houses. Customers gained entry to their yards through arched carriageways.

As in any large city, the hospitality industry was prominent for local drinkers and for visitor accommodation. Inns primarily served the latter and taverns the former. Many provided entertainment and catered about fifty clubs, national ones like St Davids, or to fire companies. In fact, like the streets and churches, they satisfied people's need to socialize. There was no shortage of public houses—in 1756 there was one (legal and illegal) for every seventy-seven to 130 residents. These public houses were relatively undifferentiated by class of clientele up to this era. The Indian King establishment, though, was undoubtedly of higher status and well regarded by visitors, such as British army officers in 1756. The Conestoga Wagon, with its large shed and yard, served wagoners from the west. The Library Tavern on Second Street catered to book lovers. The strongest concentration was south of Market Street to Dock Creek, where merchants were most prominent. The waterfront dives catered to ordinary seamen. Helltown, north of Market, and Southwark had more than their share of low-quality dram-shops. Visitors, temporary residents, and single men resided in the many boarding-houses that provided meals.

Residential population densities dropped off the greater the distance from the centre because houses were not built adjacent to one another the farther out one went. Builders worked on available subdivided land. This scattering of houses with later infilling would persist until the mid-twentieth century. With one-third more houses in 1760 than in 1749 (3,000) and even more taxpayers, the city had not only expanded outward but had intensified within. In the ten fully built-up blocks, densities were as high as 120 people to the acre—possibly higher. Residents still lived in the centre in contrast to the more recent pattern because most structures were expected to be for multiple use. In 1769 Lower Delaware Ward, which was between Market and Walnut and embraced the east side of Front and Water streets, still had 120 'houses', undoubtedly most of them for mixed uses.

Residential patterns by income and occupation were differentiated somewhat, but residents of different backgrounds and status were not quite jumbled together.

By modern though not nineteenth-century immigrant standards, many houses were crowded, but the number of occupants varied greatly. Houses that were mostly of two or three storeys in 1760, according to one recent calculation, sheltered 6.3 people on the average (though another suggests only two-thirds of that, indicating the perils of analysis).[37] Numbers per house had increased early in the century, but hardly so after 1740, signalling a limit, though in times of rapid population growth when house builders could not keep up, many must have been cramped temporarily in less space. Philadelphians probably did not experience the highest densities or crowding in that era because its blocks were not as densely filled in as those in the biggest European cities. London, for example, crammed more people in some districts; New York, Boston, Montreal, and other northern cities at the time were no doubt similar to Philadelphia. Later some of New York's neighbourhood densities would be far higher, peaking soon after the turn of the twentieth century.

Residential patterns by income and occupation were differentiated somewhat, but residents of different backgrounds and status were not quite jumbled together. Separation by class had begun, and even modern sectoral patterns (that is, wedges outward) were somewhat apparent. The residential pattern of different income groups in that era (and in the erroneously labelled preindustrial city generally) has often been contrasted to the recent American city. The stereotyped early model is this: the rich lived in the centre and the poor lived on the edge, but later the pattern was reversed. This is an inadequate and oversimplified representation of the present and is insufficient for 1760.

The 'better sorts' made up about 10 per cent of the population: prominent merchants, top lawyers, public officials, retired gentlemen, successful ship captains, and the high-status Anglican and Presbyterian clergy. Home-owners (less than 20 per cent) were only modestly segregated. Still, by 1760 the most affluent merchants (the top 100), started moving away from their warehouses and began building their homes on Society Hill because of its elevation (if slight), a practice not uncommon in other cities. They built large houses along tree-shaded streets. Building was also occurring to the west for esquires or gentlemen who would soon attend St Peter's church at Pine and Third. Many of their widows would remain in this sector, which, by 1790, had expanded to the northwest behind the State House. The grandest merchants were also building country houses, replicating London's successful, but were situated within easy reach of the city.

The 'middling sorts'—the lesser merchants, shopkeepers, tavern keepers, lawyers, doctors, and accountants—were concentrated near the centre. In 1800 when the city proper had three-and-a-half times Southwark's population, it had twelve times as many lawyers and doctors. Craftsmen were found in most parts of the city, though the higher-value producers of compact goods were more likely near the centre too. The majority lived and worked in the same houses. If 1790 distributions were also apparent thirty years earlier, some modest occupational clustering was evident. Those engaged in shipbuilding resided in the Northern Liberties near the Delaware, some butchers lived in Spring Garden, builders lived adjacent to their new constructions, and so on. Even in a city of 20,000, those with sufficient means would choose a short journey to work wherever possible.

Those at the bottom can be separated in three groups. The first group included many dependent workers—slaves, indentured servants, and hired servants (about one-tenth of the population) lived with craftsmen, while others domiciled in the houses of the affluent. In the second group were the labourers, the most segregated occupational group in the city; they may have composed less than 10 per cent of the population. Presumably they were day labourers. Only a few labourers lived in alleys in the centre of the city. Walnut and Chestnut wards had proportionately fewer of the unassessed than elsewhere. The poorest central

residential areas in the city proper may have been Helltown, north of Mulberry (Arch) near the river, where the unassessed were more frequent. Beyond Dock Creek the waterfront boarding-houses catered to mariners and 'spinsters'.

The third class of poor, 'the baser sort', which included freed Blacks, lived in shacks on the edges of the city where rents were low.[38] Many were recorded outside the city proper in Southwark, the 'wrong end of the city', and in Northern Liberties near wharves where they worked as stevedores or went to sea. Labourers probably trudged daily into the city to find work, while others helped in construction on the margins of the city. If nineteenth-century New York is any guide, some of the very poor, including some immigrants, built shacks on the edge, squatting on land owned by others or the public, with the expectation that urban expansion would wipe out their insecure hovels. The drifting homeless struggled to find a bed. Cities are always attractive not only to the most ambitious but also to the poorest of people because they think opportunities are greater; in winter charity is more likely, or the city is a more exciting place to be.

As for movement on streets and the rivers, most people walked as very few could afford to ride in carriages or on horses. The streets were where people mixed together, where differences in dress were conspicuous and defining. On Sunday a modest number walked to worship, the rich riding in their carriages drawn by sleek horses. The rich had occasion to parade their finery. The main street, the High, was the busiest; pedestrians were astir early, especially on market days. In those days when ice was a luxury for the few, housewives had to buy perishables frequently. The High was filled on celebratory days, or at times of political conflict. Wagons from the west rumbled in on the now cobbled surface, to the auditory discomfort of residents, and at times the roads became gridlocked like the freeways of today. For much of the year, Water and Front streets were beehives of swarming, sweating longshoremen and unruly cartmen. Draymen hauled goods to shops. Messengers hustled along. Dung carts clattered through the streets at night. The spatial separation of work and home, necessitating a journey to work, was not nearly as apparent as in later times, yet at the end of the work day, many of those who worked at home and those who toiled elsewhere no doubt stopped at taverns. In the evening, whale-oil lamps on the streets, at least in areas where the affluent lived and on Market Street, shed a very modest glow as watchmen made their rounds. Because the only artificial lighting in houses was by candle or (for the better-off) by lamp, and because work hours were long, retiring early was common. Some taverns stayed open late, however, and behaviour could be boisterous.

In 1760 three ferries ran across the Delaware and several over the Schuylkill, the most important at the ends of Market Street. A bridge over the Schuylkill had to wait until after 1780. The corporation's two wharves, next to the Penny Pot tavern at Vine Street and next to the Blue Anchor at the mouth of Dock Creek, provided dockage space for shallops and the less affluent travellers on a shoreline that was largely privatized. Someone allowed Jersey gardeners to land their produce at Market Street.

In summing up the spatial patterns of this city of nearly 20,000, order prevailed within the often disorderly environment. Philadelphia had a defined centre of power, if not of quite the same dimensions as those of medieval or later cities. Its density and rent gradients presaged modern cities, though a century later its triangular shape was giving way to a star-shaped pattern when public transportation began to expand. Its classes mixed in the streets, but slept in widely varied accommodations. Despite appearances to the contrary, residents as workers and consumers deferred to those with economic power, and less comfortably as citizens to those with political authority.

A Prosperous City: A Metropolis in the Making

Within ten years of the city's founding, Philadelphia had grown into the same league with New York and Boston. It slowed over the quarter of a century after 1700, as had Boston after 1640, but after 1725 the pace picked up, especially in the mid-1740s, in contrast to Boston. Encouraged by Franklin's expansionist ethos, immigrants flooded in. As suppliers of war material, Philadelphia's merchants did extremely well during the final war with the French. However, growth would again be less brisk in the decade after 1760, partly because the war years had been almost too good economically. During the revolutionary period, its economy was disrupted while British troops occupied the city. Its pace would nonetheless exceed New York's until the end of the century as the great generative organizing centre not only for its productive region but for much of the trade of other colonies. For most, the province provided the fullest and most equitable material life possible anywhere in the Western world.

As for the city itself, it has grown remarkably prosperous. As one visitor exclaimed in 1765, 'the great and Noble City … is perhaps one of the wonders of the World … the first Town in America, but one that bids fair to rival almost any in Europe'.[39] Still, in matters of governing the local environment, it left a legacy of mismanagement and confusion that would later afflict large American metropolises. Besides it was not a heaven for all. Some Europeans may have been beguiled into believing that Philadelphia's streets were paved with gold, only to be disappointed that they were muddy or at best only cobblestoned. Some may have been disappointed that the rich flaunted their wealth as they did in Europe. Certainly times of stress were experienced by many: the flood of goods from Britain, as in 1759 and 1760, created a glut that discomfited the marginal merchants and raised unemployment. For many, life was often wretched, often nasty, brutish, and short. Franklin's admonition of hard work and personal blame for poverty were hardly comforting to those left out of material prosperity. His views would remain powerful ingredients in the American ideology.

From Britain's point of view, Philadelphians were creative. Within a few years, George III would think they were too creative when gentlemen and the leather-aproned craftsmen rebelled. They had not easily coalesced as a unified front, in part because of Franklin's equivocations, but they finally did. The War of Independence was a revolution in political terms; Pennsylvanians wrote the most democratic of constitutions in 1776. That document clearly laid an imprimatur on private property rights of which John Locke would have approved, yet it attempted to balance this with a strong statement on public responsibility. The latter sentiment—republican virtue, stressing the need to act on behalf of the public good—would not, however, revolutionize economic relations. Republican virtue would prove to be second fiddle to the pursuit of gain.

Whether independence hastened economic growth is a moot question, yet economic growth, through continued areal expansion and through factories, was to be dramatic in the nineteenth century and so increased material well-being. This attracted even more Europeans. Building on its colonial industriousness, its economic leaders would turn it into the largest manufacturing city on the continent by 1860. Philadelphia would, however, lose its role to New York as the principal commercial and financial organizing centre. Lord Adam Gordon's 1765 prediction that 'Philadelphia from its Central Situation ... must become the Metropolis' of America did not hold.[40] New York appropriated the potential accorded by the central situation on the Atlantic margin.

CHAPTER 4

■

New York in the Ascendancy, 1860:
Unheard of Riches and Squalor

Early in July 1860 the *Great Eastern*, by far the largest iron-hulled steamer yet built, approached Manhattan from Liverpool. Like earlier visitors, our expectant traveller, John Brown, and his wife, were impressed with the Narrows entrance to the Inner Harbour. As the ship drew nearer, on the East River he observed a vast forest of masts of sailing ships, far more than his great-grandfather saw a century before in Philadelphia, or he himself would see later on the Delaware. The clipper ships for the San Francisco and China trades stood out. Now, too, steamers tied up at the wharves, especially on the Hudson. Smoke billowed from ferry boats traversing the harbour and from factories. Beyond the masts, he saw the church spires on Manhattan and Brooklyn's skyline: Trinity, St Paul's, and the old Dutch church, now the post office. Coming closer, Mr Brown saw a mass of buildings behind Battery Park that was much more extensive than that on brick-lined Front Street in Philadelphia of 1760, though they were still not any loftier nor different than in London (Figure 4.1).

Later he would take a boat tour around the island to observe this grand city that, together with Brooklyn and Jersey City, comprised over a million people or fifty times the population of Philadelphia or New York in 1760. The passenger was aware that the United States had surpassed Britain's rapidly growing population two decades before and now stood at 30 million, compared to only 1 million a century before. Momentous expansion had taken place.

When the ship docked at a Hudson River wharf, a throng of curious onlookers greeted this massive ship on its maiden voyage. Among the passengers who disembarked, the most prominent men, some with spouses, rode in horse-drawn coaches provided by places of business in or near the financial district, which was a few blocks away to the east. Let us put our traveller in this group. Those representing the merchants were now second in rank to financiers; they hailed cabs to reach the expanding warehouse and manufacturing districts of the 2nd and 3rd wards. Some would head to a telegraph office to wire Boston or Chicago on the deals in London or on the fashions, reports that were not dated by a month or more as in 1760 but now by only a week and a half. The hottest news had already reached New York through the telegraph lines from Halifax, where many ships made their first land call. Stevedores unloaded mail bags and merchandise from Britain and the continent.

Immigrants with money encountered little trouble from immigration officials who were less busy than they were a few years earlier. Since 1854, immigration had slackened compared to the previous decade. As before, letters of introduction from employers in Britain smoothed the way. Those in steerage were herded into Castle Clinton waiting rooms in Battery Park where health officials checked them. In 1854 the rules were tightened so that federal officials, not mayors, cleared their entry. For those with diseases deemed communicable, the wait was agonizing. Some would even be

4.1 New York City, *circa* 1855. Bird's-eye views were popular in that era. Development has rapidly expanded up Manhattan, well beyond city hall park to Madison Square. Broadway and the Bowery meet at Union Square. Castle Clinton, the immigrant reception centre, was about to be surrounded by landfill, which had already extended along both sides of the island. Wharves on Hudson River were becoming more important than those on the East River. *Source:* Prints and Photographs Division, Library of Congress, T. Muller lithograph.

quarantined and then sent back to Europe. Most, with great anticipation, stepped out into the light of the congested streets. Although poorer than others, the Irish could at least speak English, while the Germans did not. Those fortunate to have relatives to meet them eased into lodgings in residential districts with their churches providing mutual support. The unconnected and unwary discovered, however, that they could be fleeced by an 'unscrupulous army of "immigrant runners"' purporting to help.[1]

All flocked to the New World with hope for a better life. Most would pass beyond Manhattan by water and rail, some even before seeing the famous streets: Wall, Broadway, Bowery, and Fifth Avenue. Even so, immigrants and their children born in the city made up well over half the city's population. Those staying in the city would soon learn that only a few people owned the private spaces, quite unlike their vision of an America where the ordi-

nary could own land. The poor, who were the vast majority, would realize that work was arduous and not particularly rewarding in the land of opportunity. Mostly young and energetic, they would try to improve on their often dismal pasts.

New York was the commercial city of 1760 writ large, but it was also modern, even technologically. The 'age of gold'[2] through which the city had just passed placed it in the front rank of cities, not only in North America but in the world. By 1860 New York City had firmly established itself as the central urban actor, and was well on the way to catching up with London. It had surpassed Philadelphia's population around 1800, and now was twice as large, even though Philadelphia had expanded rapidly. The Chamber of Commerce could accurately boast in 1858:

Capital is attracted to this central point because it finds scope and objects for investment and distribution … The interests of agriculture and manufacturing, no less than those of commerce, share the benefits of centralization. The productive industry of the Union, thus becomes … a partaker in the prosperity of our city … and common sharers in the public weal.[3]

Let us consider first why New York rose to the top during the previous century, an era when America was still open for agricultural settlement but increasingly intensifying production. Then we will look at the social, political, and spatial results within Manhattan. Whereas Philadelphia primarily expedited the settlement of farmers, New York pulled together much of the continent. In doing so, it created untold wealth for a few, some for a growing new middle class, and promises of improvement for the poor.

The Strongest Era of Urbanization in America through Regional Development

During the century after 1760, the European occupation of the continent expanded enormously, by then encompassing most of the temperate forests and humid prairies in the West. The settlement nearly reached the limits of humid lands by then, though the semiarid plains, the Canadian prairies, and western valleys were still available. The population of the United States grew from 1.6 million in 1760 to 31.4 million in 1860, or an overall rise of nearly twenty times. In British North America, the population of the provinces that would constitute Canada after 1867 grew from about 80,000 in 1760 to just over 3 million in 1860, at a pace twice that of the United States. The ratio of roughly one Canadian to ten Americans peaked in the mid-1850s. In both countries immigration was very strong after 1820, adding to high natural increase, which was especially vigorous in newly settled regions. All this was to the detriment of the increasingly hapless Native peoples, who possibly made up nearly three-quarters of the Canadian population in the mid-eighteenth century, but

dwindled to less than 5 per cent in 1860, as their numbers were literally cut in half. In the United States their experience was even more fateful.

This extensive spread of population was increasingly balanced by intensification of economic activity as evidenced by urban growth. In America the urban share expanded from 5 to 20 per cent, to 6.2 million, a numerical gain of over sixty times. In British North America the percentage share rose too, though not quite as high. Clearly North Americans were increasingly becoming city dwellers, a pace that would slow but not falter for several decades. Besides, as a harbinger of future concentration in large metropolises, one in thirty Americans lived in New York as compared to one in eighty in the Philadelphia of 1760.

The pace of American urbanization began to accelerate between 1790 and 1810. Compared to rural growth, it actually had weakened in the downturn after 1760, and again during the War of Independence and its aftermath, and reached its lowest ebb. Then in the 1790s America began to emulate Britain's strong drive towards urbanization. After stagnating in the 1810s, the speed of growth increased markedly after 1820, reaching the greatest rate ever between about 1843 to 1857. Despite the halting progress, the nineteenth-century experience was dramatically different from that of the previous century. A host of new cities and towns appeared; overall in 1760 probably no more than ten places had populations of 2,500. In 1790 there were twenty-four such places; in 1860 nearly 400.

Regionally, stability and volatility were both apparent in the urban pattern. As earlier, the American northeast dominated. By 1810 New York was clearly ahead of all the others for good. Yet in 1860 New York, Philadelphia, Boston, and Baltimore (in that order) remained the leaders they had been in 1800. Growth was striking; over the previous seven decades, New York (including Brooklyn) averaged 67 per cent growth per decade, while the increase was somewhat less for the other cities. The long decades of Boston's stagnation had passed, Baltimore leapt ahead, especially in the early years of the century, while tapping some of Philadelphia's hinterland. Left behind in the urban race were

Newport and many other colonial ports. Satellite mill towns burgeoned in the nearby hinterlands of the big cities, so the region had the largest share of urban places, yet the region's share of large cities fell because the new West was growing rapidly.

The great western expansion of population pushed Cincinnati, St Louis, and Chicago to join the ranks of those over 100,000. In the heady days of the 1800s, 1830s, and again in the 1850s, speculators established town after town in the new West. Although most towns would fail to reach the expectations of their founders, some lesser commercial places boomed, as along the Ohio River, which developed first (Pittsburgh, which was already important industrially, as well as Wheeling, and Louisville); in upstate New York and on the Great Lakes (Syracuse, Rochester, Buffalo, Cleveland, Detroit, and Milwaukee), though few as yet on the Mississippi River, aside from St Louis, as Minneapolis was just beginning. Chicago was vying with St Louis for central power. Chicago would soon rise to the top in the region, based on the vast farming output and the forest wealth to the north.

In the South, Charleston and Savannah languished, but in the new deep South, New Orleans, which was now easily the largest with a population over 100,000, and Mobile took off. Even so, the South, following the colonial past, did not experience urbanization to the same degree as the North, even though settlement had extended across to Mississippi and east Texas (conquering Spanish-speaking Mexicans from earlier times) and in the upper South into Arkansas and Missouri. In the South, only one in ten lived in an urban place compared to one in seven in the new Midwest and more than one in three in the northeast.

Regional Development

Three general economic regions had thus emerged in the United States, each with roughly one-third of the population, the Midwest slightly less than one-third but still increasing its share, and the northeast slightly more. However, the northeast produced considerably more of goods and commercial services (over half), though the Midwest was gaining and the South was slipping. Over the two decades before 1860, the Midwest kept pace in economic growth per person, but the South less so. Regional income distribution was, however, strongly skewed in favour of the northeast where it reached over 40 per cent per capita more than the US average. The new less-developed Midwest, by contrast, still lagged at 30 per cent below the average, and the South at 20 per cent. But if only the free population is calculated, the South was a bit above the national average, hence the North was lower.

Urbanization concentrated more wealth in the hands of a few. More egalitarian rural society was relatively receding, even while still growing. Farmers on the average held twice as much real property as urban dwellers, as might be expected, since in cities most people rented. As of 1860, in the United States generally, three out of five White males reported property worth over $100 to the census takers, but in this land of opportunity, only slightly more than two in five owned real estate. The top 1 per cent owned nearly one-third of all real estate. The native-born and older people held twice as much real estate as immigrant and younger householders. Urban dwellers and immigrants improved their lot in the 1860s when urban growth and the rate of immigration slowed somewhat, and when nearly one in twenty people were slaughtered in the Civil War, allowed for a catching up. In 1860 southerners owned twice as much wealth as northerners because one-quarter of families owned slaves, and the top élite planters held an enormous share, infinitely more than property-less slaves. But in 1870 the inclusion in the census of slaves as free individuals reversed the ratio, which was obviously a truer picture. The North was richer, and the northeast more than the newer West too. One hundred-fifty men became millionaires by 1860, compared to one or two at the turn of the century, even though prices had generally fallen for several decades. New York had more than

> Urbanization concentrated more wealth in the hands of a few.

its share of wealth, but also more than its share of poverty.[4]

In what was soon to be Canada, and in Newfoundland, the urban share of population—which in 1760 in New France was the highest on the continent (over 20 per cent)—fell to half that by 1820; then, together with places in the other colonies, it rose thereafter to one in six in 1860, not far behind the colossus to the south. Quebec City, which was about half the population of Philadelphia, New York, and Boston in 1760, fell behind Montreal. By 1860 in Canada, only Montreal could claim to have reached 100,000 (including suburbs), one of nine cities on the continent to reach that population, about half the size of Baltimore and nearly twice that of Quebec City. Halifax nearly equalled Quebec in population as it was enjoying prosperity. By then Toronto was growing rapidly, but was scarcely more than half the size of Buffalo's population of 80,000. Not yet the capital of a confederated Canada, Ottawa was a small industrial town. Canadians on average probably earned incomes below the US average, as indicated by more movement to the south than to the north, but wealth was likely somewhat less unevenly distributed. American rhetoric praising the making of money may have contributed to the difference.

Although some Canadian real estate speculators expected their towns to become mighty metropolises, the intensity of speculative fever over the three decades prior to 1860 was weaker than in the United States. Certainly the cult of the self-made man was less fervently preached from pulpits. One American clergyman in 1836 (at the peak of a boom) published *The Book of Wealth: In Which It Is Proved from the Bible That It Is the Duty of Every Man to Become Rich.*[5] Poor Richard was upstaged. How could anyone with ambition resist?

The regions were linked together. Five main corridors connected the West to the East. The Hudson valley and water-level route through New York to Buffalo and the Great Lakes was the strongest link by 1860. Boston hooked onto that. To the north, the St Lawrence westward from Montreal converged with the water-level route on Lake Ontario. A third corridor crossed the mountains from Baltimore and Philadelphia to the Ohio River.

The corridors were less marked further south. Earlier, many settlers had moved through the Cumberland Gap into Tennessee and Kentucky, though this did not develop as a serious trade route. The cities on the Ohio River showed the strongest early growth in serving not only the Midwest but the upper South. The South's main inland corridor was the Mississippi River. The sea route was crucial from New Orleans and Mobile to the northeast. Elsewhere, Montreal was linked southward to the Hudson River, and much of eastern British North America was economically tied to the northeast, most obviously Boston. The presence of the long sea route around Cape Horn that connected central California cannot be ignored, as the spurt of growth following the gold rush of 1848 had an influence on eastern cities. A myriad of rivers and the sea route joined the whole of the forested North and West. The vast wilderness, or rather Indian lands, were fair game for trappers who traded with the posts of the Hudson's Bay Company, not least, for John Jacob Astor, organizer of fur-trading activities and a central figure in New York society.

What factors led to rapid urbanization and the regional patterns? Clearly an expansion of commercial activities perpetuated and raised the power of old cities in the northeast. New regional entrepôts in the West, often spearheads of settlement, replicated these coastal mercantile cities, but urban growth outran that of the eighteenth century when commerce also had been crucial. Greatly expanded and specialized trade was the major factor in urban growth before 1810, particularly the carrying trade on the Atlantic when America was neutral in the struggles between France and Britain in the Napoleonic era. Factories began to produce more goods because British goods could not reach America as easily. The continent became increasingly self-sufficient after 1820, with factory production contributing more to national output than it did previously. Other fields expanded too. Commodity exchanges expedited the enormous rise in agricultural production, the sector that still led in total productivity in 1860. Businesses in finance, commerce, transport, communications, and processing became specialized, though capital inputs in factories remained modest by later standards as machine tools were in their early development. Communications improved greatly and the costs of transportation fell dramatically. Rising real incomes, if probably less equitable, undoubtedly resulted from economic growth. Through the multiplier effects of cumulative and circular causation, higher incomes induced more production.

The West followed the northeast in intensifying; economic development in the newest states telescoped over time more than in older ones because of new technologies and organization. As in the colonial northeast where merchants were partially autonomous in the British system, midwestern and Canadian merchants were somewhat free from the control of eastern merchants. But, even more so than in the colonial era, the marketing of southern produce was in the hands of outsiders, especially New Yorkers, as we will see shortly.

Government Power

Before we focus on the power of New York, we must recognize the role of governments in directing urbanization. In this era the need for government's visible hand appeared to be less urgent than when colonies needed anchors for the lifelines across the Atlantic and for inland settlement. Yet governments were deeply involved in the economy, providing infrastructure and promulgating regulations that usually helped rather than hindered business. As always, new jurisdictions begat capitals (as in Washington), the state and provincial capitals, and the county seats. The District of Columbia, which was a compromise site for the capital between North and South, had only 14,000 people at the turn of century, but as its bureaucracy gradually expanded, its population reached 75,000. Canada would find its capital's location only later. In the united province of Canada, between 1841 and 1867 the capital rotated among five places until Queen Victoria decided on Ottawa, which, like Washington, was a compromise site.

In sharp contrast to the coastal capital sites earlier, in the new West planners centred the capitals within states, for example, Springfield in Illinois. In deference to populist rural sentiments, New York and Pennsylvania politicians moved the governing apparatus from the metropolises to the back-country smaller cities of old Albany and new Harrisburg. In northern New England and New Brunswick, more central sites won out. Capitals like Columbus and Indianapolis cost the least aggregate political travel time for legislators and citizens. State capitals were small compared to the old and new commercial entrepôts (Boston was the most notable exception), which suggests that government engagement in the material lives of its citizens was less than it was earlier. Actually, although the locational decisions catered to rural bottom-up interest (to farm populism rather than city merchants), the course of the economy was not hindered. Travel to these places may have inconvenienced city lobbyists, but the big cities continued to prosper nonetheless. Rural populism hardly limited central economic power, but it did hamper local government action in cities.

Lower in the government hierarchy, county seats, which were in many cases also centralized within their units, consolidated local authority as they had earlier most clearly in Pennsylvania. Around the Great Lakes, in the plains west of the Appalachians north and south, government surveyors provided the basis for their central role through the rectangular survey of lots and townships. Lawmakers grouped the latter into counties, which became increasingly symmetrical and standardized by size. As in Pennsylvania earlier, the decisions catered to the notion of centrality, but centrality in the economy depended more on financial power than on location itself. New York demonstrated this.

New York's Rise as the 'Heart and Brain' of America

Walt Whitman's assertion could hardly be disputed, though frontiersmen and their intellectual boosters might claim that they were the creators of the American character.[6] In 1860 New York easily stood at the top. In fact, it had pulled ahead of the others by 1810 after its population tripled over the previous decades. When the United States was created, or maybe even if it never had been, one place would almost inevitably emerge as the dominant financial and leading business centre. In the long run it would become Canada's too, as London's role faded. New York's superiority as the centre has been explained by a number of factors: rural growth; its fine harbour, which attracted shipping; its control of foreign information and trade; its monopoly of the coastwise trade, especially cotton; its domination of inland trade, transportation, and communication; and its control of finance in the private sector. In turn, the city fostered manufacturing and agriculture.

The State's Growth

The most obvious underlying indicator of its growth should be made clear. Between 1790 and 1810, New York state's population grew at a rate of 50 per cent higher than that of the United States generally. The state's growth doubled that of New England's and handily exceeded those to the south. Unlike others, its demographic expansion exceeded the new rapidly growing urban population. Manhattan, though, paralleled the state. Availability of land west of Albany encouraged early rapid settlement.

The Harbour

The high quality of its harbour has often been invoked as a blessing. In contrast to Philadelphia, and very obviously to Montreal, New York's harbour never froze over. Compared to Philadelphia and Baltimore, it was closer to the North Atlantic shipping lanes and more spacious. Boston endured more winter fogs, as did Halifax. Yet in the past, ports with inferior site qualities like Venice had overcome physical disadvantages. Situation is more important than site; Chicago may have been founded on a marshy site, but which place was better situated to capture the Midwest? To use the argument of situation, on the East Coast, either Philadelphia or New York near the central point facing the Atlantic and the interior should have

economically dominated America. New York's higher-quality harbour may have given it an edge, but centrality in economic and political conditions was of greater significance.

Transatlantic Trade and Information Flows

New York increasingly dominated America's Atlantic connections. In 1860 New York shipowners registered one-third of all tonnage in American ships, having exceeded by 1850 the combined tonnage of Boston, Philadelphia, Baltimore, and New Orleans. In 1835 over 1,000 ships sailed into New York harbour; this number tripled by 1849, putting New York far ahead of others. New York dominated import trade and led in exports. Although the volume of foreign trade in the nineteenth century rose at a slower pace than population growth, as America became gradually more self-sufficient, trade remained crucial.

Between 1790 and 1808, external trade handled by American ships experienced its golden age and the big coastal cities grew rapidly. Since Americans remained neutral in the Napoleonic wars, their merchant ships dominated the seas. Merchants of New York, Baltimore, Boston, and Philadelphia built ships and capitalized on the events. The long era of Boston's stagnation ended. Baltimore took off. However, from 1808 to 1820 or so, the times were more troubling. Peace reigned for a while, then war cut these cities off from the Atlantic and, after 1814, flooded them with cheap British goods, inhibiting the northeast's economy.

But New York took advantage of these cheap British goods; it probably already had before 1814. Merchants there captured a lion's share of the import trade. Comparing imports to exports, as in the eighteenth century, the large ports handled a larger share of imports but fewer exports than did smaller places; by 1825 80 per cent of goods from foreign countries arrived at the four major eastern ports, but a half of that already passed through New York. By 1860 New York's share had reached a staggering two-thirds. It would appear that New York surpassed Philadelphia in the 1790s; in 1791 20 per cent more foreign ships docked there. In 1805 42 per cent of the customs duties of the four major ports were paid in New York. Boston predominated in the China trade, but New York merchants began to trade there early too. Trade opened up with south Asia and Africa. All the big ports received another import: immigrants after 1817 and especially after 1825. Ships that transported timber across to Britain, for example, brought immigrants to Quebec City and Saint John, New Brunswick, on their return journey. A majority went on to the United States. New York, of course, became the most important entrepôt for human cargo.

New York cemented its control over imports because it was the first to receive European news through the packet ships from Liverpool and later the telegraph. These ships had operated in the previous century, but speed increased in the spring of 1818. The press noted that 'REMARKABLE EXPEDITION—The packet ship *Pacific* ... discharged her cargo and reloaded in the short space of *six* days...' Never before had the turn-around been so fast. The Black Ball line ran regularly scheduled monthly service with four ships. In 1822, after other companies got into the business, there were sixteen packets, and by 1840 over fifty packets ran to Liverpool, London, and Le Havre. The packets carried cabin passengers who reported news about prices, production, and fashions; mail and newspapers in both directions; luxury goods from Europe, and cotton as ballast eastward. They increasingly reinforced New York's power, at least until 1838. Intensification and control increased by 1821 when fast 'telegraph' vessels, owned by newspapers such as the *New York Journal of Commerce*, raced out to meet the packets from Britain, then landed on Staten Island (later even Sandy Hook farther south) and sent messages by semaphores ahead of the packets' arrival. Subsequently, packets sailed to and from other big ports as the Atlantic system intensified.

Steamship travel across the Atlantic began in earnest in 1838 when two ships raced one another and arrived in New York harbour four hours apart. Another decade firmly established steamships as

Packet ships and then steamships improved the transportation of people, goods, and information.

the favoured passenger conveyor, though only gradually did they replace sailing-ships as goods carriers. By 1840 steamships reduced the fastest sailing time to two weeks, and by 1860 to ten days. It should be noted that Samuel Cunard, who first seized the opportunity of regular steam sailings, began to send his four identical liners to Boston in 1840 for eight years. Subsidized by the Royal Mail contract from Liverpool via his native Halifax, Cunard probably contributed to Halifax's fastest decadal population increase between 1790 and 1860. American government subsidies did not help the rival New York-based Collins line to compete with Cunard in the short run, yet New York's economic power finally forced Cunard to move his operations to New York. The many competing ships of Cornelius Vanderbilt of railway fame helped force the issue. While between 1840 and 1860 steam tonnage exploded by a factor of six, New York's share doubled to a whopping three-quarters. We can see why Samuel Cunard could not resist. Packet ships and then steamships improved the transportation of people, goods, and information.

In the late 1840s telegraph lines speeded Atlantic activity. To reach out to Europe, companies raced to string wires through Boston to Halifax and then later to St John's, Newfoundland, to receive reports first from ships stopping at these ports closer to Europe.[7] The recently formed Associated Press made Halifax its entry point. Then, within a few years, the first (if initially unsuccessful) cable from St John's, Newfoundland, linked America to Europe. It was hailed by the *Times of London* for its 'vast enlargement' of the global economy, the greatest since Columbus (6 Aug. 1858). Some years later the *Scientific American* exclaimed that what had been inconceivable was now possible; the telegraph would create the 'kinship of humanity'.[8] Even though the United States was becoming more self-reliant, it continued to be tied to the global economy with New York providing the strongest link.

As for exports, the hegemony of large ports was, as earlier, less obvious than for imports, though their merchants organized or financed goods exported from smaller ports. American exports rose sharply in the 1790s to 1807. The Embargo Act of 1807 cut exports drastically. An indication of how important exports had been is that the 1807 value did not reach the same amount until the late 1840s, though it should be noted that prices were falling. The embargo hurt Philadelphia and especially Baltimore, more so than New York and Boston, which had developed the China trade. After the War of 1812, when exports picked up, New York took the lead, in part because vessels with imports were refilled for the return trip. The four big American ports of the northeast in 1825 handled over half of American shipping. In contrast to its dominance over imports, only in 1825 did New York achieve first place over any of the other ports in shipping exports. It handled about one-third or so from then until 1860.

Grain and wood products were important exports, but cotton rose to greatest prominence as the English mills produced more and more cloth. In 1821 cotton's share was just under two-fifths of exports, but by 1850 it was three-fifths. Exports did not cover the value of imports, but a balance of payments was achieved through revenues from shipping and British investment in America. The re-export of cotton helped New York reach first place in the export trade, thus garnering a larger share of the value of services in shipping. Earlier in the century, most cotton was shipped directly to Britain and France from Charleston and Savannah, then from New Orleans and Mobile, one leg on the 'cotton triangle'. In turn, goods and immigrants left Liverpool and other European ports for northeastern US ports, then the ships hauled northeastern goods to the South. New York's control over that trade expanded when a great deal of cotton landed in New York first. The triangle was flattened by cutting out the South's direct leg to Europe, thus enhancing New York's position.

On the St Lawrence River, the export of squared timber, lumber, and grain to Britain kept that region relatively near American levels of development. Montreal's loss of the fur trade to the Hud-

son's Bay Company was not a mortal blow. Parliament's repeal of preferential treatment under the Corn Laws in 1846 sharply cut out Montreal's advantage in grain, however, though the American exporters did not respond as much as Montrealers had feared. A sharp depression, which was not experienced in the United States, led some of Montreal's merchants to agitate for annexation to the United States. However, the depression was short-lived, and in 1854 a Reciprocity Treaty between the two countries freed up the movement of commodities to the south. Canada would remain reliant on exports, long before internal activity in the United States had far exceeded its external levels.

New York's Domestic Power: Wholesaling

If New York came to dominate foreign trade, it was partly the result of controlling commerce within the United States and, to a degree, in Canada through stronger merchant action, transportation and communication improvements, and banking and other financial intermediaries. The auction became a key institution for helping New York's merchants to rise to supremacy. When the British dumped goods in New York in the spring of 1815, far-sighted merchants used the auction system to unload goods in a hurry. Rather than mere 'good luck' at the time, it seemed they were prepared for the arrival of goods well before then.[9] Merchants elsewhere still followed the traditional method of selling on credit to back-country shopkeepers and holding title to goods until they were sold, or shopkeepers took them on consignment to be sold on commission. Since 1760 commercial methods became more specialized. More business was done by merchants on commission, who dealt only partly in their own accounts, and brokers and factors, who completely handled goods owned by others. These changes had helped, but the auction cut through this complexity, particularly when a plethora of textile goods arrived from Liverpool right after the war. The manufacturer sent the goods with his own agents, who sought out auctioneers in Pearl Street, thus bypassing the jobber or merchant. The discovery on both sides of the Atlantic that New York auctions sold more goods at lower prices turned heads in that direction. In 1817, when it seemed that other cities might move to compete, the New York state legislature reduced duties, thus reinforcing New York's auction system. Also tightening the system was the fact that only a few auctioneers (such as well-known John and Philip Hone) who were appointed by the governor or city council could sell goods by this method. This was hard on other kinds of importing middlemen for over a decade because prices were generally falling. Regulation favoured the most powerful and, it seemed, the city too.

New York's Domestic Power: Transportation on Water and Land

Increasing control over domestic transportation also contributed to New York's first place in trade. Coastal traders using sailing ships and increasingly steamships after 1810 accounted for three-fifths of American domestic movement of goods by 1853, compared to a fifth each for canals (which were then about to decline) and railways (which then rose sharply). Philadelphia and New York had undercut Boston's coastal trade by 1740; now New York outstripped the Pennsylvania metropolis. Movements among the big four intensified at the same time. While the older southern ports of Norfolk, Charleston, and Savannah languished, New York's coastwise cotton trade, which began in the mid-1780s, was on a permanent footing with New Orleans and Mobile after 1814 as cotton production expanded westward. New Orleans re-exported some produce of the northern Mississippi valley to eastern ports. Sugar, molasses, and coffee after arriving from the West Indies and Brazil, as well as hides and rice, also entered the complex coastal patterns. New York traded with Atlantic Canada, bringing in Nova Scotia coal, plaster of Paris, and mackerel. Boston continued to sell and buy more in Atlantic Canada, however, and it too increased its stake in the coastal cotton trade by supplying New England mills.

Over the decades up to 1860, inland access was enormously strengthened by the transportation improvements on a scale never seen before or since. Roads and steamboats on rivers, canals, and railways strengthened the power of northeastern cities, especially New York, even while contributing to the rise of new western metropolises. After 1760 road improvements came gradually, but spurted with turnpike construction after the War of Independence. Private companies built well-drained roadbeds outward from most cities and along main intercity routes, such as one between Philadelphia and Lancaster, and charged wagoners, stagecoaches, and riders a toll (or, one should say, they tried to—informal bypasses around toll-gates were well worn). The public purse paid for other well-used routes. The most notable links, as might be expected, were Boston to New York, and then, after ferrying to New Jersey, on to Philadelphia, Baltimore, and, after 1800, Washington. Ferries remained a necessity over the widest rivers until after 1850, though bridge-building techniques were rapidly improving. The federally built National Road west from Baltimore, winding over the mountains and completed to Wheeling in 1818, became the most important road to the west.

Travel times were reduced considerably. On the most well-travelled route (New York and Philadelphia), which was already improved by the 1760s, travel time in 1840 was only one-quarter of what it was half a century earlier. There were even more spectacular cuts in travel time on other roads as upgrading continued. As a result, between 1790 and 1825 wagon freight rates decreased by two-thirds.

Steamboats travelled the rivers much earlier before steamships plied the seas. Steamboats vastly improved river travel (most obviously upstream) and they represented the most significant transportation improvement in the West until the 1850s. Introduced on the Hudson River to Albany in 1807, and more frequently after 1810, they gave New York an advantage over other metropolises that did not have easy access into the interior. Steamboats soon plied between Montreal and Quebec, but less so above Montreal because of rapids. They were to have the greatest impact, first on the Mississippi system and later on the Great Lakes. On the Mississippi, while downstream steamboats cut freight rates by three-quarters, upstream steamboats reduced rates to only one-twentieth of what they were earlier, which was a remarkable gain.

North America was well behind Britain and Europe, however, in improving transportation over land. The great spurt of building canals in England in the mid-1700s elicited little interest on the other side of the Atlantic perhaps because instead of coal (which was the chief commodity transported on English canals), cordwood (the chief energy source for cities) could be easily transported by river, road, or along the coast. Iron making, localized by the presence of ores, could exploit adjacent forest land for charcoal. Interest in land transportation rose, however, and in the East several short canals with locks were built around falls on rivers and across narrow necks between waterways in the northeast and South. Further, as cordwood for heating and industry became very expensive by the 1820s, canal boats began to move anthracite coal from northeastern Pennsylvania to New York, Philadelphia, and New England. By the 1840s city plants transformed coal to gas for lighting.[10] In 1853 the Morris Canal Company put forward a $1 million project for dock development and, by 1858, 2 million tons of coal eventually reached New York.

The most famous canal, the Erie, was completed from Albany on the Hudson River to Black Rock near Buffalo in 1825, opening up Lake Erie and the upper lakes to New York. This canal strengthened New York's number-one status. In 1830 the Oswego branch opened access to Lake Ontario, allowing Toronto merchants to buy British imports via New York and not only from Montreal. Built in stages after 1817, the canal boosted agricultural production and urban growth from Albany westward along what the New York Central Railroad would later call the water-level route. Traffic in foodstuffs for use in the East and export gradually rose as production grew in the Midwest. The Champlain Canal connected New York to Montreal.

> ... the canals, together with the Mississippi trade through New Orleans and southern trade, contributed greatly to New York's gain of the largest share of external and internal trade on the continent.

Other large northeastern cites challenged New York's dominance in its link to the West by building canals to the west. In 1834 Philadelphia succeeded belatedly when the Main Line Canal reached Pittsburgh, but the cost was enormous, in large measure because it had to rise over the Allegheny Front three times higher than at any point on the Erie Canal, thus requiring twice as many locks and an incline railway. Baltimore merchants and others in the region finally built a canal along the Potomac, but it reached only halfway to the Ohio River. Boston considered a canal, but had to be content with the stagecoach and the Hudson–Erie route until the building of the railway. Montreal succeeded in building canals on the St Lawrence, but only in 1847, and it was hardly a match for the Erie Canal, which carried ten times as much traffic. In Upper Canada the Welland Canal, an upgraded key link in the St Lawrence Seaway, bypassed Niagara Falls, though it was many years before it lived up to its promise.

American canal mileage tripled to 3,326 during the 1830s, and connected the Great Lakes at Cleveland, Toledo, and Chicago to the Ohio and Mississippi rivers. State governments financed them through land grants and purchase of stock issued, bailed out those that failed in the panic of 1837, and so the canals were eventually completed. Thus the canals, together with the Mississippi trade through New Orleans and southern trade, contributed greatly to New York's gain of the largest share of external and internal trade on the continent. Scholars have traditionally argued that a regional triangular trade pattern developed through the system of rivers, canals, and coastal waters: eastern manufactured goods to the West, western farm and forest produce to the South, and southern cotton to the northeast. However, this greatly oversimplified the pattern: coasters brought northeastern manufactures to the South, and besides, the South produced most of its own foodstuffs, yet a system of circulation was built up. Overall, as on rivers, freight rates on canals dropped dramatically between 1827 and 1852, but canal construction in the 1840s rose only modestly and most canals were losing out to railways by 1850. Their glory was only a brief flicker when compared to the much more densely settled and earlier developed canal system of Britain.

The railway is often regarded as the greatest of nineteenth-century inventions. The 'civilizing tendency of the locomotive … so fortunately patent to all' was a wonder of the era.[11] Not everyone, however, was enthusiastic about this 'machine in the garden'.[12] Even so, the critics could do little to stop the iron horse because it sped the flow of goods and people. Also, railways added further economic power to the northeastern metropolises, especially New York. Shortly after the introduction of steam passenger rail in Britain, Baltimore merchants built the first line in 1829. Others were built quickly, at first radiating out from the large ports or connecting waterways. By 1835 entrepreneurs had linked New York, Philadelphia, Baltimore, and Washington, and added bridges soon after, so by 1840 rail equalled canal mileage, though additions languished during the depression after the panic of 1837. In the latter 1840s, while canal construction declined in the United States, rail construction increased dramatically using British capital and federal land grants that were sold for revenue, reaching 30,636 miles by 1860.

On the water-level route from New York, a line ran up the Hudson River, thence to Buffalo, and two other lines were built across the mountains from Jersey City to Buffalo, creating another corridor to the west for New York. Following this, Philadelphia and Baltimore linked with the Ohio River. In the rail boom of the 1850s, builders connected Montreal to Portland, Maine, and extended the Grand Trunk to the west through Upper Canada and took a share of the production of the American Midwest. Most larger cities in the northeast were interconnected in the early 1850s, and almost all those east of the Mississippi were by 1860. Great bridges—such as those over Niagara

Gorge, the St Lawrence, and the Ohio—sped up movement. Since the standard gauge was not yet universal and over 300 separate companies were operating, there were no integrated systems run by large corporations as yet. Nonetheless, rail lines diverted a great deal of traffic from the Mississippi system, and from the coastwise trade and the canals. The Erie Railroad, for example, hauled cattle to New York, replacing drives overland. It would not be long until Cincinnati and then Chicago would supply the beef and pork to New York. Whereas New Orleans shipped half of western exports in 1839, twenty years later it sent out less than one-fifth, as production burgeoned. Of the amount being exported overseas, New York had more than its share.

New York's Domestic Power: Information

Although people hauling goods and travellers were purveyors of knowledge and rumour, by 1860 information handling became a much more specialized activity. The telegraph can be accorded the status as the greatest invention of the era, perhaps even more so than trains, but even before 1844 when the telegram took the world by storm, communications had improved over the decades. After 1760, flows of information gradually accelerated on business deals, prices of raw materials and processed goods, the costs of borrowing and commercial transactions, legislative and executive government decisions, church and cultural matters, personal relations, world and domestic political happenings, and military events. Governments were expected to provide some information services, or at least their initial fillip. Men on horseback (pony express), in stagecoaches, ships, and boats, and on railways (beginning in 1834) carried newspapers, letters, and ideas in their heads, creating a veritable explosion of knowledge that was far in excess of the actual population growth. Some businesses used homing pigeons. Semaphores, despite a theoretical increase in speed over pony express by a factor of ninety, had not been efficient in inclement weather because operators of stations 5 mi. (8 km) apart often could not see signals.

Of all the cities, New York sent and received the greatest concentration of messages in the corridor linking it to Boston, Philadelphia, Baltimore, and Washington. Only these lay within a five-day time zone from New York in 1794, but Norfolk and Portland, Maine, also did by 1817; by 1841, coastal North Carolina, Buffalo, Montreal, and most of New England received mail within the five-day period.

Newspapers and federal government postal services expanded greatly in the United States. The number of newspapers jumped from fourteen in 1750 to 3,725 in 1860. Dailies, including cheap penny and specialized commercial editions after 1840, grew faster and increased their circulation even more. By 1860 most large cities had half a dozen—even a dozen—dailies, which were printed much more quickly with new presses. As in 1760, only in far greater numbers, publishers sent copies elsewhere by post. With its strongest connections outside and inside the country, New York's seven dailies, 100 weeklies, and over fifty monthlies carried more up-to-the-minute and fashionable European and domestic information and so were copied by those in other cities. Although Boston claimed to be more literate, New York publishers produced one-third of America's new periodicals in the 1850s.

The overland postal route network mileage expanded. nearly a hundred times over the fifty years after 1790. In the prerail era, government-financed post roads carried letters for the most part. The government reduced zonal rates in 1845, about the same time as it introduced postage stamps. These actions moved post offices towards a more modern organizational structure. New York's dominance was clear in this regard too; in 1852 letters sent and received there equalled those in its big three neighbours and New Orleans (nearly a quarter of the total), while its total share of the population of the country was just over 3 per cent. Some of these newspapers reached Canada, where local publishers reported on economic and social conditions south of the border. Canada would forever be in the thrall of American ideas and information, at least to a point. As for books, which now included novels, New York's seventeen firms published nearly one-third of the books. The first advertising agencies began to parade the need

for consumer goods. No longer was advertising simply a means of informing the public; now advertisers would persuade people to buy a variety of goods.

The revolutionary speed of the telegraph sharply increased the spatial bias in favour of New York and other large cities. The federal government sponsored experiments and the first intercity wire. From the moment in 1844 when its inventor, Samuel Morse, exulted 'What God hath wrought', private companies quickly saw its advantages; time was 'the great item with commercial men'.[13] In contrast to western Europe, the government, having taken the initial risk, allowed private firms to take over, and so the potential of a public monopoly gave way to a provate corporate monopoly. Before trains, telegraph lines linked every major city east of the Mississippi in the United States and Canada, and then quickly reached smaller places. The 20,000 mi. (32,000 km) of telegraph wire in 1852 more than doubled by 1860. Although early connections often broke down due to the heavy demand, efficient multiple lines increasingly linked the largest cities and then smaller ones, often strung out along railway rights of way. More and more lines were added to crossbars on poles.

The revolution in telegraphy initially meant that New York media, merchants, and financiers, who were the biggest early users, could reach Buffalo in hours (assuming waiting and reiteration time along the route) compared to the four days by the fastest pony express. Before long, the time would be measured in minutes. Montreal and Toronto were quickly connected to one another and to New York. In 1861 San Francisco was linked. In those early days, business messages accounted for three-quarters of the total and the press took up much of the remainder. Railways coordinated train movements by telegraph, thus vastly improving their flow and safety. Governments too began to use the code. Political parties could now operate efficiently on a national scale from Washington. Lines within cities connected businesses. All told, the telegraph 'brought about pervasive and often dramatic reductions in [regional] intermarket price differentials, information costs, and transactions costs.'[14]

Because of the anticipated high profits, new telegraph companies developed, quickly leading to cutthroat competition, especially among the large cities. Entrepreneurs of early routes reduced rates between New York, which was 'the grand focus' of the system, and other large cities; the most intense competition was between New York and Philadelphia through fourteen lines in 1852.[15] Not surprisingly, as a utility it was a 'natural' monopoly, so corporate monopoly emerged. 'Because of the importance of through traffic, telegraph companies moved through the patterns of competition, a brief era of "methodless enthusiasm", cooperation through a cartel, and consolidation in a much shorter time period than those responsible for the railroads', which are often seen as the leaders in corporate organization.[16] The speed of changes in the telegraph companies probably outpaced that in any other major field before then. Cutthroat competition gave way to cooperation by 1855. Financed mostly by New Yorkers, there were six leading telegraph companies in 1857, and three soon after. Because the American Telegraph Company lost ground during the Civil War, one—Western Union—dominated the United States in 1866. Some years later, Jay Gould, of corporate railway fame and the only one who was able to challenge Western Union, was at the helm of this monopoly! He also bought a separate Canadian company. The telegraph, due to its nearly instantaneous way of moving power through words and dollars, was a business success. It was little wonder that the most aggressive entrepreneurs moved quickly. The telegraph industry, the 'nerves of the press', also persuaded newspapers to pool their resources for transmitting stories through the near monopoly of the Associated Press and other wire service groupings, which made their headquarters in New York.[17]

New York's Power: Finance

New York is most famous for its domination of finance. The financing of government, steamships,

> Money markets grew strongly ..., promoting New York's rise to not only the first among equals, but the first without question.

roads, canals, railways, manufacturing, and trade demanded far more investment than was the case in the 1700s. Britain's role as the chief source of credit for trade had receded for the American if not the Canadian market, and by 1860 American capital was competing for commerce, infrastructural improvements, and manufacturing. To handle the great increase in money flows after the War of Independence, new financial intermediaries grew enormously. Money markets grew strongly through banking and then increasingly through the stock market, promoting New York's rise to not only the first among equals, but the first without question.

In the 1780s Philadelphia and, soon after, the other three cities founded banks as merchants gradually relinquished their banking activities, or became bankers and even stockbrokers. At Alexander Hamilton's insistence, in 1791 the federal government established the Bank of United States, which was the locus of government transactions for twenty years. New York was given a branch, which indicated its nearly equal importance to Philadelphia. By 1800 there were thirty-eight banks in the big four and other cities. By 1811 the banks in the big four held nearly two-thirds of all bank capital.

New York clearly surpassed the others in holdings by 1810. After the war, while banks in other cities fell back, New York held its own. New York banks, chartered by the state, played a centralizing role, despite the re-establishment of the Second Bank of the United States in Philadelphia in 1816. The New York bank business rose sharply around 1823. Boston's shot up too, but was already far behind. The state governments struggled over the next two decades to regulate the rapidly growing industry against fraud and to protect creditors. The big New York banks, such as the Bank of New York, the Manhattan Company, and the Bank of Commerce (1839) collectively became in effect the central bank in the United States in 1833 when President Andrew Jackson, representing western and southern antagonism to centralized banking, withdrew government funds from the Bank of United States and virtually killed it. The biggest New York banks held the balances of most of the small banks in the hinterland, including the rapidly expanding West. In 1837 nine banks held 78 per cent of these deposits, largely as a mechanism to maintain some degree of seasonal stability in the flows of money. Although there were complaints of collusion and monopoly power, western bankers had little choice but to use this system. Creation of the New York Clearing House in 1852 was a major innovation in smoothing the flows. Of 1,600 incorporated banks and 900 private banks in the United States in 1860, six in New York City stood out. Indeed, New York banks (measured by capital) held a 30 per cent lead over Boston's in 1850, which had surpassed Philadelphia during the previous decade. By 1860 New York had more than doubled the capital of Boston's banks.

The banks aided the workings of the stock market as it became increasingly important for the expanding investing public and for speculators. Banks issued call loans that were repayable on demand. Investors ploughed money into the stock market. By 1860 Wall Street dominated over other exchanges, probably because its actors were more daring than those in Philadelphia and Boston, more daring in part because they had a deeper well of capital from which to draw. Although Philadelphia, as we saw, had a kind of exchange in the 1760s, the formal beginning for the stock exchange is usually accorded to Wall Street in 1792. That year brokers and auctioneers signed a formal agreement setting up rules of behaviour, though trading remained slight. In 1817 the New York Stock and Exchange Commission put trading on a firmer footing. Securities then provided a means to finance canal building and, beginning in 1830, railway construction. The collapse of 1837 played havoc among brokers. Yet, as the strongest exchange, Wall Street in the 1840s consolidated its power against others in railway stocks. The telegraph accentuated its power in setting national prices. When the costs of plant and equipment began to outstrip the availability of investment capital from other sources in the 1850s, the market expanded its financing of manufacturing. Before

then industrialists had borrowed privately or from banks.

Ninety insurance companies operated in New York City by 1856. The expansion came after life insurance was added to colonial staples of marine and fire insurance in the 1840s, reflecting a burgeoning middle class. Because insurance companies were more conservative and less in need of liquid capital, New York did not dominate as it did in banking and stock markets. Very wealthy prominent men—such as John Jacob Astor, his son William, Edwin D. Morgan (later governor of New York State), and James Brown—dominated various sectors on Wall Street.

Despite their power, the rise of specializations, and some means to regularize flows, the financial system remained unstable. From 1792 to 1860 Wall Street experienced 'tremendous booms and busts, trustworthy issuers in need of capital and charlatans in search of lambs to shear, shareholders without power in corporate governance and managers with controls, periods of government acquiescence and others of government meddling, market leaders and market makers, winners and losers.'[18] Panics periodically swept through markets and wiped out speculative gains, as in 1819. 'Feverish activity' in money markets, excessive land speculation, withdrawal of British investment, and heavy government debts for financing canals led to another panic in 1837.[19] Failure of cotton prices in 1839 set off the severe depression sometimes called the Hungry Forties. Another panic hit in the summer of 1857 as the consequence of the overextension of credit (mostly for railway building), the drying up of British capital, an imbalance of trade showing that British manufactured goods still remained in a commanding position, and overproduction.

Calls went out for another government bank to stabilize the system. Although the federal government operated its Independent Treasury from 1840 with a branch in New York, only banks could issue bank notes. To stabilize the money markets, the National Bank Act established a federal reserve system of national banks in 1863, in large part at the insistence of New York capitalists. The federal government took over the issuing of currency from the banks. The act did not permit branch banking, however; its drafters deferred to the widespread populist fear of monopolies. By contrast, in Canada the rudiments of a chartered bank system that allowed branches emerged, starting in Montreal in 1817. The banks issued notes until the 1930s. In the long run, the American system remained more unstable, if apparently more democratic, than Canada's, but hardly prevented monopolistic power. The illusion of localism obscured New York's oligopolistic control in finance and many other fields.

New York's Power Shared: Manufacturing

New York could not claim to be the centre for manufacturing as it was in other fields. Because of its size, however, it turned out more goods and its shops and factories employed more workers by 1860 than any other city. Its role in financing manufacturing cannot be underestimated and no doubt more ink was put to paper for orders there than elsewhere. New York's inability to dominate manufacturing was directly related to the fact that most manufacturing in other cities and their hinterlands was, as earlier, for local consumption. Yet when the North, West, and even the South became increasingly affluent, suppliers created opportunities for buyers of manufactured goods, thus fostering regional specialization for national markets. The manufacturing belt of the northeast and Midwest to the Mississippi was established. Much of the St Lawrence valley, Ontario, and even Nova Scotia could be included as well, since their development paralleled that of the northern States in traded goods.

The industrial 'revolution' happened over several decades, but by 1860 it had become a mark of the Western world. Many of the key ingredients were in place: coal to make steam to power machines, thus displacing human and animal muscle power, and overhauling wood and water power (if not yet entirely). Large companies, many incorporated by then under general laws for incorporation (the American rather than the European way of organizing business to create and sell goods) had emerged, gradually replacing domestic and small-scale artisans in many fields. Much artisanal pro-

duction was probably not included in the census data for 1860, it should be noted. By the mid-1830s, corporate organization had developed to such a degree—that is, with absentee ownership, integrated production, and even control of marketing—that workers formed unions in the factories as a counterweight. The unskilled operatives of machines and the skilled workers in large organizations went on strike when wages were cut. The depression after the panic of 1837 virtually wiped out these nascent organizations, though new ones would emerge later. Another sign of a shift was the decline in the number of children in factories. Their redundancy was at least one factor for the promotion of public education.

It seems that economic development was destined to happen among North Americans of western European ancestry, as demonstrated by the participation of both the United States and Canada in this great change in methods of producing goods. Hamilton's views prevailed over Jefferson's plea for commercial farming and nice small towns. Inventiveness rose to the surface as it had in Britain; in the 1790s Yankee Eli Whitney, for example, invented the cotton gin, which increased the speed of cotton preparation on southern plantations, and Oliver Evans created a flour mill using elevators and conveyors that sharply raised productivity per worker. Several important reports by experts promoted manufacturing. Also, political and military events between 1808 and 1815 closed off British imports and thus encouraged domestic American production. Although the flood of still cheaper British goods in 1815 temporarily threw many out of work, production took off strongly by the 1820s. Overproduction of consumer goods contributed to the panic of 1837 and the ensuing depression, but was followed by a massive outburst of production until 1857, perhaps the greatest ever in relative terms.

By 1860 industrial production in the United States had still not caught up to Britain's, but it equalled that of France or Germany. The value of manufactured goods was ten times that of 1810, while the population rose four times. Production followed the pattern that was evident in England earlier, continuing the industrial character of the mercantile era and the take-off of the 1790s—that is, goods for direct personal use or intermediate goods using farm and forest products still dominated as they did in 1760. Measured by value added (that is, the value of output minus costs other than labour), the order of most production was cotton goods (first by 1840), lumber, boots and shoes, flour and meal (first in value), men's clothing, iron, machinery, woollen goods, carriages, wagons and carts, and leather. Supply and hence demand for stronger and more durable iron and other metals and machinery was, however, rising.

Regionally, the northeast dominated the country in 1860, as might be expected, with over two-thirds of manufacturing employees and of value-added goods. Massachusetts, Pennsylvania, and New York together accounted for just over half of employment, in roughly equal shares. Because they dominated in textiles, the New England states depended twice as much on manufacturing as did Pennsylvania. Although Boston led in the number of manufacturing employees in the region, Lowell, New Bedford, and Providence stood out among the many booming mill towns. Beginning in 1815 the mills put spinning and weaving under one roof, with unskilled workers operating the machines. A dozen or more places housed 2,500 or more workers by 1830. Three decades later, in over three score of places, mill owners had tapped water power and increasingly steam, and employed most local workers. In Lowell, over 13,000 hands worked in fifty-two mills. Lowell had been renowned for its female operatives, but Irish males were replacing them by this time. In addition to textiles, Connecticut and Massachusetts could claim gun and clock making, and the beginning of interchangeable parts, which was so crucial to further gains in productivity in the American assembly line system later in the century.

Industry flourished elsewhere in the northeast. The streams around Philadelphia spawned mills as they did before 1800, but in larger numbers on the Brandywine and other creeks near Wilmington and on the Schuylkill at Manayunk. Smaller cities like Reading took on factories, but Philadelphia was a bigger story. Among big cities,

Philadelphia relied considerably more on industry for jobs, twice as much as New York did, and more than twice as much as Baltimore and Boston did. Sixty thousand factory hands in 1840 increased to almost 100,000 in 1860, with over half of its workers in processing. Like all big cities, Philadelphia's manufacturing was far more diversified than that of smaller mill towns, where factory workers occupied over half of the jobs. The Quaker city's greater emphasis on manufacturing showed not only that it had fallen far behind New York as a mercantile city but also that it prefigured the next period when the largest cities would increase their share of industrial output. Philadelphia also geared much of its industry to an affluent local and subregional market. If Pennsylvania had totally withdrawn from the world after 1800, its chief city might have survived very nicely at a high level of material prosperity. Increasing specializations—in manufacturing within the city, and on resources in anthracite coal, iron making, and forestry in its hinterland, combined with its continuing strong agricultural production—supplied the world with proportionately much less, while Pennsylvania had more. Yet it was obviously not a closed system; the region's manufactures spread south, west, and north into Canada.

The New York City region grew phenomenally in the two decades before 1860: its eleven counties contributed 14 per cent (by value) of American manufactures, which was well above its share of population. The city provided over half of that; its 4,300 establishments (those that were counted), with the highest number of workers (just ahead of Philadelphia), produced more goods by value than any other city in 1860. Besides, its region boasted the large satellite cities of Brooklyn and Newark. They ranked fifth and sixth among all cities in productivity.

The falls of the Passaic River at Paterson had appealed earlier to investors in mills. Several others were of importance, not the least Jersey City and adjacent towns, and places north of the Harlem River. In 1840 two-thirds of New York City's output was for local and regional consumption, while a third was for export and for the rest of the country. The share of the latter grew over the next twenty years. Manufacturing employed a third of its workers.

In the city, sugar refining, printing and publishing, and garment making led the list by value, the latter two producing for national markets. Fourteen refineries, which used West Indian raw sugar, had displaced foreign imports, but printing and publishing were far more important than sugar for their image and influence on society. Nearly one-third of American printing and publishing were at the forefront in communication, as we have just seen. The great newspapers, the *Times*, the *Herald*, and the *World* led among seven dailies and scores of weekly newspapers and magazines. Larger presses and a massive increase in paper from wood pulp increased production. The most prominent book publisher Harper Brothers started their enduring monthly in 1850.

Clothing manufacture easily topped employment rolls, as the city already dominated men's ready-to-wear and women's apparel even more so. In 1855 over 30,000 men and women toiled over benches with their needles and increasingly with Singer's sewing machine. At an all-time peak, women made up two-thirds of all workers before decreasing to half that number in 1900. By contrast, just over a quarter of manufacturing workers were women in 1860, though less than in many New England mill towns. Large-scale wholesalers, who organized seamstresses and tailors, were making their appearance; one firm gave work to 4,000 people who worked in their homes. These large firms bought fabrics directly from the mills, cutting out the middlemen. Out-sourcing was practised; for skirts, a Bronx firm supplied hoops, for example. Quality ranged from grand gowns for the rising rich to shirts for humble labourers.

In other fields (by rank in number of employees), metal and machinery (including engines and locomotives) came next, though New York's share of the production was already falling. Furniture remained important, though its manufacture was starting to emerge in specialized clusters elsewhere in the country. Shipbuilding (also about to decline relatively), boots and shoes (which would lose out to places like Lynn, Massachusetts, and St Louis), precious metals, hats and caps, musical instru-

ments, and carriages all employed more than 1,000 workers, as did printers and sugar refineries. Thirty-three companies made pianos—obviously a sign of a rising middle-class and its modest pleasures. Bakeries and breweries continued to serve the local populace, but clothing manufacture, printing, and publishing became Manhattan's specialties.

Western cities not only replicated and elaborated on the commercial network but also followed the East in manufacturing. Buffalo and Rochester milled wheat, which by 1860 was shipped more often from the West than it was reaped locally. Pittsburgh's industrial workforce rose to within half as many as that of Baltimore, and other Ohio River and Great Lakes cities followed. Iron ore from the Mesabi Range in Minnesota began to feed new industrial complexes on and near Lake Erie ports; this was a harbinger of steel specialization for national markets. In the far West, San Francisco replicated not only Atlantic commercial structures but also shops and factories for local and regional sale. Undoubtedly, the bulk of finer goods came from the East by Cape Horn. Only in 1869 did rail reach San Francisco.

In the southern United States, the pace of large-scale manufacturing was far less marked. Craftsmen worked in the small slow-growing cities and in the countryside as earlier, especially on plantations. Southern leaders promoted manufacturing, but only Richmond came close to Baltimore's levels with its flour-milling and tobacco-processing operations. The South was in an increasingly dependent position for manufactured goods and financial investment, even while the value of its slave-produced cotton rose and the cottonbelt expanded west into Texas. The South was no match for the North in military prowess.

Canada was far more like the American northeast, if on a smaller scale. In Montreal along the Lachine Canal, a whole range of metalworks and mills developed after 1825. Textile towns sprang up. Shipbuilding remained important in Quebec City. Iron making continued nearby. As in the United States, most milling still took place in the country at water-power sites. Industrial growth in the Maritime provinces was more modest: shipbuilding (notably of great clipper ships) and a range of consumer goods emerged, not unlike New England's, to which they were partially tied commercially.

Agriculture, along with forestry and mining, undergirded the elaboration of the rapidly urbanizing economy. Agriculture led all sectors in total value of production until the 1850s. Farm productivity rose, even though mechanization was just beginning. The commodity exchange in New York and later in Chicago expedited commercialization and hence regional specialization, which became even more marked than in the previous century. Production of dairy products developed as the most important field in the northeast. Around big cities, rail milk sheds developed and market gardening greatly expanded. The wheatbelt moved westward in New York and from southeastern Pennsylvania, and almost reached the eastern semiarid plains by 1860, where it would soon become established. The cornbelt, where swine and cattle were fattened, also moved westward to make its permanent home in the more humid prairies. Grain and livestock, through associated commerce and processing operations, helped to spur urban growth in the Midwest and Upper Canada. Cincinnati was the great porkopolis of the antebellum era. However, cotton, which was the most valuable crop, did not induce the same level of industrial urbanization in the new South. Hardwood oak and northern softwood white pine forests supplied an abundance of construction materials and fuel, while the mines supplied the metallic and industrial minerals. Mining was on the verge of its great era; metals would increasingly dominate the production of goods.

If the continent remained 'a set of regional economies with low levels of interchange' compared to what it would be later, the specialized ingredients of integration—with New York at the pinnacle—had formed.[20] From New Yorkers to ranchers, North Americans had created an enormously powerful machine of growth that was integrated by communications, transport, trade, and regulations to a degree that the merchants of 1760 could not have imagined. Before the railway and the telegraph, the northeast had achieved sustained

development based on the mercantile system, especially between the mid-1780s and 1810 when New York achieved first place in trade and banking. Again after 1820, growth picked up, increasingly through internal trade in manufactured products. The growth of manufacturing in the northeast (even while Philadelphia and Boston's hinterland continued to specialize in factory production) further enhanced New York's position because its merchants and financiers were the dominant actors. Boston, Philadelphia, and Montreal did not shrink. Although they fell behind New York, their earlier established and new institutions sustained their expansion through intensification. Growth in the hinterlands of these cities continued vigorously after 1790, and through their interactions, all contributed to the super economic growth of the continent.

Let us now turn to life within New York, the greatest of all cities, to see how these economic pressures unfolded. In the process we will be able see that one other factor made New York first: in any growing nation, one place has to be created to attract the most ambitious and creative people, not least in finance and communications, otherwise their energies are dissipated. The accumulation of the nation's wealth depended on their abilities, even while they themselves, an astonishing number of *nouveaux riches*, succumbed to the lure of consumption and paraded their success. Many others were supporting actors or were in the chorus, each with her or his own sphere of dynamics. The government's role was to sustain and restrain.

Manhattan's Ebullient Society

Overturn, overturn! is the maxim of New York. The very bones of our ancestors are not permitted to lie quiet a quarter of a century, and one generation of men seem studious to remove all relics of those who precede them.[21]

To Philip Hone, the former mayor and merchant, the pace of growth in the mid-1840s was almost impossible to comprehend. The buildings of the past could be wiped out, often not just in a generation, as he said, but within a decade. He saw the social stability of the past being overturned, and so was ambivalent to this creative destruction. By 1860 there were over a million New Yorkers, including those in Brooklyn and Jersey City. According to the 1860 census, just over 800,000 lived in Manhattan, extending 4 mi. (6.4 km) or so north from the Battery, solidly to 36th Street on the east side and 50th on the west, and some beyond (Figure 4.2). Farther up the island were the villages of Yorkville, Manhattanville, and Harlem; in between were scattered dwellings and businesses. New Yorkers spilled over the rivers too. Brooklyn's nearly 300,000 people and Jersey City's 30,000 were increasingly becoming a functional part of the city. Towns, villages, and farms in the region of ten counties around the city added another half million people. By 1860 only London and Paris could claim larger populations. Unlike Hone, many New Yorkers would be less ambiguous about undoing the past; to them that opened up the future.

Manhattan had fewer than 20,000 residents in 1760. A generation of slow growth was punctuated by war, the British occupation, and then the expulsion of Loyalists. Thirty thousand lived in Manhattan in 1790, and that doubled in the following decade of rapid economic growth. It bounded ahead to 165,000 in 1825 and nearly doubled again in the years leading up to the panic of 1837. As Hone observed, the depression slowed the pace, but the number of Manhattan residents expanded over the two decades before 1860 at a rate faster than at any time since 1800 and any time after.

People were restless in this extremely transient era. Vast numbers came and went. Nine out of ten

> As New York was the chief port of entry for migrants, many sojourned there only briefly before moving on to other cities or to the countryside.

of New Yorkers were American-born in 1825, many of them attracted to the opportunities of Gotham. The number of Blacks as a share in the population dropped sharply from the highest level in northern cities earlier as few migrated and their death rate remained higher than that of Whites. Immigrants arrived in the 1830s in numbers nearly four times greater than in the previous decade, then tripled again in the 1840s, doubled in the 1850s, though the rate began to slow down after 1854. A million Irish and a million Germans landed in North America. As New York was the chief port of entry for migrants, many sojourned there only briefly before moving on to other cities or to the countryside. When they landed, fewer than half said that they expected to stay in New York, but obviously many of them did. Although Germans made up a higher proportion of Milwaukee's and St Louis's populations, those who settled in New York far outnumbered them. The Irish movement peaked in 1852, and the German one in 1854. The foreign born of New York made up just over half of Manhattan's population in 1855, the Irish leading with well over a quarter, Germans one in six, and British one in twelve. Besides the deluge of foreigners, many American young people increasingly sought their fortunes in New York. As affluence rose among Americans, tourism for more and more became possible. The city's attraction made it the chief tourist destination on the continent. Many salesmen and other businessmen visited for commerce and pleasure, but New York's established rich were restless too. In the summer they left their townhouses and country villas to board trains bound for Saratoga and Newport or steamships to Europe.

4.2 *(Opposite)* Manhattan wards, residential change, and economic activities, 1855–60. Spatial differences by class were not sharply defined, though the 15th ward was considered an élite empire. The lower east side was generally poorer. Germans were the most concentrated, while the Irish were more spread out. *Sources*: On work: Richard B. Stott, *Workers in the Metropolis: Class, Ethnicity, and Youth in Antebellum New York* (Ithaca: Cornell University Press, 1990); on ethnic groups: Robert Ernst, *Immigrant Life in New York City, 1825–1863* (New York: King's Crown Press, 1949); Kenneth A. Scherzer, *Unbounded Communities: Neighborhood Life and Social Structure in New York City, 1830–1875* (Durham: Duke University Press, 1992).

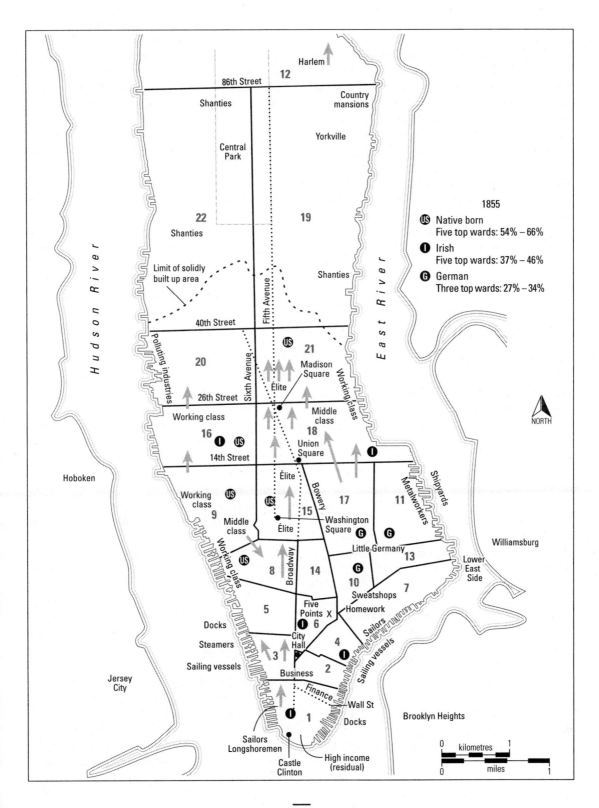

New York in the Ascendancy, 1860

Harlem

86th Street

12

Shanties

Country
mansions

Yorkville

Central
Park

22

19

Shanties

1855

US Native born
Five top wards: 54% – 66%

I Irish
Five top wards: 37% – 46%

G German
Three top wards: 27% – 34%

Hudson River

East River

Limit of solidly
built up area

Shanties

Fifth Avenue

40th Street

US

Polluting industries

Sixth Avenue

21

Madison
Square

20

Élite

26th Street

Middle
class

Working class

Working class

18

16 **I** **US**

Union
Square

14th Street

I

Hoboken

Élite

Working
class

US

Bowery

17

Metalworkers

Shipyards

9

US

Élite

15

Washington
Square

11

Williamsburg

Middle
class

US

Élite

G

G

Little Germany

13

Lower
East
Side

Working class

Broadway

8

14

G

10

Sweatshops

7

Homework

5

Five
Points X

Sailors

Docks

I 6

Sailing vessels

Steamers

City
Hall

4

I

Sailing vessels

3

2

Business

Brooklyn Heights

Jersey
City

Finance

1

Wall St

I

Docks

Sailors
Longshoremen

Castle
Clinton

High income
(residual)

0   kilometres   1

0   miles   1

NORTH

Wealth among New Yorkers was extremely differentiated; many observers commented on the opulence and the squalor. The gap between the rich and poor became even wider than it was earlier in Philadelphia. Doubtless wealth was skewed more in New York than in other cities and rural areas. A recent analysis of the ambiguous data of the 1855 census separates about 200,000 people specified by occupations into six classes by percentages: capitalists (3), small proprietors (11), professionals (6), artisans and mechanics (12), and wage workers (57).[22] Capitalists were merchants, bankers, and speculators. Compared to 1760, there were more white-collar employees and fewer artisans. The bottom three-fifths no doubt earned less than most others, though with a wide range of incomes.

The number of rich increased in this effervescent society. The new rich among major financiers, large manufacturers, great retailers, and specialized wholesalers overwhelmed the old merchant Knickerbockers, though the latter, including Philip Hone, were aggressive too and many were receptive to *arrivistes*. Because it grew more quickly than Boston, New York could not claim to have much of an old aristocracy in 1860. Constantly fluctuating, the élite were 'far less conventional' than they were in Philadelphia.[23] The share of non-business household wealth held by the top 4 per cent rose from two-fifths to two-thirds from 1825 to 1845. By then, John Jacob Astor, who made his fortune in the fur trade, shipping, and Manhattan land, was a millionaire along with nine others. A decade later there were more than twenty millionaires. One hundred thousand dollars made one very wealthy in 1852: over 1,000 were in that category and another 5,000 could be called wealthy. Affluent lawyers, accountants, stockbrokers, lesser merchants, top clerks, craftsmen of high-value goods, and the largest building contractors were included in the top ten or so. Among these were the 12 per cent who owned real estate. If comparable with other large cities at the time, the top 1 per cent held perhaps up to 50 per cent of the wealth, an amount somewhat larger than today for the country, much of it in urban land.

Walt Whitman called the men of 'moderate means' (those earning around $1,000 dollars a year) the 'most valuable class'.[24] Among the better paid above that level were lesser lawyers, retail merchants and manufacturers, and some accountants, salesmen, buyers for retailers, and commercial travellers. Tavern keepers and boarding-house operators who served the best clientele were comfortably off, as were those who operated the top-flight whore-houses, no doubt. Among the 30 per cent of white-collar workers, most were earning less than Whitman's benchmark on wages, including male office clerks and retail clerks. Some younger clerks with ambition and skill could rise rapidly, some to become owners of businesses when times were good.

Many who would have been considered craftsmen earlier were now counted among wage workers. Some received wages in kind, such as room and board. The depression following the 1837 panic had wiped out many skilled jobs, so that a majority of workers were unskilled and semiskilled wage earners. Two-thirds of these workers held jobs in manufacturing and construction, and even more worked in services as servants. The latter, who were at the bottom, were mostly young women and composed the largest occupational category—one in seven workers. The range of incomes was great. Among furniture makers, for example, were owners, master craftsmen, journeymen and helpers, or even apprentices. Without question, though, a majority of the working people—men, women, and children—were poor, many very poor, and a small minority were destitute. The numbers shrank and expanded with economic conditions and the intensity of immigration.

Most jobs paid low wages, though people generally got more for their money than in 1760. Families often did not make enough to meet the minimum standard of living set by Horace Greeley's *Tribune* in 1851 of $10 a week for a family of five. Ten dollars was actually a fairly generous income that could buy 2 lbs of meat a day, tea and coffee, butter, and cheese.[25] However, even if they were steadily employed, labourers were lucky to get $6 per week. Because of insecure employment, however, they might earn only $200 a year, but many were not even up to that level. Obviously, wives and older children had to bring in some income, the girls often through prostitution, which was endemic. Most widows and many orphans lived in penury. In winter, farm labourers, who were mostly single males, swelled the ranks of the poor and competed for seasonally short jobs. Longshoremen, draymen, and labourers, though usually in great demand for commerce, lived insecure lives. Downturns in the economy were harsh experiences for these low-paid workers in a world that defined poverty largely as the result of moral failure. Real incomes rose for many because prices fell, especially in the early 1840s, though like the early 1930s, one had to have a secure job to enjoy the benefits. Despite the bleak conditions for so many, immigrants kept coming in the hope that America would be more providential than Europe. For most, it would become that.

Not surprisingly, immigrants rather than native-born people filled the lowest ranks in the short run. They provided over 80 per cent of the manual labour in the city. If four out of ten native-born workers lived on wages rather than on salaries or entrepreneurial gain, over four in five Irish, two-thirds of the Germans, and nearly as many English and Scots did so. The Irish were more skewed towards the bottom than the others and represented a disproportionate share of charity cases. Many were cottagers, who were forced off Ireland's land after 1845 and were without urban skills or money. As Roman Catholics and victims of oppression, the poorest Irish suffered severe discrimination as they did in English cities like Manchester and Liverpool, but neither Irishness nor Catholicism would hinder their descendants from eventually rising in status. Even then, one in twenty capitalists (300) were Irish-born, as were many in middling categories. Some were secure in public jobs: one-third of policemen were Irish.

Members of other ethnic groupings were generally better off than the Irish, but varied. Germans were more numerous in the small-proprietor category in 1855 than those from Great Britain. Over half of the bakers and cabinet-makers were German. British immigrants, on the other hand, were more clearly represented among capitalists, professionals, and especially among skilled mechanics, presumably because they brought skills from industrial England. Native-born Americans were more evenly spread out across the spectrum; five of every six capitalists (most of the upper ten élite) were born in America. The native born composed half the total population, but only a quarter of the wage earners, perhaps because far fewer women among them worked outside the home.

People in various classes practised rituals. New York became a place of conspicuous display in public. The postrevolutionary republican virtue of the patricians and mechanics, which emphasized the common good, gave way to the era of the common man. That shift to populism paradoxically heightened the pursuit of status. In the 1850s the most striking ritual was the promenading of the new rich on foot at prescribed times, such as after church. Elegantly dressed women and men paraded past one another with a curtsy and a tip of the top hat, while 'recognizing and being recognized'.[26] Awed onlookers gawked at this theatre of bourgeois representation. This rigid baroque formula and the enormous work it entailed was apparently needed to define who was who and to impose some order in this the most changeable and most accumulative of societies. This gross public display after 1860 was carried over to Central Park, where the rich paraded in carriages. These public forays would eventually fade, probably when the rich spent more time in the country, but formal visiting, as on New Year's Day, was necessary to establish and maintain social standing, and would persist. But this was not enough: in a place where patricians had disappeared, celebrities were required and, ironically, they were often Euro-

> Despite the bleak conditions for so many, immigrants kept coming in the hope that America would be more providential than Europe. For most, it would become that.

peans. Charles Dickens was taken aback, exhausted, and nearly overwhelmed by the acclaim—by being turned into a 'spectacle'.[27] Copies of his bust were sold in Tiffany's. Excessive lionization of the celebrity suggests that the affluent were excessively insecure, or that Americans, who were deprived of royalty, had to invent something akin to it.

It was, it seemed, a two-class society: those among the 10,000 ('the upper ten') and all the rest—the employers and the employed, the owners of property and those without, those who consumed and the others who worked long hours to provide them comfort. Still, those in the middle ranks, now white-collar workers, though renting, defined themselves against the poorer manual workers below them in income. Among the middle ranks were the professionals and their wives dedicated to uplifting the poor to their standards of temperance and probity. The middle class went to church.

There were others. Confidence men who fleeced the unwary became the model of ambition for the Horatio Algers. Like the rich the lower classes played out rituals of display. Working boys and girls of the Lower East Side defined themselves by their leisure. Members of this rude youth culture paraded meretriciously in their finery on summer evenings, attended their theatrical entertainments, spoke brusquely to outsiders, and fought one another. Rich and poor or, to be more precise, some well-off males and not-so-rich women met. In this city of eros infested with a 'whorearchy', diarist George Templeton Strong claimed in 1853 that 'the better class of young men about town' extolled prostitutes.[28]

Unions consolidated manual workers in their workplaces, but only ephemerally. New York craft workers organized in the 1830s as never before; an astonishing two-thirds organized and called frequent strikes. Invoking the revolutionary rhetoric of liberty from the early days of the republic, they met relatively little resistance from state power. The incipient labour movement collapsed in the depression after 1837, postponing further agitation for higher wages and the ten-hour day. Many artisans lost their jobs permanently. Skilled workers,

many of them now wage earners, responded to the economic revival of the mid-1840s by organizing once again in factories and shops, only to suffer in the 1857 downturn when men paraded on New York streets, 'demanding work and bread'.[29] In this second phase of action, women were less involved in union action than they were earlier, perhaps indicating an assertion of male power and a move to a stronger but more exclusive domestic role for women. Since males were allowed the right to vote, New Yorkers (unlike European workers) did not put home and workplace together in one framework for political ends. Since they had achieved the basic democratic right, American workers went on strike only for bread-and-butter issues. In their neighbourhoods, where they had the right to vote, they supported the Democratic Party (hardly a union party), which was in New York a patronage machine. On the other hand, European workers, who were denied not only economic but political democracy, formed socialist or social democratic parties, though obviously with little electoral success until the twentieth century.

Governing an Unruly City

In a place where people were overthrowing the past, politicians faced formidable obstacles in providing services. Complicating the task was an increasingly convoluted local governing structure that resulted, ironically, from state and local attempts to make the system work better. This seems an endemic American city problem and, in the long run, has been detrimental to cities. Struggles among parties and factions often slowed action, but without parties to control the chaos, even less would have happened. Those who were excluded from economic power strove for status through politics, overturning patrician rule. From the 1840s onward, corruption and patronage pervaded politics.

Unlike Philadelphia, in the colonial era New York City's corporation was not autonomous, though it was not very open in practice. The provincial governors appointed the mayor and other officials; a limited electorate voted in others.

Not until 1834 did the state allow voters the right to elect the mayor directly. Local democracy had, however, gained ground in other respects after 1776: Jeffersonians prevailed over Federalists, who tried to perpetuate old ways. The legislature gradually overrode the earlier charter of 1731, widening the franchise from freemen to taxpayers by 1810. As elsewhere, Jacksonians had succeeded in extending to all adult male citizens the right to vote by 1840. Aldermen gave up their judicial function. But those who attempted change failed to separate executive from legislative roles by giving the mayor power over the former. City voters themselves went further by making city department heads elective through a referendum in 1849. The controller, or chief financial officer, was elected for two years, the mayor also for only two years. Separating powers may have led to some kind of crude balance structurally, but it did not overcome rivalry or gridlock. The only way out was a near one-party rule by 1860. Rampant liberal democracy opened the way to corruption. What patricians (those with old stable money) remained gave up on the city.

Attempting to control the unruly affairs, the State of New York increased its power in city affairs. The relic of charter inviolability fell by the wayside. Like all cities, New York became a creature of the state. Public and private spheres were more clearly separated. New York lost its old corporate right to own 'private' property, and thus weakened its ability to collect revenue and to provide goods. City politicians had to persuade an often reluctant and rural-dominated legislature to give financial support. On the other hand, the state imposed charter revisions on the city. To the dismay of city politicians, the legislature created a metropolitan police force under a state-appointed commission in 1857. Although New York's famous boss, William Tweed, while a legislator in Albany, succeeded a decade later to have that repealed, his victory was temporary. The state also set up other commissions, including the one to develop Central Park. These acts were reminiscent of legislative action in Philadelphia a century earlier. Only now democracy at the state level was pitted against democracy at the local level, not against an entrenched oligarchy.

As a result of these political divisions, New York state politics was 'resolving itself into the classic Republican upstate and Democratic downstate antinomy' that has persisted to the present.[30] This antinomy would contribute, as elsewhere in the United States, to a strong rural (and later suburban) bias against state financing of large cities and twentieth-century fragmentation in governing urbanized areas. By contrast, in Canada the splits between rural and urban, and between the provinces and cities, were not as pronounced. Provincial authority was not questioned, but, paradoxically, cities were able to initiate and maintain better services as a result.

Another crucial contrast that emerged in this era was the rise of American courts at the expense of legislative power. In Canada, parliamentary legislative power to write acts remained intact. In the United States, the courts increasingly made law, undercutting the legislatures of both state and city. Americans' long-standing litigiousness over property rights had already established judicial power to a greater degree than in Canada. Why? While widespread and then universal, (White) manhood suffrage at all three levels of government led politicians to pander to the voting masses. As in religion, charismatic men could exploit this to advantage. Also, the right of every man to run for office raised the real prospect of corruption. Instead of an élite that had no need to rely directly on political office for income, now ambitious men of modest means found this route to wealth through kickbacks very tempting. Ward-heelers acted like private entrepreneurs. Judges had to rise above the pettiness of politicians at all levels. Yet they ruled more and more in favour of business corporations. Freeing large business from legislative restraints weakened the possibility of economic initiatives by the state, just as legislatures curtailed city councils. A window for court power was opened when excessive state debt financing of canals in the 1830s resulted in defaults. Electors became frustrated with extravagance. More so in New York than elsewhere, judges and corporations would thus dominate American collective life. A small irony in this situation was that lower-level judges were elected, not appointed, and were therefore corruptible.

Another irony was that legislators appointed the top judges, who usually reflected the dominant ethos of the party in power. In New York, that worked against equity in the city. A greater irony was that the liberal open society—with its rampant democracy, where every White male was equal and could vote—demanded a decidedly non-democratic means by calling on judges to resolve conflicts.

Party politics played an increasingly crystallizing role in governing within the city. Professional, self-made politicians replaced amateur patricians and mechanics to create the power machine. Eventually led by Tweed, the Democratic Party bosses of Tammany Hall (named after its headquarters in the city) finally overcame the factions within the party in the late 1850s. Mayor Fernando Wood, a Democrat, was forced out even though he had strengthened local ties through ward-based schools and working men's clubs because Tammany bosses considered him demagogic, a seeker of one-person rule. Tammany Hall gradually increased its strength as a power base in the party. Through mass naturalization of male immigrants, it manufactured new voters for the cause of the Democrats, a party that had ostensibly 'established an open and egalitarian character ...'[31] Tammany Hall gained the allegiance of a majority of voters, especially in the working class. The Whigs, on the other hand, represented the rich and many in the native-born middle class. That pattern did not change when the Republicans replaced the Whigs. Reformers (some Democratic, some Whig) tried to maintain honest government, but to no avail. Other ephemeral groupings—such as Know-Nothing nativists who appealed to downwardly mobile Americans—came and went. Generally in the annual elections (which became biennial after 1850), Democrats controlled a majority of the wards that the council had increased from ten in 1808 to twenty-two by 1860. Despite considerable internal squabbling

> This was an era when corruption erupted, especially in the eyes of the American-born élite. Attempts to eliminate patronage in public jobs from ward-heelers to the mayor failed ...

between the bosses and the reform-minded, Democrats held the mayoralty in fifteen out of twenty years after 1840. Tammany won as 'men of wealth withdrew from ... leadership ... leaving charity to professionals and community to politicians.'[32]

This was an era when corruption erupted, especially in the eyes of the American-born élite. Attempts to eliminate patronage in public jobs from ward-heelers to the mayor failed, but without rules for employment by merit, a shift to the executive would probably not have helped to end patronage. 'Virtually unrestrained, the Council controlled patronage, contracts, revenues, and an array of other matters—enticements to venality as well as burdens to conscientious effort.'[33] Council members of the early 1850s, who were dubbed the Forty Thieves, virtually gave away street railway and ferry franchises, lining their own pockets with the difference between what the city received and what the taxpayers should have expected in revenues from private utilities. Health and building inspectors gave in to bribing landlords. Despite attempts by reformers and the state government to clean up city hall, 'city government was well on its way to becoming, for its critics, the greatest failure in American life ...'[34] Deals on horse-car lines were the 'beginning of the pattern of corporate abuse and political corruption so troubling to the nation and the city after the Civil War.'[35] When Tweed was caught with his hand in the public purse in 1871, many hoped the era of corruption was past, but it was not, the courts, legislatures, and reformers notwithstanding.

This image of conspicuous corruption obscured the positive role of patronage. In an era when unions and formal merit systems of public employment were non-existent, a system of patronage was needed to create jobs either directly or through contracts. Private businesses could not absorb the masses of immigrants or failed to provide secure employment. Government employ-ment was attractive to lower-income men because it paid more and guaranteed greater security than most jobs in the private sector. In 1844 city employees already numbered nearly 4,000 and would increase. Besides, the public rolls included employees in the almshouses, asylums, and pris-ons. Loyal male immigrants and many others received these jobs.

Despite patronage, professionalization and bureaucratization of the civic staff marched for-ward inexorably. Conditions in the rapidly growing city and an increasing concern about the environ-ment demanded it. In the name of efficiency and order, administration was divided into more departments, many under independently elected heads. Although these departments had the disad-vantage of fragmentation and were little auto-nomous empires, officials were able to do tasks with less council interference. In 1845 the police force of 800 men became professional under a local and later a state commission, and expanded. Vol-unteer fire fighters became paid workers. Water and sewer lines required systems. However, not until the early twentieth century in most cities did employment by merit displace patronage, at least in lower-ranking jobs.

Partly because of corruption, New York City's expenditures doubled in the 1840s and tripled in the 1850s to the dismay of the élite and state legis-lators. They had already risen twenty-four times over the first four decades of the century, while the population grew by four times, so other factors were at work, namely, civic improvements. The city came to rely more on taxes than on fees. The tax rate tripled over the two decades, though not as much as Philadelphia's and Baltimore's, and New York's debt was lower than Boston's. Taxes per capita were four times those of the rest of the state, though the rich managed to pay far less than their share. Property taxes were startlingly regressive; in 1850 William B. Astor paid less than $3,000 on an assessment of $2.5 million, though he may have been worth ten times as much. The city instituted special local improvement taxes to pay for services for a particular frontage of property, a practice fol-lowed in many cities for several decades, with

varying degrees of success. Brooklyn allowed an escape from high Manhattan taxes, but they rose there too. Higher taxes put more money in politi-cians' pockets, but it also meant jobs for workers and services for the city.

Expanding City Services to Deal with Problems of Massive Growth

Many thought the city managed its work badly; filthy streets were a major complaint. Others blamed lax regulation for the alarming frequency of steam-boiler explosions. Still, many services were progressive, with officials adopting new ideas. I will look briefly at the federal government's direct role in regulating the port. Then I will consider the following ten sets of environmental and social ser-vices that city council, other public agencies, and quasi-public bodies undertook to facilitate private economic action and improve the quality of life and the environment. The city extended the edges of the island, leased out docks, and smoothed the land surface; imposed a pattern of streets and lots following a plan of 1811 and paved and otherwise improved the streets; controlled traffic and fran-chised public transportation and ferries to private operators; installed water and sewer lines; disposed of garbage; confronted disease; and, together with private charities, ameliorated the lot of the poor. In addition, the city established schools and libraries, created open space, and ran the public markets. If city council participated less directly in running and controlling the economy than it had earlier, private interests could hardly perform outside a context of public action in numerous if often dis-jointed directions. Looking into these develop-ments will give us insight into a public at work.

Harbour Regulations

The federal government regulated the flow of com-merce in the harbour. 'Nowhere outside the national capital are the agents and agencies of gov-ernment as numerous and obvious as in a busy sea-port.'[36] As the nation's chief port, New York had the attention of nearly every cabinet minister in Washington. Custom payments provided the chief source of revenue for federal coffers. In fact, enough was collected at New York before 1860 to pay for all government operations!

Over 700 custom-house officials registered and licensed ships and collected fees in the early 1850s. Revenue cutters roamed the harbour for accidents and illegal landings. Many federal employees worked at the port: pilots based on Staten Island, lighthouse and lightship keepers, semaphore signalmen, public health officers to quarantine ill immigrants from the healthy, harbour master and staff to assign berths for docking and the like, port wardens acting as arbiters in case of damage to vessels and cargos, and so on. Foreign consuls from many countries promoted trade. The Admiralty and Marine courts convened to deal with maritime infractions. Military and naval establishments, notably the Brooklyn Navy Yard, graced the harbour. The post office was a great subsidizer of business. State officials promoted and regulated internal trade and finance. Not surprisingly, most of the federal employees who got the 'the juiciest plums' were 'deserving Democrats', since that party controlled Washington politics for much of that period.[37] These senior government services affected only a minority of citizens. For most people, local organizations, especially the city corporation itself, provided services. We will consider ten sectors of concern.

Landfill

First, to improve shipping in this most dominant of ports, the city filled the margins of the rivers with garbage and earth from excavations. It also smoothed the surface of the land. Beginning in the late seventeenth century, such action from time to time had added more than a square mile of land by 1860. The Outer Streets and Wharves Act of 1798 permitted 200 acres (81 ha) to be created in this way, and then in 1826 the city decided to add another 400 ft (122 m) more around the built-up margins of the East and Hudson rivers. (More was added after 1870—increasingly farther north— and by 1978 a total of over 2,200 acres/890 ha had been reclaimed.) On the lower East River, the landfill enabled the creation of Water, Front, and South streets; on the Hudson, Greenwich, Washington,

and West streets. As a result, the city provided new wharves and space for commerce to expand. New York was not alone, of course, in using landfill; many other cities, probably Boston especially, saw it as a cheap way to create valuable, accessible land. However, New York easily outstripped the other cities in providing wharves, nearly doubling them from 1835 to 1855 and, by 1860, building up to 300 'piers and bulkheads' for the ever-expanding number of steamships and canal boats, half of which were public. Inside the island, soon after 1800, the city also drained the Collect or Fresh Water Pond north of city hall through Lispenard Meadows along the route of what became Canal Street. The land remained spongy for many years, and only the poor occupied the area, but they were forced out eventually when the city improved drainage. Today prominent public buildings in Foley Square occupy part of the pond's site. The city and private developers levelled other less conspicuous impediments as well.

Street Plan of 1811 and Street Improvements

Second, local government's most dramatic input in shaping Manhattan's space followed the imposition of the street and lot plan of 1811. After debating over the plan for some years, the state finally allowed the city to impose a rectangular grid pattern over the island north of Houston Street. (To the south, private developers had already opted for rectangular plans on a small scale.) Surveying began immediately, though the marking out of the streets was not completed until Harlem was reached in 1868. In the process, they wiped out several colonial roads, and only a few streets, notably Broadway, which gradually extended north, cut across the grain. City surveyors laid out this rigid pattern of narrow but long east-west blocks—a pattern that Jane Jacobs would excoriate in the 1960s. Two-and-a-half times as many east-west streets than north-south avenues were planned because the surveyors expected a considerable amount of goods traffic at the ends of these streets on the waterfront, a reasonable assumption at a time when a wharf stood at the end of each occupied street. They would be proven wrong and so Manhattanites have suffered with too few north-to-south routes. Streets were to be 60 ft (18 m) wide, except for every fifth street, and the avenues were to be 100 ft (30 m) wide. (As redevelopment progressed in the old city south of Houston, the city straightened and widened a few crooked streets, though never enough for cross-town traffic.) According to the plan, each block was to contain forty to sixty lots of 25 x 100 ft (8 x 30 m), but, unlike many other new cities with grid patterns, New York would not have rear access by laneways or courtyards, which was a serious deficiency.

The city was able to undertake the task with little opposition, which suggests a widespread consensus, at least among those with power. They agreed that property rights should be overridden in the short term to expedite long-term gains, though many land developers may have wished for more flexibility. This grid template would improve efficiency in settlement and so presumably speed up population and economic growth, as it did earlier in Philadelphia and in the rural rectangular survey of the west, which was well underway by 1811. As was argued at the time, the street pattern and the right-angled lots would create 'greater economy and convenience in building'.[38]

As for street improvements, New York experimented, like other cities, with paving the streets with other than cobblestones, which was the cheapest and simplest method. Wood blocks were tried in the 1830s and later smoother trap blocks of stone. Asphalt did not become a widespread paving material until after 1860, and concrete only late in the century. Upper-class areas fared better in having their streets paved than did poor districts. Many streets remained unpaved, so mud and dust remained a fact of life. In 1823 a private firm began to provide street lighting with gas drawn from Pennsylvania anthracite coal, in the business section and on Broadway, but New York lagged behind other cities in more widespread use of street lighting. Even in the 1880s some streets in the Lower East Side remained dark and therefore 'dangerous and crime-ridden'.[39]

Traffic

The streets, especially the north-south avenues, were filled with passenger carriages, drays, carts, and pedestrians moving, it seemed, chaotically. At closing time in the shops and warehouses, workers spilled out and flowed through the streets on the way home. To an overseas visitor, the business section was a 'throng and rush of traffic ... astonishing even for London' with 'a perpetual jam and lock of vehicles ...'.[40] By 1857 600 licensed hacks and nearly 5,000 carts and express wagons increased this crush of traffic. Widening streets, suggested to alleviate gridlock, would not likely have helped. The city allowed telegraph wires and poles, a sign of progress, though to some, they added even more clutter to the congested scene. Public transportation was one solution to relieving the traffic congestion, though detractors opined that these vehicles only added to the problem.

For the most part, residents still walked, although the occupied area of the city extended 4 mi. (6 km) up the island by 1860. As in Philadelphia, the vast majority walked to work until 1890, but an increasing number of people paid fares to drivers of horse-drawn vehicles for commuting, shopping, or recreation. The horse-drawn omnibus was followed by the horse tram, and both were supplemented by steam rail and ferries. Following the lead of Paris and London, New York franchised at a fee the first omnibus (or urban stagecoach) line in 1830, with coaches carrying a dozen people. Although the first line on Fourth Avenue and the Bowery up to 14th Street was not financially successful, entrepreneurs seized opportunities, so routes mushroomed. Over twenty companies ran nearly 700 licensed buses by 1853, nearly three times more than just seven years

> The streets, especially the north-south avenues, were filled with passenger carriages, drays, carts, and pedestrians moving, it seemed, chaotically.

earlier. Converging near city hall, they ran at five-minute headways, though at rush hours they fell behind schedule and even, as Mark Twain noted, fell 'behind fast walkers'.[41] Too expensive for most people at 12.5¢ cents a ride, the omnibuses carried only a modest number daily.

The first horse-drawn streetcar operating on iron rails appeared on Broadway in 1832, running 1 mi. (1.5 km) from 14th Street south to city hall. It was an adjunct to the first steam rail line that ran northward from near Union Square. Although these vehicles provided a much smoother and faster ride than buses on cobblestones, horse trams did not start to compete seriously until the early 1850s. Omnibus owners had, it seemed, successfully lobbied city councillors against them, but when would-be tram financiers finally bribed them, the politicians franchised tram routes on five north-to-south avenues. By 1857 these one-horse-drawn streetcars carried 23 million passengers, which nearly equalled the number of omnibus passengers. Within a few years, the omnibuses were relegated to cross-town and downtown shuttle routes. The number of commuters by tram remained modest, probably less than 10 per cent of the workforce, as workers could not afford 5¢ fares.

However, on weekends, working families paid to commute to pleasure grounds. Trams took them, for example, to the amusement park in Jones' Woods along the East River, far up north at 66th Street. These trips and those of affluent women shoppers outnumbered those of commuters. (Among the upper classes, lunch near the workplace was only beginning; most people still came home for the noon meal.) The days of mass transportation were still in the future, though visionaries mused over elevated and even underground rapid transit, eyeing London's metropolitan line.

New York rail commuting began, though following Boston. A steam line for commuters reached Harlem in 1837, though the first leg from downtown was on the horse tram. Commuter rail lines extended out even to Morristown, New Jersey, by the 1850s, and to southern Westchester, thus catering to an increasing but still small number of affluent exurbanites. A freight line, the predecessor

of Vanderbilt's New York Central, ran along 11th Avenue as far as Chambers Street, though the steam engines had to stop at 30th Street while horses pulled the cars farther south.

City council franchised ferries too. Several East River ferries, charging 1¢, carried over 30 million passengers by 1860, well over 100,000 per weekday, to and from Brooklyn and Williamsburg. For many, this journey to work by ferry was more pleasant than by omnibus. Those ferry lines to Jersey City and Hoboken transported 7 million in 1855. Ferries brought rail freight cars from New Jersey to Duane and Chambers streets. At that time, the famous Staten Island ferry transported fewer commuters, though on Sundays picnickers and baseball fans enjoyed the trip. Steamships also made daily trips to regional cities. The massive daily movement across the East River began to inspire the notion of a bridge; it would not be long before Brooklyn was linked.

Water Supply, Sewers, and Public Health

Worries about fires and public health led public enterprise to supply water, sewers, and other health measures. As the city grew more crowded, owners of businesses and insurance companies agitated for a more efficient water supply. Industrialists who owned steam engines joined the chorus. Compared to Philadelphia and some other cities, New York was slow to establish a water supply and relied on public and private wells and a local private firm, the Manhattan Company, which was more interested in banking than water. The city built only a few water lines in the business district, and did not even keep track of them on a map.

A cholera epidemic in 1832 had upset many about the quality of water, but not until after a devastating fire in 1835 swept through nearly 700 houses and stores south of Wall Street did leading insurance underwriters push for a solution to the obvious inadequacy of the water service. The city and state together built the Croton Aqueduct in 1842, bringing water from a dammed reservoir north of the city to a city reservoir on 42nd Street. The cost was $12 million and caused a large increase in the debt and taxes, yet its opening was a momentous celebration. After a power struggle, the state took control of the wholesaling of the water to the reservoir, and the city took over the retailing of this water into lines for distribution. By 1848 the city gradually buried 180 mi. (290 km) of pipe that delivered 15 million gallons (57 million L) per day, though theoretically it was capable of twice that much because so much water was wasted. About then a modest number of houses, 253 steam engines, 166 manufactories, forty-one fountains, twenty-four private bathhouses, forty-three hotels, and ninety-one steamboats tapped the water system. Despite these improvements, another serious fire destroyed 330 buildings south of Wall Street in 1845, and many more conflagrations demonstrated the inadequacies of the water pressure system and also of the building code, which did not yet require brick-clad buildings.

Still, the pipes provided a more reliable and purer flow of water than wells did earlier. While outdoor spigots began to supply most people, the affluent began to install indoor plumbing in their residences. By 1856 one in twenty households had indoor flush water-closets. In the 1850s and 1860s the city extended water lines along the streets of the rich first before installing them in poor areas, just as private gas companies added lines on the same streets for street lighting and household use. Before long, these utility lines were everywhere. The sixteen largest cities in the United States operated water systems by 1860; most of them were public because private companies could not earn enough profit. Public systems could extend lines to areas with people further down the income scale, though that did not mean that landlords would take advantage of them. Health officials advocated pure water and the fire insurance industry pushed for higher water pressure, thus driving officials to act. Despite these health and safety measures, fires and water-borne typhoid were not controlled until after the turn of the century.

Just as pure Croton water led the way to better health, reformers such as Edwin Chapin and John Griscom filed a report to city council in 1843 that brought to light grim conditions among the poor. The reformers were moving in tandem with a rising tide of concern expressed by Edwin Chadwick and Charles Dickens in Britain. Two years

later, they published their findings and, among other proposals to improve health, they advocated sewers and better housing, but city council and the public did not respond very vigorously because immigrants were widely blamed for fouling their own nests. Conditions deteriorated further as poor immigrants piled into crowded houses. Subsequent studies revealed increasingly serious problems. Finally in 1859, members of the precursor of the American Public Health Association pushed for improved water, sewers, and housing, so, in the mid-1860s public health prevention was put on a more solid organizational foundation. Nonetheless, it would be several decades before premature deaths were reduced dramatically, as was the case in all large cities.

One of the ultimate solutions—sanitary sewers as distinct from drains along streets—arrived more slowly than the water system. Backyard privies persisted in older, pre-1811 districts. Elsewhere, with no back alleys, householders had to put out tubs on the street for pick-up, as they did ashes. Honeydumpers, as they affectionately came to be known, carted away the night-soil after eleven at night, dumping their collections in the river or onto boats for removal. Affluent areas received the first sewer lines before 1840, but, like water lines, their existence was not recorded carefully. These sewer lines spewed out into the river, polluting the water and air.

In 1849 when the Croton Aqueduct Department took over the sewer business, only 69 mi. (111 km) of pipe had been laid. Among the affluent, the installation of more and more flush toilets in water-closets demanded better sewers. But in 1859 three-quarters of the 500 mi. (800 km) of paved streets were still without sanitary sewers; only after 1865 did New York follow the lead of Brooklyn, Jersey City, and Chicago to create an integrated sewage disposal system. Trunk lines carried the waste to certain sites on the river, but the sewage was still untreated. Even then householders and landlords were under no obligation to connect their properties to these lines. Smells of privies and night-soil in waterfront depositories remained offensive, and were sometimes equally so around ill-constructed sewers. The system was not completed for several more decades. New Yorkers may have been olfactorally challenged, but health officials still didn't have enough evidence on diseases to persuade the taxpayers to pay for more improvements. The city slowly installed separate underground storm sewers.

Waste Disposal and Health

The disposal of garbage and manure from streets and elsewhere remained problematic, especially in poorer areas, and therefore detrimental to health. Although a street-cleaning department was established in 1802, Charles Dickens could still marvel that in this city of progress with its Broadway glitter, pigs still scavenged in certain streets in 1842. By 1854 the problem of manure disposal was still not solved. Manure was produced by pigs and cattle and the horses that pulled omnibuses, trams, and hacks, plus thousands more that pulled wagons, carts, and private carriages. Manure and entrails that piled up from slaughterhouses, private butcher shops, and public markets were often dumped as landfill on the margins of the rivers. The city council sold some of it to farmers for fertilizer. About that time, the city contracted out cleaning jobs to carters in the hope of better garbage and manure collection, though that only increased kickbacks to council members.

Obviously more action was needed in this 'survival machine'.[42] Death rates in the early 1850s were nearly double those of the early 1810s (forty-one to twenty-three per 1,000) as unhealthy living conditions and poor, vulnerable immigrants drove up the numbers dramatically. Among children under five, the death rate tripled to an astonishing 166. Many children drank unclean 'swill' milk from cows that ate distillery mash; Leslie's magazine crusaded against the unclean handling of milk, forcing the city to investigate.[43] However, even in the 1860s the death rate was still double that of the early years of the century.

> **Charles Dickens could still marvel that in this city of progress with its Broadway glitter, pigs still scavenged in certain streets in 1842.**

Disease

The affluent worried about diseases spilling over from poor areas. People became more conscious about health and fussed about diseases such as typhoid and consumption and the filth that could spread them. The cholera epidemic of 1849, when one in 100 died, led to a much stronger public outcry than had the severe outbreak in 1832. Over 1,200 physicians and dispensaries attended to the affluent's needs, and eight hospitals and infirmaries may have lessened the burden for indigents in 1855, but action was slow. General hospitals for the rich and middle class were still a few decades in the future. New York lagged behind Paris, London, Edinburgh, and Manchester, and indeed other American cities in providing public health for its citizens. The creation of the Metropolitan Board of Health in 1866 was a helpful though still inadequate move towards improved health resources in water, sewers, cleanliness, and general disease prevention.

Ameliorating Poverty

Underlying concerns about public health were the enormous problems of poverty and those associated with the 'dangerous classes', to which the city responded in a desultory fashion.[44] Although public health issues were increasingly discussed in the salons of the rich and the parlours of the middle class, few were directly concerned about the plight of the poor. To many, the poor, especially the immigrants, had only themselves to blame. Many thought the immigrants were incapable of coping. To Charles Loring Brace, they were the 'offscouring of the ... most degraded cities of the Old World ... [They were] a thriftless, beggared and dissolute population [whose] vice and laziness stimulated each other.'[45] The notion of the undeserving poor reached new heights.

A few upper- and middle-income moralists, many of them women, sought solutions that would raise the poor from the shadows into the sunshine, which was a favourite metaphor at the time. In the wake of Griscom's report, in 1843 a small band of reformers founded New York's Association for Improving the Condition of the Poor (AICP). They pushed for more action, noting that neither thirty-three charity organizations nor temporary work provided by the city during the depression had succeeded in stemming the tide of poverty. By 1853 ninety benevolent societies struggled to provide aid. For the thrifty poor, seventy-five self-help mutual benefit societies founded by churches and lodges eased difficulties.

The city, through the Almshouse Department, gave 'outdoor relief' to 10,000 people in 1854; the AICP helped too. The city was again called upon to provide relief in 1858 (no doubt inadequately) when one-seventh of the population lost their incomes in the sharp recession, even though public works projects, such as Central Park, provided 1,000 jobs. The homeless, who were then called vagrants, found shelter in police stations; in 1857 thirty shelters put up over 1,000 people every night. Following the fall of Tammany Hall and William Tweed, the city slashed relief in 1873. The able-bodied unemployed had no one to turn to. Charity did not usually extend to them.

In Foucaultian fashion, the city and charities shut away the disabled. Over twenty asylums for the mentally ill and incapable were established by 1853. A more traditional 'centerpiece' of the public charitable 'empire' was the almshouse at Bellevue on the East River north of 25th Street, adjacent to Bellevue Hospital.[46] After much political wrangling, in 1849 the bipartisan Board of Almshouse Governors stabilized administration and increased expenditures. In 1855 the workhouse under its control provided jobs for under 1,000, one in seven of inmates in city institutions, many of whom were probably not considered employable. Charities operated workhouses. The Deaf and Dumb Asylum moved north, from 50th Street and Madison to remote Washington Heights. The prisons housed mostly vagrants and, like asylums, depressed Charles Dickens.

Newsboys in bare feet tugged at heart strings. Worries over vagrant and homeless children led to the establishment of the New York Children's Aid Society in 1853. It sought to take children away from 'bad parents' to foster homes in the country.[47] Brace, the key founder, may have thought such an organization unnecessary a generation or so earlier. Now, however, despite his invective about immigrant sloth, Brace's organization represented a new and long-lasting (if inadequate and sometimes misguided) organizational approach to poverty. New York was no different from other big cities in America, but its problems were more extensive. A few institutions and modest temporary relief were not enough to deal with the degradation. The big city had more poor because it attracted so many looking for a better life. For every family disaster, though, perhaps ten families would be better off materially after initially subsisting in wretched conditions and often losing children to the grim reaper along the way.

To get rid of slums, such as the notorious Five Points not far northeast of city hall, some reformers advocated philanthropic housing schemes. In 1847, following Griscom, the AICP proposed model tenements. Supporters of the idea built two of them, but, as would often be the case with charitable housing schemes, they charged rents that were too high for most workers. To encourage working men's ownership, building associations (which were proposed as early as 1818) blossomed for a brief half decade after 1852, only to see their investors' savings sucked into the vortex of the panic of 1857. Even so, the number of cellar dwellers seems to have decreased by half from 30,000 in the decade after 1850, which suggests some progress in ridding the city of slums. Rather than direct public help or even city housing inspections, though, the weaker immigrant tide enabled those who survived the cellars to move to somewhat better housing.

Concerns over housing for the poor remained on the agenda; the first tenement house act in 1866 legislated building improvements. Interest rose to a peak at the turn of the next century during the last major influx of immigrants. However, even then neither New York City nor any other North American city was ready to support public housing, which was already underway in Europe.

Education and Libraries: The Beginning of Universality

Some reform-minded educators stressed schooling as a major solution for poverty. Just as Franklin had proposed several decades earlier, Americanization of immigrants through teaching moral uprightness was a driving motive. Others saw virtue in education as a way to teach discipline to prepare youngsters for factory work. Also, since apprenticeships no longer provided a way to practical training, schools had to take partial responsibility. Child redundancy in the workforce was the basis of these concerns. Doubtless, even without immigrants, the drive for education rose as much for children and adolescents of the upper and middle ranks as for the poor. Education gathered interest from the early part of the century onwards. The Public School Society, which was founded in 1805, provided classes, but enrolments climbed slowly; by 1840 only one in sixteen school-age children attended. As elsewhere across northern North America, the city and state regarded this as inadequate. In 1842 under the Maclay Act a public Board of Education rapidly expanded the system, so that there were forty-eight schools in 1847. Although the city rather than the state government increasingly bore the costs of education, building schools continued, so that seventy-five schools operated by 1858. The truancy law of 1853 allowed the police to arrest any vagrant child between five and fourteen years. This law may have helped to fill the classrooms; chronic delinquents found themselves in the Juvenile Asylum for correction.

By 1860 New York ranked among the leaders in promoting free and compulsory public education; school enrolments equalled one in five of the total population, while laggards like St Louis had only one in fourteen. As another indication of a rising literacy rate, Manhattan had twenty-seven libraries by the 1850s. Among many schemes for improvement of the poor, Horace Greeley, a noted newspaper publisher, suggested that ward houses provide libraries and lecture rooms. A rapidly rising number of women teachers taught children morals and citizenship, basic skills, history and geography, and a smattering of classics. For the teachers, schools provided one step on the ladder towards equal status in a society that was now creating a new middle class. Quite possibly during the middle decades of the century, the greatest gains ever were made in literacy and numeracy in this, the first and most comprehensive of universal social programs.

Critics of the new system complained that the Board of Education did not serve immigrant and Black children adequately, though these same critics were not reluctant to build showcase schools to attract children of middle-income families. Bishop John Hughes tried to persuade the board to help Irish and German children, but failed to receive public funding because of resistance to financing religious instruction, though Tweed later changed that. Parishes established their own schools with meagre funds. The public school system also had to compete with a plethora of private schools for the children of the rich, many of whom later registered their offspring in the élite, exclusive, Episcopalian Columbia College. Yet countering this, City University, founded in 1831, drew strong civic support for democratic higher education. Despite civic corruption and a dizzy economy, New York was providing (if slowly and still unequally) a measure of civility and stability.

Parks: Lungs of the City

Parks provided part of the answer to public health problems and satisfied the desire for enjoyment. Providing open space was a positive (if halting) initiative that culminated in Central Park. Open-space advocates argued that greenery would dispel miasmas thought to cause disease and allow people to escape from the ever-increasing smoke from steam engines. Even though the city owned one-seventh of the island, plans for public open spaces and for leisure and clean air were slow to be realized. In the 1811 plan, the surveyors set aside 450 acres (180 ha) for parks in eight reserves. To politicians and surveyors early on, parks were not regarded as vital since, first of all:

those large arms of the sea which embrace Manhattan Island render its situation, in regard to health and pleasure, as well as to convenience of commerce, peculiarly felicitous ...

But undoubtedly more to the point, because the prices of land are so uncommonly great, it seems proper to admit the principles of economy to greater influence than might, under circumstances of a different kind, have consisted with the dictates of prudence and the sense of duty.[48]

Private rights to property held sway. Few developers saw commercial or use value in open spaces. So, not surprisingly, only a few parks materialized: a reservoir site, Union Square, a parade on what later became Madison Square, and Tompkins Square. Originally a marsh, Washington Square became a potter's field, a parade ground, and then a park in the 1830s. By 1849 the city had only nineteen small public parks on just over one-third of the acreage that was proposed earlier. Obviously the city had failed to respond on this front; it sold off land that could have been used for parkland in the depressed early 1840s to raise revenue. Some earlier green spaces declined. Landfill expanded Battery Park, but its entertainment centre, Castle Garden, was converted in 1854 to the reception point for immigrants. Bowling Green and City Hall Park were scruffier. The poor, as earlier, spilled into the streets or visited a few commercial pleasure grounds. The city decreed that the bones of the dead should be moved from valuable sites to new cemeteries in the countryside nearby. Few churchyards, like the relic behind Trinity Church, remain today. Parklike Greenwood in Brooklyn, which was opened in 1840, was a harbinger of extensive naturalistic parks, following the development of successful Mount Auburn near Boston. These necropolises served as a sharp contrast to the living city's congestion.

However, gasworks, ships, and factories increasingly enshrouded part of the city in foul odours and smoke. The notion that parks were the lungs of the city gained currency in the 1840s, spearheaded by the landscape architect, Andrew Jackson Downing. Social reformer Loring Brace advocated wholesome, quiet, cultivated space for the poor instead of their seedy, raucous carnivals and street life. The city negotiated unsuccessfully

with the owner of Jones' Woods far up Manhattan on the East River. Then in 1851 the mayor officially put forward the notion of Central Park, which was an idea supported by the press, including the *New York Times*. The state eventually appointed a commission that bought land amounting to 17,000 building lots north of 59th Street for $5 million. It had been used as a place for quarrying, dumping, and bone rendering. It also provided space for the shack towns of the squatting poor and their pigs and goats, and even a stable Black community. The commission later added more land up to 110th Street.

Frederick Law Olmsted and Calvert Vaux designed the park and incorporated English naturalistic notions. The commissioners appointed Olmsted to begin construction in 1857 that would reshape the landscape over the next twenty years. In its early years, the park was used as a place for a 'fashionable parade' of the rich in their carriages, who were moving in just to the south on Fifth Avenue. It also had the effect of raising land values on its east side. Despite advocates who claimed that the park was created for the mass of humanity, at first it was a preserve of the rich. Only later would Central Park serve in part as a playground for working people. Even so, when completed, Central Park signalled a shifts towards leisure, a need for a change of pace in a society addicted to commerce.

Deregulation and Reregulation

Public markets faded as shopkeepers took over retailing. Therefore, the city corporation's role as regulator declined by 1860, as it did in other activities. The passing of the city corporation's traditional role of regulation thus contrasted to the increase in its responsibility for the new public services we have just surveyed. Regulation ensured a modicum of health protection and partially prevented monopolies. Public markets actually increased in number after 1760; they were in nearly every ward by the early 1800s. Community pressure led to a new Fulton Fish Market. By 1840 twelve more markets were established, but over the next twenty years, the importance of markets faded and the buildings deteriorated. By 1870 no retail market operated north of 10th Street, except in Harlem. Washington Market, which moved to West Street on landfill in 1852, became the major wholesale distributor of fresh produce. Accessible to small sloops and the like, sellers and private shopkeepers crowded Washington Market's wharves early in the morning with market-garden produce and seafood. Trains from the north hauled in goods by then as well. The total value of goods sold jumped eight times between 1840 and 1860 as rail increased the area of supply by nine times or more to supply the increasingly affluent population. In 1867 the Washington terminal supplied 5,700 grocery stores uptown, which displaced not only the markets but also many pedlars. The state would continue to ensure the regulation of public health, at least to a point; indeed, as the middle class grew in size demands for hygienically safe food increased.

As for the abandonment of other regulations, in the early depressed 1840s carters lost their monopoly rights and council preference in city work. The Irish were cheaper. Ward bosses could hand out jobs. Locally, free enterprise would reign; at the commanding heights of the economy, higher levels of government would regulate corporations, as it turned out, largely in their favour, just as local rules had helped the carters to control goods movements earlier.

The Rise of Record Keeping

In the midst of apparent political chaos, public officials rationalized and systematized their accounts. As earlier, tax records remained as certain as death records. Calculations on the quality of life in the public sphere increased during the 1840s' depression when health and social reformers undertook the first crude social surveys. The city began to keep track of its facilities, such as its water lines. The decennial censuses added more items, and the New York state census of 1855

> Calculations on the quality of life in the public sphere increased during the 1840s' depression when health and social reformers undertook the first crude social surveys.

revealed a good deal about social conditions (and hence it has been used extensively by social historians of recent times). Building up a profile of the city for its officials was a great step towards improving conditions and creating the modern bureaucratic civic order. Without numbers, tables, and maps, the world would have been left in the hands of the impressionists. Progress demanded more. Despite losing its constitutional corporate autonomy of the colonial era to New York state, city council and its officials initiated many programs. The state may have restrained excesses, but it passed many laws at the city's behest to give it power to control public spaces. In turn, the city's actions allowed private enterprise to flourish, especially in real estate.

Property: Few Developed, Few Owned

The overwhelming importance of real estate development nearly wiped out the hope for equitable public amenities. Many naively expected that the plan of 1811, by speeding up the possibility of settlement, would rid the city of monopolistic landholders and foster an egalitarian society, thus fulfilling the republican promise. However, as in the case of western rural surveys, speculators controlled the process of settling by buying government land. The public purse provided amenities to owners, surveyors, the land planners of the era, and architects and contractors, who produced housing and other buildings first for the wealthy. Private property was a central pillar of life in America from the beginning; in cities it was even more intensely pursued when cities and their economies grew in the nineteenth century. The greater the growth, the more money there was to be made. Where else but in Manhattan would urban land be so enthusiastically transformed from rural ownership and use? In 1860 the land in Manhattan belonged to a few powerful landlords, and was perhaps even more concentrated than land in the Philadelphia of 1760. To understand why, let us consider the path of land value.

Real estate values rose in spurts and then fell. The imperfect city assessments of real property are the only obvious comprehensive measures of market value, and likely underestimate the value of many-upper income houses and commercial buildings. Let us look briefly at the perils of analysing data. In 1817 the total assessment was just under $60 million; by 1850 it rose to about $230 million. On a per capita basis, the amount actually fell, from about $525 to $400! To account for what appears to be a contradiction—that real estate was in demand and indeed that the city was growing rapidly, but the assessment fell per person—we can only speculate. Prices during booms and depressions seemed to influence assessors strongly. By 1823, after a severe downturn, the assessed value fell to $50 million, but then gathered steam to an unstable peak of $234 million in 1836. Most of the gain was actually garnered over the previous two years. It seems that when opportunities for investment in productive activity dried up from overproduction and underconsumption, those with money purchased land. Speculation in real estate was irresistible for many in the mid-1830s, as it would be in the 1880s, 1920s, and 1980s. Then the inevitable panic followed by depression wiped out illusory value. The assessors dropped values markedly during the depression of the early 1840s. Assessed value of property fell to a low of $165 million in 1843 (per person from about $850 in 1836 to $475 per person), then rose only slowly until 1850, as presumably more dollars went into the production of goods than into land.

Certainly, construction continued, though at too slow a rate for New York to accommodate the great influx of immigrants, particularly in the late 1840s. Then real estate values rose again to a peak in 1857, only to fall after another panic, though far less dramatically than after 1837. The total assessment, including the more stable share of personal goods (about a quarter), reached $577 million in 1860, nearly twice that of 1836. In the long term, the trend would continue upward, with another major spurt in the 1880s.

Some landowners with deep pockets survived even the depression after 1837. John Jacob Astor was New York's most successful developer and *rentier*. Indeed, when he died in 1848, he may have been the richest American and among the top five in the world, worth an astronomical $20 to $30 million when few were millionaires. (In fact, the term 'millionaire' was coined at the time.) This peasant boy from Waldorf, Germany, had by 1800 profited handsomely from the fur and China trades, and then from banking and insurance. He knew that New York's land would appreciate, and claimed that he put two-thirds of his profits into land. Until the 1830s he bought land for development and then when it was built on, he earned rents on leases. Over the next half decade, he sold some land, often as creditor. In that speculative period as values rose, many found buying more attractive than leasing property. During the depression following the panic of 1837, Astor foreclosed on the many who failed, more ruthlessly than any other creditor, it is said. Through various means, Astor acquired nearly 500 parcels of property by 1848 when he died, earned $1.25 million in rents over the previous decade, and made more money in selling to others. He did little building himself. On his deathbed, his only regret was that he did not buy up 'every foot of land on the Island of Manhattan'.[49] He willed a division of his estate among his family, but most land went to his son William B. Astor, who was worth at least $25 million in 1860.

Others did handsomely. Second only to Astor as a landowner, Thomas E. Davies in 1840 bought 400 lots beyond the built-up area for between $200 and $400 each, and some of them rose in value fiftyfold over the decade. Ex-mayor William Brady and Amos Enos successfully kept their eyes open for land up the spine of the island, where the rich increasingly located their homes. Subdivider and builder Samuel Ruggles continued to add to his housing stock during the depression, then turned to railways. Land speculation and renting land was not for the faint of heart; only a handful of the nearly 15,000 landholders in 1854 stood on the pinnacle.

Control over property thus became more concentrated, especially after downturns that left the richest who had been cautious with more. More householders became renters; the artisan property owner of the colonial era was now infrequent. By 1853 the *Tribune* estimated that only one in eight families owned their homes, about 50 per cent lower than in Philadelphia in 1760. The power of landlords was nearly unlimited. Seeking better or cheaper housing, the poor filled the streets on May Day, topsy turvy with carts and wheelbarrows laden with belongings. The downturn after 1837 led to so much hardship that leasing landlords, as the middlemen, had to relent on rents in many cases, though Astor showed little mercy. Renting merchants and manufacturers fought to redress the balance of power in the courts, to reduce the ancient landlord's right to 'distress', that is, taking all the personal goods and tools of a laggard tenant not just his space.[50] Although the courts modified landlords' power somewhat, it did not deter the powerful.

Subsequently, New York would easily remain the largest renting city on the continent with the fewest home-owners and owners of property housing businesses. Opportunity in the big city overwhelmed the desire to own. Also, high rents limited the chance to save. With few exceptions, as we will see later, that pattern persisted: the larger the city in America, the lower the home-ownership.

Landowners were developers (to use the modern term), the primary organizers of the disposition of property. They subdivided larger parcels into building lots, the plan of 1811 making the job easy for them. In this era, they were often *rentiers*, who directly or indirectly charged tenants rents. In Manhattan a tiered system of leaseholds had emerged in the eighteenth century: developer/*rentier* to chief landlords to sublandlords to the tenant of a building, even to subtenants. In the earlier years the basic leases were for long terms (following English custom), usually twenty-one years, but even up to ninety-nine. Frequently chief landlords rather than the developers were the builders, who could trade their leases. Trinity Church vestrymen, content to collect ground rents, leased out various parcels of farm land they owned to developers for long terms, beginning in 1734. Later, critics accused the church of being a slum landlord. The chief landlords built housing or commercial buildings, then leased them. They hired carpenters and masons to do the actual construction.

After 1790 rapid growth and the loosening of earlier bonds brought change to cities as they had earlier in rural areas. The number of long leases declined. Sales of land became more frequent as land appreciated. Lessors discovered that no one would build on let land with less than a ten-year lease, so they sold rather than held onto land that had relatively declining returns. Speculative building too became more common. Instead of building for confirmed customers, some developers hired master builders, who in turn hired carpenters, bricklayers, and masons to build in anticipation of further growth. By the 1820s two-fifths of tradesmen worked on construction. Mortgage financing increasingly greased the market, one-third from companies and two-thirds from individuals in 1831. Fire insurance companies increasingly lent mortgage funds. The rich did not buy speculative housing; successful competent builders could count on putting up the mansions of the rich, with indoor plumbing and central heating after 1840.

The pressure of population and economic growth created an ongoing housing crisis. It was hard to keep up. To increase revenues, builders reconstructed much of lower-end Manhattan by replacing old wooden-clad buildings (still half the housing stock in 1830) with taller brick structures for housing. They also transformed areas from mixed residential and commercial to the latter alone. Great fires in 1835 and 1845 presented opportunities for higher buildings. To further reduce costs, builders raised the height of new structures on the edge of the city during the depression of the early 1840s; instead of two- and three-storey buildings, four-, five- even six-storey 'tenements' became more frequent, and hence more crowded with people. By 1864 possibly half the population packed themselves into these tenements, which deteriorated quickly.

Developers were so intent on squeezing in as many units on the narrow and shallow lots as they could, even in proposed high-income areas, that few developers provided open spaces to supplement the meagre contributions of the city. In 1804 the wardens of Trinity Church set aside St John's Park (Hudson Square) on the west side, only to turn it over to a freight rail depot in the 1860s. Samuel Ruggles saw the virtue of public squares for land values, and therefore created a high-income district around Gramercy Park in the early 1830s. Few others emulated him. Vauxhall Gardens was among a few private pleasure grounds that provided recreation for working people, but that area lasted for only a few years. The owner, John Jacob Astor, sensing appreciation in land value, overrode the rights of the lessee and transformed Vauxhall Gardens into building lots for elegant Lafayette Place. The failure of private developers to include open space, together with a rising interest in public health, forced the city to build Central Park. The planners of Central Park, like private developers farther down the island, displaced poor people living in shanties on what were then the margins of the city. Before we consider where people lived, however, let us consider the spatial economy: the financial and commercial district downtown, highlighting Wall Street, retail strips, especially Broadway and the Bowery, and manufacturing areas (Figure 4.2).

Downtown: Where Money Changed Hands

As in Philadelphia a century before, New York had a defined centre, though it was extremely attenuated by 1860. In keeping with my argument that government prepared the way for business, we must fix our gaze first on public buildings. The city hall of 1811 was the 'most superb building' in the country, set back in the 'verdant' triangle formed by Chatham (later Park) Row and Broadway.[51] It remained the centre while, as Philip Hone worried, most other buildings were torn down. Remarkably, it remains there today. Omnibuses and trams converged at the triangle. Earlier in the century, marchers made it their destination for speeches. Later came the Tweed courthouse (nicknamed after Tweed) of 1872, which still stands behind it. The

federal presence was very conspicuous in the custom-house at Wall and Nassau streets. Built in 1842 on the old city hall site where Washington was inaugurated as president in 1789, this 'Grecian temple' continued to impress the citizens and provide many patronage jobs.[52] Nearby stood the New York branch of the Bank of United States, which, when closed, became the assay office. In 1861 the post office was still cramped in an old Dutch church on Nassau near Liberty and Cedar. On the underside of public life, local politicians constructed new jails and almshouses beyond the city, as much out of sight as possible.

New York was, according to an English traveller in 1838, the most 'concentrated focus of commercial transactions in the world', reflecting the enormous growth in business over the century since 1760.[53] Hence, businesses diversified and specialized, occupying more space and organizing themselves in spatial clusters. Financiers and their business district became a central influence in the economy. Back in 1760 there were few 'capitalists', those who invested and lent money as their work. Merchants, as often as not, operated as bankers, lending money. Now banks and stock markets with an array of specialized money dealers became central to the economy. By 1860 many people engaged in money transactions that were intangible, remote from material goods. Control of society shifted towards these specialized financial standard-bearers, most notably on Wall Street, and spilled over into adjacent streets such as Broad, lower Broadway, and Hanover Square.

The Bank of New York, the Manhattan Company (predecessor of Chase Manhattan), and City Bank made their mark on the cityscape with their great Grecian columns. New York's stock exchange is reputed to have started under a buttonwood tree on Wall Street in 1792. The Tontine Coffee House at Front and Wall was as the locale for brokers, but the rising intensity of exchanges demanded (as it were) impressive buildings built for the purpose. Superseding an 1827 structure, in 1841 the Merchant's Exchange was built on land levelled by the 1835 fire that swept over 20 acres (8 ha) from Wall Street southward. Its great columns and dome, along with the banks, competed with the spires of churches for prestige. This was the 'very sanctuary of Trade and Commerce' where 'the wealth or poverty of hundreds of thousands—the price of stocks, the quotations of flour and cotton, the plenty or scarcity of money—all will be decided.'[54] As another sign of specialization, the Produce Exchange separated from the stock exchange and located on Broad Street in 1851. Although many insurance companies lost their records in the great fire, they too built imposing structures. The increasing number and speed of transactions, and the greater need for accurate and speedy recording, made it necessary for firms to enlist armies of clerks.

Exclusive clubs replaced coffee-houses by 1860 as the meeting places for one-on-one deals. Although the clubs charged high fees, 50,000 of the very successful and their highly paid managers belonged to about a hundred of these clubs by 1870. Some, like the Union, which was founded in 1836, moved uptown in the mid-1850s, though many remained near Wall Street. The Chamber of Commerce fostered collective support within the business community. Blessing all this heady activity at the top of Wall Street on Broadway was the new Trinity Church, which was finished in 1846. Maybe some even prayed there; during the panic of 1857 on the street, it was noted that 'every man carries Pressure, Anxiety, Loss, written on his forehead ...'[55]

The commercial city of 1760 was now far larger and more conspicuous than it was earlier, and the specialized financial sector had carved out a niche that was the heart of the economy. Merchants increased in number, many from overseas and a disproportionate number of Americans from elsewhere. New Englanders, such as Joseph Howland from Norwich and the Lows from Salem, sought their future in Manhattan, as had Franklin in Philadelphia some decades earlier. As the over-

> **The commercial city of 1760 was now far larger and more conspicuous than it was earlier, and the specialized financial sector had carved out a niche that was the heart of the economy.**

whelming importer of goods from overseas and the leading exporter, New York's more than 300 piers and bulkheads (enough to dock at least 800 foreign cargoes) spread up the East River and on the Hudson. Brooklyn, Jersey City, and Hoboken together could handle as many. Canal boats from Albany and the West and from the coalfields found berths in New York. Trains brought goods in along the Hudson, to Madison Square, and to the New Jersey side. Cotton from the South and textiles from New England were added to goods that were unloaded, the former for re-export, the latter for the clothing industries. By 1860 the hustle of New York could not be matched anywhere.

Wholesale merchant operations now specialized much more than they did earlier: nine out of ten sold a single line of goods rather than the three or four out of ten in 1790, but generally they still operated as partnerships or alone. Over 400 commercial firms—which traded on their own account in foreign trade and more than twice as many commission firms, plus British manufacturers' agents—operated in 1840 from floors of warehouses or offices. Commission agents (or factors), the jacks-of-all-trade who worked on consignment, occupied a position between the merchant who actually acquired title to goods and the broker who handled the goods of others. New York brokers predominated over commission merchants in 1860. Ancillary functions expanded. The spatial expansion of warehouses was striking, as were the specialized districts that emerged.

On lower Broadway in the 1840s, importers of merchandise such as glass and hardware, shipping agents, and transportation companies occupied old mansions. In Beaver Street were flour, grain, and provision merchants, near Delmonico's restaurant, started in 1827. Wine traders clustered on lower Broad Street. On Pearl Street were iron and other hardware traders, such as tin plate, stove and hollow-ware, and jewellery merchants. Tiffany and Young began in 1837. Early in the century, along lower Pearl in brick warehouses of three- to five-storeys, the auctioneers and textile jobbers set up operations. Textiles were an outstanding mark of the era, as an élite woman's wardrobe took up to 100 yds (91 m) of cloth, eight times what it took in 1800. Lining the stretches of Water and Front streets were the wholesale grocers and mechanics connected with shipping, along with the cotton, wool, and leather commission merchants. Farther north were the drug and paint sellers. Importers of West Indian, South American, and Chinese goods occupied the area near the site of the Brooklyn Bridge. The Schermerhorn block, on Fulton Street across from the Fulton Market, which is now preserved as part of the South Street Museum, is an impressive example of early nineteenth-century warehouse architecture.

By 1860 the warehouse district expanded by leap-frogging to the west north of Wall Street across Broadway. Dry-goods merchants began to move from Pearl Street to the 3rd ward in the late 1820s, closer to the Hudson where more goods were handled, and near hotels and retail shops. They built warehouses with palatial marble facades, replacing the old brick houses on Liberty Street. Urban renewal was, as Philip Hone rued, a fact of life. As the district expanded to the north, the first cast-iron buildings were put up in the 1850s. Some wholesalers moved across the Hudson to Jersey City. Buildings built exclusively for offices appeared. Rising land prices forced some of the moves as demand rose. Most important for the central area, the owners of land were slowly but surely transforming earlier merchant quarters into the financial district, spreading outward from Wall Street. Investment money was more important than goods and could therefore command higher-value lands.

Sumptuous Broadway and Worker's Bowery

To Charles Dickens, Broadway was already a glittering street by 1842, lit at night by gas lamps, a place for those with money to promenade in their finery. Its hotels, stores, and entertainment palaces extended northward up the spine of the narrow island, by 1840 to Union Square, by 1860 to Madison Square and beyond. Hotels served the city's huge number of visitors, who included hinterland merchants. They came, not only to buy wholesale goods but also retail novelties, and to enjoy the bright lights provided by the tourist entrepreneurs. The first great hotel was the Astor House, opened in 1836 with 300 rooms and up-to-date gas lighting, indoor plumbing, and later fans that wafted cool air off ice. At Broadway and Vesey near city hall, Astor House signalled the drift northward. Later in 1853 on Union Square, the Everett House catered to the literary and theatrical clientele who sought entertainment in the emerging Ladies' Mile. Still farther north, several hotels sprang up to serve travellers from the Madison Square rail depot. Over fifty hotels served Manhattan in the early 1850s, half of them on Broadway. Large hotels between 23rd and 24th streets were considered 'gigantic structures', such as the St Nicholas with its 800 rooms. The Fifth Avenue Hotel opened in 1856, the first building with an Otis elevator. Four hundred servants and fifty carriages catered to the visitors who absorbed its 'grandeur'.[56] These magnificent hotels 'represented a constantly advancing frontier of almost sinful luxury'[57] (Figure 4.3).

Broadway was already the prime retail street for America, as the country superseded Britain as the leading consumer society. 'Separation of the retailing and wholesaling districts' began in the late eighteenth century.[58] By 1860 residents could see clearly defined retail districts. The most obvious new features that reflected greater affluence were women's clothing and dry-goods stores. For a generation early in the century, Catherine Street, where Lord and Taylor started in 1824, was one focus of fashionable shopping, but this area was subsequently engulfed by the tidal wave of immigrants, thus creating the Lower East Side.

In the late 1820s high-class shopping was redirected to the west along Broadway, at first mostly south of city hall. Other cities had fashionable retail streets, but people from the whole con-

The above Engraving represents the interior of the most splendid establishment of its kind in the City of New-York—in the Union—probably in the world. Its decorations and furniture cost upwards of $16,000, a larger sum than was ever before expended upon any Hair Dressing Establishment since time began. The Fresco work of the ceiling, by DE LAMANO, is a brilliant triumph of a brilliant Art. One side of the Apartment is walled with $5,000 worth of mirrors. The washstand, with its statuary, costs $1,500. The marble floor $1,500. The magnificent shaving chairs $100 each, 15 in number, and the silver toilet services about $2,500. All luxury and comfort that taste, skill and money could crowd within the space, has been accomplished, and

PHALON'S

HAIR DRESSING ESTABLISHMENT,

IN THE ST. NICHOLAS HOTEL, NEW-YORK,

Is as great a lion in its way as the palace of St. Mark in Venice. The Baths, on the lower floor, were fitted up at an expense of $5,000, and are the finest in the city.

4.3 Luxury *circa* 1852: Phalon's Hair Dressing Establishment, St Nicholas Hotel, Broadway between Broome and Spring streets. *Source:* W. Roberts/Museum of the City of New York.

tinent shopped and craned their necks here in New York. Signs and plate-glass windows beckoned buyers: books, jewellery, upholstery, hats and caps, tailored suits, and millinery. Sandwich-boards advertised wares. Middle-class women could now benefit from what the *Tribune* called a 'revolution in the price of fashionable clothing', imitations of Paris for half the price.[59] Little wonder it was 'not less an avenue of business than the promenade of beauty and fashion'.[60]

Alexander Stewart opened the first department store in 1846, just north of city hall on Broadway at Chambers. Compared to earlier stores, this elegant Marble Palace was a 'real sensation' architecturally and socially, for the ladies a 'beautiful resort in which to while away their leisure hours of the morning'.[61] It had mahogany counters and a grand staircase that others would emulate. Three hundred salespeople worked on 2 acres of floor space. Stewart laid down a strict code of behaviour for the employees, who included a growing number of women. The street outside became so busy that one retailer facilitated access to his store by building a bridge across Broadway, which was often so clogged with vehicles that it was dangerous to cross on foot.

But not long after Stewart established his department store, the locus of energy moved north. Just before 1860 the fabled Ladies' Mile of the latter half of the century began to take shape north of Union Square at 14th Street. Even more so than downtown had been earlier, 'it was the place to go to, to be seen, to promenade, to be entertained, to eat, to converse—and even to protest ...'.[62] Shopping was the prime entertainment, so retail activities were central in its formation: people wanted to be seen in the finest stores. Dominating the scene were new department stores, the 'archetypal symbol of the rising bourgeois culture'.[63] Arnold Constable, Lord & Taylor, and Rowland Macy (on 14th Street) ushered in the retail strip in the late 1850s, displacing residences and small retailers. Macy introduced fixed but lower prices and cash-only sales, and advertised vigorously, all of which became hallmarks of the department stores in later generations. Detecting the change, Stewart moved his store uptown on

Broadway soon after 1860, building an iron-fronted, five-storey dry-goods palace between 9th and 10th streets. Perhaps he should have moved farther uptown than he did for, although the store was providentially watched over by the elegant Grace Church across the street, it later failed. Nevertheless, 'the largest department store complex in the world had been created'.[64]

A host of entrepreneurs set up shops and services related to dress and other luxury goods, such as home furnishings, on Broadway between 10th and 23rd as well as on adjacent streets. Clustering brought more customers. Union Square was, for example, the focus for piano sales: Steinway opened there in 1853 and was followed by others. Also on sale by then were sewing machines for home production, as were dress patterns. Shopping on Broadway was a special event, except for the élite who lived nearby. Other residential districts required grocers, bakers, dry-goods, hardware, and drugstores.

Broadway became the strip for commercial entertainment, without doubt the most magnificent on this side of the Atlantic. New Yorkers 'had developed a taste for theater in all forms—opera, drama, light comedy, variety shows, burlesque, and circuses', and minstrels.[65] Although the Swedish nightingale, Jenny Lind, sang at Castle Garden in 1850, by then entertainment was moving uptown. For the upper class, Astor Place Opera House of 1847 was followed by the Academy of Music at 14th and Irving Place. Built in 1854, the Academy of Music seated 4,600, though it was considered less plush than European opera-houses at the time. By 1860 several legitimate theatres had created the famous district that would later move farther north. New Yorkers and the flocks of visitors sought novelty, though not entirely; Shakespeare was popular. Newspapers began to hire drama critics in the 1850s. Following Jenny Lind, other 'stars' of music and theatre from Europe toured the continent, starting in New York. This area of entertainment attracted a Delmonico branch and other famous restaurants uptown.

To delight the masses (who now had a bit of discretionary money), that great master of humbug, showman P.T. Barnum, moved his museum of curiosities uptown to 26th Street from just south of city hall where he started in 1841. Franconi's Hippodrome, at Broadway and 23rd on Madison Square, attracted audiences in 1853 with steeplechases, stag hunts, and circus attractions. The Roman Hippodrome followed. At Bryant Park, at 42nd and 6th, the Crystal Palace, which was modelled after Britain's and similar to one in Montreal at the time, displayed goods from many countries. It was a harbinger of later world's fairs. The observatory nearby was the tallest structure in the world at 350 ft (107 m).

But Broadway and its environs fostered other activities. Farther downtown, on West 10th Street, the first commercial art studio was established in 1857, which marked the beginning of graphic advertising. Greenwich Village was then emerging as the bohemian district. Brothels, the 'temples of love' on streets near Broadway's hotels, catered to gentlemen of high standing. Those serving this clientele moved uptown too, where 'cruisers' hooked many visitors.[66] Landlords happily rented to high-paying whorehouse madams, who paid protection money to ward bosses and judges. On Broadway 'confidence' men led unwary farm youth down the road to corruption.[67]

Joining Broadway at Union Square and running southeast to Park Row, the Bowery was 'New York's plebeian boulevard, the workingman's counterpart to fashionable Broadway'.[68] There 240 different trades catered mainly to poor families, most of whom lived in dingy quarters and worked in shops and factories in the Lower East Side. The shops and trades also served the not-so-poor, who came looking for bargains in second-hand stores. The Bowery became a street of action at night for young people who sauntered on summer Saturday evenings. The 'b'hoy' was a 'swaggering, cocksure working-class dandy' and the 'g'hal' strolled with 'a swing of mischief and defiance'.[69] In order to afford to dress flamboyantly, young women often

> Streets were the venue for celebrations that drew excited crowds. In times of stress, the streets provided the space for political protests and parades, even riots.

scrimped on food. In the winter the young met at music-halls and bordellos. Earlier in the century, visitors frequented the famous Bowery Theatre, oyster bars, and saloons. Their numbers declined by 1840, however, because, as Walt Whitman observed, the Bowery seemed to threaten middle-class sensibilities.

In Bowery saloons, young men gathered to engage in political talk and organize demonstrations. Of the nearly 6,000 licensed drinking places and countless illegal ones, more were in the lower city than in the newer wards to the north. Many opened on Sunday, giving the clergy something to worry about. Middle- and upper-income Americans formed the temperance movement at the time, but measures to control taverns and eliminate Sunday openings failed in large part because immigrants would not countenance them. Several decades of action were needed to achieve the goal of prohibition.

Streets were the venue for celebrations that drew excited crowds. In times of stress, the streets provided the space for political protests and parades, even riots. City hall park, appropriately, was often the destination of marchers, as on the Fourth of July. After 1850 it was often the starting point (and Union Square the destination) when celebrators and protesters wished to impress the wealthy who had moved uptown. Three huge riots erupted in this era. The Astor Place riot killed twenty-two people in 1849. It was triggered by two interpretations of Macbeth! When an English actor performed at the Astor Opera House, anti-English provocateurs proclaimed the virtue of an American actor's interpretation. They, along with the Bowery Boys (a gang) and other working men, surged from the nearby new Vauxhall Gardens to Astor Place. The police and militia failed to control them. Later, a long-term feud between the Bowery Boys and another gang, the Dead Rabbits, ignited a two-day Donnybrook in 1857. State militia had to be called to help the police put it down. The third large riot was sparked in 1863 by workers protesting the draft for military service; they attacked the homes of the affluent, whose sons could buy their way out of the draft. However, frustrated by failure, before long the worst of the ruffians began attacking defenceless Blacks.

As Dickens and others noted, the social gulf between the Bowery and Broadway was enormous. 'The age proclaimed equality among all men; reality proclaimed otherwise'.[70] The rising middle class feared the Bowery. For many who toiled in making things, that street was their chief pleasure lane. Still, it did not quite signal Disraeli's two worlds: for many in a growing economy, the Bowery was a starting-point to reach upward, though few would rise far.

Industrial Districts: Sweat, Smoke, and Stench

New York produced the largest number of goods by value and its manufacturers employed the largest number of workers, as we saw earlier. Much of the organization of manufacturing had passed from craftsmen's hands to that of more impersonal entrepreneurs and managers. In the depression after the 1837 panic, half of the city's craft workers lost jobs, and the proportion of manufacturing (compared to all jobs) fell sharply. The widespread attempt at that time—to shorten the twelve-hour day in factories—foundered. Bosses made employees work longer hours than was the case in Europe, which was a curious anomaly since America was supposed to be labour short. Subsequently some workers won ten-hour days, but not many in manufacturing, especially those producing piece-work.

Every ward was a beehive of processing, due to the pervasive presence of workshops. Concentration occurred in some areas: the 2nd and 3rd wards easily ranked first and second, having lost much of their residential population. The 2nd was considered the 'great workshop of the metropolis', a jumble to outsiders, sweatshops mixed with warehousing, often in the same buildings.[71] The 1st ward at the bottom of the island still probably

had some workshops, as it ranked high in employees in 1840, though the record is missing for 1860. Substantial numbers of manufacturing employees worked in wards to the north and east, namely the 14th, 6th, 8th, and 7th in that order, though the numbers there were far behind those mentioned earlier. Like other small-scale, labour-intensive work such as manufacturing caps and hats, boots, shoes, and furniture, clothing production was somewhat dispersed, though concentrated in the 2nd and 3rd wards. Women sewed at home in the poorer wards, as did male tailors. Workshops in the 3rd and 5th wards made furniture and musical instruments. Higgins, a large carpet-making operation, located at 40th Street and the Hudson River, attracted many workers to what later became Hell's Kitchen. Just to the south was a great *mélange* of traditional polluting and space-consuming industries—abattoirs, rendering plants, breweries, and distilleries, as well as lumber yards and manure-gathering lots, all of which gradually moved north as land values and sensibilities rose.

Capital and land-intensive industries, notably in metalworking and machinery, remained strong, but had shifted to satellite industrial cities or suburbs that formed an industrial belt around Manhattan when land costs rose. At various places, developers laid out what would later be called 'industrial districts'.[72] The large factories in these areas used steam for power and billowed out smoke, while in the city only one-quarter of the shops were powered by the century's revolutionary source of energy. Among the metal and machinery operations in the city, few were large scale at the time. Novelty Iron Works, which was on the East River in 'dry dock' 11th ward, employed nearly 1,000 men to make steam engines, marine hardware, and the like.

Nearby was shipbuilding, which by then was second only to the Clyde in Scotland, but destined to fade. As the yards expanded, they moved northeast along the East River, nearly to Corlears Hook, and then around the bend to the north by 1840, in order to handle ever larger ships. Across the river in Williamsburg was the navy yard. Several large companies, like that of Isaac Webb and his son William, built sailing-ships and more and more steamships. These yards drew many workers, over 3,000 ship's-carpenters in 1860, in addition to riggers and rope makers. Crowds cheered lustily one winter day in 1851 when the Webb yard launched no fewer than three ships.

Manhattan easily led American high-value and high-pay publishing and printing by 1860, as it did the culture and information industries generally. The first penny press, the *Sun*, which was established in 1833, opened a new world that catered to a literate if sensation-hungry audience. Regularly employed skilled workers with newspapers, 'the élite of city workers', received the highest pay, while the more sporadically hired job printers received much less.[73] Unlike space-consuming shipyards, these occupiers of less but more valuable space were clustered near the founts of business and legal and government decision making. Printing House Square became the site for these specializations. Just east of city hall were the great newspaper offices and printing plants. The size of buildings increased with new, faster cylinder presses; in 1852 the *Tribune* installed a six-cylinder press that ran 15,000 sheets each hour. That was soon surpassed. Newspapermen hung out at Earle's Hotel. An indication that it was mainstream, the American Tract Society was close by. Not far to the east was Franklin Square, where Harper's prominent five-storey publishing house stood out. Job printers, with less certain work, located in the interstices, as did type-foundries and other associated services. Engravers, the predecessors of typographers and photographers, continued to operate mainly in the same areas, south of city hall down to Wall Street, at least in 1840 as in 1800. Often they lived where they worked, though by 1860 some were commuting by horse tram. As earlier, many processing operations (like bakeries) operated in residential areas. Overall, services were sorting out into the high-value national or low-value ends of the manufacturing spectrum.

The Bounding and Unbounding of Neighbourhoods

Residential densities and locations of people by incomes reflected the property and business dynamics of the city. Densities began to rise, after 1820 especially, peaking in 1854, after which immigration slowed down. The overall density in 1860, discounting the thinly populated 12th ward north of 86th Street, was 110 people per acre, about the same as Philadelphia's densest wards in 1760. The rise in density can be attributed first to the enormous population increase that pressured existing housing in a city where most people walked and therefore needed to be close to work. Second, the expansion of the business district after 1825 (in the 2nd and 3rd wards especially) forced out residential uses. Third, developers and builders infilled backyards and then, from the mid-1840s on, built upward rather more than outward. The first tenement, which was designed for multiple occupancy, may have been a row of ten five-storey structures built in 1843 in the poor Lower East Side.

Builders expanded the city northward on the constricted island, constructing row after row of townhouses and mansions for the rich. Contiguous occupation reached Greenwich Village by 1825, 14th Street and Union Square by 1840, and even farther on the west side. In 1855 over a third of Manhattanites lived beyond 14th Street, nearly as many as had lived in the city in 1840. New construction displaced market gardeners, squatters, and even outlying mansions. In displacing them, Manhattan was hardly atypical, though the speed of growth, density of settlement, and the row housing of four storeys was exceptional (Figure 4.2).

Residential densities in twenty-two wards ranged from 5.1 to 310.4 per gross acre by 1860 or, assuming 35 per cent for streets, open spaces and non-residential buildings, net densities of 8.3 to 477.0 (Figure 4.4). Some blocks housed more. There were five low-, eleven medium-, and six high-density wards. Business displaced residences in two wards, thus lowering numbers greatly. The number of residences in the 2nd ward actually began to diminish soon after 1825, so that by 1860 they represented only a quarter of the ward's earlier peak in 1825. The 3rd ward began declining after 1845, indicating the intrusion of warehouses and workshops and the displacement of residences.

The other three low-density wards—the 12th, 19th, and the 22nd—were on the periphery (north of 40th Street), indicating that occupation was still scattered, especially north of 59th, replicating earlier times. In most cities, this pattern of scattered buildings in the outskirts would persist until 1945. Established by 1825, the 12th ward was the source of most later wards.

Of the eleven medium-density wards, four were relatively new, that is, towards the periphery north of 14th Street. Seven wards were old. The population of 1st ward, south of Maiden Lane and Liberty Street, decreased after 1830, but then surprisingly doubled its numbers until 1850 before decreasing again. The other old medium-density wards were on the west side and centre. Even the 15th ward, which included the richest residents, as we will see shortly, had 139 people per acre, or 90,000 per square mile (about three or more times what are considered medium densities today). The six most densely occupied wards were located, not surprisingly, east of Broadway in the Lower East Side, even though the 14th and 16th wards in particular hosted a lot of manufacturing. Densities would generally increase even more later in the century as tenements were crowded.

By world standards, New York was thus a high-density city by 1860. Crowded conditions resembled those of large booming places of the past, such as London. Only 14 to 15 per cent of households could afford single-family houses, and only about 12 per cent owned property. Another 22 per cent lived in two- and three-family buildings, leav-

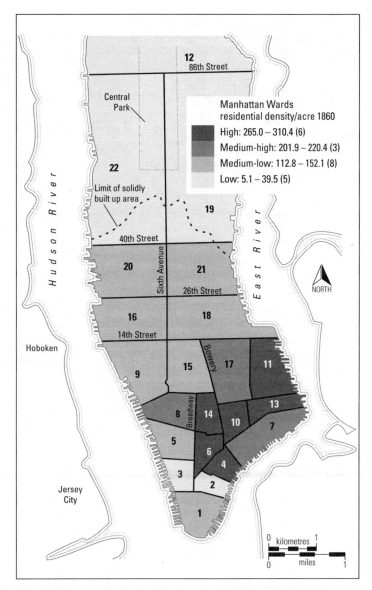

4.4 Manhattan residential densities per acre. Densities were higher than earlier, except in the expanding central business district. Occupation was discontinuous on the northern margins. The lower east side was the most crowded—on many streets overcrowded—according to reformers. *Source:* Edward Lubitz, 'The Tenement Problem in New York City and the Movement for Its Reform, 1856–1867', PhD thesis, New York University, 1970.

ing us to assume that nearly two-thirds lived in absentee-owned structures with four or more dwelling units. Perhaps about 2 per cent should be subtracted from this total for cellar dwellers and 2 per cent added for squatters' shacks.

Where one resided in New York depended on one's income, but accessibility to work obviously played a part. As earlier, microsegregation by income was still often the case, that is, clusters of rich were close to the poor and to an increasing number of the middle class. The microscale of construction was partly responsible. No builder could command large areas, so some block faces might have high-quality housing, just around the corner from low-value housing, but some districts were more clearly definable. In general terms, the 15th ward and parts of adjacent 18th and 21st wards became affluent, especially along 5th Avenue. The Lower East Side was a formidable place in most people's minds, the notorious Five Points still symbolic of squalor after several decades, as were the boarding-houses of sailors along the waterfront.

The rich—*rentiers*, merchants, financiers, and now manufacturers—could claim space that was not available to others, but they too were controlled by developers and builders and their own tendency to cluster for status. Assuming that the rich and affluent were the only ones who could afford single-family residences, in 1860 probably not more than 25,000 families owned their houses.

The pressure of central business activities mounted after 1825, so that the rich gradually abandoned their townhouses and commodious mansions downtown in the aristocratic 1st ward near Battery Park and adjacent wards. Still, some stayed near the tip until 1860, probably because of proximity to work, to the harbour breezes, and to the greenery of the Bowling Green and Battery Park. In fact, the 1st, 2nd, and even the 3rd wards were still the richest after the 15th ward as measured by personal property per head of household (unless the rich had this counted at their workplaces).

By 1800 the movement northward had begun; Chambers, Warren, Murray, and other streets west of city hall in the 3rd and 5th wards near Columbia College became fashionable. A similar spatial distinction of the rich was already slightly apparent in Philadelphia, as we saw earlier. Some later clustered for a time to the east around the high-class shopping on Catherine Street. In 1828 the top 500 resided on a hundred streets in all fourteen wards,

a strip of row houses here, another there. Brooklyn Heights appealed to those who preferred a less frantic lifestyle, though some who lived there would be roused later by the preaching of Henry Ward Beecher, who championed antislavery and women's rights at Plymouth Congregational Church. By 1855 a quarter of Manhattan's lawyers commuted to work on the ferries from their home in Brooklyn.

The move uptown was unmistakeable and more focused by the 1830s. The rich leapt along Broadway over the low land north of Chambers to Spring Street and beyond. The developers were intent on settling the rich on the healthy high ground on the spine of the island, thus giving up the margins to shipping, industry, and workers' housing. In ever-increasing numbers, the rich and affluent located just to the south of Washington Square and in St Mark's Place. Philip Hone moved from City Hall Square in 1836, predicting that the whole of Broadway would be converted to stores.

The luxury retail trade in the 1840s bordered on the fine old townhouses that were built less than a generation earlier on Bleecker and Bond streets and south of Astor Place. The rich moved on to Union Square, Stuyvesant Square, and Gramercy Park. Although a quarter of the richest 500 still remained south of the former Lispenard Meadows in 1845, over 60 per cent of New York's rich were concentrated on twelve noble avenues 'to live near friends and intimates', mostly in the 15th ward, but on some in the adjacent 8th, 9th, and 14th wards.[74] Then they began shifting from Broadway to 5th Avenue north of Washington Square, signalling the beginning of New York's fourth famous street. Churches of the rich followed, though sometimes they anticipated what was happening and so preceded most settlement.

Settlement of the rich, augmented even more by those who became wealthy during the roaring 1850s, expanded in a narrow strip between 4th to 6th avenues up to 42nd Street, beyond Madison Square, where officials and developers evicted the last squatters in 1849. In that decade, 500 to 800 high-class dwellings went up each year, mostly in this zone. The leading merchant, Alexander Stewart, lived in St Mark's Place (East Village) in the

1840s, then briefly farther south in Bleecker Street, finally moving to a new mansion on Fifth Avenue at 34th Street. The builder of the Fifth Avenue Hotel, Amos Enos and his family, who stayed on Greenwich Street until surrounded by immigrant boarding-houses, moved in 1853. After a decade or so, Union Square residences were abandoned to commerce, and residents moved northward along Broadway. North of Union Square, more modest housing with shops on the first floor lined Broadway. Along 2 mi. (3 km) between Washington Square and 42nd Street, Fifth Avenue housed 400 élite families at probably the lowest density in the built-up city.

By 1860 'Fifth Avenue was the street of aristocracy; the street of blood, the street ... "where beauty is displayed, admired, and selected...."[75] The élite may not have been as blue-blooded as those downtown or in Boston, but the conspicuous consumption of Fifth Avenue's nabobs evoked an aura of *haut* Paris and London. This area had the greatest concentration of servants, who were mostly Irish maids and some Black women. The inevitable private schools, galleries, and libraries appeared nearby. Marital ties were knotted in churches like Fifth Avenue Presbyterian that had moved from downtown. Following the trend, élite Columbia College moved in 1857 from Park Place, west of city hall, to 49th Street in 1857.

The aristocracy moved northward, sometimes pushed by business, sometimes by the encroaching poor. They moved often, though many had resisted the move from the south. Especially for the newly rich, the lure of status—the clustering of the fashionable with the fashionable, to promenade, to visit the new shops—was more vital than escape, just as it was downtown along lower Broadway before 1800 and on Broadway after 1825 near city hall. Although the rich strove to create a bounded community indoors and a little empire in the 15th ward, the swath between 4th and 6th avenues was

not totally homogeneous any more than were the older poorer wards to the south. Besides, when they promenaded, they knew others were watching. Even the 15th ward had a wide diversity in incomes and occupations, albeit many to serve the rich. In 1857 Calvert Vaux proposed elegant Paris-style apartments near Central Park. Although there was objections against them then, it would not be long before the developers perceived their value for some of the rich, who could cavort in the huge park nearby. A new era of urban living thus began.

Meanwhile, as in the eighteenth century, many of the rich built country houses on large grounds. The Gracie mansion far up the East River was one. Many of these would disappear as urban development encroached, but new ones sprang up in Westchester and other rural counties. More important for the city of the future was the first romantic exurban subdivision (the first in 1853), Llewellyn Park 12 mi. (19 km) west of New York. On 400 acres (162 ha), the project provided lots of 5–10 acres (2–4 ha) for building villas. Apparently upper-middle class men (more so than women) sought escape from the ills of the city, but not its business, as they could reach Wall Street in an hour by train and ferry. Wives more so than their husbands still sought to 'extend the spirit of the home' throughout the city.[76] Its guardhouse too presaged modern American subdivisions.

In the 1850s middle-income families, mostly native born, lived in narrow houses, some as narrow as 12.5 ft (4 m), two houses per lot, and many of them doubled up. The 8th and 9th wards on the west side between Canal and 14th streets (including Greenwich Village), and the 18th and 21st wards on the east side north of 14th were strongly middle class by 1860. Clerks were especially numerous in the 2nd and 3rd wards, (close to work), the 5th, and even the 15th wards. Boarding-house keepers, many of whom catered not only to

> The élite may not have been as blue-blooded as those downtown or in Boston, but the conspicuous consumption of Fifth Avenue's nabobs evoked an aura of haut Paris and London.

single male clerks but also even to middle-class families over long periods, operated in the 1st, 2nd, 4th, and 5th wards, often occupying mansions abandoned by the rich. Walt Whitman saw whole 'neighbourhoods' of them.[77] On the west side, families moved northward, but beyond 14th Street, the working class, drawn by industry, preceded them, so the middle class was blocked in that direction. Thus the east side became attractive to families. The native born moved upward from the 10th ward when Germans moved in, some to the drydock ward at the east end of 14th Street and then beyond. Some, including the better-off Irish and Germans, settled in healthier Yorkville between the 76th and 100th streets and commuted downtown by rail and horse tram.

As the rich appropriated the central spine of the island, and the middle class the areas to the west and east of them, manual workers and their families took over the housing they left behind. Many working-class families also lived near the margins of the island where vast numbers worked in factories and workshops. Anticipating expansion, some developers catered to them by providing new housing. Vast numbers lived in the Lower East Side. In seven wards adjacent to the rivers (a third of the population), residents owned less than 10 per cent of the city's personal property. The 'provident common class' of the native born sought 'refuge' on the western side of the city, west of 6th and 7th streets, especially in parts of Greenwich Village and Chelsea, but more or less from Canal Street to 34th.[78] There they mixed with the native-born middle class. Skilled workers in shipyards moved to the northeast in order to live near their work, but they too were located in every ward because shops were widespread. Residential construction workers sought housing near work as they would until the 1970s at least, so many lived near the growing edges.

The poorer working people lived in crowded, often overcrowded, conditions with well over one person per room. They had little choice of where to live, except that they were anxious to live near jobs, which were uncertain. The poor were packed into small houses, converted mansions, and townhouses of the formerly well-to-do (as in the 4th

and 5th wards), and into what were increasingly called tenements, those buildings that housed three or more families. Early in the century, virtually all tenants jammed into small houses, then later many moved into jerrybuilt backyard shacks or basements and sub-basements for the most destitute. Five Points, northeast of city hall, particularly the old brewery, packed in the destitute by 1820. To some outsiders, the squalor of Five Points, where a murder a night was committed, seemed to encompass the whole of the Lower East Side by 1860. In the 1st, 2nd, and 3rd wards, many resided in shabby old buildings on potentially expensive land held by speculators, who expected business expansion and redevelopment. Speculators surmised correctly from the 1820s to 1860, except in the wake of the panics of 1837 and 1857. From the 1820s to especially the 1920s, many of the poor lived on expensive land, only to be evicted when commercial expansion occurred.

By 1860 often the poor lived in purpose-built structures that were four to five or even six storeys high and covering most of each lot, the largest accommodating as many as 700 people. The number of families per building ranged from 5.5 to 9.0, or an astounding average of twenty-four to forty-three people per structure, according to an 1864 survey. The numbers of these apartment buildings would increase later in the century, much to the dismay of reformers. The city inspector reported in 1859 that nearly two-thirds of the population lived in 'dwellings' with four or more families. Although the highest concentration of multiple occupancy was in the Lower East Side, builders actively put up buildings on open land in the newer wards north of 14th Street. Lower concentrations of tenements were found in the affluent wards (the 15th and 9th), in the business-oriented 2nd and 3rd, and in the 12th ward.

Mortality rates suggest overcrowding: in the poor 6th ward, the death rate was more than double that of the 15th, in a city where the rate ran well over twice that of the state. From the 1840s onward, crowding worried health reformers who were concerned with environmental causes. Through their surveys they gradually impressed politicians and better-off citizens, though progress was slow. Conditions may have improved some-what in the late 1850s, in large measure because immigration slowed down markedly, allowing newcomers to find firmer ground. New construction may also have alleviated some of the worst conditions, such as cellars, though the cellar dwellers died in droves, but the 'dens of squalid wretchedness' remained.[79]

Wretchedness was experienced on the margins too. As a century before, the poor cobbled together their shacks, up to 24th Street in the 1830s, and up to 86th Street, including the site of Central Park, in the 1840s. As earlier, they had to move when the owners decided to build. Central Park developers suddenly expelled 300 families in the late 1850s, as they did a decade before when Madison Square was about to become a 'magnet' for the rich.[80] Squatter sovereignty was ephemeral, but they stayed on the rocky west side for several decades.

To the wealthy, squatters lived in dens of vice where prostitution and crime flourished. To the residents who put these crude structures together, their location provided the opportunity to work in slaughterhouses, to level the rocky landscapes for roads and development, to comb the garbage dumps for useful items and rags to sell, and to have a bit of ground to grow vegetables and raise a few goats, pigs, even cows. Although the density of building was far less than in the Lower East Side, a few thousand families were crowded in unsanitary conditions. Life, like their rights, was fleeting, if death rates, which were considerably higher in marginal northern wards than they were down-town, were an indication.

A Diffusing Ethnic Geography

The rich and others failed to root themselves spatially in this restless Manhattan. The affluent held a strong impression that foreign immigrants were all poor and concentrated, and that their districts ominously expanded outward. However, immigrants distributed themselves widely, in fact more so than native-born Whites in the so-called 'American' wards. Even these wards housed many immigrants. Microsegregation by block face was com-mon. Although few in number, Blacks, virtually all native born, lived in segregated enclaves in the most abject conditions. Objects of discrimination and abuse in the 8th ward, some lived in cellars filled with foul water that backed up in the sewer pipes. In 1855, when just over a half of the population was foreign born, immigrants' share of ward populations ranged from 40 to 70 per cent. Not surprisingly, the fewest lived in the affluent 15th, where many immigrant servants served their masters.

Although most of the Irish arrived in penury, not all of them did. Their vast numbers (well over a quarter of the population in 1855) meant that they were dispersed. At that time the Irish born lived in every ward, ranging from 13 to 46 per cent of residents (Figure 4.2). Their share was the great-est in the 1st ward, west of Broadway, cheek-by-jowl with the remaining rich. There the poverty-stricken among them crammed into rookeries when old warehouses and old houses opened up just as they were arriving in great numbers in the late 1840s. That transition might explain the near doubling of population in that ward between 1835 and 1850. The men found work as longshoremen and labourers, requiring access to daily calls, the women as seamstresses, and the girls as prostitutes. The Irish comprised more than 40 per cent of the population in the 4th and 6th wards, over one-third in another seven wards (three in the Lower East Side, two to the north in midtown, mostly near the rivers, and two on the shanty town fringe). Many lived across the East River.

Except for the poor on the edge, community institutions, notably the parish, provided mutual support. The church, the school, and societies gathered strength. Given the strong prejudice against the Irish for being Catholic and not Eng-lish, it was ironic that St Patrick's Cathedral would grace Fifth Avenue. Like the others, Irish leaders could not miss the opportunity for parading status. The Irish would, in the long run, be considered no different in most respects than other White people. Even in the 1850s, not all Irish were working class or poverty stricken.

The census takers reported that Germans (one in six of the population) ranged from 4 to 34 per cent per ward in 1855, so were somewhat less spread out because most had arrived more recently and did not speak English. They were mostly concentrated in Little Germany, which straddled the intersection of the boundaries of the 11th, 10th, 17th, and 13th wards in the upper part of the Lower East Side, replete with *turnverein* clubs and beer halls. They clustered by state of origin. Far fewer were located on the west side, but some lived in shack towns to the far north. Most were Catholics who had eight parishes under Irish bishop John Hughes. Before the rush of 1848, Germans had already established a district across the East River in Williamsburg. British immigrants showed some degree of concentration in the better-off 8th, 9th, and 16th wards.

The narrow configuration of Manhattan cramped expansion of the rapidly growing population, yet expand it did. The fact that it turned over and over meant it was no different than other cities, only it happened more quickly. Rich, middling, and poor people were constantly moving. The rich sought out property near their peers to display status. Their large mansions, which were in fashion for far less than a generation over the two or three decades before 1860, were left behind to others and then their sites were redeveloped. The middle ranks sought to escape the poor. They scrambled to find cheap rents in areas with access to jobs; due to the generation of economic growth in this city, some, through skill, connection, or luck, were able to move to better quarters. Neighbourhood stability was rare; May Day scattered neighbours. Greenwich Village was an exception; mixed in income and occupation and composed largely of the native born, it acquired a bohemian counterculture by 1860 that would recognizably persist down to Jane Jacobs and her friends in the 1950s. Friends and acquaintances who attended weddings often lived at a distance, many no doubt recently removed from the neighbourhood of the church. Propinquity, nonetheless, remained powerful symbolically and socially; churches, taverns, saving societies, and stores provided familiarity. Compared to over 200 Protestant churches and ten Jewish groups in 1855, twenty-four were Roman Catholic, thus obviously encompassing very large

parishes of Irish or Germans. Networks and space intersected, the degree of each depending on economic status, occupation, familial, religious, and other associational connections. Remarkably not all neighbourhoods turned over: Wall Street remained where it started, and proposals to tear down city hall failed.

Gotham: A Place of Peculiar Behaviour

Literate commentators outside and inside the city were ambivalent about the value of life in the maelstrom of New York. They praised or derided its power of creative destruction, but worried about the gap between the rich and poor, shuddered at its dark recesses of squalor, and decried its political corruption. In 1850 James Fenimore Cooper proclaimed that money from New York found its 'way into every village and to the summit of every mountain-top'.[81] As investment, this was no doubt true, but this reminds one of another perspective: Will Rogers averred that if you left a bag of gold in Death Valley, the next day it would defy gravity and land in Wall Street. Congressmen resented New York: an Indiana visitor complained that New York 'controls … with its immense monetary power, the commercial destinies of the Union'; a southerner declared that it was an 'overgrown city'.[82] In his *Symbols of the Capital* (1859), Amory Mayo, a Unitarian minister, regretted the decline of rural life, though he did not propose a return to gentler times, like many did at the time, but rather mechanization and development. New York as 'a large labor-saving machine' was a great virtue to the country that expedited trade and commerce.[83] Yet Mayo was anxious that too much power was concentrated there, that its leaders could not control destabilizing financial panics, that instead they became oligarchs of great companies and failed to root out barbarism among the masses.

Insiders worried too about the city's evils amidst the grandeur. Walt Whitman claimed it was 'the great place of the western continent, the heart, the brain, the focus, the main spring, the pinnacle, the extremity, the no more beyond, of the New

World.'[84] He later advised visitors to stay in at night, however, calling the city one of the 'most crime-haunted and dangerous cities in Christendom'.[85] He pointed to the deterioration of civic discipline, despite professional police and firemen. Some advocated buying guns or moving to the suburbs or beyond. To newspaperman George Foster, who wrote several graphic commentaries on city life, the polarization of the rich—the pretentious rich—from the poor was an evil. Yet he praised the independence, simplicity of spirit, and generosity of the Bowery 'b'hoys' and 'g'hals' that pointed to communal values in the midst of shadow, even amid the squalor of Five Points.

For many conservative businessmen, the politics of mass naturalization, which bolstered the Democratic Party and opened the way to Tammany power in the 1850s, hindered economic progress, though many were too busy making money to bother with public life. Anthony Trollope noted in 1861 that Fifth Avenue housed 'no great man, no celebrated statesman, no philanthropist of peculiar note', simply those who had made millions on shirt collars and lotions.[86] No doubt that understates inherited money. Another contemporary could find no unity of purpose in the city because its character was derived only from individuals 'seeking commercial opportunity'.[87] A recent commentator summing up impressions concluded that to many, the metropolis was 'at least vaguely un-American, that the real America was somewhere else in the midst of a glorified Nature and a sentimentalized rural life', presumably where the ordinary person supposedly could own land.[88] New York was the magnet for all who wanted to improve themselves, whether they were wretched immigrants or farm boys and girls. Despite the persistent depression of the early 1840s, the sharp downturn in 1857, the draft riots of 1863, and the death of tens of thousands of its young men in the Civil War, New York was a symbol of an America that was very much on the upswing, that was in so many ways modern and, to the distress of rural romantics, that had all the problems of Europe from which their ancestors escaped.

CHAPTER 5

■

Chicago, 1910:
The Civic Moment and the New Middle Class

Arriving at LaSalle Street Station early on a hot summer afternoon in 1910, our businessman from Britain anticipated cutting deals with certain industrialists. Being a public-spirited Tory, John Brown was very interested in the public improvements promised in the 1909 city plan. Like his grandfather a half century ago in New York, he had an eye for social conditions among the rising tide of white-collar workers and the mass of immigrants. He and his wife landed at New York harbour a fortnight earlier, and after business and pleasure in Gotham, they took a day train on the New York Central to Buffalo and Niagara Falls to see the countryside and cityscapes along the way. After a few days, they continued on to Chicago. Nearing the city, their expectations rose, like those of most other travellers.

> From Gary onward for an hour and a half was the city. They found themselves jolting over main line railroads that cross and recross at every conceivable angle, snapping up through Hammond and Kensington and Grand Crossing—to the right and to the left long vistas with the ungainly, picturesque outlines of steel mills with upturned rows of smoking stacks, of gas-holders, and of packing-houses, the vistas suddenly closed off by long trails of travel-worn freight-cars, through which the traveler's train finds its way with a mighty clattering and reverberating of noisy echoes. This is Chicago ... Chicago proclaiming herself as the business and transportation metropolis of the land.[1]

A motorized taxi carried them slowly through the congested traffic of the Loop to the new LaSalle Hotel. There they found relief from the heat with air cooled by fans melting ice. The next day they visited the prominent cultural attractions of the city and held meetings with financiers and other men of means. Our British businessman was also a guest at the Commercial Club, which was undoubtedly the most prominent club of the era and the promoter of the 1909 city plan.

'To love and render service to one's city, to have a part in its advancement, to seek to better its conditions, and promote its highest interests— these are both the duty and the privilege of the patriot of peace.'[2] Thus did prominent architect, Daniel Burnham, issue a clarion call to all Chicagoans, to 'the service of creating better surroundings'. To ensure prosperity for the city of destiny, as it was often called, all had to work for improved circulation, beauty, and social well-being to tame the 'shock' city of the previous century.[3] Burnham and his associates advocated radical surgery on the fabric of the city, with a view towards ultimate order, almost utopia, it seemed, in an 'illimitable space now occupied by a population capable of indefinite expansion'. Further, 'Chicagoans must ever recognize the fact that their city is without bounds or limits.'[4]

To uphold the 'dignity' of the city, Burnham's 1909 plan proposed far-reaching changes in the City Beautiful mode: a huge civic centre; grouping monumental buildings of harmonious classical style; relocating and grouping rail stations, tracks, and large industrial zones; straightening the south branch of the river, wide boulevards, and circumferential parkways; imposing radial streets over the pre-existing grid; widening streets; and establishing additional parkland on the lakefront and elsewhere, as well as forest preserves at a distance.

The plan was a culmination of the civic moment for the city, indeed for many cities on the continent. Burnham's great comprehensive plan would round off the rough edges of earlier rapid growth. With other reforms, it would rectify what Lincoln Steffens described a few years earlier: 'Yes, Chicago. First in violence, deepest in dirt; loud, lawless, unlovely, ill-smelling, irreverent, new; overgrown gawk of a village, the "tough" among cities, a spectacle for the nation …'[5]

The volatility of market crashes and real estate speculation seemed a thing of the past. Chicago's self-made moguls—such as Cyrus McCormick, Marshall Field, Philip Armour, and George Pullman—had died, and their companies settled into an impersonal world of corporate management that was committed to the standardization of routine. It was a world in which credentials now counted for making one's way into the stability of middle-class rectitude. Even play became ordered and, at the top rungs of baseball, commercialized. Local politics, it seemed, were less stressful in this city than they had been over the previous two or three decades as reformers modified machines. Indeed, Rudyard Kipling, had he returned at the time, might have been less inclined to denigrate the city as he did twenty years earlier: 'I urgently desire never to see it again. It is inhabited by savages'.[6] But the years around 1910 were to be only a moment of stability; within a few years, corruption and crime would reach new heights, as if the city of destiny could not deal with a world of order and slower growth.

The plan itself was part of the problem, not the solution: a plan for the 'city profitable', not the 'city livable'.[7] Behind the patina of the City Beautiful lay the expansionist urge. The plan would leave in business hands the power to control the city without limits. Our visitors became aware that the plan was silent on housing; implementation in fact would destroy the shelters of the poor, which, to the businessmen, was an act of creative destruction. But for advocates of livability, like the Woman's City Club, small-scale endeavours would have helped to make the city more like a home than a money machine. Working people would have benefited.

To assess the plan and its implementation, we will consider the relevant districts of work, home, travel, and leisure. Before that, however, we need to look at overall population and economic growth on the continent since 1860 and in Chicago. We will then look more closely at the urban environment, assessing the interplay of these four aspects in Chicago in shaping the institutions and infrastructure, which in turn altered the cityscape up to 1910, thus setting the stage for the great plan.

A Less Dramatic Pace of Continental Population Growth, 1860 to 1910

Even as the population of the United States tripled over the half century after 1860 from about 32 million to over 92 million, the rate per decade fell from over 30 per cent to 25 per cent, then down to 20 per cent. Birth rates fell, but vast numbers of immigrants could not make up for the slow-down. Canada grew more sluggishly over the four decades after 1860 before the opening of its agricultural West late in the century. The number of Canadians fell from 10 per cent of the American population in 1860 to just under 7 per cent in 1901, as immigrants to Canada and Canadians moved south of the border in large numbers. Then the Canadian population rose to 8 per cent of the American one by 1913, an early confirmation of Prime Minister Wilfrid Laurier's prediction that the twentieth century would belong to Canada; well, at least it would do relatively better compared to its massive neighbour, but not obviously so until after 1945.

People settled where opportunities beckoned. Agricultural lands still called, if less compellingly, though after 1896 they attracted more immigrants entering Canada than those entering the United States. By 1890 the American frontier was closed, according to the Census Bureau, though actually most of the humid lands with forests and tall-grass prairies capable of accommodating intensive agriculture were settled by 1860. Contributing greatly to Chicago's rise, the tall-grass prairies produced crops and livestock on a massive scale. A central-place hierarchy of trading towns that replicated the commercial East became especially well developed on these rich prairies. The short-grass plains beyond could handle more bison or cattle, sheep, and wheat than people. In Canada the western humid prairies and the short-grass plains were settled by new farmers; the greatest rush for land was between 1909 and 1913. As in the American Great Plains to the south, railway companies laid out towns along their lines like beads on strings. The farthest reaches of permanent agricultural settlement, northwest of Edmonton, were not reached until the 1920s. Irrigation transformed fertile valleys on or near the West Coast and in other arid regions.

Elsewhere, communities based on other resources sprang up in ever greater numbers. Lumber camps, particularly those based in white pine forests in a swath from New England westward through central Ontario, northern Michigan, Wisconsin, and into Minnesota, were ephemeral but obviously crucial to the increasing construction of houses and other buildings. Places farther south exploited hardwoods more gradually. As industry demanded far more metals than previously, hard-rock mineral exploitation was prominent in the Rocky Mountains and other mountain chains in the West and especially on the Canadian Shield in Minnesota, Michigan, Ontario, and Quebec. Even though the US would easily consume the largest share, Canadians would benefit proportionately more than Americans in the century to come from these developments. Coal mining towns in the Appalachian chain and on the prairies of Illinois fuelled the now greater number of factories, trains, and ships. Settlements based on oil and gas spurred growth, first in Pennsylvania, Ontario, and the Midwest, then in the prolific fields of Texas, Oklahoma, and southern California. On the other hand, electricity, which was only coming into its own, could be generated from steam in any city

and town that was located on a rail line where coal could be delivered. A few places with water power, most obviously Niagara Falls, could create the magic flow of electric power.

Urban places grew more rapidly than rural communities between 1860 and 1910 because of expanding commerce and manufacturing, though in the United States city expansion ran at a somewhat slower pace than just before 1860. The official count for American urban dwellers expanded nearly seven times, from 6 to 42 million, which was almost half the population by 1910. The number of urban places counted by the Census Bureau increased by a factor of eight—there were almost 600 urban places with populations over 10,000. An increasing number of these were industrial cities on or near the margins of the burgeoning metropolises in the northeast and Midwest. The larger and more diverse commercial cities appropriated an increasing share of the total population and also of the urban share by hosting more and more of the big manufacturing plants that produced goods for national markets. By 1910 Chicago and Philadelphia joined New York with populations of over a million each, while seven other cities had metropolitan populations of over half a million.

Regionally, the manufacturing belt that was established by 1860 filled out and expanded farther west. The margins of the Great Lakes experienced the greatest urban growth: Chicago was in the lead (with 2.2 million), while Cleveland's population equalled that of Baltimore; Detroit and Milwaukee were rising fast, and Buffalo (while slowing down) remained in the top eleven. Mid-sized industrial cities specialized in certain industries for national markets, for example, Toledo in glass, Akron in rubber, and Grand Rapids in furniture. Cincinnati and St Louis, the great river cities of the mid-nineteenth century, fell behind, but they and their adjacent hinterlands grew industrially. The New York water-level route through the Great Lakes to the west had sustained its earlier triumph over the Ohio River. In affluent Canada, Montreal reached a population of over half a million because it too benefited from the St Lawrence watershed. Toronto nearly caught up with Montreal. Toronto, which was only half the size of Buffalo in 1880, was equal to it by 1913.

Elsewhere in the West, big urban places were less frequent, but Minneapolis–St Paul, Winnipeg, San Francisco, and others served new hinterlands. Southern urbanization remained slow. Emancipation of slaves in the South led to a weaker cotton economy and little urbanization until after 1900: by 1910 New Orleans ranked only seventeenth in size; Miami, Atlanta, Houston, and Dallas remained tiny, though stirring. Another sign of change was steel making in Birmingham, and the tobacco and textile mill towns that emerged in North Carolina. On the border between North and South, Washington began its march ahead as government expanded, even while urbanization elsewhere in the northeast slowed down. But the biggest story of the era was Chicago, the key point that linked West to East, the regional centre of the Midwest. The shock city of the age rose dramatically from virtually zero in 1830 to a population of over 2 million.

> **The shock city of the age rose dramatically from virtually zero in 1830 to a population of over 2 million.**

Civilizing the shock city was not easy; young people from farms and small towns had inundated the city in pursuit of opportunity. People moved from job to job frequently. The beleaguered southern economy began to lose poor Whites and Blacks, who moved to the North. After a lull, immigrants from Europe poured in again around 1880, and virtually all of them headed for the North. In 1860 13 per cent of the American population was foreign born; the Irish, Germans, British, and Canadians stood out. In 1910 the foreign born totalled 15 per cent, in the midst of the largest immigration ever, which peaked in 1907. The 'new' immigration created much more diversity than the one before 1860, adding southern and eastern Europeans to the mix. Germans, who contributed the most in the early phases of the 'new' immigration in the 1880s, became the largest group in the US, while those from the British Isles dropped as a proportion. After 1890 Italians and eastern Europeans of various Slavic languages or of Jewish culture contributed a large share of immigrants. Interestingly, of all immigrants (in this period, two-thirds male), one third left the US around 1910, which perhaps suggests that opportunities became more limited; most returned to their home countries, though quite a few went to western Canada.

As their populations increased, large cities became the place where immigrants settled. Over a third of Chicagoans in 1910 were foreign born, while another third were native born of foreign parentage, behind only Milwaukee. Only one in five Chicagoans had native parents, and another one in ten had mixed parentage, but intermarriage was increasing that share. Those born in Germany and their children equalled the number of those with American parents, followed by those born in Russia and Austria-Hungary. Poles, who were born in all three states just noted and their children, composed a large share. Many of Russian origin were Jews, though New York hosted a larger Jewish community than Chicago. British and Irish immigrants were now considerably fewer than earlier, though those of exclusively Irish origin were still prominent. Having migrated earlier than other groups, Irish children (like the Germans) were considerably more numerous than their parents. Swedes and Italians were among the other ethnic groups.

In Canada in 1911, the census takers counted even more foreign born—between one in four and one in five. In 1913 (the peak immigrant year) Canada took far more than its continental share, one-third the number of those entering the US. Compared to the United States, those from Britain added a far greater proportion to Canada's English-speaking population. As an indication of increasing opportunities in Canada, one-fifth of migrants were American born. Southern and eastern Europeans settled mainly in the new West, making Winnipeg the most ethnically diverse of Canadian cities. By contrast, nearly 90 per cent of Torontonians could trace their ancestry to the British Isles.

Montreal became increasingly French as rural Québecois sought industrial jobs.

The religious contrast between the two countries was striking. American evangelical fundamentalism created a greater plethora of groups that appealed to salvation-seekers and the disaffected, a pattern that was already very apparent before 1850. Among popular major religious persuasions, regional variations (especially between North and South) created schisms. By contrast, nearly nine in ten Canadians found their religious identification in just four denominations: Roman Catholic (which was the largest because of Quebec), then (in order) Anglican, Presbyterian, and Methodist. Great awakenings and revivals influenced fewer Canadians, and doctrinal differences were less obvious. Reflecting the ongoing connection with Britain, Canada's Anglicans embraced a wide social range than the élite Episcopalians south of the border.

The flow of immigration would not continue. After 1910 immigration remained strong until 1913, then dropped sharply. (An eastward flow of young males found themselves in a quagmire of trench warfare.) In the late 1920s Canada hosted absolutely more immigrants than the United States. Americans wanted to protect what they possessed: in the 1920s the country, which was now the largest in the Western world, virtually closed its doors to all but English-speakers. This was a massive retreat from its historical experience fostered by William Penn. This dramatic policy shift rose from the strongest nativist impulse yet. Fears aroused by the Bolshevik revolution of 1917 triggered the closing of borders, but at the basis of this was the relatively weaker economic growth in the United States, signalled by the 1893 crash. The rejection of immigrants was an unwitting prelude to the Great Depression.

From Robber Barons to Massive Managed Corporations

Despite increasingly volatile cycles of rises and falls, the North American economy continued to surge in the late nineteenth century until 1910,

though with somewhat less vigour than between 1845 and 1860. Economic growth in Canada was stronger after 1900, paralleling the dramatic rise in population. By 1890 in the United States, the now much more intensively integrated 'common market' was robust enough to propel the country into first place among manufacturing nations, even while being the least dependent on international trade.[8] Its huge size and diverse resources allowed it to become largely self-sufficient. In the United States, output increases in the fifty years after 1860 were enormous in some fields: coal by twenty-five times, iron ore twenty times, rail mileage over ten times, Western Union telegraph messages by thirteen times. Telephone use within cities rose rapidly after 1880. After lagging, Canada raised its relative per capita gross domestic product from three-fifths in 1860 to three-quarters that of the United States in 1913. In contrast to its big neighbour, Canada continued to be much more open to trade and investment. It was increasingly drawn to the United States rather than to Britain, especially after the last great flood of British investment dissipated after 1913.

Material production was so great in both countries that it exceeded consumption, at least to 1896: after the Civil War (when inflation was high as in most wars), prices kept on dropping until 1896. The bust of 1873 accentuated the trend. Yet following the depression of the 1890s, inflation began its generally upward twentieth-century trajectory because corporate oligopolies could charge relatively higher prices. From early in the nineteenth century to 1896 can thus be termed the 'corporatizing' era, then 'corporate' era from 1896 to 1929 because they firmly dominated the commanding heights. Controlling many of these corporations were the Wall Street financiers, who signified the legitimation of the power and rights of corporations, their acceptance by the courts, and, as rural populism faded, by a wide share of the populace.

Before 1896 dropping prices induced the 'dynamics of competitive survival'.[9] As the great steel maker, Andrew Carnegie, put it, 'running full' was necessary to capture the value of the fixed capital of machines and market share. This led to overproduction, thus triggering cutthroat competition among the great entrepreneurs and lowering prices after the Civil War. Extreme volatility led the strong in this age of Darwinism to integrate vertically and then horizontally. Manufacturers sought to control resources and the marketing of their products (vertical integration). Then J.P. Morgan created holding companies that covered many fields. After the panic of 1893, Morgan and other financiers induced a massive restructuring of business through mergers (horizontal integration). Mergers weeded out the weak between 1896 and 1904. A single firm accounted for more than three-fifths of the manufacturing output in each of fifty fields by 1904. Signalling the shift to managers, many large and newly merged corporations took on abstract names, following John D. Rockefeller's Standard Oil: International Harvester instead of the name of the founder of its core operation, Cyrus McCormick; US Steel instead of Andrew Carnegie or J.P. Morgan, who took over from Carnegie to create it. In 1912 a US Senate committee learned that Morgan and his associates held over 700 directorships in over 100 of the largest corporations in all major fields of enterprise. Morgan's holding company was at the pinnacle. Although plenty of small-scale operations persisted and sprang up, often in specialized niches, many of them supplied larger companies and were thus beholden to them. Not surprisingly, the share of self-employed fell, so that corporations produced nine-tenths of manufactured goods.

Employment was concentrated in ever larger operations that were now organized by managers. In 1870 few factories employed 500 workers; by 1910 many hired over 1,000, some even 15,000 employees. The founders of great companies started on paternalistic terms with their workers and personal terms with their customers. As companies grew, the personal touch vanished as the great moguls of industry, led by telegraph and railroads, created huge empires that were organized into departments under vice-presidents, line middle-managers and their growing staffs, and foremen on the shop floor, imprinting the 'visible hand' of corporations on society and replacing or at least severely limiting free enterprise.[10]

Helping to usher in the era of corporate dominance, Republican William McKinley became president in 1896, almost as if on cue to foster Morgan's collectivization of business in a few hands. The courts favoured mergers instead of collusion through trusts or cartels among several companies because they seemed to sustain competition. Only on a few occasions did the Supreme Court use the Sherman Antitrust Act to split up giant companies. The Interstate Commerce Commission, which was established earlier in the wake of freight rate wars among railways, regulated prices in favour of large systems. Not surprisingly, prices began to rise.

Canada followed the northern American pattern, if less aggressively so. Further, American corporations took over a large share of Canadian manufacturing by establishing branch plants behind tariff barriers, though less so in mining, and far less so in transportation and communications. Canada's National Policy of 1879 fostered industrialization through protected newly emerging industries. After the great shake-out of the 1890s, the Canadian market was large enough to persuade some American companies to invest directly. Eastern Canada was far more like the United States' industrializing North than its marginal South.

The fixed capital of machines in industry, agriculture, mining, forestry, transport, and communications became increasingly widespread to cut costs and increase productivity per worker. In Chicago capitalization per worker increased fifteen times in the forty years after 1880. For energy, steam generated from coal displaced water power and much human and animal power. Because coal could be transported by water and rail, unlike waterfalls, companies could now locate near markets and labour in and around large cities to a much greater extent than was possible earlier. Use of wood as a source of energy and heating was declining after 1870 (Figure 2.5). Also about that time, trains overtook freight traffic on inland waterways. In the 1880s electricity replaced gas for lighting, and then electric motors replaced steam engines. The first great use of motors was in propelling streetcars and interurban, elevated, and subway trains. Natural gas began to replace coal for heating in some regions. Oil, along with electricity, the wonder power source of the past century, supplied kerosene for lighting and then gasoline for internal combustion engines.

New technologies led to increased inputs into capital goods to make machine tools from steel and other metals, which in turn made machines to create transport and increasingly domestic durables. Construction and then upgrading of rail lines, rolling stock, and locomotives consumed the largest share of the steel output. Implement makers introduced their labour-saving devices to farmers, in the process rendering small farmers, who could not afford them, one in a series of death blows. Interchangeable parts became the norm as goods were standardized. Rail gauges were standardized, as was time into zones. Rail efficiency and safety were served through devices such as air brakes, automatic couplings, and signalling. Laboratories, like Thomas Edison's, brought new electric inventions to commerce.

> **The drive for efficiency of workers promoted time-and-motion techniques to maximize productivity**

The great captains of industry ruthlessly cut variable costs, and not only through the introduction of machines. Suppliers often felt their wrath. But more important, industry leaders attempted to hold down workers' share of the product value. Manual work became more specialized and standardized, wiping out former skills in the whirlwind of new technologies, though those designing or maintaining machines developed new skills. As in the antebellum era, unskilled workers could operate the machines, and had to put up with repetitive and monotonous labour over long hours. To regulate their time, the factory whistle superseded the clock and the bell in the tower. The drive for efficiency of workers promoted time-and-motion techniques to maximize productivity. Early in this century, Frederick Taylor's methods were renowned, though others had preceded him in seeking the maximum human output. The industrial system, hierarchically organized with assembly-line work, produced masses of goods.

When production rose after the Civil War, even when they employed new machines, employers needed more workers to increase output. Wages for the unskilled rose, perhaps because there were fewer immigrants than in the early 1850s. Falling prices added to their real wages. Estimates of family expenditures after the turn of the century, compared with the first such study in 1875, suggest that even the lowest-income workers paid much less for food and now had far more discretionary income than they had earlier.[11] But the unskilled were vulnerable in the frequent and sudden economic downturns. When machines overproduced, workers lost their jobs and had no public or corporate social net to protect them. Those in mutual aid societies could count on a little help.

For a while in the 1880s, workers organized unions, such as the Knights of Labor, to counteract the power of industry leaders. The unions sought to increase wages, hold onto gains when prices fell, prevent production speed-ups, and shorten hours of work, even arguing for the eight-hour day. Strikes were fierce in this period until 1894, such as a widespread general strike in 1886 when workers dropped their tools in many places. Depressions, as in the late 1830s and again after 1857, decimated the ranks of workers once again. But the ten-hour day (sixty hours per week) was far more prevalent than the eleven-hour day of 1850. The skilled unions that formed the American Federation of Labor were more successful in improving the incomes of skilled manual workers than those of the unskilled. Samuel Gompers stressed bread and butter issues over direct political action.

Prices began to rise after 1896, and as immigrants (who were willing to accept low wages) arrived in greater numbers, the unskilled native born were less able to improve their position. Legislative and court actions thwarted unskilled workers' attempts to organize. Antitrust legislation led to more suits against labour than against trusts. Early in the century, less than 10 per cent of workers were organized. Flurries of strike action failed, as in 1904, 1912, and 1919. Although many worked fewer hours and Henry Ford paid well for fewer hours, even in 1920 steel workers still toiled for fifty-six hours a week, but Saturday afternoons and the few holidays a year could be enjoyed by many more than previously. Amusement parks, such as Coney Island, attracted young people to the thrills of the roller-coasters. Workers made financial and leisure gains; maybe some enlightened businessmen had finally recognized the value of their employees' spending.

The great corporations needed an increasing number of white-collar employees to run their vast enterprises. Corporations had to create offices filled with service workers, thus expanding the middle class with professionals in the upper ranks and clerical help in the lower. Large numbers of native-born industrial workers' children joined the white-collar ranks, and new immigrants replaced the aging workers. Sons and daughters of farmers left the land for the city and these jobs. Immigrants also hoped that their children would advance; they saved and scrimped to ensure this.

For middle-class workers especially, the great productivity paid off in more consumer goods and leisure time. As prices fell and goods became more plentiful before 1896, material standards rose. While workers complained about rising prices after 1896, many gained. Discretionary incomes continued to rise. While the net national product rose fivefold, and considering deflation and then inflation, real income per capita tripled between 1870 and 1914, perhaps faster than before the Civil War.

Spending on consumer durables and services outpaced basic needs. Kitchen and laundry equipment began to replace servants among the upper-middle class, though electrification of household appliances only appeared in the 1920s, and then only for a minority. Carriages and (by 1915) automobiles were mass-produced. Ice cream captivated children. In 1886 Coca Cola arrived on the scene and then Kodak cameras. Brand names began to dominate markets. Gibson girls provided a model of dress for young women working in offices. With more leisure time, more men and women joined lodges, where they could imagine themselves as grand potentates. Plays, operas, concerts, spectator sports, sermons, and phonographs entertained the élite and the masses. Many enjoyed bicycling and other outdoor recreations. As literacy approached universality, more reading material appeared. Passenger rail traffic and hotel and restaurant use increased. Because vast numbers of the middle class now earned two-week vacations, tourism took off. The more affluent could seek intellectual uplift at Chautauquas.

Specialized consumer credit organizations induced more spending to create a 'land of desire',[12] yet suppliers had to create demand. Inventive advertising captured the attention of the increasingly affluent population. As spending on advertising doubled in the 1880s, professionals standardized design procedures through such firms as Lord and Thomas in Chicago. Streetcars and posts were decorated. Billboards, disliked by many landscape purists, caught consumers' eyes. Mass-circulation magazines, such as the *Ladies Home Journal*, extolled material well-being as the route to status, and grew faster than newspapers, which were themselves sustained by extensive advertising. The great World's Fairs, which began with Philadelphia in 1876, through Chicago in 1893, then Buffalo and St Louis, bespoke the outpouring of goods that could not be resisted.

However, suppliers would have to work harder to persuade people to buy. It is doubtful that advertising increased aggregate sales; rather, its virtue was to help clever and innovative firms capture a greater market share. This would be even more apparent soon after 1910. The gains from restructuring through mergers reached their limits in many fields, though war and the production of consumer durables (notably automobiles) would postpone the economy's great fall until 1929. Even auto production became more erratic in the 1920s (Figure 2.7). Industrial employment was peaking. Even in many services, like banking, employment growth slackened. Peacetime iron-ore production hardly exceeded that of 1910 until after 1945, in large measure because railways, the great workhorse and symbol of the nineteenth century, were already in trouble when the crisis of the 1890s revealed overcapacity. In 1894 one-fifth of the trackage was in receivership, thus leading to further ownership consolidation. Abandonment of lines began, reminiscent of farmers leaving marginal farm lands in New England. By 1910 virtually all the lines had been built, though double tracking of some main lines would continue, and passenger traffic was near its historic peak. The golden era of rail was passing. Chicago, the great rail hub, would no longer command attention as the shock city of the nineteenth century.

The Metropolis of the Midwestern Cornucopia

By 1910 Chicago's population reached an astonishing 2.2 million, one of the largest in the world. Small neighbours in Cook County added another 10 per cent to that. Its economy was without question the fastest growing on the continent from its founding until the World's Fair, which coincided with the panic of 1893. Until then, the growth was so heady that leading business people believed there would be no end, and the gospel of speculation captivated many who were successful or who dreamed of success. The Board of Trade secretary sermonized in 1891 at the climax: 'Speculation stimulates enterprise; it creates and maintains proper values; it gives impulse and ambition to all forms of industry—commercial, literary, and artistic; it arouses individual capacities; it is aggressive, intelligent, and belongs to the strongest of the race ...'[13] After 1893 this rhetoric of Social Darwinism died down as growth became less spectacular.

Chicago's strength rose from its location. Situated at the south end of Lake Michigan, it was the farthest temperate location on the great St Lawrence waterway and adjacent to the rich prairie lands. The portage from the Chicago River to the Illinois, a branch of the Mississippi, was short, thus providing the easiest access to the interior. In the early 1830s smart entrepreneurs, led by William Ogden, turned the marshy site to advantage, so Chicago became the chief gateway to the West. Chicago's success as the main entrepôt, instead of St Louis, which was its chief rival prior to 1860,

was probably assured even before its first rail line was built in 1848.

The strongest route for water traffic to the east followed the lakes into the Erie Canal, with much less water traffic going into the St Lawrence. Together they outweighed the Mississippi–Ohio system. The 1848 opening of the canal along the Illinois River from Lake Michigan allowed merchants to compete farther west. Rail builders and consolidators followed these water corridors in the 1850s and later, soon surpassing the traffic on ships and barges. As rail lines had fewer constraints than waterways, more lines radiated out to the north, west, and south. Chicago boasted that it did not have to subsidize rail companies as Milwaukee and other places did because they sought out Chicago. To the east, the New York Central was the premier line along the water-level route, exceeding other main lines such as the Pennsylvania, Baltimore, and Ohio. Chicago became the number-one rail centre on the continent, with about twenty-four trunk lines by 1890. As the terminus for them all, passengers and freight cars had to be switched at Chicago.

Chicago's merchants and transportation builders made the city the main location for distributing goods: grain, livestock, lumber, and imports to the region from the East. It became the collection point for grain as the wheatbelt swept by it from the East to the West. Chicago's grain-elevator operators increased the speed of flows by altering the movement from grain in sacks to continuous golden streams that poured into ships and trains. Much of the grain was shipped east for milling in Buffalo, though Minneapolis would later be the premier site. Chicago switching yards relayed sacks of flour. Livestock—range cattle, fattened in nearby Corn Belt feedlots, as were pigs and sheep—arrived, unwitting of their fate in the packing plants. Lumber, especially the white pine from the upper Great Lakes, was sorted in Chicago. Much of it was used to build the cities, towns, and farm structures on the prairies and elsewhere in the country.

The speed of information flows increased, further enhancing Chicago's position. The telegraph and postal service became more efficient. Local business used Alexander Graham Bell's invention first and eventually eliminated intraurban telegraph. By 1910 early rapid expansion in telephones was reached (a pattern following all new technologies), but obviously growth continued at a substantial pace. Long-distance use of the telephone was slow in coming because inventors had difficulty in creating switching technology. The first long-distance call between New York and Chicago took place in 1892, yet the first transcontinental call was not made until 1914. The expense remained high and service was slow until the 1920s.[14] The telegraph would remain important until after 1945.

In finance and decision making, Chicago would remain subsidiary to New York. Indeed, the largest share of the investment capital for the West originated there, or was channelled through it from Britain. Boston contributed a lot too. Despite its regional power, Chicago businessmen did not play a leading role in financing railroads; they exerted a significant influence only on the Chicago and Northwestern line. Wall Street forged ahead as the premier financial district. Although stock exchanges operated in several other cities, including Chicago, New York increased its domination. In 1886 Wall Street had its first million-share day, leaping to a 3-million-share day in 1901. As in other fields, banking mergers occurred early in the century, concentrated in even more hands in New York, led by J.P. Morgan.

The head offices of large commercial and manufacturing firms were concentrated more in lower Manhattan than elsewhere; in 1895 its nearly 300 American head offices exceeded the combined total of the next four largest cities. The head offices of some western rail lines were located there. It became home to the largest share of the vast corporations that had merged. In midtown Manhattan, the national centres of the print media and advertising clustered close to the rising tide of commercial entertainment. The focus of America's culture, Times Square, was named in 1904.

> **In finance and decision making, Chicago would remain subsidiary to New York.**

Although second in finance, Chicagoans set up banks, insurance, and trust companies. Banks were a major source of commercial credit for the grain and other trades. A clearing-house established in 1865 expedited money flows, and Chicago became a federal reserve city in 1887, though keeping large balances with New York's financial institutions. Chicago's banks surpassed those in Philadelphia and Boston when it came to clearings and deposits. By then small regional banks became subordinate to Chicago. Only St Louis, the other early federal reserve city in the West, could command a small niche of correspondent banks out of the West, just as Boston and Philadelphia could in the East and South. The First National Bank of Chicago (1863) was the second largest in the United States by 1901. Mergers created larger banks, even while legislation maintained the fiction of democracy through many small rural banks. By contrast, in Canada legislation allowed the consolidation of banking into the hands of a few by permitting extensive branch systems, and thereby greater stability than in the United States. By 1910 big Montreal and Toronto banks were well on the way to absorbing the smaller ones.

Although Chicago bankers congratulated themselves on their 'strong, reliable, independent, and conservative banking system', like bankers elsewhere in nineteenth-century fashion, they were periodically caught up in excesses that led to the famous panics, as in 1873 and 1893, when Chicago had more than its share of failures. 'These disasters resulted from inadequate reserves, unwise investments, lax state banking laws and institutional examinations, promiscuous chartering of state and private organizations, cheating executives, and defaulting directors.'[15] Certainly these practices and the panic of 1907, which led to more bank failures, finally moved federal lawmakers to set up the federal reserve system in 1916 as the lender of the last resort. Washington (rather than New York) controlled the system—a harbinger of greater government involvement later in the twentieth century. Chicago was then the focus of one of twelve reserve districts.

Chicago's Board of Trade was a special financial intermediary for the world, it dealt in commodity futures fitting the vast resource base of the interior of the continent, where 'Chicago's greatest plungers manipulated grain and meat futures ...' In the 'pit', 'collusion and corruption abounded, but the board, empowered by the state to regulate the market, was run by the most notorious speculators; hence laws were violated, attempts to implement stronger controls defeated, and malingerers escaped unpunished.'[16] But the pit brought about efficiency, nonetheless. Because speculators sometimes cornered markets and cut corners on quality, measures were established to sort out types of wheat and other commodities and speed up movement.

Chicago's wholesalers forged into first place in clothing, hardware, groceries, and drugs; Marshall Field was the leader by 1868. Even by 1875 Chicago was the largest centre of clothing distribution in the country. Sales soared and increased tenfold in thirty-five years. Other cities also had substantial wholesaling activity—for example, Omaha, and Winnipeg for the Canadian West—though Chicago easily led. However, wholesaling declined in relative importance in the economy as retailers bought from manufacturers and opened factories themselves.

Chicago's mail-order houses were the largest anywhere. In 1872 Montgomery Ward started mailing catalogues to attract rural folk to the delights of the modern age. Sears and Roebuck came to Chicago in 1895, and later surpassed Ward and others in the country. Their success was partly attributable to cheap parcel post supplied by a government compliant to corporate interests. Catalogues sold the message of consumption to farm families as they sat around the kitchen table illuminated by lamps fuelled by John D. Rockefeller's kerosene. These mail-order retailers evoked envy in small-town dealers. In 1913 a visitor reported that 'a mail order house ... actually boasts that six acres of forest timber are cleared each day to furnish the paper for its catalogue, of which a

mere six million copies are issued annually ...'[17] Chicago was also renowned for its department stores, as we will see when we consider the geography of the central area. Unlike the corrupt actors in grain trading, department store owners were reputed to be honest. Chain stores competed too, to the detriment of small shopkeepers who sold cheap goods.

As in most other fields, Chicago's diverse manufacturing sector had reached second place to New York by 1890. The value of Chicago's manufactured goods quintupled in the thirty years after 1860. (This may understate the implied quantity since prices fell until the mid-1890s.) It nearly tripled again over the next twenty years to $1.9 billion in 1910. Within Illinois its share increased from less than half in 1870 to nearly three-quarters in 1890, thus following the pattern that large commercial cities (and their satellite industrial towns) captured more and more processing under large-scale corporate auspices. Initially its industrialists added value primarily to wood and food, but also made clothing and shoes, and then increasingly developed iron and steel and metal fabricating. By 1919 those of Chicago's establishments that had more than 1,000 workers employed nearly 30 per cent of all industrial workers.

It excelled in some fields within its diverse range of manufacturing. Displacing Cincinnati, Chicago ranked first in meat packing and its numerous by-products. Nelson Morris, Gustav Swift, Philip Armour, the Cudahy brothers, and their heirs stood out. They perfected the assembly line or, in their case, the disassembly line. A leader in vertical integration, Swift introduced refrigerator cars and opened warehouses in big eastern cities, though he had to fight resistance from local packers, who sought to control the price of livestock. After the mid-1880s they opened new plants farther west to cut costs. In the long run, Chicago would lose both the stockyards and packing plants.

Chicago was the first in manufacturing farm implements, notably those made by McCormick and Deering, which were combined into International Harvester after 1902. In 1850 Cyrus McCormick employed 150 workers; by 1916 International Harvester employed 15,000. Railway rolling stock was prominent, especially Pullman sleeping-cars. Chicago led in the manufacture of tinware, machine tools, pianos and organs, and furniture. The city produced nearly one-fifth of men's clothing in 1914, and was second only to New York in printing and publishing. In iron and steel, it was behind Pittsburgh by 1880, but would gain over the next half century; the Chicago region, including suburban Gary, Indiana, was almost first after 1905. Other steel towns such as Cleveland, Birmingham, Alabama, and Hamilton, Ontario, augmented the vast outpouring of goods, but Chicago could not sustain the movie and car industries. Even so, Chicago could claim pre-eminence in the centre of the continent:

> For her and because of her ... all the Great Northwest, roared with traffic and industry; sawmills screamed; factories ... clashed and flowed; cog gripped cog; beltings clasped the drums of mammoth wheels; and converters of forges belched into the clouded air their tempest breath of molten steel.[18]

By 1910 the pace was less frantic when the corporate era had its finest hour, less impeded by rural populism, small producers, and unions. With the economy in mind, let us look at the inner inward workings of Chicago.

Divided Chicago Politics and Society

What were the chances for fulfilment of Burnham's plan, given the power relations in the city? On the plus side, the plan followed a decade and a half of reform. The rhetoric of the reform era reflected high expectations in remaking the city. To Frederick Howe of Cleveland, the city became the hope of civilization. In 1905 Chicago's Charles Zueblin, another great publicist for cities, saw 'spectacular evidences of civic progress'.[19] Indeed, 'civic progress of the last decade is greater than that of all our previous national existence.'[20] Not long before, Englishman James Bryce had chided Americans for having the worst-governed cities in christendom, but to Zueblin, cities were no longer a 'political

failure'.[21] All the great improvements pointed to a 'civic awakening'. He described the White City of grand classical proportions at Chicago's 1893 World's Fair as 'a miniature of the ideal city', inspired 'by a common aim working for the common good'.[22] What was needed was comprehensive planning, 'the ripest expression of the new civic spirit'.[23] Then too, these reformers linked education directly to planning.

Flushed with their successes in designing skyscrapers and monumental buildings, thereby establishing their professional influence, Burnham and the other architects of the time widened their focus to encompass the city. Burnham was the principal figure in coordinating the design of the White City (as contrasted to the smoke-darkened grey city) on the World's Fair site of 1893 to celebrate Columbus's arrival, the wealth of America, and the spectacular success of the shock city. He had participated in the revamping of L'Enfant's Washington in 1900, which triggered a decade of enthusiasm. His plan for the great civic park bounded by monumental buildings in Cleveland was underway. He presented a plan to San Francisco the year before the earthquake, but it was not followed. Inspired by Burnham's admonition 'to make no little plans', architects in other cities proposed similar schemes. Actually, 1909, the year of Chicago's plan, was momentous in other respects for city shapers. It marked the official start of the urban planning profession with the founding of the American Planning Association, the first university course in urban planning at Harvard, and the first legislation with an urban planning label in Wisconsin. Urban planning reached exalted heights for uplifting and standardizing America and even Canada, to bring order out of the chaos of the previous century, under a joint business-civic banner.

Reform action in Chicago was not isolated from those in other cities on the continent, though the city had its own distinctive political culture. As in business and labour, countrywide municipal-interest bodies were formed during this era. The National Municipal League was founded in 1894, followed by a similar organization in Canada. Civic officials, such as engineers and health advocates from both sides of the border, exchanged ideas at their conferences. A journal that began in 1899 promoted urban improvement and good government, as did other similar magazines. The American Civic Association (1904) mobilized professionals to the cause. The City Club in New York brought together reformed-minded businessmen, as did Chicago's Commercial Club. Following New York, many cities had a Bureau of Municipal Research or similar bodies that promoted scientific analysis of governing.

The prospect of fulfilling the plan and other reform measures through these mostly business and professional groups seemed promising. But social divisions and conflict in Chicago and the concerns of other groups, some of whom could be considered reformers too, suggested that a consensus would be difficult. Fabulous economic growth in the gilded age created great fortunes, thus dividing the rich from the poor, probably as much so as in 1860 New York. Chicago's early successful men were largely self-made people who seized the opportunities in the new open, booming environment until the 1850s when risk was at a minimum. Eventually, though, as in eastern cities, the road to riches was reached increasingly through inheritances, though continued rapid growth until 1893 engendered more great fortunes. In that era of Social Darwinism, to quote Theodore Dreiser, 'all life was … the strong preying on the weak'.[24] The handful of New York millionaires in 1860 increased markedly. By 1892 there were 278 millionaires in Chicago, though some lost ground in the depression that followed. When he died in 1906, Marshall Field, the retailer, left a fabulous fortune of over $150 million. Potter Palmer, who was also a retailer and land dealer, seemed like a pauper by comparison because his worth was only $8 million. Machinery and plumbing-fixture maker, Richard Crane, died with more than that in 1912. The great utility magnates, Samuel Insull and the unpopular Charles Yerkes, also amassed fortunes. Yet making money was not enough for some: Cyrus McCormick became preoccupied with genealogical research into his roots and craved the attention of European nobility. By the turn of the century, managers in large companies had replaced most of the old rich entrepreneurs as companies

sought to standardize methods, workers, and products. Managers and the *arrivistes* scrambled to rise into the élite and signed up for a listing in the social register. No doubt, even after the gilded age, the urge to socialize was a greater pursuit than 'to love and render service' to the city, as Burnham put it.

The middle class grew in Chicago as it did elsewhere. In the United States while the number of farmers and the self-employed fell, the salaried 'new middle class' quadrupled from 1870 to 1910.[25] Immigrants sought entry into their ranks. Children and grandchildren of the 'old' German and Irish immigrants achieved this status by 1910. Professionalization in many fields was stabilizing the upper end of the middle class. Engineers, who were in increasingly in specialized niches, architects, doctors, dentists, nurses, and teachers greatly increased in numbers, and all formed associations to tighten up credentials and exclude the mavericks. Building managers organized associations and, like the others, published trade journals. Even the boisterous real estate industry moved towards rules through boards after being burned by the excesses of the 1880s.

Service employment doubled in Chicago between 1890 and 1920, which expanded both the upper- and lower-middle ranks. As the professions expanded, the number of clerical office workers grew disproportionately more than many other fields, partly as the result of the creation of large corporations. Salesmen for wholesalers drummed up trade in the city and from town to town in the hinterland. New gadgets for the home and new industrial technologies gave rise to a host of new technical workers such as electricians. Retail stores expanded opportunities, though working as a clerk in stores could be nearly as wearying as making garments. A quarter of women worked for wages, as in New York earlier.

Chicago's industrial semiskilled and unskilled waged workers laboured in the now often extensive factories. As in New York earlier, many of these workers were foreign. Separating the industrial workers from the residual poor (the underemployed underclass), as social analysts intent on classifying tried to do, was not easy in a fluid and growing open environment. Men still worked long hours in the factories under trying conditions, as Upton Sinclair made clear in his 1906 exposé of the meat-packing plants. Women huddled over sewing machines in sweatshops and in the dim light of their homes. Boys hawked newspapers. Intent on reducing costs, most business people intensely opposed unions for unskilled labourers.

In the gilded age, violence—even murders—characterized thousands of strikes and lockouts in Chicago, a few of which reached epic proportions in 1866 and 1894. After police killed four workers at McCormick's plant during the Haymarket protest, which was organized by a German workers' newspaper, someone threw a bomb that killed some policemen. Other officers responded by shooting into the crowd. Businessmen and the newspapers sought law and order, raising among the middle class the fear of anarchy and communism, even of the apocalypse. Some members of the élite encouraged the execution of the presumed anarchistic immigrant perpetrators. Eight years later, the élite supported George Pullman and the suppression of workers through federal military action. Pullman's workers struck against pay cuts and, despite national support from Eugene Debs's rail workers' unions, the workers lost and the depression continued. Debs went to jail. Strikes occurred even in the quieter corporate period after the depression, as in 1905 and later in 1910 when garment workers marched in the streets. Although Chicago had fewer conflicts than elsewhere in 1912, antagonisms had not dissipated.

In electoral politics, most of Chicago's working people, who included most immigrants, voted Democrat, while businessmen voted Republican. The polarization might have been even more severe: despite the strife, labour and socialist parties had only brief moments of attraction for workers, even in 1912 when Debs ran for president. In rural areas he won 7 per cent of votes, and only 13

> **Businessmen and the newspapers sought law and order, raising among the middle class the fear of anarchy and communism, even of the apocalypse.**

per cent in Chicago, which was less than in smaller cities. He won only a minority of working men, and even fewer in the burgeoning middle class. Compared to Europe, third parties were ephemeral. The promise of America and its middle-class vision of a voter democracy engendered resistance to the European notion of labour parties. The level of popular participation declined generally in the North after 1896, as if more and more workers were resigned to Republican corporate dominance.

Locally, the conflict between Democrats and Republicans was sharp enough; mayoralty elections were often close in Chicago, though Democrats won the post more often than not. However, division over reform issues was not always along party lines. The progressive era brought out several strands of reform that never tied together. Various social and environmental movements sought to effect change in Chicago. How Chicago would be governed was contentious in this era, and obviously had a bearing on public action.

To many citizens, Chicago's city government was unwieldy. The mayor held little legal power, compared to mayors in other cities. The city council was often divided, and bosses reigned in wards. Besides, several separate autonomous and independently financed commissions did not even have to report to city council. Three park boards, the sanitary board, and the school and library boards were independent, though the mayor had the right to appoint members to the last two. Other officials were elected too, so the voters had to scan a long ballot. A sign of state imposition of its will on this urban place, the county fixed assessments and levied general taxes, then turned funds over to the city. To city politicians, this was a sore point.

Mayors thus had to balance the concerns of different reform interests and those of ward-heelers. After 1900 Carter Harrison, Jr, was the most adept at doing so, bringing in civil service examinations and therefore weakening patronage to a degree. His successors, Democrat Edward Dunne and Republican Fred Bosse, were less able to perform the balancing act. Various groups in Chicago sought to clean up corruption and rewrite the charter. These included the Citizens' Association, which was possibly the first on the continent and a model for other cities, and also the Municipal Voters' League, the Civic Federation, and the City Club, all of them largely élitist. More democratic in intent, the Independence League and the Municipal Ownership League supported structural reform too with more ordinary people in mind, but the latter failed to municipalize the street railways. The Civil Service Reform League succeeded in pressing for employment based on merit. In 1909 the indefatigable Charles Merriam was successful in bringing about budget reform. The City Club set up the independent Bureau of Public Efficiency as a watchdog, yet several attempts to rewrite the charter of city government failed, notably in 1907.

In 1906 businessmen and professionals (mostly Republicans of varying stripes) set up a charter convention. The convention, which was stacked with their own crowd and included only a handful of labour and social advocates, finally approved a list of reforms, though these did not strengthen the mayor's role and did not advocate vigorous home rule to loosen state control. The weak product was handed to the state legislators for approval. The legislators, who were mostly rural Republicans, altered it further, notably to gerrymander wards in favour of city Republicans. Combined with left-leaning, non-partisan bodies, the United Societies, representing ethnic voters, mounted a vigorous attack on the charter, with the support of the Hearst paper, the *American*. The city electorate then turned it down by a two-to-one margin, leaving the editors of the right-wing *Tribune* to fume. Obviously, not only Democrats but many Republicans and moderates were disgusted with what businessmen had created. Upstate-downstate tension continued; Illinois farmers and small-town lawyers preferred to support the economically powerful Chicago Republicans, apparently under the illusion that their political control was the equal of economic power. The city was left with its fragmented governing system.

New York, as noted earlier, was in sharp conflict with the state too. In 1898 its charter was rewritten to strengthen the power of the mayor. Advocates thought that this might control Tammany Hall, but Tammany nonetheless continued to hold sway. By contrast, Chicago was not dominated by a citywide machine in this era, nor would it be until the 1930s when *de facto* if not *de jure* centralization was finally achieved.

Until 1930, most aldermen who represented the thirty-five wards were not enamoured with structural reform, nor were mayors with leadership skills. Controlling their turfs was more important. They made alliances on issues. Central-city Democratic aldermen held power for many years. The renowned Ward 1 aldermen, Bathhouse John Coughlin and Hinky Dink Kenna, were untouchable, at least until 1910, as was Johnny Power in Ward 19 west of the river. Saloon-keepers paid them to protect them from prohibitionists, as did madams of the bordellos in the Levee and gamblers from other moral crusaders. The police turned a blind eye, even abetted in these protection rackets. Even so, in exchange for votes, the ward politicians provided working people with municipal jobs, which generally offered more security than industry or construction—and the proverbial turkeys at Christmas. An alternative political system provided an alternative ward economy and welfare system. Chicago's politics had, though, already reached their corrupt nadir in the early 1890s. In 1897 a British visitor reported to the world on the effort of the Municipal Voters' League: 'It is thought a great step forward that there are now actually one-third of the members of the municipal body who can be relied upon to refuse a bribe.'[26] Actually, reformers increased this share of non-partisan council members within a few years and briefly controlled the city council. As for Burnham's plan, both Democratic and Republican ward politicians may have supported it officially, but one doubts whether it was high on their agenda.

In other large cities, mayors enjoyed greater executive power, which state governments conferred on them, as they sought to undercut the alternative governments of ward bosses. Elected with business support, Cleveland, Detroit, and

Toledo reform mayors went further than elsewhere by embracing reforms that helped lower-income people, but in doing so, they alienated the upper classes. Milwaukee, though, consistently voted in socialist mayors. Many American cities failed, however, to maintain the momentum of the civic moment after 1910. Most Canadian cities, which were never as expectant, never rising to exultant heights, retreated less markedly from more modest gains of reform; their paid officials were less subject to the whims of party politics and therefore capable of providing an orderly system of management. Yet even these merit-based civic services could not prevent nepotism.

Aside from structural reform, six thrusts for social change can be identified: more efficient and humane social welfare, improved workplace conditions, expanded education, playgrounds, better housing, and clean-up of vice in the city, which were seen as progressive to many in the rising middle class. Then too, physical reform was at least partly successful in supplying clean water, installing sewer lines, and combating disease. The drive for efficiency fitted nicely with Burnham's underlying motive of centralizing power through landscape design. People of various classes and interests wanted change; consensus was reached on some issues, but irresolvable conflicts remained.

Reforming Society to Create a Middle-Class City

Social Welfare

After the World's Fair in the fall of 1893, the beginning of the most severe depression since the 'hungry forties' provided a moment for liberal reformers to attack inequities. An estimated 40 per cent of unemployment moved a reporter to exclaim: 'What a spectacle! What a human downfall after the magnificence and prodigality of the World's Fair which has recently closed its doors! Heights of splendor, pride, exaltation in one month, depths of wretchedness, suffering, hunger, cold in the next.'[27] The British social gospel minister, William

Stead, hectored an audience: 'If Christ came to Chicago ... He would not find Christian sympathy ... but callous indifference, ... a growing separation of classes, and an ominous spectre of class conflict.'[28] Underfunded organizations, Stead asserted, could hardly improve conditions on their own, so Chicagoans with power must act more forcibly through public housing and direct welfare. They would act, but only up to a point, and not to the extent that European countries did at the time.

Even free-enterprise Chicago had from an early date provided social assistance, parsimonious though it was. Cook County provided relief, though 'without notable generosity', and almshouses continued ancient practices, but responsibility for helping the poor still lay largely in the hands of churches and philanthropists.[29] Ward bosses maintained their power by trading votes for help. The moguls, or rather their wives, the charity matrons, engaged in philanthropy. Before 1860 wealthy Christian gentlemen established the Chicago Relief and Aid Society. Their benevolent wives canvassed for funds, and were very involved in creating institutions like orphanages and other asylums.

In contrast, the gilded-age generation was not as generous, embued as it was with the notion of survival of the fittest. *Noblesse oblige* fell by the wayside; relief payments to the employable were cut, as the distinction between the worthy and unworthy sharpened. The élite pursued club life and a listing on the social register. Rather than direct aid to the poor as earlier, they retreated to arm's length charity balls. They hired those aspiring to be social work professionals to do the work while they (to social analyst Thorstein Veblen the leisure class) consumed and wasted conspicuously. They gave far more generously to cultural institutions that enhanced their own status and the image of Chicago as a great city than to the programs for the poor. High culture, one visitor proclaimed, was 'pouring into Chicago as rapidly as pork or grain ...'[30]

However, reforming professionals, who were gathering in strength and numbers, sought a more hands-on approach and were not as interested in distinguishing between the deserving and the undeserving as earlier moralists. Seeking to overcome the 'growing separation of classes', in the 1880s the Charity Organization Society (COS) promoted direct 'scientific' work among the poor, though it competed against the élite's Relief and Aid Society for funding.[31] Pursuing this aim, Jane Addams and Ellen Starr founded Hull House, a community centre (or a settlement house, as it was then called) in 1889 on Halsted Street. The depression marked a return by some of the affluent towards much more direct action through the Civic Federation, but businessmen's enthusiasm waned quickly when they sided with George Pullman in the fierce 1894 strike. However, Jane Addams and her cohorts persisted, forcing businessmen to listen. The stuffy Relief and Aid Society merged with the more forward-looking COS. Philanthropists and churches established eighteen more settlement houses in poorer areas of Chicago by 1909, which were similar to hundreds throughout the continent.

Following the practices of people in other cities, Robert Hunter, Edith Abbott, and others undertook scientific house-to-house surveys in poor areas, then publicized the wretched condition of the poor, calling for environmental and social improvement. Charity districts were created for 'friendly visitors', social workers. In 1914 the Chicago Council for Social Agencies took systemization, centralization, professionalization, and survey work one step further. Centralized funding campaigns, predecessors of the United Way, began as they did in most other big cities in North America. Jane Addams and her associates ameliorated the immediate condition of some of the poor and helped to create a social work system. Systematizing took the bite out of concerted action, however: in the 1920s the élite kept more for themselves, the professionals controlled welfare, and the poor, many of whom were still living in poor housing, became just numbers in surveys. In churches, the social gospel waned and reactionary notions held sway.

Workplace Reforms

Liberal reformers advocated legislation to help when they realized that charity donations for working people were inadequate. On labour issues, Addams and others successfully pushed for arbitration of industrial disputes and workmen's compensation. Businessmen eventually agreed to the latter because they would be relieved of pesky lawsuits. Child labour laws were also passed. The more optimistic of reformers sought to reduce class tensions by encouraging management and labour to discuss their differences rather than submit to industrial confrontations, but these attempts of Addams and others never achieved much. Only unions or the *noblesse oblige* of industrialists could have a serious impact; some gained the eight-hour day, but many still did not.

Education and Libraries

The most successful positive social reforms were those that helped to create the middle class and to raise the poor into that class. Support for the expansion of education, libraries, and playgrounds was strong in Chicago and elsewhere. These concerns coincided with women's increased public participation and their agitation for voting rights.

Schooling was an entry point to the middle class, and probably the most significant social gains of the era after 1860 came about through education. It was done under the banner of 'saving' the children and thus redeeming the city.[32] If reformers thought that most poor and working-class adults were beyond redemption, their children were not. As we saw earlier, the drive for widespread free education began in the 1830s and 1840s. In the heroic era of expansion that followed, the public share of education spending shot up from less than half to four-fifths in the United States. As in Canada, schooling became not only free but compulsory, gradually up to age thirteen, then up to age sixteen around 1920, at least in the industrial North and West.

By 1910 in Chicago, nine out of ten children, aged seven to thirteen, attended schools, though attendance for many was irregular. Roman Catholic schools and others were hard-pressed without public funds, but sought to catch up. Attendance varied among ethnic groups. High schools became tuition-free, so enrolments climbed, by 1930, to include almost all young people up to sixteen and many beyond that age. Kindergartens were introduced, and by 1910 125 of them were established, many in poorer areas. Vocational and technical schools built after 1900 became necessities and catered to low-income young people, thus reducing the drop-out rate in the process. Women taught home economics and hygiene to young women. Sunday schools—such as those operated by Dwight L. Moody, the great evangelist, and by most churches—were less needed for secular learning, though evangelical groups continued to set up inner-city missions. In general, between 1860 and 1910 or soon after, nearly everyone had some knowledge of reading, writing, addition, and subtraction. The most important basic skill goals had been addressed.

The salvation of children was motivated by necessity and idealism, so additional goals for bettering children's lives were important. School enrolments rose after 1860 when it became apparent that child labour was increasingly redundant in cities. Machines altered employment. Child labour legislation, fostered by a humanitarian impulse, in turn helped the shift to machines. Telephones in downtown businesses, for example, restructured messenger boys into students, as did pneumatic tubes for handling cash in department stores.

Teaching basic skills and providing child care alone could hardly justify the vast expenditures in education. Schools had to serve also as locales for socialization into the virtues of the middle class, that is, to protect children from the temptations of the world; teachers prepared children for the discipline of the factory clock or, increasingly, for the routines of the office. Not least, schools sought to make citizens, especially to head off the communism that had seemed imminent in the crisis-laden 1880s and 1890s, and Bolshevism after 1917. As fewer families needed children's wages to make ends meet, by 1910 most parents could recognize the virtues of advancement through education.

Although Italians in Chicago still resisted state and church intrusion on education, which was traditionally a family matter, by the 1920s they saw, like others, that their children could move up the ladder through education.

'Child-saving' was the goal of progressive educators, and they created the first great dimension of the welfare state that marked the century to come. The school systems were clearly successful in harnessing children. Of course, they could not redistribute incomes, so undoubtedly many children, on becoming adults only to work long hours in factories or stores for low pay, would hardly believe they had been saved.

Ironically, though, building a successful education system was costly not only in funding but for democracy. To educate children required bureaucratic—even draconian—measures. Lockean preoccupations tended, as one scholar has put it, following Tocqueville, 'to promote Hobbesian solutions'.[33] Strong-willed superintendents and a school board, which was usually dominated by businessmen appointed by the mayor, laid down tough rules for behaviour. Compulsory education created the position of truancy officer. Except for a few years after 1910, the system remained so centralized that the board allowed teachers to have little input, let alone middle- and low-income parents. Through their federation, teachers struggled for recognition as professionals while seeking smaller classes and higher pay. Young women dominated the ranks of elementary teachers, and boards thought these women would likely be married soon, so they kept salaries low until after the inflationary Great War era. Still, like office work, teaching was an entry point to middle-class life.

Postsecondary institutions, such as the Rockefeller-financed University of Chicago, which had many renowned scholars, professionalized and fulfilled middle-class aspirations. Credentials helped to standardize and articulate ranks within the mid-

> 'Child-saving' was the goal of progressive educators, and they created the first great dimension of the welfare state that marked the century to come.

dle class. John Dewey's celebrated experimental school, which emphasized child-centred pedagogy, marked a clear turning-point from the boisterous unself-conscious growth of the nineteenth century. Promoting individual worth, Dewey's concept fitted well with the liberal child-saving philosophy of reformers, though businessmen who were intent on discipline resisted. Wittingly or unwittingly, attention to personal value enhanced consumerism and the status of the middle class, and would eventually undercut centralized power. Christian denominations set up seminaries; McCormick endowed Presbyterians and Moody started his Bible college as social gospellers fought conservatives and the great debate between modernists and fundamentalists over biblical veracity heated up.

Libraries and associations were also signs of the rising middle class. As in the education system, the mayor appointed trustees to expand the library system. In 1873 they built the first free library. By the end of the century, Chicago's book collection was on a par with European cities, even before Andrew Carnegie showered his fortune on large and small communities all over the English-speaking world. The reform impulse democratized the library system further as its professionals decentralized access in branches throughout the city. Libraries, like the earlier mechanics' institutes, provided adult education. The élite established historical and scientific societies, as someone said, for 'scholars and genteel dilettantes'. As with libraries, reforming professionals sought to democratize the arts. Certainly members of the now expanded upper-middle class could indulge in the symphony and opera established by the élite. Hull House endeavoured to teach art to immigrants, hoping to enhance their self-esteem.

Parks and Playgrounds

The reform impulse gathered collective action for parks and playgrounds. Until the 1890s the aesthetic impulse was the strongest and hence parks appealed to the upper half of the income scale. Following New York's creation of Central Park, Chicago instituted a park system in 1869. Three park boards, which borrowed money guaranteed by the city, built elegant parks and boulevards.

Along the waterfront, parks began to occupy parcels of land in part created by landfill. Enthusiasm for more arose, and Burnham proposed additions, including what would become Grant Park. Urban analysts admired Chicago's efforts, though women activists promoted the idea of many small beaches accessible by streetcars. Chicago was able to maintain green public spaces on most of the central waterfront because shipping and industry had already moved inland along the river and canal and to the Calumet region to the south.

Reformers pushed for more parks, in part to wean young people from tawdry amusement parks at the end of streetcar lines, or at least supplement them. As a result, Chicago shared in the continentwide movement to establish small parks and playgrounds, particularly in working-class areas where vacant lots were rare. Although some were started by ginger groups such as Hull House, by 1915 the city itself ran fifty-five supervised playgrounds. The playground movement argued persuasively that organized play would reduce juvenile delinquency and gangs and, therefore, like schools, save children. Organized sandlot baseball became very popular to keep boys occupied and out of trouble.

Then there were spectator sports. Baseball was professionalized and commercialized—the White Sox and Cubs in new parks and minor league systems. Baseball served to socialize immigrant males into thinking as middle Americans (and Canadians, for that matter, as cricket was virtually overwhelmed). The relatively non-violent competition of the game was a respite for many male workers from the daily grind of a numbing industrial routine. The professional players, who were under the thumb of owners, hardly had the security of engineers and architects. Organized hockey, which began in Canada, eventually reached Chicago.

Housing

Housing was left to the private market. Aside from a few municipal boarding-houses and miniscule philanthropic experiments, there were no direct public inputs into housing in North America during the early twentieth century, except briefly after the Great War. The issue of public housing, which was debated vigorously among a few professionals and union leaders who learned that European cities were acting on this issue, did not capture the imagination of business people. The few socially progressive reformers of the era could not dent the thick, ideological *laissez-faire* armour, and they alone were not capable of challenging the prevailing order of property rights. As we will see later, the slums remained, if gradually thinned out when those with the means to move out did so.

Morality Reform

Along with the campaigns to save the city through positive action—that is, by recognizing that poverty and behaviour were a consequence of environmental conditions—other reformers sought to cleanse the city of vice, 'negative environmentalism', as it has been called.[34] As earlier, these advocates blamed individuals for their behaviour, but they believed, like those who helped the poor, that changing the environment would help. Now the rising clerical middle class joined many professionals and businessmen in demanding a clean-up of drinking, prostitution, and gambling. Many business leaders 'attributed the spread of urban vice and corruption to immigration and popular sovereignty'.[35] If law and order could be brought to bear on union unrest and the threat of communism among immigrants, then they could also come down on the ward-heelers who maintained the systems of vice. A clean and planned city could not tolerate dens of degradation in Ward 1.

Over several decades, the forces of temperance gradually gained strength as the virtues of self-improvement and self-reliance caught the middle class's imagination. However, unlike rural Illinoians, a majority of Chicagoans were neither native born nor Protestant, so they were not easily moved to accept prohibition. Many Protestants were enamoured with the notion of prohibition. The debate was not simply between pietists and European Catholics. The temperance issue was a sharp dividing line in Chicago, and the 1907 charter reform foundered in part on the issue. Finally, in 1919 prohibition was achieved.

Prostitutes and pimps were harder to rout in a city of a thousand brothels, but after 1909, even the hitherto protected prominent bawdy-houses,

such as those of the famed Everley sisters, were forced to be less conspicuous. Open gambling, such as at the Washington racetrack, was restrained. The great moral uplifter and evangelist, Dwight L. Moody, had he still been alive, no doubt would have seen these as victories for the middle class, who were 'the back-bone and sinew of any city'.[36] Indeed, Moody went beyond liberal reformers in saving adults (mostly the alienated native born) for middle-class rectitude and the recovery of rural virtues.

But the outcome of the negative cleansing campaigns would create new problems. Indeed, the dirty trades were driven underground as the ward bosses lost control. The illicit booze trade became an opportunity for organized crime in the 1920s. Prohibition thus created a new kind of free enterprise. The extent of Al Capone's venture into the 'creative destruction' of capitalism should not have surprised anyone, but then America and Chicago were known more for constant fluctuations and change than for stability and regulation. Clearly negative reform had its limits. Positive reformers, on the other hand, sought to bring order out of chaos and to improve the lot of the poor, though they could hardly overcome the great inequalities in income and wealth that persisted. Having achieved some social ends, the push to do more petered out in the 1920s.

Shaping the Physical Environment: The Heroic Era of Servicing the City

The élite politicians and the reformers, despite frequent conflict, managed to deliver many public services, far more so than New York did in 1860, a 'triumph', as one writer has claimed.[37] Indeed, the plan of 1909 capped considerable civic accomplishment in Chicago and in other great cities of the era, but the limits of improvement were reached before all residents were adequately served. We will consider the interplay of public and private action on Chicago in several respects, all of which had social implications: annexations; water and sewers; other public health measures; gas, electricity, and public transportation. They provided a context for the actions of speculators and developers. These issues lead us to a closer look at some of the salient spatial results on the ground.

Annexations

On 29 June 1889 the City of Chicago annexed an enormous area on its edge, quadrupling the space under its jurisdiction from 43 to 169 sq. mi. (111 to 438 km^2) (Figure 5.1). It continued to take over small areas, so the total reached 190 sq. mi. (492 km^2) in 1910. As in most other successful cities in the mid to late century on the continent, Chicago's leaders sought to enhance the status of their city by annexing suburban subdivisions, organized communities, and even large swaths of farm land, much of it held by urban speculators. When it was incorporated as a city in 1837, Chicago's area was a mere 2.5 sq. mi. (6.5 km^2). By 1869, after receiving permission from the state legislature, it expanded to 35 sq. mi. (91 km^2). No more space was added until 1887. During this time a large number of new governments on the urbanizing fringe replaced the rural townships. The state incorporated a few of these townships, the largest of which included several communities. The state also incorporated urban villages. In the mid-1880s the pressure to provide services persuaded outlying residents, speculators, and their politicians that the surest way to well-being was to let the city do the job. City dwellers and leaders were also enthusiastic about quadrupling the space governed by the city corporation. Lake, Hyde Park, Jefferson, and Lakeview were the largest additions. Annexation contributed to the doubling of Chicago's population in the 1880s; its 1.1 million population in 1890 equalled that of Manhattan, which spurred the municipal expansion of New York in 1898.

Most North American cities annexed large parcels of land, with state or provincial government approval, in the latter half of the nineteenth century. For example, Philadelphia took in the whole county, and Buffalo expanded, as did Montreal. Boosterism, engendered by rapidly growing populations and a booster legacy of the Jacksonian era, inspired annexations. Growing public demand to protect property in some nearly ungovernable marginal areas led to takeovers so that police could bring these areas under control. Undoubtedly the strongest reason was to apply economies of scale to the mounting costs of providing increasingly sought-after water and sewer lines, as well as paved streets and sidewalks. Speculating subdividers were anxious to shift the costs of services onto the tax base of the central city, even though many arranged earlier for incorporation of their areas into towns to lift the burden from their own shoulders. Many central cities were willing to pay these expenses. Some annexations were the result of marginal communities that defaulted on their debts, often because they were too generous in subsidizing manufacturing plants. The annexation process was halted in some other places, as in Boston and St Louis, even before Chicago's great annexations of 1889. In 1899 Mayor Harrison shied away from further annexation because Chicago was 'so spread over an extensive and unproductive territory ...'[38] After 1900 the pace slowed markedly because of resistance from suburban (particularly affluent) communities that were provided with services, often through special-purpose metropolitan bodies. Some central cities balked at further costs. Toronto stopped annexation in 1912, and Los Angeles also stopped virtually by 1913. Chicago added only smaller parcels of land; Evanston, Oak Park, and Cicero (the other large places in Cook County) remained unannexed.

Public Works: Water and Sewers

The vast public works that were undertaken earlier in the century, as we saw in New York, increased extensively later in the century, the great era of urban environmental improvements. Annexations encouraged the spread of these works. American cities, followed closely by most Canadian cities,

5.1 *(Opposite)* Chicago annexations, 1833–1910. Like most cities, Chicago added pieces to the city. Resistance to annexation grew after 1900. Chicago's biggest expansion year (1889) was just before the depression of the 1890s. Builders then did not put up houses for many years on premature subdivisions in most of these large parcels of annexed territory. *Source:* Adapted from the Chicago Community Inventory, University of Chicago.

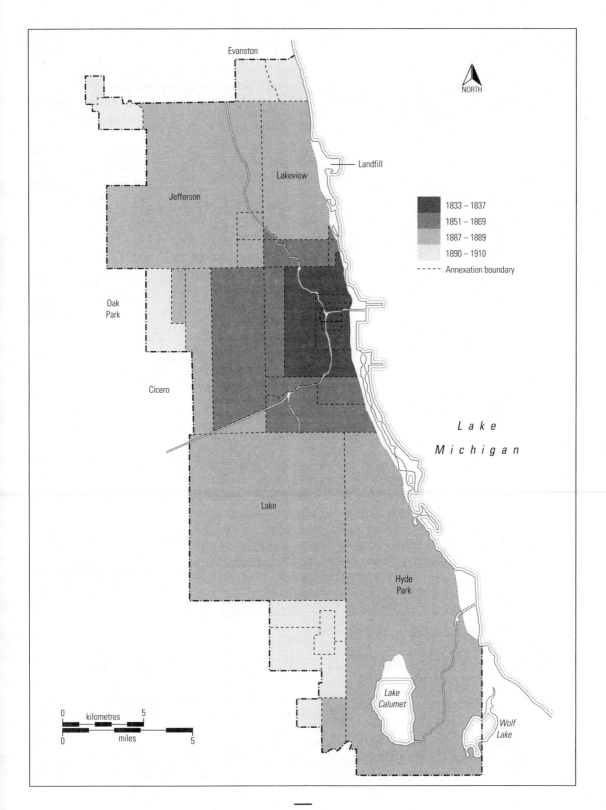

Evanston

Landfill

Lakeview

Jefferson

	1833 – 1837
	1851 – 1869
	1887 – 1889
	1890 – 1910
- - - -	Annexation boundary

NORTH

Oak
Park

Cicero

*Lake
Michigan*

Lake

Hyde
Park

*Lake
Calumet*

*Wolf
Lake*

kilometres
0   5

miles
0   5

vaulted ahead of western Europe in quantity (if not quality) of services by 1900, and if not in social services. This suggests that despite the ferocity of politics, local and state governments and political parties were sensitive to needs, and that there was money to construct and maintain these works. Corruption may have siphoned off some money, but cities made improvements yearly, thanks to experts who were tenacious and rose above the petty squabbles. Indeed, these improvements contributed to the sense of civic pride and to what I call the civic moment early in this century. The rapid increases in population put pressure on cities to act, but also the awareness of germs highlighted the problem of water quality. Likewise, the rising preoccupation about children's well-being, as shown in education, was also reflected in concerns for public health. The fall in death rates, if erratic and regionally varied, indicated success.

Water remained a central issue for policy makers and engineers until nearly 1910 when most cities established systems to provide adequate quantity, high pressure, and improved quality of water for fire-fighting, industrial, commercial, and residential use. Waterworks were already in place in many cities by 1860, but, like New York before 1835, still relied on private investors. Professional fire-fighters replaced volunteers by the 1850s in New York and also in Chicago. Chicago's great fire of 1871 encouraged insurance underwriters to press for more sophisticated equipment and higher water pressure. Still, even early in this century, demand mounted for more water pressure, up to and beyond that needed for the great fires in Baltimore and elsewhere. Even by 1900 American cities were better equipped than European cities, though continued construction of wood-clad buildings susceptible to fire meant a greater need for more water. In the wake of the 1871 fire, city council prohibited wooden buildings in Chicago's central area, though it had trouble in enforcing this. In eastern cities, brick was required more frequently.

By the 1870s in large cities, it was apparent that to ensure wide availability of water, water systems should be public. Apparently businessmen who calculated the risks decided there was too little profit in universal systems that would serve all people. When Chicago became a city in 1837, engineers laid a pipe into Lake Michigan. After the Civil War, city engineer, Ellis Chesborough, ingeniously expanded the system; for gravity flow, a steam engine drew water up into the famous water-tower. Although the system was no match for the great fire, the clean-up provided an opportunity to establish more water lines. Between 1890 and 1902 Chicago nearly tripled the mileage of the water system. By then, American urban dwellers used about four times as much water per capita as Europeans, but working-class homes were far less likely to have the amenities of a tub and water-closet than those of the rapidly expanding middle class. Immigrant home-owners and slum landlords resisted paying for plumbing. Slowly following the lead of Europe, working-class citizens resorted to public bathhouses. Public beaches in the summer helped the cause of hygiene, once they were cleansed of sewage.

The quality of Chicago's water, like that of many American cities, fell behind Europe's, however. This exemplified the American penchant for quantitative improvements as opposed to quality. Polluted water remained a problem, as the death rate in Chicago remained above European levels until 1900. In 1893 a British engineer bluntly stated that residents should shun Chicago's water, but, since obviously people could not, recommended that they boil it. Water filtration and then chlorination lagged behind European systems, though by 1910 American and Canadian cities were finally catching up. Treated water flowed to one-third of the American urban population; cholera had disappeared, and typhoid nearly so. The exception was Pittsburgh, where mortality rates remained high because the city still had a polluted river supply. But as cleanliness had risen in status next to godliness, cities strode ahead.

Much of the water pollution was the result of dumping raw sewage into waterways, and American cities were behind European cities in treating sewage. Sewer lines that ran into waterways supplying water exacerbated water quality downstream. In Chicago, inshore pollution necessitated the extension of water-pipes farther into the lake. Chicago extended sewer lines under its streets; in fact, it had more miles of these lines than New York by 1902. It banned privies in 1894, though new solutions came slowly in working-class areas. Following Boston, water and sewer districts were created. To cope with the rapid increase in population, Illinois created a metropolitan sewage commission, whose Sanitary and Ship Canal alleviated much of the problem, at least for most of the city when it opened in 1900 to the acclaim that it was the grandest public works undertaking ever. Thereafter, the Chicago River flowed southwest more often than into Lake Michigan, but towns downstream on the Des Plaines River had to cope with untreated sewage from stockyards. By 1910 Chicago and some other cities began to treat sewage in aerated holding basins in response to lakeshore pollution, though politicians in most cities were mindful of costs and therefore saw less urgency in sewage treatment if incoming water was adequately cleansed. At the time, only a tenth of the American urban population enjoyed some kind of sewage treatment. Even by 1930 the proportion rose only to one-quarter of the urban population. Some large cities, such as Montreal (which has a natural flush down the St Lawrence), have yet to provide full sewage treatment, and others still have none. Industrial toxic wastes, in the age of new chemicals, were left for future generations to worry about.

Public Health

Other measures to improve public health were pursued with greater vigour than earlier. Vaccinations and inoculations became more common. After being pressured by reformers, including public health officials, cities began inspecting slums to clean up garbage and enforce the tenement law of 1902 that specified water hook-ups, inside toilets instead of privies, and a certain amount of cubic sleeping space per person to prevent overcrowding. Hull House advocates made an effort to ensure enforcement of these measures. Their surveys testified to the conditions. By 1910 considerable gains were made in indoor plumbing, but landlords often bribed housing inspectors to save on costs. Whereas earlier hospitals were havens for sick indigents after germs were discovered, large general hospitals that were partially financed by philanthropists began to treat middle- and upper-income people. In the wake of Sinclair's *The Jungle*, federal governments passed pure food acts in both countries.

Air pollution in cities from the burning of bituminous soft coal was severe; in 1905 H.G. Wells complained that Chicago's smoke 'has a reek that outdoes London'.[39] Belching coal dust from the thousands of trains and factories blackened buildings; the White City of the World's Fair could hardly cover up that fact. Without the use of cleaner hard anthracite from eastern Pennsylvania for domestic heating, conditions would have been worse. Chicago conducted its first study of pollution in 1912. Campaigns to alleviate the pall and noise were, however, largely unsuccessful in an era when smoke signified progress.

But waste became a bigger problem. European visitors complained as they did earlier about the lack of cleanliness in the streets, though New York was doing better. Cities looked after street cleaning and garbage and ash disposal, but safe disposal of mounting waste from an increasingly affluent city resulted in more and more dump sites. The problem of street paving was finally resolved in the 1890s when asphalt began to replace planking, cobblestones, granite blocks, and brick. For sidewalks, planks gradually gave way to concrete.

Gas, Electricity, and Public Transportation

In contrast to water, sewers, and streets, gas, electricity, and public transportation remained in private hands in most American cities, but not without a great deal of controversy over whether regulation was adequate to ensure good service or whether public ownership would be needed over these monopolies. Some degree of public control of these major services, which were increasingly seen as necessities and no longer just luxuries, brought the business reformers into political action because they depended on these services. Gas was fair game for political corruption. In 1865 when Chicago's city council let them, two companies colluded to set rates for gas, which was then used widely for street, business, and upper-income residential lighting. Later in 1887, investing Philadelphia capital, Charles Yerkes organized the Chicago Gas Trust. After 1897, when Samuel Insull took over, Peoples Gas Light and Coke Co. monopolized the field. Controversy over gas rates continued for some time.

By 1892 Insull, financed by Chicago capital, took control of electric lighting through Chicago Edison. For street lumination, arc lights displaced coal gas jets, and were then almost completely superseded by Edison's incandescent bulbs by 1910. By then electricity was replacing gas in homes. Gas was increasingly used for heating and industrial purposes. As in every city, through the 1890s electric trams replaced horse trams. Insull's success was assured when he was able to replace dynamos in business buildings with lines from his coal-burning central station on Fisk Street in 1903. To do that, he had to overcome the costs of cyclical daily and seasonal uses of a commodity that could not be stored. He was able to work out a two-tier pricing system for business and residential use. Use of domestic appliances and industrial electrical motors grew gradually, creating 'The Electric City'.[40]

While Insull's name became synonymous with private power, magnates in other North American cities also saw great possibilities for revenue. The Buffalo World's Fair of 1901 highlighted the virtues of Niagara power to consumers. Like all utilities, competition would be short-lived. Almost inevitably, electricity became a monopoly. While most Americans accepted private but government-regulated monopolies, Canadians outside of Quebec were far more amenable to gas and water 'socialism' emanating from Britain where Joseph Chamberlain and his Conservative colleagues created municipal ownership. While telephones and gas remained private but tightly regulated, Toronto and other cities gladly agreed with Ontario Conservatives on the state ownership of power in 1906. The private monopolies that supplied power were routed. Local commissions retailed provincial Niagara hydro.

> Like all utilities, competition would be short-lived. Almost inevitably, electricity became a monopoly.

In 1921 Toronto also put street railways under municipal control. Chicagoans would have to wait until 1947, though in 1906 they nearly owned their trams. The 'upper stratum' held Yerkes 'in contempt ... because of his recent arrival, the unsavory connection between traction franchises and municipal fraud, notorious stock-watering, and inefficient and unsafe service.'[41] Politicians lined their pockets with kickbacks in the 1890s during the era of electrification. Boodle was everywhere, Steffens reported. Businessmen and other citizens perceived long-term franchises, but there was insufficient revenue sharing with the city, inadequate guarantees for maintaining roadbeds on streets, and haphazard service schedules and crowding during rush hours, as well as long waits at crosstown transfer points.

In 1905 Edward Dunne was elected mayor on a platform of placing transportation under municipal control. A majority of citizens agreed. However, he eventually temporized, to the disgust of advocates. Preoccupied with other issues, Dunne agreed to a compromise: Yerkes agreed to several improvements, such as more and newer equipment, and provided more frequent service, thus preventing the municipal take-over. In 1913 Yerkes unified the street railway system with the Chicago Surface Lines, thus ending the bitter franchise fights and boodling by politicians. Yet the rapid transit elevated lines and suburban commuter rail services remained separate and private, and not integrated with streetcars through transfer points. When we consider circulation later, we will see that the various transit systems worked better than they had earlier, but by the end of the Great War, inflation pushed them into financial difficulties, and in the 1920s the auto would undercut their ability to survive.

Taxes

Paying for the new larger scale of services was a fiscal challenge for politicians. Cities relied on five methods to pay for ever increasing projects.

- First, general tax revenues (chiefly on property) became the main method. Unlike most large cities, in Chicago the county assessed, collected, and allocated funds back to the city.

- Second, the city charged special or local improvement taxes against the abutting properties that benefited directly from paving and sewer lines. This was the main development in Chicago between 1845 and 1860. The pace of development was thus put in the hands of private owners who could choose to pave or put in sewers. This tax system failed because poorer owners and landlords resisted paving, and people other than adjacent owners used the streets, hence maintenance costs had to come out of general funds. It was recognized that most works were public in nature and served the whole city. As elsewhere, the city continued to use special assessments, though as an increasingly minor supplement to general taxes because a rational, fair system was hard to figure out. In a volatile land market, some owners went bankrupt, so taxes were unpaid.

- Third, for capital projects, cities borrowed up to the limits that states would permit. The bonds were eventually retired through tax revenues. In the 1880s, they were reluctant to borrow, even though pipes were being laid. Politicians and local electors were more favourable to bond issues after 1900. There was a continual struggle over balancing direct and indirect beneficiaries and immediate versus long-term expenses.

- A fourth method—fees for permits to draymen, liquor outlets, and others—brought in some money, though relatively less so than a century and a half earlier.

- Finally, more so than many other cities except Boston, the state created special-purpose agencies in and around Chicago to operate parks, sewers, and other services; these agencies had power to borrow and could therefore avoid local politics. Thus the per capita debt of the city itself was lower than elsewhere, but led to fragmented governing.

Residential properties carried a proportionately larger share of the tax burden than commercial properties, according to critics. Assessors vastly undervalued the properties of larger businesses and, according to one estimate, assessments were usually less than 12 per cent of their actual revenue-making power. Comparing residential and commercial values is, however, like trying to compare apples and oranges. Businesses claimed that they generated wealth and income, while residences did not, so that lower commercial rates were not, they averred, really subsidies. Thus debates over using market value of properties for assessment—that is, the presumed selling price compared to nearby properties—would continue to the present.

Land Speculation, Development, and Building

The heroic era of servicing did more than open up land for development and redevelopment of commerce, industry, and residences. Speculators (who sold and resold) and developers (who subdivided property into lots for building) shaped the horizontal dimensions within the template provided by original rural surveys of mile (1.6 km) by mile, then later half-mile by half-mile squares marking arterial roads. Within the latter, three east-west streets usually crossed by seven north-south streets. On the edges of the city, the developers often ran ahead of services, so they sought annexation. Thousands of builders, few of whom were subdividers in this era, cashed in on the construction. Architects did well on the high-value projects. Before looking at the spatial patterns, we need to consider the process of creating the cityscape.

Usually well beyond the built environment, speculators subdivided farms and fields, more so in fast-growing Chicago than elsewhere, at least until 1891. Often subdividing went through two or more stages. On land that was initially subdivided, market gardeners often grew high-value vegetables and flowers for consumption in the city. As building on this land came closer to realization, the same speculators (or more often other developers) often resubdivided their parcels again into small building lots. They roughed in local streets and demanded services from the city or suburban communities if developers were willing to pay special local improvement taxes. They sold lots to small-scale realtors, and advertised to the public and auctioned off properties if they were in a hurry to sell. In some cities, as in Boston, developers owned street railways and extended them into their distant subdivisions. Yerkes obliged some prominent developers in Chicago by extending his lines out to new subdivisions. Commuter rail and elevated lines in Chicago opened up some new areas that were far from the city centre. Some developers, as we will see in the discussion of the spatial order, developed industrial suburbs; big industrialists even developed satellite cities, while others established suburban retreats near commuter rail stations. In the older areas, notably in the central district, financiers and owners redeveloped properties for offices, stores, hotels, and the like. On the edges of downtown, speculators held run-down slum residential lots for commercial and industrial redevelopment.

Speculative real estate was like a roller-coaster, which was also an invention of the era. To ride the peaks and troughs was thrilling, unless one fell off on the way down and many did. As elsewhere, major peaks were reached in 1836, 1856, 1869, and 1873 when the pace was frantic. Expectations of future growth during the astronomical boom of the 1880s, which peaked in 1891, resulted in the laying out of about 125,000 lots, some of which were even beyond the 1889 boundaries. In 1890 about half of the city was already subdivided far beyond immediate needs. Ordinary middle-class and even working-class men and women were drawn into the maelstrom of speculative madness. Drummers visited offices and factories to persuade the gullible, and there were many. Samuel Gross, who was the greatest developer and builder of the era, rented trains to bring prospective buyers to some of his numerous projects (Grossdales) on the edge of the city. So popular was Gross, 'the Napoleon of house builders', that he was the candidate for mayor under the Labor banner in 1889.[42]

Real estate values fell dramatically after the land boom, as they did after the earlier peaks, and presaged the 1893 financial panic in stocks and commodity futures. All this activity underlined the excesses of the gilded age. Banks went broke after lending to excessive levels in non-productive real estate when industry could not absorb their capital. Many people in the thrall of speculative power saw the value of their suburban dream investments vanish. The city and suburban municipalities had to take over the properties of people who defaulted on taxes and absorbed the cost of unused services. An unwilling landowner, the city then unloaded its sheriff-sale lands at fire-sale prices. Only in 1910 did aggregate assessed land values reach the 1891 level.

After economic growth renewed, Chicago speculators shied away from subdividing because there were so many excess lots. Around Chicago in 1910 the surplus of lots was still twenty-five to fifty times the foreseeable need. The speculative bug was so slow compared to the earlier frenzy that many thought the excesses of the previous century had passed into an era of greater stability. Yet in new and rapidly growing cities, particularly in western states and provinces, speculation continued until the economy levelled off in 1913. Saskatoon, which was perhaps the most extreme case (at least in Canada) had enough subdivided land for over sixty times as many people as its population!

Land-use planning as a recognized discipline arose partly because city officials and professionals were worried about this 'premature' subdivision. They had sufficient power to persuade legislators in Wisconsin, Ontario, and other states and provinces to pass subdivision-control laws over lands on the margins of big cities. Actually when the economy slowed down soon after, it appeared in retrospect that these laws were no more effective than changing the weather. In cities that experienced substantial growth in the 1920s, speculation fever returned once again. In 1926 a new peak in subdivision and land values was reached on the margins of Chicago, and even higher around rapidly growing Detroit and Los Angeles. The cautious predicted yet another panic, the greatest of them all in 1929.

On the edge of settlement, builders often operated far behind the subdividers, but followed roughly the same cycles. After 1891 building began fifteen to twenty years after subdivision. Residential builders operated almost entirely on a small scale, constructing only one or a few houses a year, mostly on speculation. The vast production of housing for the increasing population was made easier by balloon framing, using white pine and cheaper conifers when the pine became scarce. The affluent could afford custom-built structures; builders often followed pattern books, which resulted in considerable standardizing of appearance and size. They usually subcontracted out certain jobs. Falling prices of materials until the mid-1890s led to larger houses for the rich during the gilded age, as well as for the rising professional classes. Thereafter, smaller units, including bungalows, increasingly became the order of the day, while in Chicago, apartment structures for the middle class were in demand the most. Costs also rose with the addition of new necessities: gas and then electricity conduits; furnaces for central heating, which necessitated basements with concrete floors and walls; and plumbing for toilets and bathtubs.

If speculative builders constructed most of the houses, including subdivider Gross, many owners of small plots put up their own houses with the help of neighbours. Developers who held too many lots in subdivisions laid out in the 1880s were willing to sell them cheaply. In the best-known case, on the margins of Toronto after 1900, British immigrants built their own small and unserviced 'unplanned suburbs' in what the press derisively referred to as 'shack towns', even though, unlike the squatting Irish in New York earlier, they held freehold title to the property.[43] Self-builders put up one-third of houses between 1901 and 1913, many in areas that were subsequently annexed to the city in that period. The poor built in some locales beyond Chicago's boundaries. Speculative and self-builders' small-scale operations often resulted in scattered construction in clumps or even single houses in the newest subdivisions. Building was spread out over many years on many subdivisions, and depended as much on business conditions as on the process of subdivision itself.

These development and building processes give the impression that the sequence of speculation, subdivision, and building moved more or less gradually outward from the city. This picture has to be modified. Some large companies built separate satellite industrial towns on the margin, a few 'community builders' created satellite residential communities, and many earlier outlying places added housing.[44] The railway companies also built extensive railway-yards, some with adjacent industrial areas, so the overall landscape had certain peaks beyond the city. People of means also redeveloped the centre of the city in dramatic fashion. Let us turn to the geography of the city, first the central area, then circulation, manufacturing clusters in satellite cities, and finally residential patterns. Burnham's specific proposals will lead us into Chicago's geography, starting with the centre, the focus of greatest attention.

The Plan and Chicago's Dominant Centre

By 1910 Chicago was the archetypal spatial model of urban places of the era with a strong centre, the Loop, from which radiated roads, rail, and transit over a nearly flat plain. The star-shaped city replaced the triangular or elliptical walking city. It would not be long before the star shape gave way to a circle when the car permitted the infilling of the interstices. The 1909 plan was designed to cope with what Burnham and others thought were inadequacies in the spatial order, imperfections that they envisioned could be corrected through what might today might be called an economic strategy. I will assess the plan against the actual layout of Chicago around 1909.

Although interpreters of Burnham's plan have stressed parks, beautification, and circulation, one can argue that the plan was primarily concerned with the power of the centre, to maintain the Loop's business dominance over the city and the region. Indeed, Burnham imagined a grand civic centre at Congress and Halsted, sitting on the 'center of gravity' in the city[45] (Figure 5.2). A wide bridge would help break the river barriers. Simi-

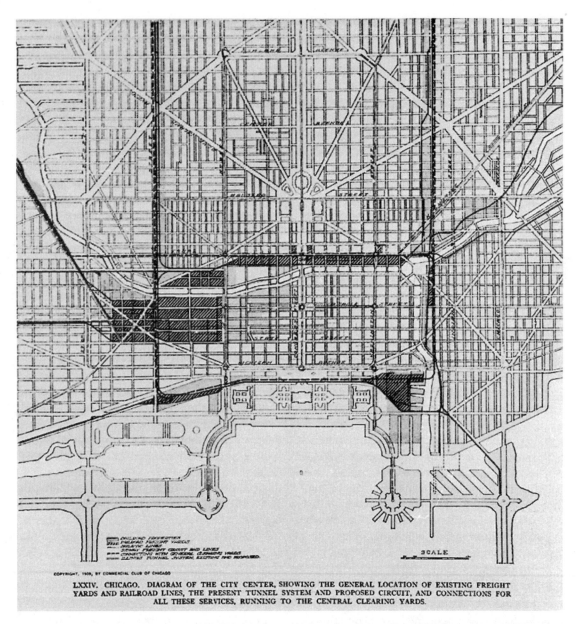

LXXIV. CHICAGO. DIAGRAM OF THE CITY CENTER, SHOWING THE GENERAL LOCATION OF EXISTING FREIGHT YARDS AND RAILROAD LINES, THE PRESENT TUNNEL SYSTEM AND PROPOSED CIRCUIT, AND CONNECTIONS FOR ALL THESE SERVICES, RUNNING TO THE CENTRAL CLEARING YARDS.

5.2   Burnham's plan for the Loop area. Daniel Burnham proposed a grand city centre of Parisian style with elegant parks on the waterfront and symmetrical diagonal streets to cut through slums; he also suggested straightening the river and relocating rail stations. *Source:* Daniel H. Burnham and Edward H. Bennett, *Plan of Chicago*, edited by Charles Moore (New York: Da Capo, [1909] 1970.

larly, moving most of the passenger terminals to a row along the south side of 12th Street (Roosevelt) and thus removing acres of tracks would allow the central business district (CBD) to expand to the south. Improved bridging of the main stem of the river would also allow the commercial centre to expand to the north. Straightening the south branch of the river would expedite water traffic. Diagonal arterial streets would enhance access to the centre, as in Detroit, more to promote the newly invented automobile than to improve transit. In short, the plan's key concern was to strengthen the centre to encourage office development and alleviate congestion. Little wonder, it seems, that prominent businessmen were enthusiastic about promoting the plan to citizens and schoolchildren.

Certainly, the plan has been seen as the culmination of the City Beautiful movement foreshadowed in the White City of the 1893 World's Fair, as expressed in the monumentality of the classical square and of the baroque formal gardens on the waterfront. After all, a cardinal quality of Chicago's great buildings was their size. The fact that the plan was exhibited first in the Art Institute attests to this. The plan also advocated the City Efficient through its dramatic proposals for rail and highway traffic. Socially, the plan was the high point in the élite's attempt to direct (if not control) the city: the top architects and businessmen saw their interest and that of the city, or the CBD, as one and the same. Ward 1 boss, Bathhouse John Coughlin, approved the plan, which underlines its compelling case (though perhaps he doubted whether its chief proposals would ever see the light of day). The plan was thus an embodiment of the City Controlled, even if its proponents did not say so. To them, improvement was for the public good.

The CBD was distinctive: more than in any other city, the 'steel girdles' of rail lines hemmed it in. Unlike New York and London, Chicago's CBD did not split into two. The plan would open the way for an even more extensive central district, not a new major secondary one.

By 1910, more clearly so than in 1860, a hierarchy of central power as expressed through land values emerged. Powerful redevelopers recognized opportunities. Central districts of large cities were composed of clusters of several main activities. Here we consider seven such activities, and then circulation. In this era of corporate dominance, government continued to claim centrality; Burnham would enhance its symbolic presence. Financial institutions, corporate headquarters, and city and downtown-serving professionals occupied Chicago's office buildings; they were a powerful force in shaping the centre. Far more so than

Broadway earlier, retailing occupied more blocks and block faces. On the other hand, as in Manhattan earlier, wholesalers were pushed to the periphery of the district and even beyond because they were unable to pay the mounting land costs that peaked around 1910. Department stores and offices, on the hand, could afford the costs. City and regional entertainments managed to expand in part by sharing space. A wide range of hotels accommodated visitors, most of whom arrived by rail. On the CBD's margins were the great train stations of regional and national significance. Also peripheralized was what writers at the time called vice: prostitution and gambling. We will not consider manufacturing until later because it was much reduced downtown. The great fire of 1871 had eliminated housing. Let us consider the major dimensions and the shape of the downtown, and how Burnham's specific suggestions would alter the scene.

Government

Government buildings were bold and prominent as they were in most cities at that time. The Federal Post Office and Customs Building opened in 1905, replacing the one built on the site in 1879. Its monumental size, crowned by a great dome, complemented the classical city hall and County Building four blocks to the north at Washington and Clark, which was built between 1909 and 1911 to replace an 1885 structure. Municipal government was in the throes of expanding its office complement, as the paperwork entailed in promoting technical efficiency grew. The central library, which was previously in city hall, was housed in 1897 in an impressive new structure facing the lake on Michigan Avenue, a sign that literacy was more widespread. Except for the courts, the state of Illinois was then visually inconspicuous downtown, though obviously Springfield was often on the minds of civic leaders and politicians who had to ask for enabling legislation.

Burnham's great civic square at Congress and Halsted would have accommodated buildings for all levels of government. In keeping with the civic high point, his city hall would be pre-eminent, flanked by the federal and county buildings. Not only that, the dome, in the artists' drawings, would have rivalled the Capitol in Washington in size and height. It may seem curious, then, that the city went ahead with a new city hall at the time of the plan, on its former site far from Burnham's centre, as if to deny Burnham his goal. Perhaps the architect was musing about the future; Chicago was in the habit of tearing down the old to build the new. In the early 1930s, the post office would in fact move into a huge building straddling Congress, but the civic square site would become a freeway interchange, which was symptomatic of a later age not inclined to civic rectitude. In fact, new civic squares were few and far between after 1910.

Skyscrapers: Chicago's Glory

Chicago's reputation in the world rested in large measure on its skyscrapers, which by 1910 reached up as high as twenty storeys, overshadowing the new public buildings. The great architects—William Jenney, John Root, Daniel Burnham, Louis Sullivan, and others—became internationally renowned by designing office, civic, and retail buildings that captivated the attention of élites elsewhere between the 1870s and 1900s. They became known as the Chicago School, claiming precedence over New York by emphasizing function over form and appearance. Not quite 'modern' in style, their office buildings still reflected classical lines, with flat roofs and cornices as if they did not dare to supersede the spires of churches over which they now towered. Oddly, however, Burnham's artists suggested classical office structures of lower uniform height, more like Haussman's Paris, which undoubtedly inspired his plan. Burnham himself, though, designed taller ones with little sense of uniformity in height.

The Chicago architects began their odyssey in new designs after the fire of 1871. It provided an opportunity to break out of the 'district's prison

> **Chicago's reputation in the world rested in large measure on its skyscrapers, which by 1910 reached up as high as twenty storeys, overshadowing the new public buildings.**

walls' to build higher (six even seven storeys), surpassing the only earlier tall buildings, which were the flour mills and grain elevators.[46] Over 300,000 sq. ft (27,870 m^2) of office space were added in 1872 alone. After slowing down, construction sped up in the 1880s until 1892. The architects transformed space using new techniques. The Montauk with ten floors opened in 1882. Burnham's Monadnock building of 1889 with its sixteen floors reached the limits for load-bearing walls, which were 6 ft (1.8 m) in width at the base. Chicago architects and engineers were innovators of foundations for an area that was still spongy from the glacial era: concrete cushioning was first used in 1885. Modern Portland cement came into use at the time.

Two crucial innovations—the steel frame and fast, electrically driven elevators—led to higher buildings. In 1885 William Jenney employed the steel frame in the Home Insurance Building. Following various precedents in cast-iron and wrought-iron framing in Europe and America, Jenney broke the barriers to achieve the design of the modern skyscraper. Indeed, the steel frame 'completed the most radical transformation in the structural art' since the Gothic in the twelfth century.[47] The quality of steel was strengthened sufficiently so that beams could bear the weight of those above. Unlike the mills of old or even the Monadnock, the walls no longer had to bear the weight; they became curtain walls, thus permitting more windows and light. The elevator was nearly as important a people mover as the transit lines that brought workers and shoppers to the centre. Electric power in the 1880s superseded the steam and hydraulic lifts that were used as early as the 1850s in New York hotels and Chicago in the 1860s. Another improvement was fireproofing to protect workers.

Behind the architecture and the engineering was the demand for space. Suppliers willingly catered to the great American desire for displaying wealth to gain prestige. Corporate bureaucratic organizations emerged in the late nineteenth century, and armies of male clerks (who were now using Hollerith calculating machines) were augmented by hosts of young women sitting in front of Underwood typewriters. Increased office work created the need for buildings exclusively for office use, and the quality of the workforce changed. Still, only a few corporate offices occupied entire floors at the time. Many others crowded into the Loop's new office buildings: lawyers and accountants who served corporations, business schools that trained young women, as well as top realtors, doctors, and dentists occupied much of the space within the elevated loop rapid transit of 1897 from Lake to Van Buren, and from Wells to Wabash. They also used the many postfire buildings still standing. Demand does not clearly explain why buildings had to be built higher and higher. In Europe, businesses expanded their offices without resorting to skyscrapers. In Chicago the office construction could have extended across the river to the west as Burnham envisaged, and did in fact to the north after 1920.

The developers were a driving force; they had the money to hire architects, who fulfilled the expectations of successful people and institutions for power and image. Among the prominent clients of architects in the 1880s were Peter and Shepard Brooks of Boston, who financed the Montauk, the Mallers, and the Monadnock. The Women's Christian Temperance Union and the Masons also engaged architects. The Masonic Temple, which had mostly offices, rose to twenty-two storeys. The 1880s to the early 1890s was a period of intense speculation, not least in office construction, that satisfied property owners' and non-profit organizations' drive to power. Upper floors were less valuable at first than they were later when elevators were faster. Prestige was a commodity.

Considerably overbuilt in the speculative era of the late 1880s to the early 1890s, skyscrapers had high vacancy rates through the depression until after 1900. Although office complements kept growing, the pace of construction was less frantic than it was earlier when land values soared

between 1885 and 1892. In 1910 the amount of floor space added, 1.2 million sq. ft. (111,524 m²), exceeded that of 1892 for the first time, bringing the total amount to nearly 10 million sq. ft. (929,000 m²). A new peak in construction was reached between 1911 and 1913. During that time the Loop enjoyed the highest land values relative to the rest of the city.

After slowing down, construction of the temples of commerce increased again in the frenetic 1920s, peaking in 1927. In 1930 the total office space had reached three times that of 1910 and ten times that of 1872. With buildings north of the river, such as the Gothic Tribune Tower and the Wrigley building, architects now dared to point towards the heavens.

In Chicago, the last building of the overbuilt era was the Field building of 1934, which was forty-three storeys. Its austerity presaged the next great round of office construction in the modern style, beginning in the 1950s. In the meantime, New York had beaten Chicago in the race upward, first with the Woolworth building in 1913 and the Empire State in 1939.

Focused in 1910 on LaSalle south of Washington, the core high-value financial district appropriated many skyscrapers. In 1865 the Board of Trade, the anchor of the financial sector where commodity futures were traded, moved south from Market to Washington, then leapfrogged farther south to Jackson in 1881. In the late nineteenth century, dramatic shifts like this occurred in many CBDs. Banks, brokers, and lawyers followed. Then the expansion of business forced new construction northward along LaSalle. The Board of Trade building was superseded in 1889 by the third Chamber of Commerce building at Washington and LaSalle. There in the 'pits', men gambled on wheat, corn and pork futures, and telegraphers sent the results of this 'bedlam' of commerce to the world.[48] Across LaSalle was the Chicago Stock Exchange. By 1910 the financial district was anchored, remaining there until the present, though in some new buildings.

Retail Stores

The Chicago architects also designed the great department stores in the Loop (some as high as office buildings); they dominated retailing. Chicago followed New York in spawning famous 'palaces of merchandising'. Stewart's 1846 Marble Palace on Broadway was followed in Chicago by Potter Palmer's in 1852. Every big city had as a trade mark the familiar names of prominent shop-keepers: Wanamaker in Philadelphia and New York, Hudson in Detroit, Eaton in Toronto, and so on. In affluent Chicago until 1910, twenty depart-ment stores vyed for the dollar almost exclusively along State Street, the 'great street' near Madison. That corner had the highest land value in the city. State replaced Lake after 1867 when Palmer, who was often said to be 'the first merchant prince of Chicago', bought out the owners of several blocks and persuaded the city to widen the street and raise the grade to overcome the high-water table and permit sewers. By persuading the Field, Leiter & Co. department store to move to State and Madison, and by building the first Palmer House, Palmer influenced other retailers to locate on that stretch, where retailing in the Loop has been located ever since. The department stores had lav-ish buildings with domes, skylights, and great staircases. Now they created irresistible fantasies in their display windows, thanks to designers like Frank Baum, best known for his *Wizard of Oz*.

Marshall Field expanded his store several times to take up a whole block; eventually its six-teen floors had 50 acres (20 ha) of floor space. One commentator in 1907 saw it as 'more than a store. It was an exposition, a school of courtesy, a museum of modern commerce.'[49] 'To give the lady what she wants', selection was among the largest anywhere.[50] Marshall Field had 'a switchboard large enough to take care of the entire business of a city the size of Springfield, the capital of Illi-nois'.[51] Field provided a school to teach stock-boys and stock-girls their fractions, and a children's playroom that kept children occupied while moth-ers shopped.[52] Until he was at least seventy, Field, 'the Master', was relentlessly in charge, 'seated in his glass cage, like a spider in his web'.[53] Carson, Pirie, and Scott also expanded. Its 1904 addition, which was twelve storeys high, was one of the last landmarks of the Chicago School and Sullivan's swansong. Burnham's company added another

addition in 1906, following Sullivan's scheme. It has been considered one of the masterpieces of 'modern commercial architecture in the world'.[54] Small shops, especially those that sold women's clothing, huddled close to the big stores. Arcades, the predecessors of today's enclosed malls, graced a few properties.

By 1910 department store magnates had refined their businesses and struggled for market share. They added an increasing variety of goods, far more than Stewart had back in 1846, and pur-chased directly from manufacturers (or ran their own plants). They sold for cash, though some had short-term credit available. They advertised exten-sively in newspapers and with flyers and bill-boards, and had seasonal discount sales, low mark-ups, and rapid turnover to ensure profits. They offered an increasing range of services within the store, employed many women, and delivered using horse-drawn vans and then trucks. Service on a first-come basis created the illusion of egalitarian democracy. The magnates themselves greeted shoppers, but paid the clerks little for ten-hour days on their feet. The customers came—women mostly: first the rich, then the ever expanding mid-dle class, all treated with respect. The consumer culture of selling and buying was at a high pitch.

The department stores dominated the down-town retail scene, focused at State and Madison. By 1900 chains like Woolworth challenged the lower end of Chicago's retail market after starting in smaller cities. Many office buildings within the Loop, many of which were from the 1870s, housed retailers on the first floor. Along Michigan Avenue, two blocks east of State, the carriage trade was housed in old upper-income residences or on the first floors of new office buildings or hotels.

Mail-Order and Wholesale Workhouses

Less of a direct challenge to retailers in the city itself were the formidable mail-order catalogue operations. By 1904 cramped quarters in the ware-house area downtown led Sears to move west onto 40 acres (16 ha) near Kedzie and 12th, where he built vast warehouses. Smaller but still successful, Ward built on the north branch of the Chicago River.

By buying directly from manufacturers, large retailers cut into wholesaling, which was the activity in the central district. Wholesalers continued to expand, however, with the growing population and affluence. Indeed, wholesalers in the 1870s and 1880s built more structures, many with elegant cast-iron facades. In 1887 Marshall Field's new warehouse occupied an entire block. Even so, the dominant era of wholesaling had passed. They now had to locate on the margins to the west rather than in the core. Clothing, jewellery, and other light manufacturing occupied space in these buildings. Along South Water was the congested wholesale produce market until it was forced out by Wacker Drive redevelopment in the 1920s.

Arts, Entertainment, and Accommodations

Downtown entertainment palaces constituted an expanding sector of the centre's activities. Chicago led the way in creating other prominent buildings in the central area for entertaining the élite and the rising middle class. Largely funded by the public purse, the Art Institute (1883) and Orchestra Hall (1904) east of Michigan Avenue signalled the beginning of land reclamation in that area. Nearby was Sullivan's mixed-use Auditorium where the Civic Opera performed for many years from 1889 onward. The Coliseum hosted political conventions and prizefights, and was well placed for Ward 1 Democratic parties thrown by the long-standing aldermen, Bathhouse John Coughlin and Hinky Dink Kenna. Downtown vaudeville theatres, especially on Randolph, provided a variety of entertainment from magicians to serious skits, and even some movies by 1910. Live theatre, such as the Garrick, catered to a more literate clientele. Like department stores, they all made the city more 'down-towny', according to Henry James.[55]

Hotels serving visiting businessmen and tourists expanded in size. In 1889 when Chicagoans were heady about growth, Kipling stayed at the Palmer House where he 'found a huge hall of tessellated marble, crammed with people talking about money and spitting about everywhere'.[56] Just before 1910 the Blackstone and the LaSalle graced the downtown skyline. A few hotels were found in better residential districts. Less pretentious hotels catered to small-town merchants and travellers, while shabby hotels on the periphery were for the down and out.

Rail Terminals

The six rail terminals were landmarks in the downtown area. Earlier terminals and tracks defined the limits of downtown as early as the mid-1850s, creating the 'iron girdle'. To the south, the tracks limited entry to the Loop to three streets. Until the mid-1880s additional trunk line companies sought to enter the city, so cooperative deals had to be made in this congested environment to accommodate all twenty-four companies in six depots. The Illinois Central and three other tenant companies entered along the lakeshore; in 1892 the IC moved intercity trains to the new Central Station to the south, using the station at the foot of Randolph for commuter traffic. To the west, Dearborn Street (1885) served seven companies, then LaSalle (1903) served three and Grand Central (1890) served four. Along the south branch of the river was the old Union Station (1880) hosting five and, just to the north, the Chicago and Northwestern Station (CNW), which replaced its 1885 structure in 1911. Together the stations handled nearly 70,000 intercity passengers daily and over 123,000 commuters daily in 1912.[57] This daily flux may have been the greatest in the world. The train stations were 'a great crucible into which the people of all nations … are being poured.'[58]

Burnham sought to free up more land for the central district south of Congress by unifying at least four of these terminals along the south side of 12th Street. This would also have untangled the mass of rail lines that led into the stations, and undoubtedly travellers changing trains would have benefited. The Chicago and Northwestern Station could not have joined in as it ran northwest, nor could the Milwaukee Road. Still, all the intercity lines to the east, south, and southwest were invited to unify their terminals, but they declined to act. Only the new Union Station in 1925 altered the pattern. The inertia of property interests prevailed over public collective business planning.

Circulation: Water, Road, and Rail

Daily movement in the downtown area created considerable congestion. Retail clerks, typists, secretaries, stockbrokers, bankers, clerks, and shoppers converged there daily, a swirling mass of humanity. As we will see later, in 1910 most came by some form of public transit. A few of the rich came by carriage, but many of the less affluent walked, mostly from the near west side, and draymen drove their horses and wagons in each day to haul goods. In 1910 perhaps 750,000 people worked and shopped there daily. Within the downtown area, horse-drawn cabs and omnibuses (connecting rail terminals) operated, though they were in their last days. Internal combustion-driven taxis would take over.

The Loop was renowned for its street and alley congestion. One commentator noted that 'New York does not for a moment compare with Chicago in the roar and bustle and bewilderment of its street life.'[59] A New York visitor remarked that during rush hour, 'certain streets in Chicago are so packed with people as to make Broadway look desolate and solitudinous by comparison'.[60] Or, 'inside the Chicago loop are several dozen Thirty-fourth streets and Broadways.'[61] One photograph showing Dearborn and Randolph in 1912 has been reproduced several times with captions referring to 'excessive' congestion. The lined-up streetcars, though, may well have been in a staging formation waiting for rush hour. Even though planners of the era worried about what resembled clogged bronchial tubes, and recent writers have been aghast at the sight, people and vehicles were nevertheless able to move.

Policemen and then, in the 1910s, traffic-lights regulated traffic and therefore facilitated the flow of more cars. In their desire to alleviate congestion by widening streets, planners unwittingly helped to promote the car and therefore fostered even further congestion. However, they could not have foreseen this, as commuting and shopping by automobile was only just beginning in 1910. Some

> **The Loop was renowned for its street and alley congestion.**

of the surface freight traffic congestion was actually obviated by little underground trains that hauled goods from freight rail depots, ashes from office furnaces, and the like. This underground tube system with 62 mi., (100 km) of track went 'far toward clearing the streets of forty thousand turbulent teamsters'.[62] Still, the Loop remained congested, and vehicles crowded together during rush hour. Chicago was not the only place with this problem: all cities large and small had central areas where the main business action took place. Efficiency in movement would have cancelled out the value of concentration. Most people might complain, yet would they have wanted it otherwise? Dramatic changes later in the twentieth century suggest that maybe Americans (or at least the developers with power to alter the landscape) did, but we will consider that matter in subsequent chapters. Here we will consider the overall regional transport systems: on water, road, and rail.

Water Traffic

Chicago's location, as we saw earlier, was optimal for its access to and from the centre of the continent. Despite its initial marshy site, it was relatively easy to organize the space into a transport pattern. The waterways were adaptable and created only minor complications for other traffic, except adjacent to the Loop. The road pattern was a direct result of the rural grid survey pattern. Although the railways did not have to bear the cost of great bridges, like the Eads over the Mississippi in St Louis, the railways managed to create a complex pattern of tracks as they sought access to the city's centre, beginning in 1848.

Early traffic on the lakes created the harbour in the mouth of the Chicago River. Storage and manufacturing crowded along the main trunk of the river. As the city grew, activity spread up the north branch and more markedly along the south branch, and then along the Michigan–Mississippi Canal after 1848. Heavy water traffic impeded land crossings across the river, especially the main stem. Bridges slowed movement from the north to the

Loop, and therefore slowed down residential development. By 1870 the river's capacity for movement was strained, and adjoining businesses suffered.

Luckily Chicago had alternative options for water access from the lake and inland. As in the case of other large infrastructural undertakings on water, the federal government built the new harbour at the mouth of the Calumet River and opened up Calumet Lake for industrial development. This greatly reduced pressure on the river downtown and obviated the need to build a harbour on the central waterfront, therefore permitting parkland. On the canal that connected with the Mississippi, traffic peaked in 1882, but the new Sanitary and Ship Canal gave water traffic a new lease on life when it opened in 1900. It was 28 mi. (45 km) long, 24 ft (7 m) deep, and cost $60 million. The engineering experience there demonstrated the feasibility of the Panama Canal started in 1906. The Port of Chicago handled more goods as a result of these changes—over 10 million tons in 1911—one of the largest amounts on the continent. However, by then railroads were taking a far larger share of the transport. Water traffic was increasingly relegated to low-value bulk carriage and modest ferry traffic across Lake Michigan to Michigan ports.

Street Traffic: The Heyday of Streetcars

The road pattern was on a simple grid, as generally elsewhere in the American Midwest, to ensure efficiency. In and around Chicago, half-mile arterials and local streets provided an easily understood system for traffic and land development. Most Indian trails were wiped out in the process, though early diagonals to the northwest and southwest more or less followed them. Topography, railways, and developers, if ever so slightly, complicated the pattern. The city hired Frederick Law Olmsted to design boulevards in 1869 that would join together large inland parks. Outside the Loop, traffic was only locally heavy here and there with wagons and transit vehicles before 1910, but by 1910 10,000 autos presaged a new era to come.

Burnham's plan would have introduced far more diagonal streets, not only in greenfields but also through already built-up areas to expedite flow, thus creating a lattice of major arterials and minor connectors on top of the grid pattern (Figure 5.2). In the inner city, the diagonal streets would have wiped out slums in the process. Planners at the time in many cities never tired of pointing out that diagonals (the hypotenuse of a right-angled triangle) would save time and energy from having to travel along the other two sides. Burnham's suggestion was eventually followed in only one case within the built-up city, and in only a few cases in other cities, largely because diagonals would have been too costly or too disruptive, even in poor areas, if they had been implemented. Freeways later, of course, would not present this problem. Only cities whose early plans incorporated diagonals, such as Washington, Buffalo, and Detroit, would enjoy their efficiency. Burnham also proposed widening streets. Some were widened later, notably 12th Street, which was extended on a viaduct over the tracks, and Michigan Avenue to the north. Like diagonals, widenings in built-up areas cost taxpayers a lot of money, so few were done anywhere. The city did transform Water Street into Wacker Drive in the 1920s. This may have created a stretch of elegance, but forced the tatty yet important wholesale produce market to relocate to the southwest. The 1909 plan also argued for three circumferential bypass highways at various distances from the city. In the 1950s two of these appeared as Interstate 290/355 and 294.

Public Transport Focused on the Loop

Public rail passenger transport was of four types. Three of them ran by electricity: streetcars or trolleys, elevated rail, and electric interurban lines (large streetcars), as well as steam commuter trains, all converging on the Loop.

Streetcars superseded horse trams after 1890 and cable-cars by 1906. Chicago had the former in common with most large, medium, and even small cities after 1850, the latter with only a few. Expanding ever farther outward for thirty years, the horse trams had contributed to the expansion of the city, and established public transit as a necessity for more and more of the middle class. Increasing amounts of horse dung was one factor that encouraged the pursuit of new technologies. Chicago used cable-cars extensively after 1881 with 86 mi. (140 km) of track in 1894. The cable-cars increased the efficiency of transit times.

The use of electricity for power as well as light led to the trolley as the best adaptation. After several years of publicity, apparently Windsor, Ontario first adopted it in 1886. Within a few years, almost all cities on the continent said goodbye to horse trams. The streetcar doubled the speed of its predecessor, at least during non-rush hours. During those hours, passengers crammed in cars that inched along; though they often complained, they had little choice. As Yerkes observed, 'the straphangers pay the dividends'.[63] As service improved after the 1907 agreement, ridership rose and reached nearly 500 million in 1910, or each person rode the streetcars on average over 300 times a year, which was double that of the rate in 1890. New York was first in ridership, while Chicago vied with Toronto for second place among large cities.

The 1913 unification created the longest system in the world, running on nearly 1,000 mi. (1,610 km) of track, and ridership was still climbing. Streetcars ran straight through the downtown area, which made it more convenient for some riders. Although the system remained focused on the centre, new cross-town lines and free transfers now encouraged riders. The quality of cars improved. However, Chicago lines were not connected to the lines operating in adjacent municipalities, so commuters had to transfer from one car to another at the city boundary. Tunnels under the river at Washington and LaSalle in the 1890s enhanced access to the Loop. Internal combustion buses, which replaced omnibuses of earlier times, began in 1917 and presaged the eclipse of the streetcar in Chicago as in most cities, an issue we will consider in the next chapter.

Trolleys alone could not carry the massive daily flow of commuters. The Loop itself was a creation of the union of the elevated lines (els) in 1897. The first el was run to the south in 1891, arriving at the World's Fair site on 63rd Street just in time in 1893. Subsequently, three branches extended service, one into the stockyards. Two

lines ran to the west and one to the north; all these lines had branches, including one to the northwest. These were run by separate companies that initially created their own loops downtown. To overcome the confusion for riders, in 1897 cooperation led to one loop eight blocks long and five blocks wide. Although it was criticized as creating another steel girdle around part of the downtown and leading to a deterioration of land values of adjacent properties, undoubtedly it was efficient. The els carried about one-quarter as many passengers as the streetcars, peaking at 229 million riders in 1926, the year ridership on all transit began to decrease. Slow to follow New York and Boston, Chicago did not replace the els with subways until the 1940s. To accomplish that, the streetcars, buses, and the els were brought under the Chicago Transit Authority, which indicated that private ownership of this vital public utility had failed to be profitable in the long run. To build subways, public money and control were necessary, as in New York and Boston.

The electric interurbans (or radials) provided a modest share of commuter traffic to the centre. Conceived in the 1880s as a cheap, electrified intercity alternative to steam railroads or as a complement to them in connecting Chicago with small towns, construction of the interurbans exploded all over eastern North America after 1900. Their founders dreamt of great networks. Seven lines served Chicago by 1912, but the lines had trouble gaining direct entrance to the centre, though the Chicago, Aurora, and Elgin lines did. The Evanston line from the north eventually ran on el tracks, and the one into Indiana came into the Illinois Central station. In many other big cities, the owners had to be content to stop at city boundaries, where commuters had to transfer to streetcars, but then that was the case with most transit companies, which were either restricted by local governments through franchises or their own unwillingness to cooperate. In the overall scheme of things, this may have been an inconvenience to only a few people in the short run, but it undercut the functioning of the whole urban area as a unified whole.

Steam rail lines also facilitated commuter flow to the centre from 60 mi. (97 km) out. As in several eastern cities beginning in the 1850s, the Illinois Central, the Rock Island, the Burlington, the Milwaukee Road, the Chicago, and the Northwestern ran commuter trains to and from small villages. The Illinois Central was the largest carrier. One day in June 1915, when tram workers were on strike, ridership to the four stations handling commuters peaked at 625,000, about five times the usual number. Most of them worked in the central area.

The Grand Nexus of Rail Systems

In this greatest of railroad cities, rail lines carrying intercity passengers and freight dominated the overall landscape and altered the settlement pattern (Figure 5.3). Twenty-four trunk companies converged and drove into the centre along seven corridors, to the terminals noted earlier. Between 1848 and 1885 agreements on running rights among competitors led to cooperation, but the task had not been easy. The last main line to arrive, the Grand Trunk from Michigan and Ontario, had to run far to the south of the site of Gary, then far to the west through Harvey, then north, and finally east to gain access to the city. Four belt lines, owned by major companies and connecting these trunk lines, nearly encircled the city at various distances outward. The farthest one out was the Elgin, Joliet, and Eastern, which was owned by US Steel and extended into Indiana. Switching and terminal rail companies also operated. The Chicago Switching District, which was at its maximum extent about 1920, embraced a vast 400 sq. mi. (1,030 km^2) within a line from Des Plaines to East Chicago (thus not crossing the state lines), with 160 freight yards, seventy-three freight stations, and 5,717 mi. (9,200 km) of track.

Because trains from the West did not pass through to the East and vice versa in this great break-in-bulk centre, classification yards in Chicago were the most important switching points in the country and perhaps in the world. Although they had to maintain inner-city yards for freight and express destined for the local market, as traffic grew in the late nineteenth century, the companies built new yards farther out to open fields. By 1910 classification yards were located roughly between 7 to 10 mi. (11 to 16 km) from the Loop. In the 1920s new yards up to 21 mi. (34 km) took up greater swathes of the landscape. Burnham's plan would have rationalized the systems to cut waste by creating one massive freight yard south of Cicero. It would be linked to the downtown area, the harbour nearby, and to Calumet harbour by an underground rail line. Not surprisingly, the rail companies did not respond to this scheme, which was an even more utopian idea than his proposed rearrangement of central passenger stations. From the 1920s onward, except during the Second World War, the excess of rail facilities would become more and more obvious. Today, derelict rail yards litter the big cities of the continent; only a few have been redeveloped.

A plethora of rail lines divided the city. The divisions were accentuated when the tracks were elevated, channelling traffic on arterial streets through underpasses and blocking many local streets. Beginning with the Illinois Central at the entrance to the 1893 World's Fair site, tracks were raised on embankments to permit increased train speeds and ensure safety for pedestrians. Prior to that, the death toll was enormous at level crossings because people dashed between trains, which passed every few minutes. The embankments defined the city as never before into cells.

The Suburbanization of Industry in Satellite Cities

Outside the central area, as the scale of industry expanded, larger concentrations of workplaces emerged, some with adjacent residential areas. Earlier, manufacturing and storage facilities sprang up along the Chicago River. Activities crept inland upstream on both branches, and then along the canal to the southwest. Also, rail sorting yards and rail lines encouraged the establishment of manufacturing plants, mostly to the west and south. Lands adjacent to the Calumet River and Calumet Lake became huge industrial districts, as did the sand dunes in Indiana to the east. Even though the Loop intensified office and retail activity, concurrently with the great increase in the city's popula-

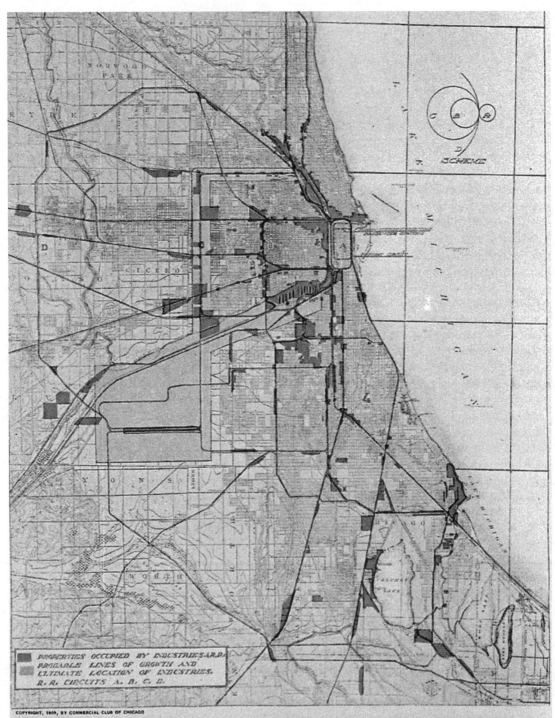

LXXIII. CHICAGO. DIAGRAM OF THE CITY AND SURROUNDING COUNTRY, SHOWING RAILROAD CIRCUITS, B, C, D, AND E, WHICH ARE, OR MAY BECOME, TANGENT TO THE INNER CIRCUIT (A).

The diagram also shows the existing industries, and the probable trend of growth away from the center of the city.

tion, the larger-scale organization of manufacturing and the encouragement and support of governments led to the creation of these new suburban landscapes.

Since Chicago assumed the role of the premier break-in-bulk centre in the middle of the continent, it is not surprising that grain elevators and lumber piles were among the first visible signs of commerce alongside the main stem of the Chicago River, beside warehouses, breweries, distilleries, abattoirs, and McCormick's reaper plant. By the 1850s industry appeared on both branches of the river, and along the canal opened to the Des Plaines River. Reshaping the margins of the river in 1857 created slips and 10 mi. (16 km) of frontage.

Meat-packing operations so increased in size that the Union stockyards were created in 1865 beyond the built-up area on the south fork of the south branch in Lake Township. Large packers, such as Gustav Swift and Philip Armour, built plants next door. By 1884 the stockyards alone covered 360 acres (146 ha) laced with 40 mi. (64 km) of tracks to unload livestock.

The great fire of 1871 reduced industrial activity on the main stem of the river. McCormick built a new plant on the south branch near Ashland Avenue. It continued to expand: McCormick and Deering merged after 1900 to create International Harvester, the largest implement maker in the world. Inland and farther west in Cicero, Western Electric, which was a subsidiary of American Telephone and Telegraph, set up its massive Hawthorne works in 1903. Its research laboratories hired inventors and technicians, another sign of corporate times. By 1910 Cicero housed 15,000 residents.

Developers created large manufacturing districts with small structures for a wide range of light industries, including incubator operations. The Chicago Belt Railway started the central manufacturing district just north of the stockyards in 1890. Slowed by the depression of the 1890s, the same corporation started the clearing industrial district

south of Cicero in 1898. These districts, all of which were connected to rail and water, presaged zoning of industrial parks. Although the centre remained powerful, the outward expansion of the city was well underway.

This was the exalted era of satellite industrial cities. As we saw earlier, large-scale industrialists began moving to New York's suburbs by the 1850s, though earlier mill towns in the eighteenth century could be classified as satellites too. Now no longer constrained by water-power energy (as in Lowell and other early textile towns), companies established themselves in the city. However, greatly expanding heavy industries sought to escape the congested constraints of the city and workers' unions, and sought cheap land for massive works fuelled by coal. In the Chicago region, the first satellite industrial city was Calumet, which became the most important harbour complex on the Great Lakes. In 1871 dredging, breakwaters, and piers opened the channel, creating a real estate boom. Grain handlers built elevators and rolling mills there. In the town of South Chicago, iron ore from the Mesabi range in Minnesota, coal from central Illinois, and local limestone were combined in Bessemer and later open-hearth steel furnaces. Boats and railways shipped out the products from the finishing mills of Illinois Steel (later Carnegie-Illinois and subsequently US Steel).

On the west side of Lake Calumet, George Pullman developed his sleeping car works and renowned model town in 1880. Although industrialists had previously started company towns in England and in America, and owners generally sought to limit workers' rights, Pullman undoubtedly had the strongest utopian vision of any industrialist. To create a happy populace, he provided workers with homes, schools, churches, clubs, parks, and a hospital, but forbade saloons for their own good. He hoped to keep workers away from the temptations of the city and unionization. Men were paid good wages in his plant, but they had to walk to Kensington a half mile to the west for a

5.3 *(Opposite)* Burnham's plan for the region. Burnham proposed a vast rail yard and industrial park south of Cicero, circumferential highways, lakeside parks, and forests. *Source:* Daniel H. Burnham and Edward H. Bennett, *Plan of Chicago*, edited by Charles Moore (New York: Da Capo, [1909] 1970).

drink. But then Chicago annexed Pullman in 1889, thereby limiting Pullman's paternal power by controlling services to the community. In the wake of the panic of 1893, the great strike of 1894 put an end to his vision. As the market for sleeping cars fell dramatically, Pullman reduced wages but not rents. He had to concede that linking work, home, and community in his style would not work. After the depression, sleeping car production rose again erratically and after 1910 with much less gusto. Employment fluctuated between 1,500 and 15,000, as conditions were volatile. To paraphrase a worker, employees no longer had to live in a Pullman house and be interred in a Pullman cemetery, but most probably still felt that when they died, they would end up in a Pullman hell. Other factories emerged near Pullman, some supplying his works.

Begun by US Steel, Gary was the other conspicuous industrial satellite town to the east across the state line. South Chicago had inadequate space, so beginning in 1906, the company built a new harbour, great steelworks, other steel-producing factories, and a new residential city on sand dunes. Said to be the 'greatest single calculated achievement of America's master industry', Gary followed South Chicago, Homestead near Pittsburgh, Lackawanna near Buffalo, Fairfield near Birmingham, east Hamilton, Ontario, and others.[64] The steel plant was of the most modern design for efficient handling of raw materials and finished goods. By 1910 it employed 14,000 workers, most of whom walked to the steel mill, often through a pall of smoke. Still growing, Gary had reached a population of 16,000. Warned by the failure of Pullman's utopian scheme, US steel set up a real estate division to sell home lots to developers and builders, who sold to settlers. Other plants using steel sprang up nearby, such as manufacturers of tube, sheet metal, wire, and railcars. Between Gary and South Chicago were other satellite industrial towns—Indiana Harbour, East Chicago, Hammond, and Whiting—the latter known as the distributing centre for oil, the wonder fossil fuel of the twentieth century.

Developers created other satellite industrial towns and industrial areas, such as Harvey and Chicago Heights to the south. Thirty miles (48 km) out to the west in Elgin and Joliet, industries altered the economic base away from agricultural servicing. To the north on the lake, only Wilmette and Waukegan hosted a modest amount of manufacturing. Access to water and the prime rail orientation to the east induced large industries and their satellite towns, mostly to the south and southwest of the city.

In addition to these satellite industrial towns with their residential districts, there were also some outlying residential communities. A few large community builders built upscale towns and tried to control the style and size of buildings through covenants. Revered in the planning literature are Riverside (1879) west of Chicago, Kenilworth to the north (established by Joseph Sears as a model suburban town), and also Forest Hills on Long Island (following Lleywellan west of Manhattan), all of which focused on commuter rail stations. They resembled necropolises with their curving streets and weeping willows. After the turn of the century, large projects became more common on the margins of most cities. In the 1920s community builders refined neighbourhood planning, providing a wider range of commercial and community services. However, like other residential developers, even the community builders could not fill up their lots quickly, especially when badly timed, yet they foreshadowed the post-1945 system when corporations provided almost all new housing. For the rest of this chapter, we will focus on residential areas inside the city. There apartment living was the most striking development, and a large majority of Chicagoans were tenants.

Tenancy and Home-ownership in Chicago Compared to Other Cities

Apartments

Single-family houses may have dominated the urban scene generally in North America, but more so in smaller places. In a few large cities where land was expensive, apartment structures became the norm. Following New York and Boston, Chicago's first apartment building appeared in 1878, and others followed quickly in the 1880s. After the turn of the century, builders in Chicago preferred to construct apartments rather than houses. Between 1905 and 1910 apartment units comprised 80 per cent of new residential construction. In 1910 Chicago dwelling units (apartment buildings counted as one to census takers) housed nearly nine people, more than double that of low-rise Indianapolis (the lowest of twenty-eight large American cities), and nearly twice as much as Baltimore, Philadelphia, and most other cities. Chicago roughly equalled Boston, Newark, and Jersey City in its share of apartment buildings, but had fewer than Brooklyn, the Bronx, and, not surprisingly, Manhattan, which easily topped the list. Apartment construction continued through the 1920s in Chicago and many other big cities until 1928. In the 1920s the average size of individual units was smaller than they were earlier and catered to middle-income earners. Many of the apartments were in small buildings, often in duplexes or triplexes; in 1930 (when the data was better recorded than earlier) nearly one-third of Chicagoans lived in two-unit structures. Probably half of the city's population lived in taller ones—four, five, even up to twenty storeys, the tallest followed the Marshall Apartments of 1906 in palatial style. The predominance of apartments led to a prediction that the 'long-time tendency is definitely toward the multifamily dwelling'.[65] Corbusier's vision, it seemed, won over Frank Lloyd Wright's.

The following comment on Chicago in 1908 seems to contradict the trend to apartments. 'The chief ambition of every citizen seems to be to build himself a house standing in its own grounds, and so the town grows in extent at a prodigious rate.'[66] Not counting self-built homes, single-family housing was not the reality for most residents. It may have been the ideal, as many writers claim, though only a few large cities followed this path. In Philadelphia and Baltimore, single-family row housing dominated; in Toronto, which severely limited the number of apartments, semidetached houses were common. In these cities, many people doubled up and took in roomers and boarders, especially among the working class, because single-family units in Chicago and other expensive big cities cost more and rents were higher than in smaller centres. Was the abundance of apartments the result of supply or demand?

Home-ownership and Tenancy

Owning a home has been regarded as a cardinal measure of middle-class status. In Chapter 3, I stated that farmers on their freeholds were, in fact, the first middle class. In recent times, blue-collar workers have, it has been believed, joined the middle class when they owned their property. Having owners (instead of landlords) responsible for their property has been widely considered a virtue over the past century. In fact, academics have claimed that it was becoming the norm, as people had a 'will to possess'[67] that translated into a 'cult of homeownership'.[68] Figures from the United States only modestly support this view overall, as owner-occupied homes of urbanites remained roughly constant between 36 to 38 per cent from 1870 to 1910, after which the level grew slowly to 46 per cent in 1930. Even then, over half of the urban population still rented. Why have historical geographers and historians argued that ownership was seemingly natural? They are assuming that post-1945 levels have always been the norm in people's minds, and that the earlier drive to own rural land freehold (which is indisputable) was carried over to the city.

Chicago and some other cities hardly sustained the will to possess, it seems. In 1910 the census takers reported that 26.2 per cent of 'homes' owned, based on a number about equal to the number of 'families', were counted as owned at the time. This proportion was lower than 28.7 in 1890 and slightly higher than in 1900 (after the depressed 1890s). The level was slightly higher again in 1920 and it reached 31.1 per cent in 1930. In short, after forty years the proportion had changed little. One-third or more owners rented out a 'home' on their second floors in 1931, which was more than the proportion of those who occupied single-family houses.[69] Renting a single-family home was also not exceptional. Apparently relatively few were able to take advantage of the subdividing and house building of Samuel Gross, other builders, or even their own. Like Chicago, levels in other cities do not square with the desire to possess a home.

New York, Boston, and Montreal had low levels of home-ownership. New York hardly shifted from its 1860 rate. Even low-rise Philadelphia was reported at the same level as Chicago, though by 1920 it rose considerably higher. Buffalo and Detroit had relatively more owners. Small New England cities had lower rates than those in cities of similar size or larger cities in the West. Generally home-ownership in some Ontario cities went up dramatically above the American level in the first decade of the century: in fast-growing Toronto, ownership jumped an astronomical 73 per cent between 1901 and 1907, raising the total to near 50 per cent. The picture obviously varied, but does not clearly sustain the belief in the drive to ownership—at least half of the households rented.

To understand why some cities seemed to foster home-ownership more than others is not easy. The built form of residences provides a piece in the puzzle. In Chicago most households lived in rented apartments or duplex units. High land values probably encouraged suppliers to construct apartment units, especially when labour and building costs rose after 1896. Land prices may have made owning prohibitive for most. The case for New York, Boston, and Montreal is similar: units in tenements, three-deckers, and walk-up houses were distinctive, so households were counted as tenants. In Toronto, and maybe in other cities with few formal apartments, single people and families lodged in the house-form units, helping owners to make ends meet. They would more likely be counted as members of the owners' households and not in their own right, thus obscuring the tenancy level. As for low-rise and slower-growing Philadelphia, the picture is not clear: costs relative to wages were considerably higher than in Chicago, at least in 1909.

Financing may have had some effect on ownership. Building and loan societies operated in all cities, though more vigorously in Philadelphia and Baltimore, which may account for the house-form of development. Although mortgaged homes in Chicago rose by 50 per cent between 1890 and 1920, the ownership level actually dropped. It is possible that regulations varied across states and provinces and that custom played a part in determining mortgages, the type of building, and the degree of self-building. In Ontario municipal voting was restricted to owners, which may have encouraged self-building. In Australia public policies favoured self-building on large quarter-acre suburban blocks where working-class families could produce a great deal of their own food.

Finally, it is hard to conclude that the larger the city, the higher the rental level, or that the speed of development was a certain factor. It would seem that new cities in the West were more likely to have higher ownership than older ones in the East. Perhaps as discretionary incomes rose, home-ownership became a greater possibility and more likely in new environments. In the East, older customary practices were still followed. In the last analysis, particular cases need to be assessed against one another. As we will see in the next chapter, the drive to higher ownership levels after 1945 was contingent on massive government aid.

As for classes of people, in Chicago many people in the middle class occupied apartment units, and hence were tenants. Many of the rich moved into palatial units, leading the way for the affluent and then the lower-middle classes. Non-British immigrant manual workers, on the other hand, may have sought home-ownership with far greater determination than others, as has been argued elsewhere. A sample study for Chicago in 1907 suggests this, though Edith Abbott, a prominent Hull House community surveyor, warned against drawing any conclusions from the small number in her sample.[70] In the 1930 census, 41 per cent of foreign born owned their homes (10 per cent higher than the city as a whole), which supports the contention that many poor immigrants made 'extraordinary efforts to purchase a home', even putting this ahead of schooling.[71] Quite possibly immigrants preferred not to deal with English-speaking landlords, who were probably more likely to make life more difficult for them than for the middle class, or maybe non-British immigrants had more to prove in the new land than others. Even so, a majority of immigrants and their families still rented; many of them were poor and lived in the worst old houses, that is, apart from those inhabited by Blacks, who increased greatly in number between 1910 and 1930.

Where They Lived

To some visitors, Chicago seemed like a spread-out, 'straggling' place,[72] but the straggling had a star-shaped pattern. The star shape resulted from streetcar lines and els that ran in corridors outward from the centre, with construction in the interstices only later. The failure to establish cross-town lines with free transfers until 1913 accentuated the pattern. Beads of commuter settlements along rail lines also encouraged the maintenance of the star shape. Later widespread use of the car would gradually fill the interstices.

Overall, Chicago's gross density in 1910 within its 190 sq. mi. (492 km²) was about 11,500 people per square mile. After the great annexations of 1889, it was only 6,300, and would rise (with only minor additions) to just 16,300 in 1930 when the city space was more or less fully occupied. Brooklyn's density was twice as high as Chicago's, and Manhattan's was far more. What do these contrasts signify? Earlier eastern cities were denser, though Baltimore and Philadelphia were less so than others because of low-rise row housing; newer western cities, except San Francisco, were markedly lower in density. Great Lakes and Mississippi valley cities were in between, though Toronto was denser than others. Greater collective affluence and higher personal incomes allowed more space per person and therefore reduced densities, helped along by public transport, which became a necessity for many people. Another factor was crucial: the vast industrial works within city boundaries consumed enormous amounts of land, thus reducing the overall density.

As one might suspect, Chicago's highest densities were adjacent to the downtown and to industrial districts because many people still walked to work. Chicago was pretty typical at the time, with the highest gross density (75,000 people per square mile) located on the Near West Side (Figure

5.4). There and to the north, only nine mile-square blocks contained over 40,000 people and another thirty-one blocks had over 20,000, out of a total of 190. (I write this in an area of central Toronto that has a density of about 25,000.) At the other end of the spectrum and towards the northwest, southwest, and south, over fifty mile-square blocks housed fewer than 1,000 people each, even though most blocks were subdivided by 1891, which was the peak of the subdivision boom. This low number testified to scattered construction, the excess of the previous century, and to disappointed investors.

Residential zones and sectors reflected class and ethnic distinctions. As noted earlier, within the city, increasing numbers of the rich lived in high-rise rental apartments, though the gross densities of their neighbourhoods were not high. The middle class increasingly resided in medium-density areas in low-rise, smaller rental apartments and in flats in houses. Many people with manual occupations owned new single-family houses, though most still rented flats in older houses. The poor, who were mostly recent migrants, including Blacks, rented old single-family dwellings in high-density areas that were often overcrowded. Chicago was thus like what New York was earlier, with most people in rental accommodation and with more and more in apartment structures. However, unlike New York, the rich were less inclined to ownership and more inclined to live in apartments. The poor were the same and lived in houses, including alley houses, if not purpose-built tenements. The middle class, which had increased since 1860, could live comfortably in apartments.

The rich tended to live close to one another so they could enjoy the social benefits of propinquity, just as they worked together on LaSalle Street and socialized with one another in their clubs. In the gilded age, as their wealth rose, they built ostentatious, massive brick and brownstone houses in the asymmetrical Romanesque style of the architect, Henry Richardson, along boulevards and adjacent streets. The houses of the rich exuded flamboyance; Potter Palmer's 'castle' was the most elegant of the lot. Their Episcopalian, Presbyterian, and now the *arriviste* Methodist churches displayed

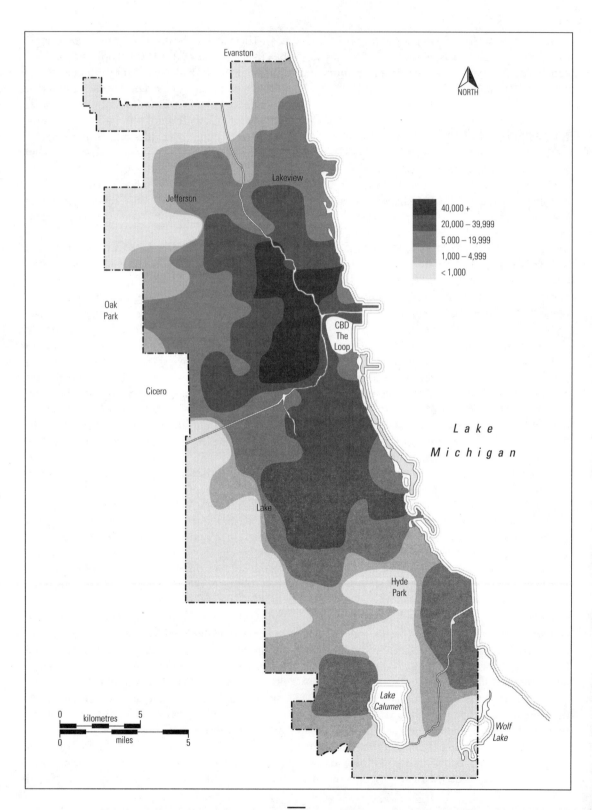

Evanston

Lakeview

Jefferson

Oak
Park

Cicero

Lake

CBD
The
Loop

Hyde
Park

Lake
Calumet

Wolf
Lake

L a k e

M i c h i g a n

NORTH

40,000 +
20,000 – 39,999
5,000 – 19,999
1,000 – 4,999
< 1,000

0   kilometres   5

0   miles   5

their success. They bought coaches to display themselves on the boulevards from which freight traffic was prohibited, but as in New York, such public displays were passing. Increasingly, the well-to-do sought privacy and separation from others. Their conspicuous consumption was displayed more and more to one another, though the *hoi polloi* could still catch glimpses of elegance as the wealthy arrived at the opera-house. Listing one's address, visiting days, and clubs in the social registers was a must.

In the city, those with high incomes lived in well-defined sectors (south, west, and north), though these changed markedly (Figure 5.5). Early on, the wealthy were attracted by the lakeshore and boulevards to the south; after 1870 Prairie Avenue south of 17th Street became the most prominent address. Near the centre of the city along Michigan Avenue, George Pullman's mansion led the way southward in 1873. The sector expanded in that direction for the next twenty years, reaching the area near the World's Fair site and the University of Chicago. Directly to the west of downtown, the affluent moved towards Garfield Park along the Madison Street axis, then to Austin, and on to Oak Park outside city limits, a place of 20,000 in 1910. To the near north, because road traffic was constantly held up by swing bridges that accommodated the heavy river traffic, the élite were slow to move.

Up the shoreline, however, new communities formed, such as Evanston, which had access by rail to the Loop. Evanston was the largest city in Cook County, other than Chicago, with 25,000 people. Places like Irving Park and Norwood Park to the northwest, and Washington Heights west of Pullman, would serve as nuclei for later upscale communities.

The rich had a propensity to move to enhance their status by following the newest trends in housing styles and conveniences. After all, as was asserted in 1905, 'our homes in America are mere extensions of clothes; they are not built for the next generation. Our needs change so rapidly that it is not desirable.'[73] So after 1893, in the southern sector down to nearly 39th Street, they abandoned their mansions to institutional or lower-income multiple occupancy. Union Park was also left behind. North of the river then became the main area for the top élite. By 1890 the Gold Coast became a fashionable residential area. Increasingly the rich preferred Gold Coast palatial apartments. By 1910 few showed the inclination to build large houses. From then on, the rich Gold Coast sector would remain fixed, though in the long run, many other wealthy neighbourhoods would turn over. As indicated in Figure 5.5, in 1930 the top quintile in income (as measured in rent paid) overlapped considerably with the area of below-average home-ownership, most markedly near the lakeshore and in Austin to the west.

The middle class grew enormously. By 1910 most of them lived as tenants. For the middle class, a strong selling-point for early apartment units were expensive amenities, such as central heating and electricity, which were available sooner than could be afforded in single-family dwellings. The material lot of the middle class improved because of dropping prices between 1870 and 1896, more amenities in homes, and extensive public works. By the 1920s the middle class lived better electrically with appliances like vacuum cleaners and radios, though more slowly with refrigerators, coffee-makers, and the like. They lived in apartments that extended inland in the north and west, adjacent to the rich in the south. Beginning in 1874 Irving Park to the northwest became an upper-income Protestant commuter enclave. By 1910 it had predominantly middle-income people, including northern Europeans, many of whom lived in new housing of multiple units. Even so, after annexation to the city after 1889, Irving Park

5.4 *(Opposite)* Chicago densities, 1910. Like most cities, the highest densities were adjacent to the central business district where many workers lived at high density to be near work. The CBD had expanded so as to exclude residential uses. Densities fell off towards the edge, except in the industrial clusters in the south and the southeast. *Source:* City of Chicago, *Population of the City of Chicago Per Square Mile 1900–1970* (Chicago: Department of Development and Planning, 1975).

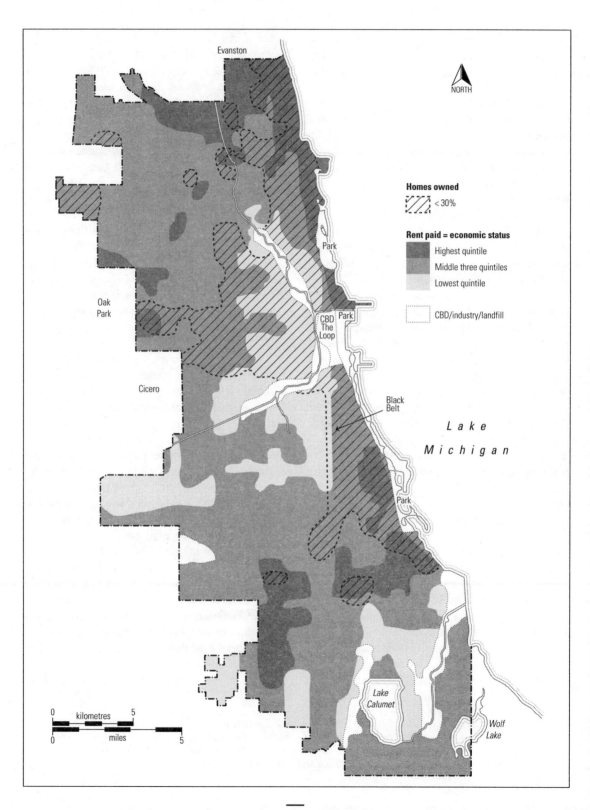

maintained a strong community identity thanks to its Women's Club and the Improvement Association. In 1930 the middle-income 60 per cent of the population coincided to a large extent with above-average tenancy (Figure 5.5). (The map obscures, though, second rental units in owner-occupied homes.)

The bottom fifth of the income scale included most industrial workers, most of the recent immigrants from eastern and southern Europe, and a smaller number of native-born poor. In 1930 a large number were owners, especially those adjacent to the industrial sector to the southwest (Figure 5.5). Many had been on the move from the rental units in the inner city. Around 1910 outside observers saw a wide swath a mile west of the Loop—'a vast wilderness of shabby houses—a larger and more desolate Whitechapel that can hardly have a parallel for sordid dreariness in the whole world'.[74] It extended to the southwest of the Rock Island line, then southwesterly to the Back of the Yards and beyond, immediately west of the Gold Coast, and along the river to the northwest. Another observer described it this way: 'Plunging in the electric cars through the river subway, and emerging in the West Side, you realise that the slums of Chicago, if not quite so tightly packed as those of New York or London, are not a whit behind them in the other essentials of civilised barbarism.'[75] To a New Yorker, 'The long trains of elevated cars are slipping along their alley-routes, skirting behind long rows of the dirty colorless houses of the most monotonous city on earth ...'.[76] In the inner city, these wood-frame single-family houses built after the fire usually housed more than one family, especially in the dense Near West Side, in overcrowded conditions. Many lived in decrepit alley houses. Had the observers looked more closely, they would have found a lively community life nonetheless.

In 1910 the highest gross density (75,000) was within the square mile of 12th, 22nd, Halsted, and Ashland, which was the chief immigrant reception area. This was the highest of any district in this century, approached only in 1950 by the Black Belt on the south side. Assuming one-third of the area was for uses other than housing, the net density would have been well over 100,000 per square mile, about 175 people per acre, which was only two-fifths as high as New York's densest ward in 1860 and parts of Manhattan at the time. Undoubtedly some blocks had considerably higher densities; one near Hull House, according to a survey, had 412 people per acre. Overcrowding could be just as enervating in an old, poorly serviced three-storey house or a backyard hovel in colder Chicago winters as in a six-level tenement. Hull House sought to uplift the denizens of this slum. Many of them would leave.

Even by 1910, when Blacks made up only 2 per cent of Chicago's population, they had little mobility and were increasingly concentrated. By 1930 nine out of ten lived adjacent to one another. Before 1900 the few Blacks were often in poor alley dwellings adjacent to Whites, while others were servants who lived near the rich, as in the South. By 1910 the renowned Black Belt was already the largest of a small number of increasingly Black city neighbourhoods and suburban communities in Chicago. Hemmed in on the west by the Rock Island Railroad and by resisting working-class Whites, as well as on the east by the middle class, the Blacks began expanding in a narrow band to the south, as far as Garfield by 1910, embracing in part the old rich sector. Escaping from the South's bleak economic conditions and Jim Crow laws, Blacks faced limited access to jobs and housing, though Black mothers replaced Irish colleens as living-out servants. Even beaches were segregated; a Black youth who crossed an invisible colour line in the water touched off the bloody race riot of 1919. Few Blacks could become middle class, that

5.5 *(Opposite)* Chicago's social geography, 1930. Home-ownership and economic status as defined by rent in Chicago. The city average of owner-occupancy was 30 per cent; relatively more renters lived east of the heavy line. Economic status does not correlate with ownership; note suburban (ethnic) working-class home-ownership in the south and west. *Source:* Adapted from Social Service Research Committee, 'Census Tracts of Chicago 1930: Economic Status of Families' and 'Census Tracts of Chicago 1930: Percentage of Total Homes Owned' (Chicago: Social Service Research Committee, University of Chicago, 1930).

is, with white-collar jobs; preachers had relatively higher status in Black communities than they did in White communities because they provided more support. European immigrants often displaced Blacks in semiskilled jobs such as barbering. When the immigrants stopped coming in 1914, the Blacks occupied the very bottom rung of the income ladder and lived in the shabbiest rental housing. Landlords charged them relatively higher rents than what they charged White working-class residents to the west. Thus in 1930 their economic status (as suggested in Figure 5.5) was considerably higher than it actually was.

Chicago is said to be a city of neighbourhoods, but then all cities make this claim. Its rail lines defined cells, which helped to sharpen this image. Neighbourhoods connote stability with at least a core of people anchoring the local institutions. Cutting across the stability, of course, was the constant movement of people and of networks of relationships outside neighbourhoods. One of Chicago's best-known, stable, and very ethnic working-class neighbourhoods (even now) is Back of the Yards, which is just west and south of the stockyards and packing plants. Home-ownership there was considerably higher than in the slum areas on the Near West Side, though half the owners took in boarders. Someone said that they would 'starve their families to buy'.[77] Well, not quite. Though overstated, many residents in a better area of the neighbourhood lived in 'comfortable single and two-family houses, neat yards with well tilled gardens'.[78] It was an area where saloons were crucial social institutions; those near the plants drew workers in at lunch-hours. In the evening, neighbourhood taverns, like clubs, attracted men, especially single immigrants. Churches (mostly Roman Catholic) were community centres with many clubs for men, women, and children. They and the public school board improved the lives of children. The University Settlement House helped to make life slightly easier.

As in Manhattan in 1860, the reality was of considerable ethnic mixing, though newer immigrants had a greater tendency to concentrate.

Young people went to the dance halls and, on weekends, visited an amusement park called the White City, which emulated the World's Fair site. People from several ethnic groups (Irish, German, and several Slavic groups) resided there, though in somewhat clustered groups. Ethnic saloons provided a wide range of community help. In fact, in this area of about 250 residential blocks, there were enough Poles for three parish churches. The church farthest from the stockyards and packing plants to the south served the highest-status Poles. Those that were nearest had to put up with the stench of the yards and plants and of Bubbly Creek, the receptacle of offal.

In looking at ethnicity across the city, community analysts' maps have tended to show ethnic homogeneity (partly because of the difficulty in showing different kinds of peoples mixed together or in showing multiple uses, and because outsiders have always tended to generalize): the Germans to the north; the Poles to the south in districts vacated by Germans in Little Poland, and also to the southwest in several areas and in South Chicago; the Jews in the near westside ghetto; the Italians to the north; the Czechs just to the west in Little Pilsen; the Swedes north of the Germans, and so on.

As in Manhattan in 1860, the reality was of considerable ethnic mixing, though newer immigrants had a greater tendency to concentrate. In 1910 members of all larger ethnic groups were found in all thirty-five wards. If one examines the number of foreign born of each group in these wards, the level of concentration can be ascertained. According to the census numbers, in the top five wards for each were three-quarters of Chicago's Italians, two-thirds of the Poles, half of the Austrians, Russians, and Swedes, but only one-third of longer-standing Germans and Irish. By 1930, 60 per cent of Poles, who were the most concentrated, lived near other Poles, a level far below that of ghettoized Blacks. Others among the most recent European groups were less than half that percentage.[79] People had spread out a great deal from the original core areas, yet churches in the old areas continued to attract people on Sundays, as did schools and shopping areas (which were somewhat concentrated ethnically), thus giving the impression of neighbourhood continuity. In new districts with sufficient concentration of one ethnic group, churches (the Roman Catholic ones were often enormous to make, it seems, a symbolic statement to Protestant America) and schools continued to be organized on ethnic lines. In 1930 over half of Catholics still attended ethnic churches, and over half of the pupils attended ethnic schools. The pull of ethnicity remained strong.

Almost half of the Roman Catholic pupils did not attend ethnic schools, on the other hand, which suggests that dispersion to new suburbs and ethnic mixing was widespread. Besides, many (such as the Germans and Irish to the north) intermarried and were now in the middle-income class, living in apartments. Public elementary schools and high schools undercut ethnic identities, yet sustained the sense of belonging to neighbourhoods. As in other large cities, the Chicago School Board built many elementary schools from 1880 onward. The elegant Carl Schurz High School far out on Milwaukee Avenue near Irving Park, built in 1909, likely had a catchment area of 3–4 sq. mi. (8–10 km^2). Cutting across neighbourhoods were large technical schools that covered even wider areas. Outlying shopping strips emerged along streetcar lines and focused on accessible intersections. Department stores opened branches in some of these shopping areas. Because housing types were often microsegregated by block faces rather than over wider areas, even the neighbourhoods defined by Chicago sociologists, such as Little Sicily in the 1920s, remained mixed—with the exception of the Black Belt.

The Civic Moment Dissipates

Burnham's city plan was the high point after two decades of the civic moment. It was 'one powerful attempt to revive Chicago's reputation as an inspirational white city.'[80] However, the planning commission that was set up to implement the plan made only modest improvements. Grant Park and others came cheaply on landfill. The widening of 12th and Michigan Avenue, the Ogden Avenue diagonal, and the relocation of the produce market were expensive but miniscule compared to what Burnham had originally proposed. In the far regional reaches, officials did set aside forest reserves. Burnham was not, of course, concerned with the housing of the poor except to clear them out of rookeries. The plan was an impossible fantasy, rising from Burnham's White City of 1893, prototype for Disneyland. It was utopian like Pullman's vision or, for that matter, like most communitarian villages of the nineteenth century. Burnham's attempt to shape the city failed.

One can wonder why so much excitement was generated by the 1909 plan. Burnham and Charles Zueblin no doubt genuinely felt that the tide of reform was on their side. Prohibition was on the way out, and vice was coming under control—purifying and standardizing the city and its citizens. Also, as we saw earlier, structural changes reduced corruption and patronage. Class strife was less virulent, and corporations were riding high. In 1911 James Bryce, who two decades before had excoriated cities as America's 'most conspicuous failure', could now exclaim 'the sky is brighter, the light is stronger'.[81] In their innocence, these leaders underestimated the power of property, even as they touted the great money-making value of the great plan.

The will of the supporters of reform petered out. The plan was the finale of the fleeting moment of civic rectitude. After 1910 the reform impulse began to fade. Businessmen became less interested in political action and, as a result, professional reformers without business leadership had less influence. Business sought stability, predictability, and security; reform programs delivered a measure of these with regard to street railways and other services. Experts in the large and nearly autonomous bureaucracies had virtually finished the vast infrastructural undertakings. Women activists largely gave up on reform. Many of them tired of trying to make the city livable. Writers, drawn earlier to the exciting city, departed.

Political conflict returned. Most reformers were Republicans. The election of a Republican mayor in 1915, Big Bill Thompson, led to even worse corruption and scandal than among the boodling Democrats of the 1890s, and pointed to a failure of the élite, as did Al Capone. The élite obviously no long had any positive control over politics. The civic moment had passed; Chicago became a 'shameful disgrace' once again.[82]

Ironically, Chicago politics became even more unstable. The speculative world of Chicago in the previous century carried over into an unsettled culture for the élite and others. Of course, their activities on LaSalle Street were hardly more honest than ward boodling, but the press did not report their dealings. The inheritors of wealth were less confident than their fathers, those great moguls of industry and finance. Symptomatic was the political career of a *Chicago Tribune* heir:

> Joseph Medill Patterson was temporarily a great champion of the underdog. Haunted by guilt over being an 'idle-rich young man,' in 1906 he resigned as reform public works commissioner and joined the Socialist party. Patterson now attacked amassers of great fortunes, like Field, and their legatees and advocated the labor theory of value, redistribution of wealth, and nationalization of industry. Within ten years, however, he was vice-president of the family paper and a staunch reactionary.[83]

This about-face, if rather exceptional, can be taken as a sign of inherent instability among the élite leaders, an insecurity that undercut public responsibility and the possibility of *noblesse oblige*. The *Tribune* remained reactionary (interestingly

like Thompson), and diverted citizens' attention by blaming England for Chicago's woes. Like rich heirs elsewhere, many of the third generation dissipated wealth. 'Even before the Roaring Twenties and cafe society, divorce, drugs, drunkenness, and death began to undermine the upper class in Chicago as well as in New York.'[84] Even more telling, Chicago's élite took off in a 'massive flight to distant places, a source of erosion spared' big eastern cities.[85] Maybe that could have been predicted; Burnham himself lived in Evanston to escape the taxes and the politics of the city he proposed to improve. Also, the city probably lacked the infusion of new entrepreneurial blood as its economic growth weakened. If the health of a community depends on the stability and participation of its successful business people, then Chicago became worse off than other big cities.

Even while many rural midwesterners remained dependent on the value-added largesse of the city, and many of their offspring sought opportunities there, their perception of Chicago remained that of 'The Grand College of Vice ... pouring a cloud of immorality and pauperism' over the region.[86] Chicago failed to measure up as the centre of what Frederick Jackson Turner and Frank Lloyd Wright had asserted, the 'heartland of the national ideology', the Midwest.[87] After 1910 Chicagoans acted as though they believed this too—despite the civic enthusiasm of the previous decade. Even in 1910 our English visitors may have been impressed by the still-apparent verve, but came away harbouring doubts that the élite had a deep commitment to public life, wedded as they were to the bottom line. From a longer perspective, Chicago could not seem to handle the inevitable flattening out of its economy—nature's revenge on nature's metropolis.

CHAPTER 6

■

Los Angeles, 1950: The Working Class Thriving on Military Largesse

As their plane prepared to land at Los Angeles Municipal Airport on the late afternoon of a warm fall day, our now aging British couple hoped for a restful vacation in the Mediterranean climate of southern California and, with their usual enquiring bent, to learn about life in this renowned place. As their Constellation glided towards the airport, they could see a mix of metropolis and rural areas, a 'repetition of squares and rectangles ... curiously mixed with citrus groves, dairy lots, hills spiked with oil derricks, tank farms, and precise truck gardens'.[1] Disembarking from the plane, they enjoyed the breeze and clean air from the nearby ocean.

They decided to explore the city by starting at its centre, though they knew that to understand the whole region, they would have to travel long distances to many communities. The taxi ride on the wide boulevards to the Biltmore Hotel downtown seemed to take an eternity. Indeed, as a frequent flyer said of the taxi ride downtown, 'now we begin the hazardous traveling'.[2] On arrival downtown, they wondered whether their lungs would survive the blanket of smog enshrouding the city. For a while, they thought maybe another ride back to coastal Santa Monica might make sense, since they had heard that one centre in the region was like any other, but they resolved to stay at least briefly downtown.

Their contacts here had been telling them about 'one community', yet they were somewhat confused by the plethora of different communities. They were not the first. In 1931 another visitor had

questioned: 'It struck me as an odd thing that here, alone of all the cities in America, there was no plausible answer to the question, "Why did a town spring up here and why has it grown so big?"'[3] In 1938 journalist Westbrook Pegler had referred to Los Angeles as a 'big sprawling, incoherent, shapeless, slobbering civic idiot in the family of American Communities'.[4] Other big cities had identifiable landmarks, but an Angeleno, puzzled when asked about his town, replied: 'I don't think of anything ... except, maybe, my own garden.'[5] Perhaps he should have applied Gertrude Stein's view of Oakland that 'there is no there there'.

Others would retort that they could explain the city. Like Chicago before it, Los Angeles was a place that 'knows it has a destiny ...'.[6] 'Climate has a cash value ...'[7] was obviously an answer; the place was the Riviera of the West. In the 1920s it became the Detroit of the West, and by 1950 it was the chief arsenal of America. Los Angeles was also the legacy of retired, White, small-town and farming midwesterners who established the community on the site of an earlier Mexican settlement. They attempted to create a White, Protestant, prosperous, and orderly America in the communities of the basin.

But Los Angeles was riven with contradictions. When the economy failed in the 1930s, despite the ballyhoo about free-enterprise, its leaders embraced government aid. When government subsidies became permanent in 1950, they denied the reality. Earlier they tried to run the city as if it were a military base. Ironically, in the end, the military establishment of Washington ran it during the near-corporatist era. Despite the boosterism of Los Angeles' 'one community' and unification by bureaucracies and infrastructure, this massive, car-addicted urban region was fragmented socially, politically, and hence spatially. Early on, those who did not fit in were left out. The rich spatially excluded others as never before, and the middle class did the same to those with little. When industrial workers made unprecedented material gains, even they, who had suffered extreme deprivation during the depression, excluded the poor. To the English visitors, the underlying presence of government, autos, fragmentation, and exclusion stood out in a city that had all kinds of wondrous things their compatriots at home could only dream about.

North America's Demographic Decline and Baby-Boom Recovery

In 1941 demographers in California and elsewhere predicted that 'the day of immigration is over and the time of small families has come' as the United States 'approaches a stationary population'.[8] The 1930s had been such a shock to population experts and everyone else. Even in 1950 projectors were chary about forecasting strong population growth; the recent wave of births 'probably has not brought about any long-term swing'.[9] However, the United States and Canada were, in fact, growing again more vigorously, thanks to the optimism after the Second World War. The baby-boom generation actually began by 1943. In 1940 only half of American women ages twenty to twenty-four were married, but two-thirds were in 1950. The population swing reflected a volatile half century in the economy, and one that saw a slow-down in the 1920s and a more marked one in the 1930s. In the forty years after 1910, the American population as a whole increased by only 65 per cent to just over 150 million; Canada, on the other hand, gained more (94 per cent) to reach 14 million.

California was a newer place than most. The state's population rose four and a half times between 1910 and 1950 to 10.6 million, faster

than other large states or provinces. It experienced particularly rapid growth in the 1920s and especially the 1940s, at least by the slower standard of the early twentieth century compared to previous centuries. Southern California was even more ebullient. In the 1920s Los Angeles outdid any other large city, double that of Detroit, though much smaller Miami, Houston, and Dallas were rapidly catching up. Los Angeles county's population went up even faster. After the slow-down of the 1930s, over the next decade (especially during the war), the city grew markedly to 2 million, and the rest of the county to 2.1 million. One-third of the county's population was in other cities, the largest of which was Long Beach, followed by Pasadena, Glendale, Burbank, Santa Monica, South Gate, and thirty-nine others. Ironically, the population of Los Angeles and its region was markedly older than most of the other populations on the continent, and hence could not reproduce itself. By the mid-1940s family size in America had dropped from 4.6 at the turn of the century to 3.5, but to 3.2 in Los Angeles. With a surfeit of single persons and the elderly, this playground of America had fewer children.

Like big cities earlier, the rapid population rise was a consequence of migrants. Beginning in the 1880s, the migrants were not Europeans but mostly White Americans and Canadians who left the countryside because of low farm incomes and rural boredom. Americans born outside of California contributed to three-quarters of the growth from 1910 to 1930. Midwesterners were predominant in California throughout the 1920s. In the 1930s three-quarters of the migrants came from west of Mississippi, many of them indigent and gaunt from malnutrition, forced out from Oklahoma, Texas, and other marginal farming areas by drought and foreclosers. Although southern Californians had previously laid out the welcome mat, many of these poor migrants were discouraged and often rebuffed. Even so, the county's population increased by half a million people in the 1930s. The grapes of wrath then turned to a harvest of incomes for those who had stuck it out. From 1940 to 1944, 780,000 newcomers to the Los Angeles region arrived, again mostly from the Midwest and southcentral states. This exceeded the number of migrants who went to the East and the Midwest, many of whom came from the South. Although unemployment rose after the war to 10 per cent in 1949, migrants (if in lesser numbers) were still attracted to Los Angeles because personal income in manufacturing was 15 per cent higher there than elsewhere. Then the Korean War, beginning in 1950, attracted even more people as the region's military economy became entrenched. Fewer Blacks travelled to the West than to the North, but Los Angeles was not very accommodating to the over 200,000 Blacks who lived there in 1950.

The number of foreign-born people declined generally on the continent. In Los Angeles county, they counted for less than one in five in the late 1940s, much less than New York or Chicago early in the century. Back then, Los Angeles did not attract many. In the mid-1920s federal legislation reduced most of the European and Asian influx. Not surprisingly, Mexico provided the largest number of immigrants—indeed, by 1925, Los Angeles was the largest Chicano city besides Mexico City. Of course, Spanish-speaking people had predated the American arrivals, though their culture was largely overwhelmed quickly. The British Isles, Canada, Russia, Germany, Scandinavia, and Italy provided most of the other foreign immigrants.

The Roller-coaster Economy after 1910 Saved by War

For four decades, the American and Canadian economies were more closely tied than they were earlier and fluctuated wildly. Workers experienced great rises and falls in income and employment. Bursts of production in peacetime until 1913, in the 1920s, and during the two world wars were interspersed by a weakening—even stagnating—economy, particularly during the Great Depression, which was the worst ever. Corporate concentration failed to stabilize the economy. Government had to step in as never before.

> **Corporate concentration failed to stabilize the economy.**

Let us briefly track the experience to underline the need for stability. Early twentieth-century growth continued until 1913, very rapidly in Canada as the western prairies were being occupied. Between 1910 and 1913 Canadian economic gains narrowed the gap with the United States to 75 per cent of gross domestic product (GDP) per capita, which was the highest since the 1850s. Growth stalled between 1913 and 1915, falling a sharp 8 per cent in the US, only to rise again primarily because of the First World War. Growth stalled more so in Canada because of its deeper and longer involvement. In 1919 growth dropped off briefly in the United States by 6 per cent. For Canadians the fall was much sharper—apparently a whopping 27 per cent decline from 1917 to 1921 (assuming the figures are accurate), as wheat, mining, and manufacturing sales fell. Compared to the US, Canada's per capita GDP slumped dramatically to 63 per cent. The Canadian economy took nearly ten years to recover to the 1917 level of productivity per capita. Production in both countries rose to historic peaks in 1928 and 1929, but the economy in the 1920s was not as stable as it was before 1913. Agriculture in both countries was troubled, land speculation rose dramatically, then fell, and consumer financing had to sustain even the widespread use of cars and household durables in middle-class homes, as testified by the creation of General Motors Acceptance Corporation. By 1928 stock speculation began its rapid rise as the only place to put money because overproduction in many fields discouraged investment, with the exception of skyscraper offices. The stock market spiralled upward without any direct relation to the actual economy. Wealth and income became more concentrated at the top end of the scale; the top 10 per cent of the income scale held 36 per cent of the wealth in 1929. The bottom half lost a larger share to the top. It was a society in which its leaders believed in 'the foolhardy assumption that the special interests of business and the national interest were identical'.[10] It was a society in which advertisers produced ads showing an executive, phone in hand, looking out from his high office window, like a duke of old from his ramparts, surveying the landscape—'the master of all he surveys'.[11] Herbert Hoover's 'welfare capitalism', supposedly to improve the lot of the workers, had wrought these distortions. The twenties did not roar for everyone.

Then the stock market crash in October 1929 triggered the deepest and most prolonged slump ever since European settlement. This signalled the end of the short era of corporate hegemony. The organizational restructuring through mergers after the 1890s depression had stopped yielding. If the incomes had been higher for those in the bottom half of the income scale, 1929 might have been postponed, and perhaps the economic downturn might have been weaker. The crash meant that the rich lost; 'large incomes simply had farther to fall than small ones'.[12] As is well known, bank failures in the United States peaked, resulting in a bank holiday at the depths early in 1933. Chicago's commodities exchange was hit hard. Only because they were big, the seven chartered banks in Canada survived. Montreal's stock market, which was also beset by corruption, lost industrial shares to Toronto, thus pointing to a regional shift of private power in Canada.

Companies laid off workers in dramatic numbers, a quarter or more by February 1933; in many industrial cities, the ratio was much higher. Young women left the offices, not for marital bliss but to return to the kitchens of their mothers. Or their quotas for producing garments increased. Businessmen, such as Henry Ford, who paid $5 a day earlier for mind-numbing assembly-line work and non-unionization, could not fathom that their own workers needed money to buy goods. Although some businessmen could preach that this was a 'chastening of the spirit', to most people Ford fell from his pedestal as a folk hero.[13] Investment dropped like a stone, new capital issues by 90 per cent. Consumer durable goods production was hit hard, their value of sales dropping by two-thirds from 1929 to 1933, and those of cars fell by four-fifths. Producer durable goods dropped even more. New private investment in construction nearly disappeared: housing construction in the United States was less than 10 per cent of the 1925 peak,

though in Canada it fell to only 30 per cent, cushioning the blow somewhat. Over a quarter of office floor space was vacant. The excesses of the chaotic land industry of the 1920s had contributed to the length and depth of the depression. Not surprisingly, non-durable goods, including food and services, did not fall as dramatically. In all, American and Canadian productivity dropped by one-third. The excesses of the 1920s, when corporate power rode high under Calvin Coolidge's dictum that 'the chief business of the American people is business', could not be checked by weak unions; too little government regulation of speculation and banking, and maldistribution of income led to overproduction and underconsumption. A crash was needed, it seemed.

Recovery came slowly after 1933, though the American economy took a sharp drop in 1937 once again. Canada's recovery was slightly stronger, so that its per capita GNP reached two-thirds that of the United States in 1939. American productivity levels reached 1929 levels in 1940. Until then unemployment remained high, if not as high as it was in 1933, even though those with jobs bought more household durables. Office vacancies still remained high at 17 per cent. Construction lagged behind the mid-1920s' rate, though housing starts reached two-thirds of the 1925 peak, thanks to government public housing and mortgage guarantees that began in 1935, on which I will amplify later. The 1939 World's Fair in New York cheered people up with promises of a glorious future. But only war 'brought all of American industry out of the crisis of the thirties'.[14]

Production reached new heights in 1944 as the machines turned out more goods than ever before, this time for deadly weapons. Production jumped from 1939 by a whopping 83 per cent in the United States and 56 per cent in Canada. Incomes rose by a quarter during the war, especially among industrial workers; it was the strongest gain ever, setting the stage for fulfilling the workers' material dreams and having, as we will see, a profound effect on cities. That the top earners lost 10 per cent of their 1929 income level after the crash and especially during the war was a good sign for the rest of society. The federal government employed four times as many people in 1945 as it did in 1940, in addition to those in uniform. Factories worked at capacity and more. Because full employment was achieved, large numbers of women, symbolized by Rosie the Riveter, joined the workforce. Although workers could not buy consumer durable goods before 1945, they could save for postwar spending through war bonds.

But once again, the return to the civilian economy led to a slow-down, much more sharply so in the United States than in Canada, in part because of overproduction by 1944. President Roosevelt cut military spending substantially in 1944, and then President Truman did so again in the fall of 1948, apparently in the hope of returning American defence to the militia. So in 1949 American per capita production was only three-quarters of what it had been in 1944, while Canada's was over nine-tenths, in large part because its mines were turning out strategic materials for American stockpiles to provide for a war to come. Despite a return to building houses and cars at 1920s' levels and more, many workers found themselves out of work once again; in 1949 unemployment was well over 5 per cent in the United States, which was double the Canadian rate. Wages dropped off.

In 1950 the Korean War provided new opportunities for military production, so employment picked up. It seems the cold war, which became fixed during the Korean conflict, was necessary to prevent a roller-coaster economy. It certainly stabilized growth and incomes, though at the price of inflation. Even so, the US would not recover the 1944 levels of per capita productivity until 1964, as the postwar drop of 26 per cent was a lot to overcome, just as Canada's recovery from decline had been after 1918. Canada made it back to the peak of war levels as early as 1951. In the process Canadians reached 71 per cent of the American standard of living, which was the highest since 1913, presaging Canada's sustained greater gains until the late 1980s.

Governments were thus drawn dramatically deeper into the economies, first when they responded to the crisis of the 1930s by providing relief and public works for employment, then indirectly by granting rights to industrial workers, establishing housing programs and serious social security measures, and later still through military spending. In the United States, Roosevelt acted quickly in launching the New Deal to provide direct relief, in contrast to Hoover, who had only reluctantly stepped in to sustain voluntary Red Cross measures. States and especially local governments had already faced reality: their spending on welfare had jumped ten times from 1927 to 1932, then doubled again within two years and continued to climb. Under Roosevelt federal welfare support quintupled, then receded as unemployment slowly fell. In 1938 help was needed again. As Steinbeck made it clear, Okies in California enjoyed peace and comfort only in government camps.

Under the First New Deal, Roosevelt's administration flirted with national economic planning. It financed public works because municipalities and states, which were hitherto largely responsible, could no longer carry the financial load of considerable expansion in the 1920s. Local spending on public works peaked in 1928. Special tax measures to fund borrowing for such projects failed soon after when property tax delinquencies in the early 1930s dramatically undercut municipal finances in both countries. Indeed, states and provinces had to bail out many bankrupt communities that were also saddled with providing relief on a massive scale.

In the past, the federal governments had, of course, financed roads, canals, harbours, railways, and industry directly and indirectly, but now the crisis called for more than that, not least the restoration of employment. In the teeth of the free enterprise ideology, the Great Depression forced action (if not through full-scale economic planning) through methods that were later justified by Keynesian demand management. Among English-speaking countries, New Zealand led the way with social programs in the 1930s, though the US then created a social safety net and recognized unions, with Canada lagging. At the end of the war, all Western countries committed themselves to full employment, though English-speaking countries would not stick with the idea for long.

By 1950 the regional economies in North America were just at the threshold of unprecedented change. In the United States, the West and parts of the South began to contribute far more to the economy, with Los Angeles at the leading edge, largely because of government initiatives. In Canada Alberta's oilfields would have a profound effect on the economy. But these changes were just underway. The manufacturing belt of the northeast and Midwest in the United States and adjacent Canada, which was established by 1860, still dominated goods production, though it was badly shaken in the depression. The American part of the manufacturing belt would fall behind the West, South, and central Canada over the next generation. Maritime Canada's industry had already fallen into permanent depression in the 1920s and New England textile production declined as plants opened in North Carolina.

Mining in the western United States and on the Canadian Shield continued to yield vast mineral wealth for the factories of the northeast, while the Appalachian chain and prairie states yielded coal, and Texas and California produced oil. After 1918 farmers bought tractors, trucks, and cars, but many fell deeply into debt. The northern spring wheatbelt, which was firmly in place only by 1913, was troubled soon after the First World War. The region suffered greatly during the depression, as would the winterbelt farther south. In the South, cotton production peaked in 1926. The more marginal farmers left the land in great numbers, in part because of greater mechanization and low prices. To survive during the depression, many city dwellers returned home to farms.

Contributing to the economy's revival after 1933 were the electrification of farms and in particular the power dams and navigation improvements of the Tennessee Valley Authority; finally, after many decades, the government in Washington paid heed to southern woes. Consumption was aided by the forty-eight (in some cases, forty-four-hour) week for those with work; people had more opportunity to spend. During the war, the northeast hummed once again because Washington's military spending, including aircraft procurement, went to the industrial belt. New York and Michigan led, though California rose to third place, with Los Angeles as the main military manufacturing centre. The Pacific war turned attention westward. The South also received a larger share of military spending than it did in earlier wars. The shift of economic power to the West and South was beginning. In finance, Wall Street would never be the same again. The financial world had been changed: 'federal government spending … diminished the power of New York capital'.[15] The corporate era gave way to a time of cooperation among corporations, government, and even workers to a degree. American society was, it seemed, becoming corporatist.

The LA Economy: New Opportunities and Excesses

By 1950 Los Angeles was well on its way to the top rung of the military industrial establishment. It was already the premier city of the West. From 1920 to 1930, it rose dramatically to become the third city on the continent in manufacturing output, and second in value-added. Military spending began in the late 1930s. The war gave the city a great boost as a major production focus. Although the postwar economy was wobbly with the return to civilian spending, the demands of the Korean and the cold wars got it moving for several decades.

Before 1920 the local economy had been dependent on agriculture for national markets, investment in land by affluent migrants, and industry to serve the immediate region. High-value agriculture took off in the 1880s with refrigerated shipments by rail. In 1892 growers combined into the cooperative Southern California Fruit Exchange using the Sunkist label. By 1910 Los Angeles was the first county on the continent by value of farm products, a position it held until urbanization overran the orchards and beanfields. Bountiful electrical power and local oil facilitated production and consumption.

During the First World War, California's manufacturing for a national market began with shipbuilding. Although this lapsed quickly, it signalled new possibilities. Aircraft production rose, and Los Angeles supplied the country and the world with motion pictures. In the 1920s, however, most of its manufacturing activity remained regional. It became the Detroit and Akron of the West. Los Angeles's auto and rubber factories were only the most obvious of its many branch plants. By 1929 it reached ninth position industrially among cities. Sports clothing production took off because of a rising influx of tourists. Beginning in 1921, the All Year Club, led by the *Los Angeles Times* and subsidized by local governments, successfully promoted tourism by selling California's climate. Real estate speculation was rife in the early 1920s; in 1923 it exceeded the peak years of the 1880s. However, with a slowing population growth, housing production eventually weakened too. More dollars went to building apartment blocks. Even before 1929 many had lost savings in real estate speculation and wildcat oil ventures.

The Great Depression hit the region hard, if less markedly than it did in the industrial East. By 1933 Los Angeles's industrial activity fell to three-fifths of what it was in 1929. Wholesalers, banks, insurance companies, and oil firms went bankrupt; these defaults were the highest anywhere. Suicides were high; an Arroyo Seco bridge was a common leaping site. In 1934 30 per cent of Californians received relief, two-thirds of those in southern California. Los Angeles police took up posts at the state borders to stop poor migrants from entering the city. 'The famous Los Angeles hospitality had suddenly been turned off for those unable to pay for it.'[16] As a result, for two months the city managed to save on its relief funds. Needless to say, there was less marketing of California's temperate weather in the 1930s. 'Decades of Ballyhoo had at last caught up to the boosters.'[17]

Even so, people kept coming to Los Angeles in the 1930s. Industrial activity recovered to the 1929 level in 1937. Film production increased. Several auto plants were added, and oil was exported to Japan, at least until 1939. Because the federal government started the war effort well before 1941, particularly in aircraft production for allies and American preparations, economic improvement occurred rapidly from 1939 onward.

Employment in manufacturing rose from 205,000 jobs in April 1940 to the peak of 560,000 in October 1943 in nearly 500 new plants and 1,000 more expanded operations. Defence production by October 1941 took up nearly half of this expanding manufacturing sector by value; by 1943 this share rose to 80 per cent, and provided one-quarter of all employment. Aircraft production accounted for three-fifths of that share. A steel mill east of Los Angeles facilitated production in aircraft and shipbuilding. Increased payrolls helped the movie industry to divert more people from the worries of war. Southern California's oil fuelled the planes and ships in the Pacific. Among American cities, Los Angeles had achieved third rank in manufacturing.

Despite rising fears about the erstwhile enemy and then ally, the Soviet Union, demobilization of the economy came rapidly after August 1945 to the detriment of Los Angeles and the GDP figures. Although the population continued to grow, manufacturing employment fell. Business failures exceeded new starts. The movie industry declined markedly from its peak in 1946. The military share of aerospace production dropped even more dramatically before rising again sharply in 1950. The cut in defence spending in 1948 hurt two aircraft companies that were virtually dependent on government contracts. They laid off nearly one-quarter of their workers before rehiring them the next year. Although indebtedness to governments remained high, in 1949 some suggested that government support of urban development to stabilize the economy would be timely.

The economy expanded in other directions, cushioning the declining government influence. New firms flocked to Los Angeles. Carnation, Rexall, and other companies moved their headquarters to the city. Civilian aircraft production rose. Sports clothing, food processing, and furniture production rose, so that Los Angeles stood third in these fields. Nine assembly plants produced 650,000 cars a year, boosting postwar sales (Figure 2.6). A quarter of married women worked in many fields, despite the decline in war work. Los Angeles became a leading city in book publishing. Manufacturing employment thus remained far above the 1939 level. Construction increased in the late 1940s as well, with twenty times the number of building permits recorded in 1950 as in 1939. Housing construction by 1948 nearly reached the peak year of 1923, though it did not meet demand. Oil production exceeded the wartime peak. San Pedro remained the biggest fishing port on the West Coast. Besides the naval base, the combined harbours of Los Angeles and Long Beach provided several thousand jobs, handling exports and imports of lumber and other goods. The two harbours were so important that the federal government declared them a Foreign Trade Zone, together with New York, New Orleans, and San Francisco.

As a sign of affluence, beginning in 1947 Angelenos bought television sets, and then at an accelerating rate, spent more on laundry and cleaning, hired more domestics, and boasted a higher proportion of professionals compared to people elsewhere. The number of government jobs rose. Incomes were 50 per cent higher than they were elsewhere on the continent. The hype of boosterism was strong, but the reality was different than those who had trouble finding jobs. Government spending in massive doses was still needed. To highlight the above, let us consider three strong areas of economic activity: petroleum, the entertainment media, and aircraft and space.

Petroleum

In 1892 Edward Doheny drew oil near the La Brea tar pits. Soon after, Collis Huntington drilled successfully at Huntington Beach. Then in the early 1920s Santa Fe Springs and Signal Hill oilfields boomed, but peaked in 1923 at 700,000 barrels a day from hundreds of wells in the 'black gold' towns.[18] This abundance enabled affluent Angelenos to fuel cars cheaply, and spawned an equipment industry. The dazzling profit also induced speculative excess. C.C. Julian swindled many in 'the biggest and phoniest of promotions'.[19]

The oil industry was less glamorous by 1950 because gushers were less frequent. Three years after the opening of the Wilmington field in 1936, sixteen fields produced less than half those of 1923. This amount, which was mostly refined

nearby, was more than enough to supply much of the West Coast and the Japanese for a while. Although it was predicted at the time that reserves would last 'less than ten years', new oilfields and more efficient techniques of extraction increased production during the war. In 1949 production tripled that of a decade earlier. The area refined three-fifths of California's oil needs, and the chemical industry expanded. In fact, at the time markets could not absorb all that was produced. However, because the number of cars in the county (already over 2 million) was expanding rapidly, some predicted that finding oil would become more difficult. In the long run, the region would indeed become dependent on non-American sources, and on Texas and western Canada for natural gas. Even by 1950 oil imports in the United States were beginning to rise; oil consumption surpassed that of coal.

Entertainment

Entertainment—motion pictures, radio, and television—were a second strong field. After Edison's invention in 1894, movies were made in many cities, including Chicago. By 1910 the movie industry concentrated in Hollywood as producers sought to escape the monopolistic Motion Pictures Patent Company in the East. The *Count of Monte Cristo* and then David Griffith's *Birth of a Nation* put Hollywood at the top. Charlie Chaplin, Mary Pickford, Douglas Fairbanks, and the DeMille brothers became famous in silent pictures when 'the industry entered upon an era of exaggerated profits, preposterous salaries, soaring production costs, amazing expansion, and general extravagance that relegated the wildest California real-estate and oil booms to the level of sober business enterprises.'[20] The industry turned to Wall Street in 1919 to fund even more extravaganzas. In 1926 sound 'brought devastation to many parts of Hollywood. Screen favorites who enjoyed world-wide fame and commanded fabulous salaries disappeared in a few months from public notice …'[21] Yet for the industry, sound was an enormous boost. Big electronic companies, such as RCA, were helped greatly. Innovations in lighting and photography led to construction of studios outside of Hollywood.

The depression failed to slow Hollywood as it did other sectors; output rose by a quarter. Sixty films were produced in 1935 and distracted people from the anguish of the depression. Although this golden age resulted in some memorable movies, only a few like the *Grapes of Wrath* addressed social issues. Viewers wanted to be distracted by the romance of cowboys killing Indians, not social action. In 1939 Hollywood spent nearly $9 out of every $10 on making films in the country. Wages for over 30,000 industry workers were reputed to average $4,200, which was four times the local average income; this was skewed, of course, by the outrageously high salaries of stars and directors. Outsourcing multiplied activity and wealth. Industry profits provided investment for offices, hotels and apartments, racetracks, and oil wells. Film production continued during the war, much of it as propaganda for the Allies' cause.

The great fall after 1946 created panic in the industry as other diversions emptied movie theatres. Payrolls fell by one-third. By 1950 European producers offered more daring or substantial films, and television presented a challenge. Yet Hollywood was still producing three-quarters of the world's movies; over 20,000 people worked in the industry, and others managed the estates and affairs of the stars. Seven major and ninety other studios could not be written off, and recovered to a new phase when they started to produce for television. As we will see shortly, the social influence of the industry was as profound as its economic impact.

Radio stations began broadcasting in 1920 and, though the region did not produce sets, it did produce programs. The Warner brothers operated a station in 1927, though another entertainment enterprise, Aimeé Semple McPherson's Four Square Gospel, beat them. Specialized music stations emerged. Twenty radio stations were on the air by 1950. Networks also developed in the 1920s; NBC broadcast the Rose Bowl nationwide in 1927. In the 1930s NBC and CBS set up their western headquarters in Hollywood. Early in the 1930s they started production in Hollywood; Jack Benny and Gracie and Fred Allen stole the attention of Sunday night audiences away from preachers. Because of local talent, Los Angeles had clearly established itself as second to New York in broadcasting.

Television became viable only in the late 1940s. Although the networks broadcast some programs in Los Angeles, New York was the prime centre. However, Hollywood recovered from its slump to carve out a share. Helped by AT&T's microwave and coaxial cables across the country, Hollywood could broadcast its programs to audiences in the East. The movie studios both small and large learned to respond to the public's interest in television by making films for the new medium, and thus recovered from the postwar shock by 1955.

Aerospace

The aerospace industries became even more crucial to the local economy, drawing a large proportion of government spending. Aircraft production began at various points on the continent: by the early 1920s in Cleveland, Dayton, Detroit, New York, San Diego, Seattle, St Louis, and Wichita. Donald Douglas started production in Santa Monica in 1922 with the support of financiers like Harry Chandler (the *Times* publisher) and a navy contract. The Lockheed brothers arrived. Military aircraft production predominated early, then shifted more to commercial output in the next decade when the Lockheed Electra and especially the Douglas DC3 began flying, the first large plane to make money by carrying people. By 1937 nearly nine out of ten aircraft added to airline fleets were DC3s. Half of American aircraft construction workers lived in the area, thus creating, to put it ponderously, a 'functionally interrelated system of agglomerated production activities'.[22] Suppliers provided parts and services, even eastern companies supplied engines and some parts, and universities turned out engineers, creating a culture of aeronautics. By the end of the decade, Douglas and Lockheed, along with Boeing in Seattle, were developing four-engine passenger planes. Other aircraft companies set up in the region: Hughes, Consolidated, North American, Vultee, and Northrop. They did so in part because the biggest story of the latter part of the decade was government contracts.

The British and French began to buy planes in 1938, and then the United States began to expand its own demand in expectation of war.

Whether the Second World War was won on air power is still debated, but the enormous output of planes in the United States and in Canada can hardly be denied. Four of the top five companies that produced 300,000 planes during the war were based in Los Angeles. The industry rose to first in dollar value in the US. The Big Six in Los Angeles expanded their floor space six-times between 1940 and 1944. The network of outsourcing rose dramatically, and big competitors cooperated with one another. Employment jumped two-and-a-half times to between 190,000 and 240,000, which was half the number of manufacturing employees in the county. (Counting varied, depending on direct and indirect activities.) Shipbuilding employed up to half the number of workers in the aircraft industry, and helped to push Los Angeles into the top rank of industrial centres.

After the war, however, orders decreased by 90 per cent, so aircraft employment dropped to 40,000. By 1948 employment had shrivelled to less than 15 per cent of industrial employment, which was about the prewar level. Commercial production may have risen sharply as scheduled commercial flights more than doubled those of 1939, but this could hardly compensate for the loss of massive military production, even though other fields of manufacturing (such as auto production) took up much of the slack. Douglas was easily the largest commercial producer with its DC4s, then DC6s and DC7s, though Lockheed competed with its renowned Constellation. The Korean War, of course, meant even more in military spending, though commercial production continued. North Americans could have butter along with guns without the rationing of the 1940s.

Other developments in aerospace were yet to blossom. As early as 1929 Theodore Von Karman at the California Institute of Technology experimented with jet propulsion. By 1943 military jet aircraft production began, though their practicality was still problematic in 1945. Not until 1956 would commercial jet production be perfected. Jet propulsion also led to missile experimentation

and a contract in 1945 for Douglas, then other companies.

Major spin-offs from aircraft and jet propulsion developed in electronics. Compared to the northeast, these electronic spin-offs were very limited in the area before 1940. Then radar systems and the like were perfected. Los Angeles thus became a premier high-technology city with high-paid staff. Indeed, in the 1950s its share of government-supported programs rose dramatically as it became the anchor of the sunbelt.

Still, the Los Angeles economy was vulnerable and therefore dependent on Washington's dollars. In fact, the economy had long been subject to the whims of financiers in the East before the explosion in military expenditure. Regionally, its financial institutions were weaker and subservient to San Francisco's. Much of the oil production was controlled by eastern companies. Movies in the 1920s were funded from New York, as were the newer high-technology fields. Savings and loan associations imported money and, by 1950, overtook other institutions in providing mortgages. The higher demand for mortgages in postwar Los Angeles was reflected in interest rates that were higher than elsewhere in the country, largely because the demand for housing finance was greater than elsewhere. Three-quarters of its houses were mortgaged at rates that were 10 per cent higher than the country. But then most well-paid Angelenos could afford that. By 1945 the San Francisco-based Bank of America's assets were three times that of Los Angeles's largest bank. The local California Bank was only fifth. After A.P. Giannini opened Bank of America branches in the region, other banks had to follow suit. The assets of California banks were collectively greater than that of banks elsewhere in 1950 and reflected affluence, but they were still dependent on outside financial support. Giannini was changing that: 'As J.P. Morgan had personified banking for the wealthy classes, so A.P. symbolized banking for the masses.'[23]

The federal government, though, was the main key to prosperity. Consider the benefits to developer and industrialist Henry J. Kaiser. Before the military build-up, the Hoover administration supplied funds for the Hoover Dam. Supported by

Bank of America's Giannini, Kaiser controlled the building of the dam. Washington bought up Los Angeles municipal bonds earmarked for the project in 1933 when the open market collapsed. Under the New Deal, especially under the Second New Deal, a coalition of western investors, including that of the Republican Kaiser and the unions, now with official sanction to organize, persuaded Washington to give the western states more money for dams and great bridges, to Kaiser's benefit. When war threatened, Kaiser and others plotted further on how to build up the West's industry with federal money. One result was the construction of Liberty ships at three shipyards, including one in Los Angeles, so that Kaiser became America's largest shipbuilder ever.

Steel was needed for this huge enterprise, and the shipping costs for steel from eastern mills were high. Only two small plants operated in Los Angeles at the time. US Steel built at a Utah coal mine when Defense Production Corporation paid the bill. Kaiser acted too. Borrowing from the Reconstruction Finance Corporation, he established his steel mill at Fontana in 1942. Coal was brought all the way from Utah and iron ore from nearby Eagle Mountain in the Mohave Desert. As elsewhere, Kaiser, who had adopted a corporatist philosophy, respected the rights of unions, paid high wages, and provided health care for employees, many of whom lived in the area.

The sprawling plant was the region's 'de luxe war baby', totally independent of Wall Street and the eastern banks because of government and some Bank of America financing.[24] The spin-off of secondary steel and machinery plants in the area led Kaiser to boast that his development would focus a great industrial empire for the West, providing for the expected increase in mass consumption after the war. Kaiser also got into the housing business, providing prefabricated units for war workers. Success and *noblesse oblige* rode on government largesse.

Just after 1960 a commentator noted that in California, the

... income elasticity of demand for climate, retirement, vacations, and specialty foodstuffs is high,

and they are among the underlying elements in California's rapid growth. However, the recent growth ... is also associated with the high rate of increase of defense-related spending....[and] is based upon national social and policy decisions rather than upon the intrinsic character of private demand ...[25]

Federal spending ran higher than Washington's revenue from the state. By 1958 California's share of the national debt per capita was six times that in the rest of the country, even though its per capita income was 25 per cent above the national average. The United States had 'gone far toward socializing research and development—if not in the organizations carrying out these activities then certainly in the source of demand for these outputs.'[26] The single buyer was Washington. However, what would happen 'the day the federal money stops?'[27] Few would hear the question, let alone try to answer it.

LA Society: The Contradictions of Fantasy Land

In 1913 a critic exclaimed that Los Angeles was the 'stronghold of transplanted midwestern provinciality, prudery, and spiritual aridity.'[28] Indeed, 'enjoyment is considered the first step to perdition ... creating a city devoid of lenience and cosmopolitanism', said another.[29] Frivolity was limited to the state picnics of expatriate Iowans and other midwesterners. Even charity balls were held without liquor. A clergyman moved to Los Angeles in search of 'a city without tenements, a city without slums'.[30] Unlike New York and Chicago, which attracted many aliens, Los Angeles sought to make the city like a small midwestern community with openness and no extremes of poverty. To maintain this innocence, people had to deny and ignore the realities around them. Let us consider the challenges or contradictions to the comfortable, egalitarian way of life that so many expected or hoped for: the pursuit of gain, limits to discourse, Hollywood, religious hucksterism, the Great Depression and unions, the rise of a national élite, and excluded minorities.

The Pursuit of Gain

The first challenge to the egalitarian way of life was embedded in the culture of the early American settlers themselves. Buttressed by Helen Hunt Jackson's *Romona* romanticizing the Mexican past, the southern California's 'sunny Arcadia' would mythically provide 'days of quiet content', far from the madding crowd in the East.[31] The clever and the ruthless pursued moneymaking through real estate speculation. Accumulating property for development enhanced power, so these men became the leaders. In principle and rhetoric, these leaders remained egalitarian, but in practice they created a society with extremes of hucksterism in economy, culture, religion, and income. Los Angeles reinforced the tradition that 'commercial élites in America have received esteem and deference from a society in which acquisition or wealth and success in business are reigning values.'[32]

During the land boom of the 1880s, this group of leaders arrived from the East and the Midwest, some with inherited wealth; others who had failed earlier sought their 'last chance'.[33] General Harrison Otis, later his son-in-law Harry Chandler, Henry E. Huntington, Eli Clark, Moses Sherman, William Garland, and Alphonzo Bell would dominate the local economy, even if they were dependent on outside capital for many of their exploits. They and others combined into a Chamber of Commerce and later the Merchants and Manufacturers Association to promote the city through an image of clean, Protestant free enterprise. Most of them had diverse operations and were very involved in creating a public environment amenable to their goals, so they did not hesitate to use public money for their own ends. Water provision was foremost for their land development because of the semiarid climate. Some controlled local transportation and other utilities. Few Catholics and Jews were among the power brokers.

> The clever and the ruthless pursued moneymaking through real estate speculation. Accumulating property for development enhanced power, so these men became the leaders.

Limits to Discourse

Otis and Chandler personified the early twentieth-century élite through their wide-ranging exploits, but primarily through their voice in the *Los Angeles Times*, which Chandler took over from Otis in 1917. His son Norman replaced him in the 1930s, and tried to perpetuate the same views. The *Times* was the ideological mouthpiece of downtown power, expressing the civic ideology of growth and the importance of protecting the innocents from alien influences.

The second contradiction to the egalitarian way of life, therefore, was to make an extraordinary effort to exclude thoughts and actions that did not conform to this ideology. 'The Otis-Chandler dynasty of the *Times* did preside over the most centralized—indeed, militarized—municipal power structure in the United States.'[34] That Otis had been a general in the Philippine campaign symbolized the nature of things. The city was to be run like an army and the innocents would fall in line. While other papers also boosted the civic ideology, the *Times* could be counted on in all debates to take the hardest, most uncompromising free-market line. In 1937 Washington journalists said it was the third 'least fair and reliable' American newspaper, just behind the Chicago *Tribune* and the Hearst papers.[35]

The *Times* maintained the myth of free enterprise by resisting unionization. Los Angeles was the citadel of the open shop. The bombing of the *Times* building in 1910 dampened public support for unions and socialism when two unionists confessed to the crime. Later General Otis and the Merchants and Manufacturers Association organized the hiring of off-duty policemen to break up strikes. Street railway workers who attempted to organize were beaten up in 1919. Los Angeles remained a low-wage city, except in Hollywood. The Russian Revolution and particularly the Communist Third International in 1919 provided another opportunity for repression. Bolsheviks induced fear. The *Times* persuaded mayors to appoint police chiefs like James Davis to sweep the streets of politically incorrect talk, not only action. Davis organized the Red Squad to ferret out the slightest deviation by non-conformists. 'Fanatically anti-labor', Captain Red Hynes hounded union leaders and organizing drives in the 1920s.[36] The third degree (that is, torture) was practised. Nonetheless, throughout the era, the 'lawless' police were involved in booze, drugs, gambling, and prostitution rackets.[37] Los Angeles was no cleaner than Chicago, though Capone received more publicity than LA gangsters. The innocents did not know where to draw the line, so it was all or nothing; prohibition, the ultimate hurrah for moral purity, provided an opportunity for the unscrupulous.

Corruption did not disappear because the downtown élite abetted (if not fostered) it. In 1929 the *Times* ran a special supplement on the 'The Forty-year War for a Free City … for industrial freedom', praising 'the independence of working-men free of the shackles of unionism'.[38] Nine years later, the *Times* lost one of its few serious ideological battles, calling the victorious mayor, Fletcher Bowron, 'the unwitting dupe of the CIO [Congress of Industrial Organizations], the Communists, and certain crackpot reformers'.[39] That rhetoric failed, at least for a few years.

Hollywood

Hollywood's entertainment industry included people with immense prestige and wealth, thus creating a parallel power structure to downtown's business élite. It also changed behaviour. Since Hollywood's markets were national, even international, its influence on the local environment was less by design than what was sought by the downtown élite. Hollywood's effects powerfully undercut middle-class rectitude, thus posing a third challenge. Jews, who were otherwise excluded from mainstream society, rose rapidly to fame and fortune. From very humble origins, Louis B. Mayer, Samuel Goldwyn, the Warner brothers, and others would eventually dominate the tinsel world

of make-believe. Mayer received the highest pay in town, maybe anywhere. With their stable of stars, they formed an unprecedented west side élite. Beverly Hills almost seemed their creation in defiance of downtown's centre of power, and was also a way of protecting their own uncertain status. The movie élite might fight and squabble, but their prestige became legendary throughout the world. Goldwyn might have said that so many people attended Mayer's funeral in 1954 to make sure he was dead; more to the point, so many people came to be identified with the glitter and prestige. Budd Schulberg, the son of a successful producer, may have punctured the pushiness in *What Makes Sammy Run?* (1940), yet just added something to chatter about.

The early 'cheap, tawdry, vulgar, ostentatious, vapid' films of the 1910s and early 1920s were reined in by censorship through the Hays office.[40] Cecil B. DeMille's dictum of 'giving the public what it wants' through racy sex was slightly transformed through religious extravaganzas like *King of Kings* and *Ben Hur*. Sex unleashed in 1920 remained standard fare, but was mostly toned down to sentimentality, despite the glamour of Garbo and Lamarr during the golden age of the 1930s and on through the 1950s. No one reined in guns and physical violence.

More important than films was the lifestyle in this 'incredible fairyland', which was also 'the most unhappy city in the world' with its 'own royalty in ermine robes', who looked down on a 'humble peasantry'.[41] It was 'John Bunyan's town of Vanity, magnified to huge proportions, streamlined … the manipulator of the … creed of narcissism.'[42] Hollywood and Los Angeles had the highest suicide, drug, and divorce rates in the country. The police impotently stood aside from the scandals, rapes, and murders, apparently too busy trying to prevent unions. Prudish southern Californians were converted to chasing fame and fortune, or at least hedonism, by the excesses of the fabulously wealthy. For the innocent, Hollywood inspired even higher levels of consumer indulgence in clothes, homes, and relationships, and epitomized 'frenzied recreation and pecuniary accumulation'.[43]

Religious Huchsterism

Religion was the fourth contradiction. Although many of the Iowans stuck with their Methodist and Lutheran roots, their churches and voluntary associations 'failed to recreate the cohesive community so crucial to their personal aspirations'.[44] Others became bored, 'tired of oranges', averred Nathaniel West in *The Day of the Locust* (1939).[45] They might have been vicariously tititlled by Hollywood's antics, but direct religious experiences peddled by 'the high-pressure salesman of salvation' were more exciting. Southern Whites did not take to old-line northern churches and started their own congregations. Blacks, as always, had to rely on themselves. As for Roman Catholics, Archbishop John Cantwell from 1917 onward kept the poor local Latin population in line through verbal assaults. He ranted against 'atheistic regimes in Mexico City and Moscow'.[46] Cantwell supported Hitler and Mussolini, attacked the New Deal and condemned Hollywood's immorality, but he did not refuse money from oil magnate Edward Doheny for a seminary, money that was tainted in the Teapot Dome scandal. Doheny had bribed the Secretary of the Interior in Harding's cabinet.

The Great Depression and State Support of the Working Class

The Great Depression was a shock to innocent Angelenos. In fact, many middle-class professionals and business people lost jobs and their faith in capitalism. The system no longer delivered in paradise. People sought other alternatives. F. Scott Fitzgerald and others read Karl Marx. Others, reputedly including Walt Disney, were drawn to fascism. Movies and religion diverted those with jobs. Many looked to secular panaceas in this age of 'crackpotism', and turned to everything from the Technocracy Utopian Society in 1933 to Roy Owens's 'ham and eggs' campaigns in 1938.[47] Francis Townsend's old-age revolving pensions and other similar odd schemes helped to create a climate that led to national social security, which was a major social gain of the 1930s. Most seriously, under the banner of 'End Poverty' in California, socialist Upton Sinclair nearly won the state governorship as a Democrat in 1934. He probably would have succeeded had not Roosevelt and moderate Democrats shied away from his advocacy of use instead of exchange value, of cooperatives over profit. Too many people preferred to revolt against taxes.

To the dismay of the innocents and the *Times*, the other great gain of the 1930s was the legitimization of unions. Although local department stores (with the help of the business associations) opposed the Teamsters as late as 1937, the tide had turned: transit workers unionized in 1934 (decades after other cities); the federal Wagner Act of 1935 allowed workers to organize and strike as never before; movie unions were strengthened; and Douglas gave in to the United Auto Workers (UAW) in 1937, the same year that General Motors lost its battle with the UAW. Most aircraft companies unionized, as did most large manufacturers all over the continent.

Los Angeles was no longer the open-shop city. Enough industries were organized to keep wages up, so by 1940 it became a high- rather than a low-wage city. All of this helped the working class, at least the White industrial working class. Wages during the war rose even more, though workers in large industries had to strike in 1946 in Los Angeles and elsewhere to ensure their share.

Crime and harassment of critics stopped. In 1938 Mayor Frank Shaw was recalled by an irate electorate and Judge Fletcher Bowron was elected. He abolished the Red Squad and appointed a trustworthy police chief when Angelenos finally recognized that the police were corrupt terrorists. Illicit gambling enterprises left for Las Vegas. The patronage racket that had 'utterly demoralized the city's civil service organization' ceased.[48] Soap-box speeches returned to Pershing Square.

The Rise of a National Élite

During the later 1930s and the war, there was confirmation of a new élite that was supported not so much by downtown or Wall Street as by the federal government in a massive denial of free enterprise. The new élite included corporate leaders such as the Lockheed brothers, North American Aviation's James H. Kindelberger, and Donald Douglas. They added prestige to the city, though they generally avoided socializing with the older élite. Howard Hughes and Henry Kaiser contributed to the diffusion of power in the region. Given the enormity of the change wrought by the consequences of the Great Depression and the war, local élites there and elsewhere no longer wielded as much power as they had previously. Washington was now the nexus of power as never before, but the innocents in Los Angeles would not quite concede.

By 1950 most old families had lost power, as did the early Hollywood moguls. Except for the Chandler family and a few others, kinship no longer counted for status, as in some earlier great cities. 'Disruption at the top reflected and intensified the rootless, inchoate atmosphere of Los Angeles' that had fragmented the scene for half a century.[49] Society life was guided by 'gossip columnists and press agents' rather than by the society matrons of the nineteenth century.[50] The corporate élite of the early twentieth century increasingly failed to maintain productivity and control the country, as finally revealed by the Great Depression. Government became the glue holding together the economy and society.

> The corporate élite of the early twentieth century increasingly failed to maintain productivity and control the country, as finally revealed by the Great Depression.

Los Angeles still held on to its middle-class innocence, though its citizens were now caught up in a consumerism that would try anything new. The chairman of the Brand Names Foundation asserted in 1946 that Angelenos would try new items first, then if these products were successful, they would be marketed in other cities. 'If anything, that is the magic which has made Los Angeles great'.[51] The Security-National president said that impulsive politics and exotic religious cults helped: 'it means that people coming out here are open to new ideas'.[52] Even in its decline, Hollywood served up pap like *The Philadelphia Story* to anaesthetize audiences.

But the dark side of this consuming society was that it was still resistant to new ideas. Odd,

religious charismatic leaders were now hardly a threat. But the gains of the depression and war were in retreat, as the effort to limit free speech returned. After 1945 the innocents mounted yet another purification campaign, doubling back to pre-1938 days. Norman Chandler may not have been as tough as his father, but the *Times*, having been set back by the clean-up of the police force, readily pursued the purge of communists, those who had dared to preach any degree of collectivism during the 1930s. The decline of Hollywood led to a search for scapegoats on the local scene. Richard Nixon could win a congressional seat by calling Jerry Voorhis a communist; in 1950 Nixon won a Senate seat by referring to his opponent, Helen G. Douglas, as 'the darling of the Hollywood parlour pinks and reds'.[53] To draw attention to the pursuit of Reds under every bed, in his play, *The Crucible*, Arthur Miller graphically portrayed the outcome of charges in the Salem witchcraft trials of 1692. Only those with closed minds failed to see the parallel between those witchcraft trials and what became known as McCarthyism.

Unions were forced to retreat. One-third or more of the workforce carried union cards in the country in 1948, remarkably up from less than one in ten in 1935. The auto, steel, and other industrial unions affiliated with the CIO made the greatest gains, and white-collar unions in government, transportation, and movies tripled their membership to over one in six workers. But the Red scare compelled the expulsion of communists. They turned to bread-and-butter interests rather than broader social concerns (like public health care) that had not been put in place during the New Deal. A truncated corporatism would not favour unions in the longer run.

The Red scare was an opportunity to shed some New Deal social gains, though Los Angeles could not revert to the open shop because the large industrial magnates did not agree. The *Times* might have liked the reactionary 'right to work' retreat from the Wagner Act in 1947, but that was, in the short run, only persuasive in the South and in the less industrialized states of the North. The Canadian solution—the Rand formula—was ingenious in preventing the open shop: you didn't have to join a union, but if there was one, you could not be a free rider. You had to pay dues.

Military rule during the cold war became permanent in 1950. Unlike earlier conflicts, during the Korean War most people hardly noticed the pervasive military presence. But the innocents still worried about those with heretical views. Police oppression returned. In the early 1950s under Chief William Parker, they ferreted out corruption as one of 'a "few good men" doing battle with a fundamentally evil city'.[54] Parker increased the fire-power of the force to Marine levels. In 1960 he said that the police were an 'embattled minority' against the crime of Black and Chicano gangs and their communities.[55] The Church reinforced this notion. In 1948 Francis McIntyre succeeded Cantwell as archbishop. Having learned to use strong-arm tactics against those priests who were sympathetic to unions while he was an assistant to Cardinal Spellman in New York, McIntyre was even more rabid, extolling McCarthyism and defending Chief Parker up to and including the Watts riot in 1965.

Excluding Minorities

Why would Parker see the police as an embattled minority? The city was socially and spatially divided by class behind a façade of classlessness—the final contradiction of the egalitarian, open society. Certainly the façade was carefully constructed. High school culture ignored the world outside, and so maintained the myth; Friday night football (now with lights) and the soda fountain of the corner drugstore attested to the democratic quality of life. California led the world in providing postsecondary education in junior colleges, state colleges, and universities. Opportunity was there for all to grasp. Incomes were higher than anywhere else in the world. The city had the highest proportion of proprietors, managers, and officials. Lending

libraries were set up in neighbourhood shopping centres. It was, indeed, a place with a wide distribution of wealth.

But all these advantages were in peril from enemies without and within. The enemies within were not so much the communists as the minorities, the masses who did not fit. The innocents could maintain their status only by excluding others—the vulnerable who challenged the clean-cut. Nixon, the Quaker boy, had learned this. The poor of Los Angeles were largely hidden from view and therefore did not exist. John Steinbeck may have earlier been the 'the poet of our dispossessed', but that mass of White dispossessed now was being materially fulfilled after 1940.[56] The minorities were the problem.

Visible minorities—Japanese, Chinese, Mexicans, and Blacks—made up only a small share of the population, but to the innocents, they were not supposed to be there. Attempts had been made to exclude Japanese and Chinese immigrants by 1906. In 1913 the Anti-Alien Land Act tried to prevent the Japanese (some of whom were already successful market gardeners) from owning agricultural land. The Immigration Act of 1924 effectively shut off the flow of the 'yellow peril'. By 1941 most Japanese were citizens; nonetheless, they were shipped out of the area eastward to internment camps. At the end of the war, Mayor Bowron, the most enlightened of all mayors, apologized to the returnees for having supported the move, but subtle exclusion remained.

Mexicans had been the original residents before being displaced by midwesterners. Other Mexicans arrived again in large numbers during the First World War. By 1925 Los Angeles had the largest Mexican urban concentration outside of Mexico City, but they were poor. Even though public health measures reduced infant mortality by 1930, the Mexican rate was one-third higher than that of Whites. Composing only 10 per cent of the population in 1930, they were overrepresented as labourers, many of whom worked seasonally in the fields and as servants at the bottom of the income scale. Women worked in low-paid clothing and food manufacturing. Employers discriminated against them. Agricultural workers struck in 1928 in the Imperial Valley's 'factories in the fields', in the early 1930s when organized by communists, and again in 1941 near Santa Paula.[57] These events unsettled the innocents. Los Angeles leaders organized a repatriation campaign for these supposedly unassimilable folk. Some (though obviously not most) left in trauma. All the while, family life sustained the cultural integrity of Mexicans and their ability to survive hostile, anti-Mexican sentiment. Even in 1941 one in six pupils in city schools was Chicano, yet denied entry to high school. Four in five were residentially segregated. The war brought greater job opportunities for all. Even so, the June 1943 Pachuco Zoot Suit riot, in which frustrated White males attacked defenceless Mexicans, embarrassed the school superintendent sufficiently that he began a 'good neighbour policy' to promote integration. Still, the fear of conflict between Whites and minorities remained.

Blacks were far fewer, but in 1950 they suffered discrimination even more severely than Mexicans, as they did in Chicago earlier. Earlier, even Black film stars were segregated. Jack Warner kept them away from his lot. Blacks held only 3 per cent of the jobs in Los Angeles in 1930, but were disproportionately labourers, janitors, and porters. During the depression, they were shunted aside. The war improved their job opportunities; Roosevelt outlawed discrimination, yet police continued to harass them. Black writers like Langston Hughes and Chester Himes could speak of racial hell.

Spatial segregation distinctly showed that protection of freedom for the majority overrode the espoused ideology of freedom for all. Much of what follows in this chapter shows how society worked to increase freedom for many while denying it to others. To focus on the latter, consider the views of some writers. Will Rogers may have been folksy, but he was the first mayor of the most famous of all exclusive communities, Beverly Hills. *Film noir* novelists captured the dark side of this fascinating place. To Raymond Chandler, 'modern Los Angeles ... [was] the consummate symbol of cultural pretense ... an empire built on a spurious foundation, decked in tinsel, and beguiled by its own illusory promises ... a metropolis of lies ...'.[58];

'pare away the layers of pretense and you only find more delusion'.[59] Nathaniel West, Evelyn Waugh, William Faulkner, Aldous Huxley, and Ray Bradbury all saw the deep contradictions—the simple Midwest juxtaposed with the bizarre, the crime, the opulence, the poverty, and the exclusion. The ambition of one of Huxley's characters was to make Tarzana, far out in the San Fernando Valley, 'the Living Center of the New Civilization'.[60] America had always been so. On the western edge, this was the final attempt to create utopia, but midwesterners could not escape the past, only pretend that they had. Was it not Groucho Marx who said, 'if you can fake sincerity, you've got it made'? Most could not fake sincerity, to their detriment, but admired the successful who could. Most were listening to neither the critics nor Marx (Groucho or Karl). The middle class, joined by the industrial working class, pulled itself together again after the depression. Life took on an aura of satisfaction and rectitude once again. In its profile of the city in 1950 *Holiday* neutralized the critics: 'Typically American and prodigiously proud ... offering something for almost anyone ... What but our national character writ large?'[61]

The Politics of Social Exclusion and Bureaucratic Inclusion

Fragmentation of local government has been the dominant image of twentieth-century American metropolises, not least Los Angeles, but the word 'fragmentation' is too abstract and lacking in social content. It also hides the degree of unity that has been necessary to govern these ever-enlarging places fed by migrants. To counter the drive towards autonomy and local democracy driven by exclusion, large bureaucratic structures emerged to supply the apparently neutral services, the services that no one could survive without: water, power, transportation, and others.

Titanic political struggles led to the creation of systems that became bureaucratic empires. These were all sorted out in Los Angeles with little help from the state of California, at least until the demands of war and then for a freeway system came to dominate action during the 1940s, and soon after for even bigger water systems in the state. The governing process was dominated by mistrust but implicit trust: elected politicians were not trusted, nor were bureaucracies, but only on the surface. In the end, residents complained only sporadically when their ideological nerves were struck or when their immediate interests were threatened. Most of the time most people were fatalistic. Those who had great economic interests never slept. In 1950 in Los Angeles county, forty-five cities (of which Los Angeles was by far the largest) and the county government competed with one another for ruling power, yet had worked out some ways to complement one another.

Autonomy and procedural checks on politicians characterized one side of governing structures. 'Home rule', a sacred principle in twentieth-century America with deep roots, denied Dillon's rule that municipalities were simply creatures of the state. Unlike Illinois and New York, California ignored Dillon as early as 1879 by letting local communities decide whether to incorporate or not. After the turn of the century, California allowed cities to establish permissive home-rule charters. Los Angeles and some other cities thought it was necessary to control their own destinies. Annexation and incorporation rules were localized. Most remarkably, California allowed counties home rule, so they could act more or less like municipalities and govern areas referred to as 'unincorporated', that is, not incorporated into cities. (California did not have urban towns and villages.) In 1913 Los Angeles county was granted home rule, thereby setting up a long-term struggle with the city of Los Angeles that has never been resolved through a comprehensive metropolitan structure. However, compromises and cooperation gradually reduced the stress. By mid-century a balance of power had been achieved, partly because the county also provided services to smaller cities. By then, too, the metropolitan population was expanding into adjacent counties.

City politics were reformed early in the century. Like other cities, but earlier than most, Los Angeles created a civil service in 1903, thus reducing patronage. Breaking the political monopoly

controlled by the Southern Pacific Railroad was not easy, but the progressive reformers managed it by instituting open primaries and thus weakened parties. John R. Haynes, the most noted Republican reformer of the era, also led the way to recall and referendum, the first in the country. The progressives distrusted the electorate, not surprisingly because working people's votes could be bought. Free enterprise racketeering by politicians did not fit the progressives' image of corporate integrity. In contrast to many cities, the reformers managed to limit the mayor's power, though he could still appoint commissioners and the police chief. The reforms cleaned up visible politics, but they had their downside. As we have seen, the police were ominously autonomous in the 1920s. It seemed that when corruption was eliminated in one place, it cropped up in another, as in Chicago.

Eliminating wards meant that only those who had citywide influence could prevail. Instead of local doorbell canvassing, those with large campaign funds could buy billboards and advertisements in newspapers. The change also enhanced the power of the citywide press. The wider constituency meant that 'public relations, not party discipline [remained] the key to electoral success'.[62] The losers were radicals, minorities, and neighbourhood politics. The *Times* slates of candidates usually carried the day. Indeed, even after 1938 when the corrupt police chief was removed, it was claimed that a *Times* reporter controlled conservative members of council by signalling members on which way to vote. Adding women to voting lists hardly made any difference, it seemed, nor did restoring (enlarged) wards again in 1925.

A bigger story was that of the rise of municipal enterprise in Los Angeles, which was more extensive than in most American, if not Canadian, cities. The reformers combined with the downtown businessmen and great land speculators to create the Department of Water and Power (DWP), which is said to be 'the most powerful municipal agency in the United States'.[63] It became nearly autonomous from the politicians. Although the mayor appointed commissioners and the city guaranteed its bonds until 1947, by and large the department was a law unto itself. It was able to do so because it delivered the goods.

Corporate reasoning by the powerful won out. If Los Angeles was to grow, it needed water, and water was too important to be left with the private sector. As we saw in earlier chapters, private water systems gave way to public enterprise. In Los Angeles, private enterprise could not have financially undertaken the great projects. General Otis, who would never dream of socializing the economy otherwise, pushed vigorously for public water. In an apparent odd contradiction, only socialists opposed a proposed massive enterprise because it would, they observed correctly, line the pockets of great land speculators. In 1907 the voters approved the funding of a very daring scheme to bring water by viaduct from Owens Valley on the east side of the Sierra Nevada Mountains. Constructed under the aegis of legendary engineer, William Mulholland, the viaduct conveyed water down a sluice at the north end of the San Fernando Valley at the opening in 1913 while bystanders cheered, just as they did in 1842 when Croton water reached Manhattan.

Hydroelectricity was a by-product, but not as important to Otis and Chandler. Although Southern California Edison consolidated most power operations in the rest of the county, the city took over the largest private system in 1914. John R. Haynes and other reformers were able to withstand the *Times*' and Merchants and Manufacturers Association's pressure to prevent municipal control or their later attempts to privatize the system. The Chamber of Commerce and newspapers other than the *Times* supported the mayors and council, who upheld, if sometimes under pressure, the widespread view that public power was cheaper and more efficiently delivered than by a private utility. In Canada this view was even more widely shared.

The reformers continued to be successful in the 1920s, even when enthusiasm for progressive public action had waned elsewhere. More water came from the Sierras, though the residents there carried on a virtual war against Los Angeles, and then the voters approved water from the Colorado River and power from the Boulder (Hoover) Dam on that river. Harry Chandler resisted that initiative

because he owned irrigated cotton lands in Arizona, a state that obviously had a stake in the river. The city bought out the last private power (and gas) company in 1936, just as Boulder Dam power arrived. Colorado water would come soon after in 1941. The massive DWP enterprise was firmly established as a permanent fixture and praised for its foresight. Except for some minor corruption and a strike of some workers in 1944, the department was a model bureaucracy; between 1929 and 1943 it quadrupled deliveries to the burgeoning city. In 1947 it could borrow in its own right. Its success gave lie to the view that only private enterprise could be efficient. The Department of Water and Power was restricted to Los Angeles because the state itself would not undertake public ownership; attempts to do so were denounced as 'socialistic'.[64]

By 1913 all factions favoured growth of the city based on public water. Instead of selling water to other jurisdictions within the county, the city embarked on an annexation program that made it the largest city in area in North America by 1927. State rules said that annexed areas had to be contiguous to a city. The first major moves were made before 1913 to create a municipally run harbour to the south at San Pedro and Wilmington. This 'socialist' action, which was advocated by Otis, helped to undercut the Southern Pacific's harbour at Santa Monica. To do so, the city annexed a narrow 'shoestring' and then the towns mentioned earlier between 1906 and 1909. This annexation was not more extensive because of another permissive state rule: owners could resist annexation if they had enough power. Fortunately, Otis and the city had enough power to ensure contiguity. The city continued to expand: since the city controlled water, Hollywood in 1910 and many more areas were annexed. The arrival of Owens Valley water in 1913 and the insistence of land speculators led to the gargantuan acquisition of the San Fernando Valley in 1915, thus doubling the size of the city, even though much of the water would irrigate oranges and beans for years to come rather than more lawns. Other annexations based on the promise of water brought the total area to over 450 sq. mi. (1,165 km²) in 1927, virtually to its current limits. Thus Los Angeles exceeded the grand dimensions of New York and Chicago.

The year 1927 marked a watershed: Owens Valley water was inadequate for further major expansions. Compared to other big cities, Los Angeles was able to prolong the process of annexation later into the century, while most others had stopped by 1900 or 1915. By the time Colorado water arrived in 1941, it was too late. The drive to local autonomy and exclusion of minorities and lower classes was too great; further fragmentation would occur.

Before Los Angeles began to annex early in the century, several outlying communities had incorporated, twenty-nine of them by 1913. Some cities had held Los Angeles at bay: Long Beach, Pasadena, Santa Monica, Glendale, and Beverly Hills had sufficient water. Some indulged in even more spatially odd strip annexations, though often with good reason. For example, Long Beach acquired part of Signal Hill's oilfield. Given the permissive state that allowed easy local initiative, other places were incorporated as cities; by 1930 there were forty-four in all in Los Angeles county. The depression and war shut down further action with one exception. A revival awaited 1954 when a new rash of incorporations brought the total to seventy-three within a few years (Figure 6.1). It was obvious that many cities were incorporated as 'zoning devices … to preserve a desired land-use pattern'.[65] In a few cases, as in Vernon, this protected industry from higher city taxes. Often local control was cited, but in most cases the exclusion of undesirable poor people, whose needs might drive up taxes, was the main reason. Or alternatively, very rich places like San Marino had such high land values and relatively high taxes that the *hoi polloi* could not afford to buy there anyway.

Paradoxically, fragmented exclusiveness would raise the county's status as a major governor in the metropolitan region. Many of these cities, espe-cially those with weak assessment bases, could not survive without county support or cooperation. The county's willingness to service some areas held off the city's annexations. In 1915 the county provided public health for unincorporated areas, and in 1932 for smaller cities. It expanded library, police, and fire services. While doing so, its civil service grew by eight times between 1914 and 1950. With only five council members (called supervisors) and three other elective positions, its bureaucracy increased in power, efficiency, and effectiveness. The big city itself gladly gave up the administration of weights and measures and especially charity to the county, though relief recipients in the 1930s complained that the county was not very generous. The depression and war slowed fragmentation and deals. Contracted services with the county reached a crescendo after 1954 when the 'Lakewood plan' was instituted. The developers of Lakewood (a new postwar suburb started in 1950) encouraged incorporation. Its new officials, who were reluctant to tax and anxious to set up boundaries against alien influences, made a deal with the county to provide virtually all services. The county could provide these at relatively low cost to these 'minimal' cities, equipped as it was with economies of scale.[66] By 1961 it was contracting combinations of forty-two services to all but one of seventy-three cities. Of a potential 3,066 service arrangements, the county contributed over 40 per cent to these 'cities by contract'.[67]

Annexations and incorporations left out bits and pieces of the county; some were actually inside cities, adding to the fragmented space, for which the county was responsible. This was in part possible because, as noted earlier, groups of landowners could resist annexation, just as incorporation was easy for those with enough power to pull together a sufficient number of people in support. The county officials in certain circumstances, on the other hand, could refuse requested incorporations.

6.1 *(Opposite)* Fragmentation and exclusion in the Los Angeles basin. Until 1927, the City of Los Angeles had annexed large parts of the county, including the shoestring to the south to encompass the port, Hollywood, and the huge San Fernando Valley. It had failed to annex many places like Burbank. Subsequently, the state allowed many communities to incorporate as cities, leaving the county to govern leftover pieces such as the Sunset Boulevard Strip east of Beverly Hills (darkest shading). *Source:* Winston W. Crouch and Beatrice Dinerman, *Southern California Metropolis: A Study in Development of Government for a Metropolitan Area* (Berkeley: University of California Press, 1963):Figure 1.

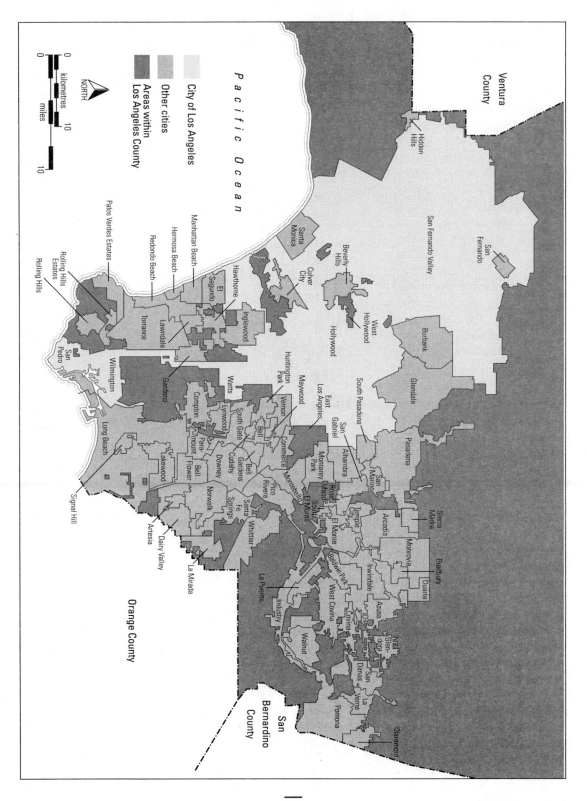

Los Angeles, 1950

Legend:
- City of Los Angeles
- Other cities
- Areas within Los Angeles County

Pacific Ocean

NORTH

kilometres 0 10
miles 0 10

Ventura County

Hidden Hills

San Fernando Valley

San Fernando

Santa Monica

Beverly Hills

Culver City

West Hollywood

Hollywood

Burbank

Glendale

Pasadena

South Pasadena

Manhattan Beach

Hermosa Beach

Redondo Beach

El Segundo

Hawthorne

Inglewood

Palos Verdes Estates

Rolling Hills Estates

Rolling Hills

Lawndale

Torrance

Gardena

San Pedro

Wilmington

Long Beach

Signal Hill

Watts

Huntington Park

Maywood

Vernon

East Los Angeles

San Gabriel

Alhambra

San Marino

Sierra Madre

Arcadia

Monrovia

Bradbury

Duarte

Compton

South Gate

Lynwood

Bell

Cudahy

Commerce

Montebello

Monterey Park

Rosemead

El Monte

Temple City

Baldwin Park

Irwindale

Azusa

Covina

Glen-dora

San Dimas

La Verne

Paramount

Bell Gardens

Downey

Pico Rivera

South El Monte

West Covina

La Puente

Industry

Walnut

Pomona

Claremont

Bell-flower

Lakewood

Norwalk

Santa Fe Springs

Whittier

La Mirada

Artesia

Dairy Valley

Orange County

San Bernardino County

Sixty-nine 'unofficial' cities existed in the urbanized areas of the county in 1950. Some were affluent, but places like East Los Angeles and Willowbrook south of Watts were poor. West Hollywood—the Sunset Strip—stayed out of Los Angeles so that its businesses could escape higher fees.

The county thus provided services to its 'unincorporated' areas and to many cities. This would not have been possible, however, had not the big city cooperated. It too provided services to some cities: sewers for Vernon and later Culver City in 1909, and electricity for Pasadena, Burbank, and Glendale. The city's school district, which was another nearly autonomous power, embraced many other communities. As elsewhere, a superintendent with strong powers ran the system, though he was theoretically responsible to an elected board of seven (usually upper-middle class) people.

Most vital was the city's willingness, through its Department of Water and Power to help organize the Metropolitan Water District. Recognizing in 1927 that Los Angeles could not afford to annex further using Sierra Nevada Mountains' water as a lure, its politicians and bureaucrats decided to pursue Colorado water in cooperation with the county and other cities. Representatives from twenty-one municipalities in four counties served on its board. In the 1940s, the district gathered in further communities. Even San Diego joined in. This major expansion of water servicing doubled the district's area.

This huge hydrological unit led some experts to hope that metropolitan government was imminent, but this desire for 'one community' was only wishful thinking. Without a state government, which had 'absented itself so long from the leadership of local governmental affairs' to lead, restructuring was not possible.[68] Only in a region that was more socially unified and politically coherent was this possible, as in Toronto in 1953, as we will see in the next chapter.

> **The big developers saw planning as a way to grasp control of the property rules and hence to improve their profits in an era of rapid growth.**

Several other special-purpose bodies amalgamated communities, though many joined reluctantly because their tasks were less essential than water and power. Several sewage districts were established in 1923, as were sewage disposal plants, though voters, who were wary of higher taxes, resisted bond floats for improvements. On the other hand, the more pressing need for flood control was organized at the county level. Smog was another governing problem; finally in 1947 a county board was established after years of debate. Later the state became active in making the smog issue a problem for national concern.

Urban and regional planning loomed large in governing the cities and county during this era. The county level would count most, largely because the automobile dominated policy interests from 1920 on, overtaking water supply as the major tool in promoting annexation. The urge to exclude unwanted people and activities brought zoning and subdivision controls onto the civic agenda in the 1920s. Planning concerns reached their zenith just before and after 1940; by 1950 it seemed that advocates were marginalized.

Through their voluntary City Plan Association formed in 1915, progressives first convinced the city in 1920 to set up a planning commission (which was composed of major real estate developers, downtown business interests, and a few intellectuals), and create a city department. The big developers saw planning as a way to grasp control of the property rules and hence to improve their profits in an era of rapid growth. The realtors themselves had organized an association in 1917 to establish professional credentials and so 'run on a scientific basis'.[69] Establishing credentials became a way to mark the boundaries in many fields and eliminate amateurs. Planning would follow when Harvard set up the first academic planning program. As in Chicago in 1909, business was intrigued with the possibility of control through planning. New York businessmen would also set up a regional planning commission soon after. Like Burnham, the West Coast advocates proclaimed: 'We are the ones who should "dream dreams and see visions"—visions of a better City to be'.[70] Like Chicagoans, they had to promise that 'the benefits to be gained … must equal or exceed the cost'.[71] As it turned out, while some of the commissioners may have dreamed, the city department spent most of its energy on more mundane zoning issues.

The Los Angeles County Regional Planning Commission, the first public regional body on the continent, would have a better shot at visions, particularly about roads and freeways, which were the central concern. Both commissions also fussed with public transit, but with weakening resolve. The county commission was composed of representatives from cities and the county itself. It was enjoined to coordinate with the city of Los Angeles and an increasing number of other cities, thirty-seven by 1950. It also had responsibility for zoning and subdivision planning in the unincorporated areas of the county. Both commissions expected to produce master plans that they hoped the city and county officials and politicians would accept. These bodies were powerful enough to persuade the legislature to pass a planning act in 1927, and in 1929 the state required counties to have planning commissions, though with modest powers. While Secretary of Commerce, Herbert Hoover also helped the cause with a model zoning act in 1923 and a model planning act in 1928 for states to follow. The Supreme Court upheld zoning. All of these initiatives signalled that property issues were too important to be left to the vagaries of the market alone. Corporate powers saw it was in their interest to control urban land uses.

The New Deal cemented their power. Although President Hoover brought together realtors and builders to find ways to speed up construction, under Roosevelt the Federal Housing Administration headed by realtors exercised far greater power over planning than any voluntary and modest state powers had previously. The key to power was mortgage insurance that 'virtually eliminated the risk for lenders'.[72] The banks and other financial institutions and the big developers were

pleased with this guarantee of windfalls. Eventually too, would-be home-owners would be pleased with the massive outpouring of housing built under these rules. To be eligible for federal support, communities had to plan. In California a governor with real estate credentials signed a stronger planning law in 1937.

It seemed the United States was headed for a planned utopia. A federal planning and state planning boards were established, supposedly to integrate local efforts into a wider cause for subsidized order and growth. The federal government built some low-income housing itself. In Los Angeles the county commission undertook many surveys of activities and land use (some subsidized by Washington), solicited opinion from citizens and groups ranging from the Chamber of Commerce to the Welfare Council. This led to an official plan for the region with thirteen different master plans covering a range of concerns, including airports and shorelines, that looked forward to the future. Planners and intellectuals saw great promise in planning. Said one: 'Los Angeles, with its youthful vitality and because of its minimum of encumbrance from the past, should be one of our first cities to point the way to this future—the city beautiful.'[73]

Private rights would be subordinated to the public good. No longer would Americans and Angelenos have to put up with the cities described in 1915 by Frederick Howe as 'inconvenient, dirty, lacking in charm and beauty because the individual land owner has been permitted to plan it, to build, to do so as he willed with his land'.[74] Government and big land firms would make sure of that.

This was the high point when the highest degree of consensus about planning was reached. War impinged with its planning priorities. Those controlling property were able to convince Congress to end the federal planning board, and likewise convinced the states to end theirs because they seemed Soviet. A new federal housing act was passed in 1949 to spur on new housing, some of them public. By then California and other states had passed redevelopment laws even though they were not clearly meshed with housing proposals, nor newly established redevelopment authorities with housing authorities. Many people active at the time confessed to conflict over goals—most of the time, property rights won over those with social goals.

In 1949 a well-known planning activist, Mel Scott, presented a popular justification of planning for Los Angeles high school students and citizens of the region. Even while calling for a planning body to cover a wider area than the county and the many blessings planning could bring, he wrote defensively: 'True, some citizens among us hold the opinion that planning for a whole metropolitan community is undemocratic—that it smacks of totalitarianism ... in contrast to our ideal of action from the grass roots upward.'[75] Although 'plans reveal community desires', he had to invoke the bottom-line argument, like Burnham earlier, that 'with permanent coordination ... we would get more for our money'.[76] This argument was too late. Those with the most to gain had already gotten what they wanted: power to reshape the city through zoning and rules to build subdivisions to their liking. It may have been that planning could reach a 'no higher end than contributing to the development of human beings who passionately want a full life for others as much as they do for themselves',[77] yet the lines of exclusion had been drawn: the industrial unionized working class could now enjoy the pleasures of the middle-class innocents and have decent housing and services, but the poor would still be neglected and ghettoized for the benefit of the rest. Few would take Scott any more seriously than they did the *film noir* critics. Right-wing politicians helped clear social advocates out of the way through wild accusations of communism.

The first major result of the politics of social exclusion and bureaucratization was the balkanized region with many small cities in addition to the big one and the big county. These small cities varied from affluent to modest. The poorer ones were mostly exclusive, emulating the rich communities. Their main goals were to hold down taxes, buying only necessary services from the county, and keeping out undesirables. Later we will consider one of these modest cities. Many unincorporated communities were badly under-

serviced, yet that was also true of poor areas in the city.

The second major consequence was, paradoxically, the creation of powerful, nearly autonomous departments run by commissions and boards. It has been argued that they were independent of '*all* political control'.[78] That they were large should occasion no surprise, given the long drive to provide essential services. That they were not accountable to elected politicians was likely too, given that they worked so well, especially for those with economic interests, and given that most citizens deemed them outside of politics. By 1950 even the *Times* could see this. Although free enterprise in theory, it had no trouble accepting these large municipal enterprises, even electrical power eventually, any less than it found it worthwhile to support massive federal funding for the region's economy.

> These 'new machines,' more monolithic by far than their ancient brethren ... are entrenched by law, and are supported by tradition, the slavish loyalty of the newspapers, the educated masses, the dedicated civic groups, and, most of all, by the organized clientele groups enjoying access under existing arrangements.[79]

Similarly, in both countries, all cities developed little empires of departments, boards, and commissions. Yet in Canada these bodies were responsible to politicians. The equivalent of a DWP could not borrow in its own right, nor could local municipalities exclude people. Service delivery was decidedly more equitable.

Services Delivered, But with Priorities

> Water and power—these are as vital as climate in selling southern California ... [T]he boosters created a metropolis of strange and lopsided proportions—in commercial benefit, a giant; in public improvement, a dwarf ... [S]chools, sewers, and public transportation do not have the kind of cash value that stirs booster action.[80]

Thus did writer Remi Nadeau draw a stark contrast between services for the economy and those for society, but as we saw earlier, big cities had hardly been paragons of service delivery. On the agenda of businessmen and public officials, water and power, like oil and Washington's defence spending, were higher than support for the marginal people and the environment. In Los Angeles, water was of greater urgency than elsewhere and as we will see under regional planning, arterial streets and then freeways were high priorities. A cost-benefit analysis of enormous complexity would be required to analyse how effectively the city, county, other smaller cities, and (in some cases) private non-profit enterprises, delivered their services; of course, even if that neutral exercise were possible, in the end political judgement would prevail. Consider the delivery of water and power, sewers, garbage disposal, flood and smog control, education, leisure and recreation, welfare, and then the sad tale of public transportation and the happy one of cars.

Water and Power to Sustain Growth

Water was the highest policy concern until the 1930s. On the ground, the Sierra run-off was joined by a flow from the Colorado River, stored in holding reservoirs in the mountains to the east. By 1930, virtually all the subterranean basins were severely depleted, though percolation from irrigation kept the pumps going, so cities had to rely increasingly on the water district set up in 1928. By 1950 most of the cities relied on the joint systems. Even Los Angeles, which consumed double the amount of water it had in 1920, began to rely more and more on the district system. Certainly the future was worrying some people: as per capita consumption rose, there was talk of limiting certain luxury uses. Bringing northern California water to the south was being planned, Columbia River water was a vision, recycled sewer water was a possibility (as had been done elsewhere for irrigation), and desalination a more remote idea of engineers. There was no cause for alarm: 'The aridity of the climate no longer limits the future of this desert metropolis.'[81]

At the household level, however, many of the poor had to rely on outside spigots. As in Chicago earlier, landlords and poor home-owners were reluctant to install plumbing for proper bathing facilities and toilets. In fact, in St Louis in the mid-1940s, young White women still resorted to bathhouses because their old homes lacked hot water, tubs, or showers. There only half of rented units were completely serviced. Meanwhile, southern cities were even less serviced. In northern and Canadian cities, most houses were fully serviced. In Los Angeles nine in ten units had full plumbing, though somewhat less in rented than in owned. The few bathhouses that remained catered to gay couples or single men in flop-houses and on the street.

Until the 1940s hydro plants supplied virtually all the electric power. Southern California Edison, serving the county, built generating plants in the southern Sierra Nevada Mountains and the city's DWP tapped the aqueduct's generating power. Hoover Dam power arrived in 1936 in time to alleviate potential black-outs that might have occurred earlier had the depression not slowed consumption. However, war production demands began to push the limits, so suppliers built power plants fuelled with oil and gas. Few houses were without lighting by 1950, and many used power for cooking and heating, though natural gas through pipelines from west Texas continued as a competitor.

By the 1940s sewage disposal was a serious problem. As already noted, voters were reluctant to spend money on proper sewage treatment. At the city's large Hyperion operation near the airport (on land annexed in 1917), raw sewage flowed into the ocean. After a reluctant electorate agreed, a new plant was built in 1924, yet two beach walkers in the late 1930s, Aldous Huxley and Thomas Mann, observed the 'malthusian flotsam' at their feet.[82] Pungent odours were more repugnant than condoms. In 1940 the State Board of Health ordered a clean-up in one year, then quarantined 10 mi. (16 km) of beach, to the detriment of the decaying resort community of Venice. But despite the pleading of the city engineer, politicians and voters kept turning down bond issues; there were other beaches. The war slowed environmental concern.

Finally in 1948 taxpayers heeded the state's entreaties. A new plant was opened in 1951. An impressed Huxley declared in 1956 that the waste after treatment was '99.95' per cent pure.[83] Sludge was turned into organic fertilizer. Methane was supplied to a nearby power plant. Smaller treatment plants were upgraded, and many smaller cities joined the city's sewer system. Various sanitation districts combined into a county treatment plant south and east of the city. However, insatiable growth would place strains on these systems in the future.

Garbage disposal became an increasingly vexing problem as the urban area grew more wealthy, thus wasting more. At the same time, residents became less tolerant about careless disposal. In 1902 the city began to collect wet garbage to feed hogs. In 1921, after paying off city councillors, Fontana pig farmer, A.B. Miller, received carloads until 1950, but then feeding raw garbage to pigs was outlawed. Dry solid waste was either incinerated in backyards or, if reusable, it was picked up by scavengers, at least until 1912 when the city started collections, as it did of dead animals. In 1924 it sold glass and metals to a company that recycled material; the rest was dumped in landfill sites in valleys in the hills. Backyard incinerators belched smoke until the 1950s when legislation finally eliminated these contributors to smog. Without hogs and incinerators, and a burgeoning of packaged goods in the ever-consuming society, larger and larger stringently regulated sanitary landfill sites took all the garbage, except what went down the drain into the sewers after being ground up in kitchen garburators.

Natural Hazards

Los Angeles, more so than other large metropolises, was concerned about floods, brush fires, smog, subsidence, and, of course, earthquakes. Countywide control agencies were eventually established. In dry, Mediterranean-type climates, winter rainfalls are sometimes torrential and the run-off induces landslides on steep slopes that are often denuded by brush fires. In these climates, a great flood happens about once in a century, major

ones about six times a century. The floods of 1914, 1933, and 1938 were exceptionally severe. The first one led to the county's construction of dams. In the 1930s federal engineers built more dams, some downstream with large holding basins. These had the added positive effect of allowing percolation into the ground for water supply. Lining stream walls allowed water to flow to the sea, though the above measures reduced the amounts reaching it. The costs were great. Wildfires can occur in the fall after dry summers, especially when drier Santa Ana winds from the desert (adiabatically heated) blow down mountain slopes. Volatile chaparral bushes are easily ignited. Fire departments developed various techniques to control the blazes, but they often failed. Public authorities continued to allow developers to build on the slopes among the tinder-dry chaparral.

Smog became a serious problem only in 1943 when the first great photochemical blanket covered the region. Inversions of temperature (that is, higher atmospheric levels that are warmer than lower levels) trapped various hydrocarbons near the ground. Blame was attributed to oil-burning industrial plants (coal would have made it worse) or to backyard and dump incineration. The car was identified as the major offender only after 1950. Before then, oil companies, auto assemblers, bureaucrats, politicians, and drivers were locked into a psyche of denial. Removal of groundwater and oil caused the land to subside at a number of places in the basin, most markedly at Wilmington. The costs of environmental protection were quite high.

As for earthquakes, after the 1933 Long Beach earthquake, building codes were more strict and children were taught safety measures in schools, but subsequent earthquakes revealed that these measures were inadequate, especially for those unlucky enough to live on landfill. Needless to say, the migrants kept coming regardless of earthquakes and other environmental threats. State and national taxpayers covered much of the cost of correcting and preventing these environmental problems through emergency funds, indirectly through the massive transfer of funds to fuel the Los Angeles military economy.

Human Capital and Democratic Schooling

In education, by 1950 the 'human capital' argument was taking a strong hold of policy makers and planners, who believed that prosperity depended on education. In 1949 the massive Los Angeles City School District served pupils within 850 sq. mi. (2,200 km²), which was nearly twice the size of the city, with 315 regular elementary schools, thirty-three junior high schools, and forty-five senior high schools, plus schools for disabled children and adults and four junior colleges. As elsewhere, secondary schooling increased greatly after 1910. After 1945, given the strong commitment to democratic public education, continued immigration, and the early years of the baby boom, pressure on space for schools was great. Large classes and split days were common, but more space was on the way. The superintendent promised ten new classrooms a week; by 1949 contractors were building three classrooms per day. Outside the city system, poorer school districts with low assessments had to do more with less; in fact, 'fantastic variations' in spending also persisted within the city system.[84]

The ten institutions of higher education were expanding because of veterans' demands: UCLA, still seen as a branch of Berkeley, enrolled 14,000 students, while USC had 16,000. The great expansion of the multiversity was imminent, as was that of state colleges. Meanwhile, the basin's private colleges catered to the élite and upwardly mobile. Aiding the burgeoning aerospace-electronics economy was the California Institute of Technology, which had done so since Robert Millikan brought Theodore von Karman from Chicago in the 1920s to engage in atomic and jet research. Together with the research capabilities of the industrial firms and research bodies, the military was well served by the region's engineering élite. MIT and Stanford may have been more visible in the country because the status accorded to the Boston and Bay areas was higher than that given to Los Angeles, but Los Angeles became the top centre for high-technology nonetheless.

Tourists and Mass Leisure

Los Angeles had already developed a vast array of recreational facilities. In the private sphere, there was Hollywood and Vine for the gawking tourists, Knott's Berry Farm for modest thrills and jam, some beaches, and backyard pools for the affluent. One city block in Beverly Hills, it was noted, had more pools than the whole of Cincinnati. Affluence and keeping up with the Joneses would bring more, though the YMCA and school pools (in better areas only) would still be used. Yacht clubs expanded. In quiet Whittier, middle-class high school girls and boys kept busy with Job's Daughters and DeMolay while their parents were active in Masonry and Eastern Star. Certainly there was 'more leisure than ever before' and 'for it to be … veritably a national and international playground is nothing short of phenomenal'.[85]

In an increasingly privatized social environment, public facilities lagged. The city had rested on its oars since early in the century. Planners worried that a 'large segment of voters looked upon public recreation for adults as a new kind of governmental nonsense.'[86] The city planning commission pointed to the need for neighbourhood playgrounds, district playing fields, major sports centres, and youth camps, though large parks like Elysian and Griffith (philanthropic gifts) had long been available to the masses. The public library system expanded through regional level branches. The old élite gradually built up culture, such as Huntington's garden and library. Not until 1965 would the city have a prominent art gallery and Mrs Norman (Buffy) Chandler's Music Center.

Social Welfare:
The Undeserving Poor Linger On

Like the society matrons of old, Mrs Chandler was active in charity, especially on behalf of hospitals. But the city and county lagged behind big eastern cities in its provision of hospital beds and counselling and psychiatric clinics, though they were on the verge of great expansion like education. Providing confessionals for the rich of Bel Air lined psychiatrists' pockets. Although early on the city had the usual orphans' home, it ignored (more so than other cities) the challenge of *noblesse oblige*. Still, like other cities, Los Angeles created a Municipal Charities Commission in 1913 to regulate the solicitation of funds for charitable purposes and to endorse worthy social agencies. In 1918 the Alliance of Social Agencies was set up, which led to the Welfare Council, a voluntary social planning body for the metro area. The Community Welfare Organization and a fund-raising Community Chest in 1924 unified groups as never before.

The Great Depression took a toll on voluntary efforts, and the city needed an enormous amount of relief. Only state and federal help could stem the tide of misery, as we saw earlier. The war made welfare work much easier when the factories and the military hired more; full employment then as always was the best cure for poverty, idleness, and crime. Women who did not work in plants ran the Hollywood Canteen and the like. By 1948, volunteers' enthusiasm weakened, so the Welfare Council established a volunteer bureau in the hope of encouraging more women to help those deemed less fortunate. In the affluent society, the poor were not very visible. Industrial workers were now well off, so it seemed no one should be poor any longer.

Transportation: Cars Win, Transit Loses

In his famous book on Los Angeles, Reyner Banham praised its 'freeway pilots', who exercised disciplined freedom on the freeways.[87] Los Angeles was the archetypical automobile city, evoking such notions as sprawl and placelessness. But the car did not conquer the streets without a struggle, though it almost seemed as if this 'democratic alternative' to transit would.[88] By 1950 cars were so prevalent that voices for public transit were hardly heard and only marginally heeded.

Earlier the yellow streetcars of the Los Angeles Street Railway (LARY) rumbled along the arterial streets. The big red cars of Henry Huntington's renowned interurban Pacific Electric (PE) linked the towns in the region to the centre. Angelenos rode to the beaches on Saturdays and Sunday. Both systems would pass into public ownership by default in 1951, and disappear from the landscape in a decade. As early as 1913, LARY was in trouble

> **Los Angeles was the archetypical automobile city, evoking such notions as sprawl and placelessness.**

when ridership decreased. Although the system recovered somewhat during the war and for a few years afterward, ridership gradually declined during the 1920s. On weekends, more and more people drove to shops and to entertainment, and on workdays, more drove to work. The PE began to abandon lines. By 1924 only half of those commuting to the downtown area came by transit, down from four-fifths earlier in the century when walking was the main alternative. By 1931 the number of transit commuters fell to less than one-third and would continue to fall markedly during the depression, despite the introduction of streamlined, smooth-riding streetcars. Revival of transit during the war only prolonged the agony. When the state utility commission refused a fare increase but demanded upgrading, the red cars were sold to a bus company, which in turn gave up.

The competition against transit was stronger in Los Angeles than elsewhere, but this was the story all over America; streetcars disappeared in small cities first, then in larger ones. Many systems survived the 1920s by combining transit with electric utility companies, but a 1935 federal law outlawed these trusts, apparently in the hope of resurrecting competition. Bus companies took over some transit systems, municipalities did so with others. Cities recognized that not everyone could or would drive a car to work or for pleasure. Many streamlined streetcars ended up in Toronto. Public transportation did not disappear because buses gradually replaced streetcars, though with far fewer riders in most places. Most systems, which were largely in public hands by 1950, operated at a loss, so they reduced service, making matters worse. Often the transition to buses in American cities over the decades from 1920 to 1960 has been attributed to a conspiracy in the late 1920s among the big corporations with a strong interest in buses and cars. Whether General Motors, Firestone, and Standard Oil did so is probably beside the point. They may have speeded up the demise of streetcars, but their interest was hardly the core reason. Public policy gave way to the suppliers, but also to the public demand for space to run autos. Only a few relatively dense cities (notably New York, Chicago, and Toronto) could maintain low-subsidy systems. Most of the older cities also continued to rely on commuter rail; this was not the case in Los Angeles even though the new Union Station of 1939 provided an elegant downtown entrance to the city.

Car traffic increased. By 1920 Angelenos were buying cars at a rate almost four times that of any other city. By 1930, after a decade in which car registrations rose a phenomenal eight times, each auto served one-and-a-half inhabitants; in Chicago it was one car for every eight people—an enormous difference. Only Detroit itself would come close among the biggest cities. The Great Depression slowed car production, the war stopped it, but production was in full swing again by 1950. In fact, virtually every industrial worker was able to buy a car and commute in it too. People had no choice other than car-pools. In 1955 the county had more cars than all but four countries in the world!

The policy and planning battle was won by the proponents of the car in the 1920s, but not without a series of fights focused on the downtown. LARY lines converged on the centre, like streetcars elsewhere. Unlike interurbans in other cities, PE routes also radiated out from the centre. When downtown commuters gradually switched to cars, they competed with streetcars in the narrow streets. Ironically, Los Angeles then had less space on its streets than other cities, a legacy of earlier surveys.

Downtown interests were pitted against those in favour of dispersion. The Central Business District Association (composed of realtors who specialized in the area), retailers, financiers, and companies with headquarters downtown sought to retain their power. Their view was that transit should be given precedence and improved and stiff parking regulations should be instituted, but they were up against the powerful Southern California Automobile Association (SCAA).

City council had to balance these interests as best as it could. As early as 1919 it struggled to bring in parking rules, especially during rush

hours, to regulate recalcitrant motorists. This helped for a while, but the drivers kept coming. A traffic code included signals, banned left turns, and created one-way streets. Private operators put up parking structures. These control measures only encouraged more traffic. Meanwhile, the streetcars, which lacked manoevrability, slowed down the traffic.

City council supported two plans—one for highways and one for transit. In 1924 Frederick Law Olmsted, Jr, and his associates proposed the first significant road plan for the region since the original pattern was laid down. The Major Traffic Street Plan recommended 200 specific proposals such as widening and extending arterials, opening new ones, and, in the process, distinguishing a hierarchy among several hundred miles of streets. Council 'enthusiastically' put the scheme to the voters, asking for $5 million to float bonds. By then 'the power of consensus' was so great that an overwhelming number of voters said yes, many of whom now shopped in outlying stores rather than downtown.[89] Supportive legislation allowed the projects to proceed and the county cooperated. Still, many more millions would be needed, partly from abutting landowners through special local improvement taxes to fulfil the plan.

Yet planners and some businessmen still advocated for transit. City council agreed to hire experts R.F. Kelker and Charles De Leuw, who designed Chicago's Rapid Transit Plan. In 1925 Kelker and De Leuw argued that 'the county would continue to rely on the interurbans and streetcars to an even greater extent in the future'.[90] To achieve efficiency in the central area, they suggested rapid transit that was either elevated or below ground. Although Los Angeles was a low-density city, autos would continue to create congestion downtown. Most downtown groups and public bodies agreed that transit should remain.

But, under the banner of the SCAA, the car lobby was stronger. Long-standing antagonism to the street railway monopolies, as elsewhere, did not help the transit cause. Transit commuters had long felt that they were the source of profit for transit companies that intentionally crowded cars. Besides, riders resisted higher fares. Some also remembered the boodling of the 1890s when the owners of new electrically driven routes bribed city councillors. The truth was that even before 1920 the privately owned street railways of Los Angeles and other cities were not earning a profit and began to reduce service. As for the proposal itself, few were enthusiastic about underground or elevated lines for the PE. The *Times* complained about the possibility of 'four miles of hideous, clattering, dusty, dirty, dangerous street-darkening overhead trestles'.[91] Others pointed out that more and more commuting was from outside the centre. The car won hands down: it was not even political in the sense that trolleys were. The streets were public, while cars were private but not monopoly-owned and therefore very democratic. By 1930 cars were used in four out of every five trips in the city.

Tangled up in this debate were attempts to deal with the problems of access to the centre across steam rail lines and rail yards, which were mostly east of downtown, and of the obvious need to rationalize rail lines. Main line railways were not very cooperative, nor were the downtown élite consistent in their position. Finally, some overpasses were built and the rail yards were rationalized in the 1930s. Few people commuted downtown by rail.

The massive shift to the car did not solve the congestion problem downtown, nor even in the decentralized shopping areas. Department store owners and other storekeepers were complaining by 1930. The city was 'literally gagging on its own traffic congestion.'[92] Outlying businesses called for more arterials. Freeways were needed. In 1930 Olmsted mooted grade-separated, limited-access 'parkways' without signals. New York regional planners were advocating them and Robert Moses had already organized their building on Long Island.

The depression dampened the urge to improve traffic flows, though affluent Angelenos continued to buy cars. Yet gradual recovery raised spirits to

the point that in 1937 the SCAA submitted a motorways plan. At the same time, the county planning board was conducting a wide range of surveys of activities from airports to manufacturing in the region. The SCAA lobbied Sacramento for motorways inside cities, not only between cities. *Times* writer Ed Ainsworth opined that Los Angeles was 'ideally situated … for the first great experiment in loosening the strangling noose of traffic'.[93] (Of course, something similar had been asserted in support of 1924 arterials.) By then, even downtown businessmen, as well as the *Times*, desperately supported motorways, hoping that they would improve access to the centre and therefore halt its decline.

They had lots of support. The Transportation Engineering Board report of 1939 stated that 30 per cent of travel time was spent waiting at intersections, and that rail movement was slower than ever, holding back traffic. The board drew up a map of proposed routes (Figure 6.2). The city agreed to establish the first two routes, the Arroyo Seco parkway to Pasadena and a stretch over Cahuenga Pass to the San Fernando Valley. Engineers built parkways elsewhere on the continent, but outside cities. At the time, open Los Angeles could build urban parkways with little disruption. By 1941 both the city and county planning staffs and councils said the long-term plan was largely consistent with their master plans, believing that the freeways would relieve congestion, reduce the notoriously high accident rate, and stabilize land values by drawing excess traffic away from arterial roads.

The first coherent freeway plan envisaged eight routes: 108 mi. (174 km) in the first stage, 103 mi. (166 km) in the second, and finally 613 mi. (987 km) in four counties. Price tags were put on the first two stages. Although tolls were debated and would be used elsewhere, Angelenos rejected the idea. The war intervened to reduce car travel and stop planning. Gas rationing and car-pools

6.2 *(Opposite)* Los Angeles freeway plan, 1939. The Arroyo Seco was already built, and the Hollywood was started. Many freeways eventually constructed did not exactly follow the routes shown here. At the time, rapid transit was still a possibility, but Angelenos had to wait several decades. Other cities already had plans then too. *Source:* George W. Robbins and Leon D. Tilton, *A Preface to a Master Plan* (Los Angeles: Pacific Southwest Academy, 1941).

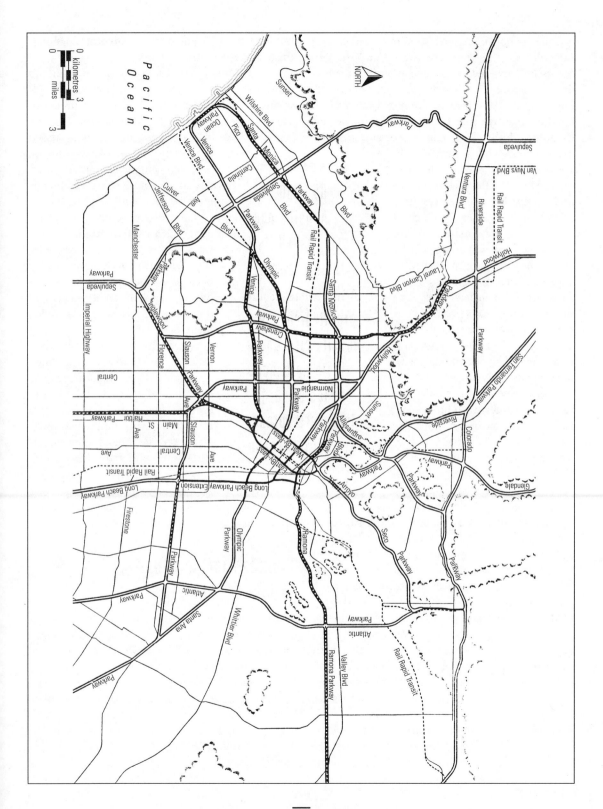

became the norm, yet congestion remained severe. After the war, the SCAA, the county, and the city successfully lobbied the state legislature for subsidies. Governor Earl Warren signed the Collier-Burns Act in 1947, which committed the state to building freeways. The Los Angeles region would receive two-fifths of the money to be amortized by gasoline taxes and licence fees. In 1947 a ten-year program for 165 mi. (265 km) of the 1941 plan was undertaken. Because progress was actually slow by later standards of construction, only one-fifth of the original scheme was completed by 1958. Then the pace quickened in large part because Congress finally empowered the federal coffers to subsidize up to 90 per cent of construction costs in 1956. Legislators, who were previously reluctant, were persuaded on national security grounds.

Because newly opened stretches became clogged at rush hour, some planners still held out hope for transit. In a wide-ranging set of planning essays written in 1940, a transit advocate pointed to several plans for transit after 1925, and recommended that electric rail remain in place, arguing that half the commuters still relied on it. For some planners and civic leaders in 1949, transit was 'the only long-range solution to the problem of transporting the masses of people to and from work daily, especially if the population ... increases to 6,000,000 or more.'[94] The Rapid Transit Action Group and the Chamber of Commerce sought rapid-rail lines in freeway medians, but 'Rail Rapid Transit—Now' campaign failed with suburban voters, some of whom condemned it as socialistic—a sure way to kill a public initiative. Sacramento supported the new municipal transit authority in 1951, but did not provide funding. The vision of transit was kept alive and would eventually be modestly realized, though the old systems were discarded in the meantime. A financially successful system needs not only a big city

> By 1950 Los Angeles had established its mythological place in the world as the prototypical low-density city where single detached houses on large lots were spread over the landscape.

but one of sufficient density. Advocates ignored this, as Lewis Mumford lamented: 'by allowing mass transportation to deteriorate, and by building expressways out of the city and parking garages within, in order to encourage the maximum use of the private car, our highway engineers and city planners have helped to destroy the living tissue of the city.'[95] Why blame planners or even engineers rather than developers or bankers or aircraft moguls? Who had the power?

Could it have been otherwise in low-density Los Angeles? Freeways were a response to low density. The car won too mainly because Angelenos believed or were persuaded that it would free them from constraints, the fulfilment of the liberal culture. They tolerated gridlock at rush hour because on nearing home, they could move faster than they would have had they walked from the tram or bus stops. For non-rush hour movement, the car was obviously speedier, more convenient, and more comfortable. Detractors may have criticized the vast subsidy for cars through low gasoline and licence taxes and government subsidies for streets, but few listened.

Planning for Communities to Clarify the Sprawl

By 1950 Los Angeles had established its mythological place in the world as the prototypical low-density city where single detached houses on large lots were spread over the landscape. Yet to boosters, it was also a collection of communities composed of homes, schools, churches, shopping, and other services. Planners took a benign view that a 'subdivision is not merely a means for marketing land .... [It] is far more, a process of community building.'[96] Consciousness of community and neighbourhood was greater, it seems, than it was earlier in the century. We need to explore how planners and developers contributed in ways other than the car in creating the form of the city.

In contrast to other times, this was the golden age of detached family houses. Between 1946 and 1956, virtually all the houses completed were single-family homes. Not surprisingly also, home-ownership increased, yet Los Angeles went through a phase in the 1920s when, like Chicago, apartment construction was strong and almost exceeded that of bungalows. After 1960 ten-unit 'dingbats' would wipe out many bungalows, so the golden age of single-family housing was a historical exception in Los Angeles and in many other large cities.[97]

As for housing unit starts since 1910 in the United States as a whole, construction of these units remained strong until 1917, then rose in 1921 and peaked in 1925. The years 1926 to 1929 were still strong years for new housing, though nearly one-third were multifamily units. After New York and Chicago, Los Angeles had the largest number of apartments containing five or more units. The collapse of residential construction in the early 1930s was followed by vigorous recovery to 1941, then after the war, new heights were reached in 1955. In Canada, housing starts were weaker after 1912 following its earlier golden years of single-family house building, though apartments went up in the late 1920s too. Starts continued at a higher pace in the early 1930s than in the US, but production would not reach the same level until 1945, when another burst of construction almost matched that of the United States.

American home-ownership rose gradually from 1910 to the 1929 level of just under half of non-farm units, thanks in part to somewhat easier mortgage financing. Apartment construction in the 1920s held back home-ownership in Chicago, as we saw in the previous chapter. In Los Angeles, 37 per cent of housing units were owned as of 1930, somewhat higher than in Chicago. During the depression, many lost their properties to foreclosure until they were protected in 1933, so the level dropped to one-third. From 1940 the share of home-ownership in the United States increased from 41 per cent to 53 per cent in 1950, then to 61

per cent a decade later, and higher in Los Angeles. Because of the weaker economy after 1913, home-ownership in Canada only equalled that in the United States in 1930, then fell, but rose to the slightly higher level of 56 per cent in 1950 and continued to rise. In these respects, the two countries were similar: both had a volatile house-building economy; some of the big cities retained high tenancy rates, but now both were committed to home-ownership as never before. Even though Los Angeles ran only modestly ahead of other cities and housed many in apartments, it perpetuated the ideal of the single-family bungalow.

The land-development industry evolved from fragmented activity to building large communities. Planners for developers created models of a hierarchy of neighbourhoods within larger communities. The federal government stepped in during the depression to bolster the mortgage market with ironclad guarantees for housing, thus promoting home-ownership to more of the population, including the industrial working class. One might conjecture that the baby boom, which began about 1943, was based in part on the expectation of a greater supply of houses and undergirded by an optimism about victory in war after a decade that depressed procreation.

This brought about the rise of the 'community builders', the developers who not only subdivided as speculators of the nineteenth century but established rules for building and built themselves, or at least organized the subcontracting of building.[98] These developers had to persuade governments to draw up rules that favoured them and disadvantaged small-scale speculators and builders. It was these developers who encouraged and then benefited from government support. Other sectors had been corporatized, aided by government regulation. Now it was real estate's turn.

Controlling the subdivision process was essential for power over property. By 1915 prominent realtor and early community builder, J.C. Nichols of Kansas City, 'advocated a public-private partnership in the preparation and execution of private urban land development ...',[99] but this view encountered considerable resistance. Regulations in California were imposed much later than elsewhere; only in 1893 did subdividers have to register plans. Subsequent acts tightened the rules, usually after periodic downturns left vast areas prematurely subdivided. In 1915 the state's Map Act required subdividers within 3 mi. (5 km) of municipal boundaries to have their plans approved by local authorities. California was catching up: Wisconsin, Ontario, and others legislated this a few years before. Locally, large-scale developers Colonel William Garland and Harry Culver pushed for more controls, including master plans, zoning, and planning agency reviews. They and others organized realtors into a profession and pushed successfully for the city and county planning boards. However, even many large developers were reluctant to hand over too much regulation to the public sphere, hence the modest gains in regulation.

In the 1920s the boom in subdivisions exceeded the wildly speculative 1880s. In 1923 over 1,400 subdivisions were registered. 'Fly-by-nighters', who were able to evade regulation, laid out unrecorded subdivisions. The struggle was about profits. Petty developers wanted to maximize the number of lots on their land for sale to those with modest incomes. These 'land butchers' drew lot lines as narrow as 25 ft (7.6 m), 'creating conditions for future slums'.[100] (Residing on a 25-ft lot myself, I question the inevitability of this.) The petty subdividers often sought buyers who wanted low taxes and limited services, and who in some places jerry-built their own shacks, which was what happened on the margins of Chicago and Toronto earlier. The big community developers, on the other hand, were selling well-serviced large lots to the affluent and creating well-defined, orderly communities, though they claimed to hold an areawide concern. They were anxious, of course, to socialize the costs of infrastructure by asking the cities and the county or subsequent owners to pay. As we saw in the previous chapter, the struggle over who pays for services had been going on for several decades.

An example of resistance in the 1920s was the drive by certain substantial rogue realtors to use zoning to their advantage. To planners, zoning was a device to stabilize land values in the context of master plans, as we will see shortly. However, realtors on the west side of Los Angeles had the bright idea that zoning could be used to *raise* values. In 1921, when half the subdivided land in the city was not yet built upon, 69 per cent of the lots were advertised for single-family homes and 75 per cent of existing houses were single family. Yet, at the behest of powerful land holders, city council zoned less than 10 per cent of the lots for single-family houses, all of them near posh housing! Then city council designated 59 per cent of the lots for 'residential-income' areas (for rental apartments and duplexes) when these totalled only 20 per cent of existing housing.[101] Not only that, but 13 per cent of the lots were defined as commercial frontage when this sector actually composed only 6 per cent. By the end of the decade, when the hapless acolyte planners completed the mapping, 600 mi. (965 km) of street frontages had been assigned to commercial—enough for 14 million people in a city that has a population of 1.2 million at the time! Someone thought there was enough to serve 'the entire population' of the country.[102] So when real estate slowed down, land adjacent to arterials remained vacant, except for billboards (Figure 6.3). To be sure, a substantial number of apartments went up late in the decade, raising the share of multiple units to roughly 30 per cent, those in buildings with five or more families to one in six, though the number of multiple units was well below those of New York and Chicago. In those cities also, zoning, as a kind of ersatz planning, defined unnecessarily large areas for apartments and commercial use. It was obvious that realtors and their friends at city hall were carried away in the heat of the speculative 1920s. The top community builders found their task had been subverted by others with power.

But there was more: 'spot rezonings', which were considered a 'vicious practice' by planners, also subverted community builders' aims.[103] A daring developer, A.W. Ross, was the prime example. Opposed by downtown interests, Ross persuaded city council to redesignate pieces of property he had assembled on Wilshire near La Brea to create a commercial zone that he foresaw would cater to the affluent of Beverly Hills and nearby.

6.3 The overzoned city. In the 1920s the city zoned an excessive amount of land for commercial frontage, as indicated by the vacant stretches. It also favoured multiple housing over single-family units. *Source:* Air Photo Archives, Department of Geography, UCLA.

Others scoffed at 'Ross's folly', but it became the Miracle Mile, and Ross a hero in the late 1920s, as department stores with large parking lots built there.[104] Free enterprise was still stronger than corporate control.

In 1926 Harry Culver, who was then president of the California Real Estate Association, was determined to push for more state rules. He had the support of some of the larger brokers, bankers, and title companies. Even so, they were not successful in their efforts; the planning and map acts of 1929 were still too weak in his view because too many realtors were not convinced of the need for orderly development, even though subdivision activity was declining and everyone was aware of the collapse of the great Florida land boom. In 1929 more than half the subdivided lots in the county remained vacant.

The Great Depression led to the final solution,

a victory for the big operators under government auspices that would pay off for millions of ordinary people who wanted good, inexpensive housing. Thousands of lots and houses went into sheriff's sales, in some places on the continent far more than in Los Angeles. Even there, uncontrolled land development had left, even in 1937, 40 per cent of the county's surveyed lots empty. Elsewhere, as around Detroit, 95 per cent of the lots grew weeds. 'Metes and bounds' (that is, unregistered subdividing without a plan) was coming under fire from banks and utility companies that were increasingly unwilling to support chaotic subdivision and building. For several years the pace of building fell to a crawl.

California community builders won the war with the free enterprisers in 1937. New planning acts and map acts, signed by yet another realtor governor, eliminated metes-and-bounds subdividing, strengthened municipal enforcement, and gave power to the state's Real Estate Department to enforce map filing and subdivision sales regulations. Realty licensing was tightened. Politicians too had tired of servicing costs for subdivisions that were undeveloped for years. In both countries, local improvement taxes were mostly jettisoned in the 1930s. Thus developers would either put in their own servicing or pay municipalities up front. If developers were to organize building immediately, the long gap after subdividing would disappear.

More important was the federal government's initiative. Setting the stage, President Hoover's Conference on Home Building and Home Ownership concluded that the federal government, together with private real estate and construction bodies, could facilitate the transition from subdividing to homebuilding. It was only a short jump under Roosevelt to the home-owner act of 1933 and the 1934 act that gave the Federal Housing Administration (FHA), especially its Land Planning Division, the mandate to promote housing. It was staffed with successful realtors. Government greenbelt towns fostered community building, as had large private projects like Radburn, New Jersey, in the late 1920s. The master key to speed and efficiency was mortgage insurance, thus 'virtually eliminating the risk for lenders'.[105] In 1934 up to 80 per cent (90 per cent in 1938) of loans were insured for twenty years. Insurance allowed homeowners to borrow from savings and loan associations and builders to borrow from commercial banks at reasonable rates. Stability of prices was also guaranteed through FHA appraisals, which was 'tantamount to price setting'.[106]

To protect this massive investment, the FHA was careful. To ensure sound loans, it developed a standardized appraisal procedure to assess borrowers' material circumstances, the quality of properties and neighbourhoods, and municipal regulations. This allowed the FHA to exercise enormous control over real estate development, and over design and engineering features. It provided free advice and organized conferences to encourage standardized suburbs, to an extent that Daniel Burnham could have only dreamed. Guidebooks

on how to plan neighbourhoods properly set the pattern as never before. To maintain the illusion of egalitarian democracy, FHA officials were 'always very careful to cultivate and preserve the image of voluntarism'.[107] Obviously, though, the carrot was used as a stick; they would 'not propose to regulate subdividing ... but insist upon the observance of rational principles of development ....'[108] This double-talk favoured large-scale developments and the big developers. Corruption was minimized as inspectors were 'generally very tough and honest'.[109]

Near the end of the war, the government added the Veterans' Administration (VA) to the FHA. Low down payments, low interest, and twenty-five-year amortization made it possible to invest vast resources into home-ownership housing. 'Without FHA and VA loans merchant building would not have happened ...,' asserted Ned Eichler, one of California's great builders.[110]

Housing starts increased: with the renewal of building in 1935, the FHA guaranteed 6 per cent of the units in the country; by 1939 it guaranteed one-third, then four-fifths of the units in 1943. After the war, even though the FHA scaled back housing starts (in 1950 including VA units), federally insured housing starts still totalled just under a half of all in the country, most of them single-family dwellings. The scale of building, which was just what the community builders wanted, had concentrated dramatically; by 1949 4 per cent of all American builders put up nearly half of the units. Costs of servicing fell with more rational planning. Developers reshaped old premature subdivisions from the 1920s. Building followed closely after subdivision.

By the late 1930s California led the way in construction, double the next state in insured mortgages. Because of its military industries, it was favoured by the FHA. In 1939 the FHA handed out over 30 per cent of its mortgage insurance in the Los Angeles area for small workers' houses, far more than any other metropolitan area. Home-ownership was thus given a shot in the arm. The FHA also enabled 1 per cent of builders to build 15 per cent of all the houses. Los Angeles's merchant builders thus built many more houses annually than they did earlier. Although most of them could rarely produce 300 houses a year before 1950 because control over organizing mass production was still unwieldy, they later perfected further techniques to build more quickly. Added to what was called the 'California method', (predetermining the number of pieces of lumber and other materials that were needed) were Levittown mass-production methods.[111] Sawmills were set up on site. Eichler, a San Francisco Bay area builder, stated that Los Angeles's producers, like Fritz Burns, were 'the hardest driving, most competitive, and most imaginative men in the nation …'[112] The community builders won the war over the petty producers, thanks to their own organizational skills and to pliant governments that were anxious to generate growth. For the masses, many of whom had built their own houses in the 1920s, house prices were lower in Los Angeles than elsewhere, even if mortgage costs were a bit higher.

The Federal Housing Administration also directly financed the construction of some projects. When the war seemed imminent and the production of war material was urgent in 1938 and 1939, developers who were paid from the public purse built projects near plants. Los Angeles was the major beneficiary. Wartime meant direct government construction, particularly in 1943 and 1944. After the war, many more FHA project units were put up, so by 1950 they represented one-quarter of all those insured. The government acted in a fashion that not even the most fervent housing advocate of 1910 would have thought possible.

The Canadian federal government also was reluctant to take public action. Like the United States, but unlike European countries that were deeply involved in social housing, Canada had only briefly dabbled in public support for municipal commissions set up at the end of the First World War. During the depression, there was less urgency to act because housing starts did not fall as dramatically as they did in the United States. By 1939 starts had recovered to late 1920s' levels in urban areas. The Canadian government passed housing acts in the 1930s, although it subsidized only the upper third of income earners to help fill up subdivisions. The housing acts were designed to increase employment, not to solve social problems. The 1938 act specified public housing, but the rules were such that willing municipalities could not pry any government money loose for building. As a consequence, unlike the New Deal, Canada did not build any public units. During the war, it depended far more on doubling up in units than on direct action, though towards the end of the war, Canada began to build a small number of modest houses for defence workers and then for veterans until 1949.

The Canadian housing acts of 1944 and again in 1949, like those in the United States, greatly expanded the role of insurance and of regulation through the Central (later Canada) Mortgage and Housing Corporation. A well-known American housing expert, Miles Colean, thought that 'Canadians were a cooler lot than our people', and that they 'proceeded with a firmer sense of direction and less confusion …'[113] If Canada seemed nearly as intent on 'keeping to the marketplace', governments guided changes more firmly.[114] In the next chapter, I will discuss the struggles over public housing in both countries, though I will briefly record the early experience in Los Angeles shortly.

In both countries, after long debates, housing policies were geared primarily for large-scale single-family projects. Because the developers realized by the 1940s that they could make more money by building on larger lots, this meant that they needed minimum lot sizes of usually 5,000 sq. ft (465 m^2) with frontages of 50 ft (15 m), which was twice the size (or more) that of earlier lots. This would in turn mean that people would rely more on the car than on transit in all cities of the continent.

The Counterweights to Dispersion: Stability, Protection, and Separation

Property developers wanted to increase the value of their land, while most people wanted some degree of order and security in their lives, to settle down, have children, enjoy the use of their homes and yards, and not have to watch the real estate market closely. Most people agreed with the big developers that they wanted more space. As the community builders had earlier recognized for their own upper-income clients, volatility in land markets (prevalent from 1880 to 1930) would not serve others either. Most families sought diversion in community facilities. Most planners were anxious to make this kind of community available to everyone, they said. Community planning would provide for central shopping areas, schools, libraries, churches, and other facilities, including decentralized municipal offices. Planners argued for small neighbourhoods (ideally of 10,000 people) organized around elementary schools and play spaces. Schools and shopping would be hierarchically organized on principles of centrality, scale, and the range of a good—that is, the area needed to have sufficient numbers of consumers of various services and goods. Replicating earlier high-income enclaves, streets would be curvilinear. By the late 1940s these notions were widely accepted. To avoid competition with one another, churches (at least mainline Protestant denominations, if not the fundamentalists devoted to free enterprise religion) cooperated in planning. Stability and order: these neighbourhood ideas had been worked out in the 1920s by experts such as Clarence Stein and Clarence Perry, though they still had to be promoted in the 1940s.

Although the very act of planning these communities promised long-term stability and protection of property values, in 1920 they were only a promise. Because massive growth had led to the abandonment and decline of many residential districts, many in the middle class sought other means of protection. We have already noted that the incorporation of some cities was a step towards protecting communities and values. Private restrictive covenants and public zoning were the two other major devices used to exclude certain activities and unwanted people. Laws of nuisance had been in place since the beginning of cities in North America. Building codes later provided some degree of order. Developers of affluent communi-

> **Private restrictive covenants and public zoning were the two other major devices used to exclude certain activities and unwanted people.**

ties placed covenants (or deed restrictions) on properties they sold to protect neighbours from undesirable appearance, uses, and people. The problem with covenants was that they expired after a set period, and they could also be broken. The power of the state was needed; legislation would grant the police power to enforce rules.

The start of zoning is often attributed to California cities that prohibited Chinese laundries from certain residential areas in the 1880s. Defining districts (not just specific properties) for residences only, for factories, or for commerce came about gradually, following German practice. By defining what is permitted within an area, other activities would be excluded. This was a step towards the planners' theory of using zoning as the legal basis for fulfilling plans. It also meant an expansion in the state's power over private property. Zoning meant control over density, setbacks, and height of buildings, as well as use of 'private lands for the common good ... aimed at an orderly development', and 'the protection of property investments'.[115] At the local community level, this was virtually the equivalent of planning.

In Los Angeles early on, the city restricted some activities such as oil drilling, to certain areas. Following the 1907 financial panic that halted another spate of land speculation, Los Angeles was the first city to pass a comprehensive zoning law at the behest of realtors and progressives. The ordinance defined seven exclusive industrial areas and three large residential districts, in which it prohibited a wide range of manufacturing and heavy commercial activities. The ordinance did not even tolerate activities that were already in place, at least in one case; it forced a non-conforming brick maker to move from his clay pits and kilns. Unlike Los Angeles, New York's widely studied 1916 zoning scheme permitted or excluded commerce from certain blocks, defined others as residential, and also set limits on the shape of office buildings to ensure some sunlight on the streets.

Zoning to segregate uses (and thus stabilize land values) increased about 1920; it was 'the heaven-sent nostrum for sick cities'.[116] Under Secretary of Commerce Herbert Hoover, planners and realtors drew a model comprehensive zoning law in 1923. By the late 1920s most states had passed legislation along its lines, thus enabling municipalities to act. The courts in California and many other states upheld them. The Supreme Court, ruling on *Euclid v. Ambler Realty* in 1926 (a Cleveland case), agreed that a municipality could override private property rights for the common good. The developer could not do what he wanted and did not need to be compensated for any expected loss of value. Police power, not eminent domain, was ruled; that is, governments could not expropriate land for uses it had not defined for public use. Protection of private property was the issue.

The first task of the Los Angeles Planning Commission in 1920 was to draw up a new refined 'scientific' zoning scheme. Its staff created five categories: single-family residential, multiple residential-income, business, non-obnoxious industry, and unrestricted (meaning industrial but still allowing residential and commercial). The planning staff drew maps for large areas that were already subdivided, initially for those west of downtown. The top downtown realtors and retailers wanted these areas defined first, and pushed for the designation of single-family residential for most of the Wilshire and Hollywood district. Their aim was to slow commercial development outside the centre, and fulfil the goal of a bucolic city, but (as we have already seen) they lost out to west-side realtors who sought to raise land values, not stabilize them. Even though their nice idea was turned upside down, the prominent realtors and planners did not concede to city council's overzoning.

After the near total collapse of the land market in 1929, city council agreed to a new zoning ordinance, added new categories, and asked the staff of the commission to rezone. Yet even then, the Realty Board, pushed by the maximizers, persuaded the city council to require that no zoning change to a more restrictive category take place without the consent of 65 per cent of adjacent owners. They did not want the council to go too far

the other way in reducing apartment and commercial potential! City council tightened the processing rules, though these did not prevent some politicians and planning commissioners from operating a 'zoning variance racket' to seek pay-offs for spot rezonings.[117] The result was that requests for rezoning went before the Board of Zoning Appeals under reform mayor, Fletcher Bowron. In those relatively calm political years after 1938, this worked.

More important, the Federal Housing Administration helped considerably by requiring zoning laws to protect whole neighbourhoods of middle-income families, not just those of the wealthy. Cities that persisted with the extravagant overzoning of the 1920s, thus promoting speculation, would get fewer mortgage guarantees. Needless to say, cities fell in line. In 1946 Los Angeles brought all 453 sq. mi. (1,174 km^2) under zoning control; the most recent was much of the San Fernando Valley, which was still sparsely occupied. Other cities did likewise, and Los Angeles county zoned unincorporated areas. The federal government had finally wrung free enterprise out of development on behalf of big developers, and stabilized neighbourhoods as well.

But covenants and zoning could not prevent change. Home-owner associations, some of them organized by developers, attempted to put up roadblocks, especially against minorities. Smoke from factories and unwanted commerce paled in comparison with the fear of loss of property values if Mexicans and especially Blacks infiltrated communities. In the extremely racist 1920s west-side communities limited the number of Jews and Black celebrities. It was easier to put up 'barbed-wire social fencing' around an incorporated area, such as Beverly Hills.[118] With or without covenants, voluntary associations acted as vigilante groups to resist change. The Ku Klux Klan moved in. An Anti-African Association was formed; signs, such as 'Keep Slauson White!' appeared on lawns.[119] Later Whites resisted Blacks who tried to move into parts of the San Gabriel and San Fernando valleys. The strongest line of defence was the realtors. If they did not show minorities houses in certain areas, they could at least limit change. *De facto* practices would maintain segregation to a degree.

Blacks who moved to Los Angeles after 1940 to work in factories had to live somewhere, so some communities adjacent to areas that were already Black had to give in, however reluctantly. After the war, the racial clause in covenants and zoning became embarrassing. The state abolished the racial clause in 1947. Blacks would continue to expand, despite resistance.

In Canada, there could be no such federal requirement for zoning because of provincial power. The provinces had little interest in zoning. From 1900 to the 1920s some cities passed by-laws of a zoning nature, mostly at the behest of high-income neighbourhoods. Like Los Angeles, neighbourhood ratepayer associations were active against threats. Some lived with restrictive covenants that excluded minorities. Jews were the main target. However, Toronto did not have a comprehensive zoning by-law in place until 1954. This was partly a result of the better condition of cities in Canada as compared to the United States. The urge to expand was not as strong, and owners of older properties maintained theirs by and large.

Invasion/Succession: A Theory That Implied Threat

Around 1920 sociologists in Chicago devised a theory of city growth that would dominate thinking about the city for several decades. Robert Park, Ernest Burgess, and others argued that as cities grew and expanded outward, the social composition of neighbourhoods would change in a predictable, almost natural pattern. New residential areas of single-family houses on the edge of development, 'the light bright zone', would attract affluent families with children from older suburbs. Others would seek apartments not far inside the edge. The old suburbs would appeal to upgrading immigrants who would leave the more central areas that had the oldest and poorest quality housing. This zone of slums closest to the downtown would continue to attract poor immigrants (especially those from overseas) and poor, rootless Americans, replacing those who moved up in the world—the 'motor' of the model.[120] The slums

would also house the stranded poor. The central business district would encroach on this immigrant reception area, forcing people and the zone outward. The size of immigrant settlement would also expand the area into the newer adjacent area, giving rise to the notions of invasion and succession. Thus, if this process worked, the older the housing, the more it would filter down the income scale. This, the sociologists thought, was an inevitable 'ecological' process, creating 'natural areas', newness on the edge, dilapidated housing near the centre, and decay progressively expanding outward in waves. No neighbourhood could avoid change; those closer to the centre were imminently vulnerable.[121]

The scholars modified this simple zonal pattern, recognizing that in Chicago the rich maintained the Gold Coast sector to the north along the lake, and that the Black Belt had emerged in the south. Other modifications, such as the rooming-house belt and particular ethnic clusterings were added to the zonal pattern. Because the apartment house had been 'growing in popularity' in the 1920s, one of the Chicago group argued in 1933 that this would dominate housing construction in the future.[122] His assertion contradicted a strong movement to 'prohibit' apartments and to favour single-family housing.[123]

Homer Hoyt altered the picture in the late 1930s by arguing for sectoral patterns in which different income groups would expand outward in wedges; this was already obvious in the two specific cases of the Gold Coast and the Black Belt. People like those in Back of the Yards would expand outward from that area to the edge. In some cases, Hoyt noted that sudden shifts in income occurred at certain lines. He drew maps for many American cities to show his sectors, though obviously Chicago was the prime example, having the added virtue of being spread over a nearly featureless plain.

Despite the alterations to spatial theory, one common element stood out: the inner area with the oldest housing was the most decrepit and was where the poorest people lived, with a few exceptions of old, poor suburbs. As noted in the Chicago chapter, commentators described a wide swath (as much as 3 mi./5 km) of greyness and decay. The Chicago sociologists observed social disorganization within that swath. As this was less obvious in some cities than in Chicago, the expectation that housing would filter down the income scale became deeply entrenched in the minds of policy makers and planners.

If this city development was an inevitable natural ecological process, what could be done about it? Suburbanization and the end of the influx of poor immigrant Europeans in the 1920s led to population decline in the inner cities especially, presaging a vacuum. Central business district expansion also slowed down in most cities. Ghettoized Blacks moved into some of the worst housing, but their movement from the South slowed after 1918 when fewer job opportunities were available in the North. Without new migrants, older housing and nearby commercial buildings became more vulnerable to decay. To make 'suburbanization a nat-ional priority', as President Wilson's Secretary of the Interior suggested in 1919, would only exacerbate the city's problem.[124] In Los Angeles, even if the decay of the inner city was less marked, the suburban drive was as powerful as elsewhere, probably even more so.

> ... the expectation that housing would filter down the income scale became deeply entrenched in the minds of policy makers and planners.

Renovation and Redevelopment Proposed

By the 1930s there were two possibilities for renewal of these bleak zones: renovation and redevelopment by tearing down and building anew. As for renovation, this depended on money in the pockets of the poor or their landlords. After the bottom of the Great Depression was reached in 1933, the Roosevelt administration persuaded Congress to set up the Home Owners Loan Corporation (HOLC) to refinance mortgages of properties that were 'in danger of default or foreclosure', to grant loans to many who had already defaulted, and to allow a much more extended amortization period.[125] A huge 40 per cent of eligible Americans sought aid over the next two years, and one-tenth of home-owners received assistance. Excessive borrowing in the 1920s was a major reason for so many applications. Like financing for cars and household durables, borrowing became easier because the institutions with money were anxious to lend it. Many overextended lenders and borrowers had thus been made vulnerable in the 1930s. HOLC action stabilized home-ownership, but it did not deal directly with deteriorating structures.

In 1934 the Federal Civil Works Administration, using a uniform appraisal scheme of four categories, undertook a real property inventory of 3 million housing units in sixty-four cities. Appraisers, who were mostly from the real estate industry, surveyed block by block. They generalized by dividing cities into numerous small districts, and drew coloured maps: green for the best areas and red for the worst, the old, and decaying. Based on the ecological assumption of natural areas according to the Chicago school of thought, they displayed those adjacent to the CBD and those with renters and Black people ('the colored element') on maps as red, and most of the others close to the centre (or declining) as yellow.[126] They discovered that nearly one in six units was overcrowded or lacked private indoor toilets. Nearly a quarter were without bathing facilities, and nearly one in ten were without running water. In St Louis, one of the worst cities, 40 per cent of the poorest housing units in slums were without indoor toilets, and the tuberculosis morbidity rate was very high. Most suburbs were green or blue, except for the surprising number of shack towns. Although the maps were a 'gold mine' of data, they were not published.[127]

Initially, at least in Newark, New Jersey, HOLC supported poorer areas with loans: nearly one-third of category three (yellow) and also nearly as many in the red 'hazardous' areas. After HOLC provided stabilization, owners could then put some of their money into renovation. HOLC's task was massive, and loans to poorer areas dried up quickly. The FHA proceeded to 'redline' most of the poorest areas, ending any serious direct attempt to improve older stock. Higher incomes in the 1940s led to private renovations, though in 1950 many older areas were still not renovated. Then over the next decade, as analysed in the next chapter, conditions deteriorated further.

By contrast in Canada, the federal government provided nearly 126,000 loan guarantees under its Home Improvement Program between 1936 and 1940. At one loan per unit, over 5 per cent of houses were aided. Toronto and a few other cities had already started their own programs; city staff photographed houses before and after renovations. Many inner-city houses were upgraded and, in the process, provided the construction industry with jobs. In Canadian cities the task of improvement was formidable, but far less so than in the United States. A 1941 survey of large cities showed that nearly 100 per cent of houses had toilets (if not all exclusive), running water, and bathing facilities, thanks to medical health officers, who aggressively pushed landlords to provide these to tenants. A 1943 planning report for Toronto showed that areas totalling less than a square mile could be defined as a slum or blighted area largely because invasion/succession was far less pronounced. Even so, planners who were indoctrinated with Chicago ideas defined four-fifths of the city that had pre-1914 housing as declining areas or vulnerable to decay. Instead of declining, these areas were subsequently improved, only a small portion of it through redevelopment, as we will see in the next chapter. Unlike Americans, Canadians seemed less inclined to denigrate 'undesirable' elements.[128]

Redeveloping the slums and derelict downtown areas seemed a more probable solution to blight than renovation, though writers in the 1930s were pessimistic that central businesses would rebuild in zones of transition. With the New Deal, though, some American officials and planners were eager to tear down and rebuild, but the path to the redeveloped city and particularly public housing would be tortuous. The Public Works Administration, the US Housing Authority, and some states and housing authorities in municipalities built over 300,000 public units by 1943, mostly on green rather than grey fields. This was far below the 8 to 10 million public units that some planners said were needed. Wartime housing was built in places with defence industries, though much of it was shoddy. After a hiatus, the numbers built rose in the late 1940s. The 1949 housing act promised to finance over 800,000 units annually for six years, to be built by local housing authorities, but, as we will see, they achieved nowhere near that figure, despite the rhetoric of a decent home for every family. And what was built was a terrible imposition on the poor: they were ghettoized even more than they had been in high-rise apartments.

The problem of acting on public housing was the same all along: the real estate industry opposed it. Unlike European countries, where at least a partial consensus was reached on providing housing as a 'public utility' for a large percentage of the population, America was hamstrung.[129] The private sector asked for subsidies, but every calculation showed that direct government building was the only way to provide housing that was cheap enough and decent enough for as much as one-fifth of the population. In sponsoring the 1949 act, Republican Senator Robert Taft drew on his sense of *noblesse oblige*: 'the rich have a special obligation to show the poor how to take care of themselves.'[130] This view was not shared by most members of his own party, nor by southern Democrats, nor in those terms by those who supported the bill.

Redevelopment for purposes other than housing was a different matter. Businessmen were keen to have the government bear the cost of buying up and tearing down slums and derelict commercial properties so that they could build. Many public officials were enthusiastic about new, imposing public structures. Modest funding was provided in the 1930s for new public works, like the federal courthouse and the post office in Los Angeles. Following the regional plan of 1929, New York undertook redevelopment under Robert Moses, using far more federal funds than any other city. Although Moses attacked New Deal politicians and planners because they 'destroyed property values', he did not hesitate to take this money, while wielding as much power as any Soviet commissar.[131] But New York was desperate: the failure of Wall Street revealed how weak the great metropolis had become.

Los Angeles authorities built a modest 3,500 public housing units in the late 1930s. The 1940 census revealed that 106,000 dwellings in the county were substandard, half of them in the city. Architect Richard Neutra stated that most Mexicans and Blacks were 'in need of a slum clearance and rehousing program'.[132] As studies elsewhere had shown, blighted and slum neighbourhoods cost the city a great deal of money in fire protection, health, policing, and foregone taxes on depressed properties. Officials identified 20 sq. mi. (52 km^2) of the central area of the city as blighted. The census in 1950 reported that around 50,000 (one in thirteen) units were dilapidated or without running water (or not reported). Even in 1980, over 2 per cent of units in the county lacked full plumbing.

Instead, the planners and others were more interested in redevelopment than in housing the poor. Mel Scott asserted in 1942 that 'we must actually replace much of the community that exists today. We must create many things still in the imagination. Blocks and blocks of buildings must be cleared to make way for freeways.'[133] Likewise, the Regional Planning Association said that what was needed was a 'geometric wave of entirely new buildings, replacing virtually everything standing downtown'.[134] Despite their hope of accommodating the poor in new housing, this visionary talk only increased uncertainty among slum dwellers and those who feared that slum dwellers would eventually invade their districts.

California passed a Community Redevelopment Act in 1945, as did other states about then; this allowed local agencies, whose members were appointed by mayors, to buy up blighted properties (by eminent domain if necessary), clear the land, and resell or lease it for rebuilding. Los Angeles set up a redevelopment authority in 1948. Its local housing authority drew up plans for a dozen sites for public housing, hoping to mix people with a range of incomes and ethnic backgrounds. When it came to implementation, however, the state blocked that goal through legislation that prohibited the sale of redevelopment agency land to the housing authority. One dismayed advocate concluded that any public housing would simply recreate ghettos of the poor, increasing 'mass segregation of economic classes'.[135] Citizens rejected public housing in a 1952 referendum, fearing that it represented a socialist threat. Reform mayor Fletcher Bowron, who had presided over the city's most open political era and was a supporter of public housing, left the political scene. His successor cancelled many projects. One such project was Chavez Ravine; once cleared of housing, the city handed it over to Walter O'Malley and the Dodgers. The redevelopment enthusiasts would later tear down old housing on Bunker Hill for grand civic projects and upscale private schemes, and freeway builders drove through low-income Mexican neighbourhoods in east Los Angeles.

One Community, Segregated Space

Famous architect Richard Neutra told befuddled Europeans in 1939 that Los Angeles was composed of 'cottage suburbs and satellite garden cities ... [that] seemed to extend amorphously' over 300 sq. mi. (777 km^2), mostly over what Banham imaginatively referred to as 'the plains of id'.[136] Another described the city as 'centrifrugal'.[137] Sprawl springs to mind, but it is hardly an adequate word to describe Los Angeles's space. Even if by 1930 it had the lowest density of any large city, it was articulated by activity and classes of people. It was only superficially amorphous. Earlier we considered protective devices for neighbourhoods that countered the spread somewhat, though they mostly had the effect of drawing sharp boundaries. Anxiety levels remained high, except during the less self-conscious war years; fear that developers or unwanted people would create insecurity and the loss of property values was pervasive. Let us consider several regions: the central business district, industrial zones, and residential areas, concluding with some specific cases.

Central Business District

The Los Angeles central business district (CBD) has often been the butt of jokes. Despite its nadir in the 1950s, it has remained the centre of the region. After 1920 it lost a large share of retailing and office activities. Contributing to this decline was its decreasing financial power, even though its high-status banks remained downtown. Building regulations to limited earthquake damage prevented costly office buildings from towering skyward, at least before 1950. Regional newspapers remained downtown, though their ranks were reduced to five by 1950. They still had to compete with over twenty local dailies in the region and a host of weeklies. City and county government offices firmly planted there continued to expand employment. The electric rail lines still converged on the centre in 1950, and the earliest freeways under construction met at the 'stack' just to the northwest. Soon after 1950, the downtown élite still had sufficient power to ensure that Dodger Stadium was built just to the north, and that new civic buildings were situated in Bunker Hill. Centres never disappear, no matter how derelict they might become. The downtown power élite remained in the CBD even though the major economic decisions for the region were made in Washington after 1930.

Within the CBD, various nuclei migrated from the original centre at the plaza to the south, which was a pattern consistent with the movement of CBDs in other cities. The dominant retail stores moved down Broadway early in the century. The department stores, with adjacent women's clothing shops, at Seventh and Broadway anchored the sector by 1920. The locus of the largest hotels showed the same pattern of movement, though the Biltmore, overlooking Pershing Square since the 1920s, finally faced competition from the Hilton a few blocks to the west in 1950. The expanding office sector moved also until it reached Ninth Street. At the far western edge of the CBD was the most conspicuous office building, the late 1920s' Richfield building with its metal tower, black

masonry, and gold terra cotta. It symbolized the power of the oil industry in the region. The financial district was focused on the stock exchange at Spring and Sixth, with the clubs of the top élite nearby. Its activity in 1950 had recovered from the devastating 1930s, but it was a minor player in global finance. Pedestrians competed with cars and trams at downtown intersections.

The government stayed at the north end of the CBD in an increasingly shabbier section of old offices and lofts that were abandoned to lower-value activities. City hall, which opened in 1927 and was the most conspicuous building in the city, stood on an elevation and symbolized civic pride. Reaching a height of 464 ft (141 m), it easily exceeded office buildings that were earlier limited to 150 ft (46 m). (Increasing numbers of office workers led Buffalo and others to erect grand city halls.) Just to the north was the conspicuous courthouse and post office, which opened in 1939 and housed many federal departments. The county Hall of Justice and Hall of Records were close by, as was the State Building, like the federal monolith, a public works measure. Since 1910 governments continued to expand despite the ideology of free enterprise. The elegant central library was located farther south in 1926. The *Times* stayed in the CBD also and opened its third prominent building in 1935. Forever defiant on behalf of free enterprise, the Chandlers had placed on the roof their bronze eagle, a survivor of the bombing of the *Times*'s 1910 building.

Downtown retailing certainly took a beating after 1920, more quickly in Los Angeles than in other large cities. The big downtown department stores managed to remain solvent by decentralizing beginning in the mid-1920s and having the good fortune to operate in a city where income levels were 50 per cent higher than elsewhere in the country. Despite this advantage, the share of downtown retail sales fell in the 1920s, then even more in the 1930s. By 1939 Hollywood, Long Beach, and Pasadena showed similar declines, but virtually all of them had higher levels of sales than the retailers downtown. The affluent Wilshire Miracle Mile hardly fell at all, doubling its sales over 1929. Downtown sales continued to erode in the 1940s, yet in 1950 middle-class shoppers from Whittier and elsewhere still shopped there before Christmas. By then, planners and others worried about the rise of retailing in tawdry goods that hardly yielded the same business tax revenues. Assessed valuation dropped dramatically during the depression, and also because many office functions were weakened. The quality of buildings could not be maintained. Headquarters drifted out, and after the war, new arrivals like Carnation and Prudential moved to the Miracle Mile. By then, farther west in Beverly Hills, Wilshire sported the 'swankiest' stores, Saks Fifth Avenue and I. Magnin.[138]

On the margins of the CBD, Chinatown was to the north and Little Tokyo was to the east. Below that was Skid Row, which expanded in the 1930s, retreated in the 1940s with full employment, and then expanded again by 1950. The missions to the homeless could not do more than provide a meal. Just to the east were the prisons. Farther to the south were the wholesale food and flower markets, wholesalers, and apparel manufacturers. Such a pattern was more or less typical of other cities, though elsewhere food terminals were beginning to decentralize. The city had closed all but one of its local public markets; a private farmers' market opened near the Miracle Mile to satisfy a bucolic urge among professionals and tourists.

Industrial Zones

The central area lost not only retailing and offices but also many manufacturing and service activities, such as wholesaling and utilities. In 1924 three-quarters of wholesaling firms were adjacent to the CBD core, extending down into the city of Vernon, which was purposely incorporated for industry only. This share of wholesalers fell to a quarter by 1960, and processing even more so. Even so, manufacturing in the region was not scattered, but like many service industries showed definite zones of concentration as late as 1960. It was still strongly correlated with rail lines, even though many plants no longer used rail sidings (Figure 6.4). Inertia was powerful, though slipping. The larger the plant, the more primary the activity, and the stronger the association with rail. Owners had located their plants outward along rail lines since 1924, especially to the south, adjacent to the Los Angeles River, but also to the west of downtown and around the harbour. Only the service industries along the Wilshire corridor and Beverly Hills were independent of rail.

Clusters of plants in certain districts were evident. Of the nine auto plants in 1950, the earliest were clustered in a corridor south of Vernon in Maywood and South Gate, or in Long Beach, with one in Van Nuys to the north in the valley. These plants produced about 650,000 cars and trucks a year, maintaining Los Angeles's position as second to Detroit. Tire, parts, and accessory plants nearby permitted almost just-in-time service. Unlike Vernon, residential areas developed nearby.

The aircraft industry occupied even more space adjacent to airports. In 1950 Douglas had three plants in Santa Monica (the first) in Long Beach, and in El Segundo, next to the municipal/international airport, as was North American in Inglewood. The latter company operated a plant well to the east in Downey, while Lockheed was to the north in Burbank. Closer were Northrop in Hawthorne and Hughes in Culver City. Jet aircraft were assembled out in the desert at Palmdale. Hundreds of parts plants close by supplied the aircraft plants.

As for Hollywood, the need for spacious studios to make talkies led to decentralizing in the late 1920s. In 1950 half of the major studios operated outside Hollywood itself: MGM in Culver City, 20th Century-Fox west of Beverly Hills, and Universal and Warner Brothers in the San Fernando Valley, where Walt Disney and Republic also had their operations. Some had back lots or ranches there. The take-off of television in the late 1940s led the big networks to expand beyond their radio facilities in Hollywood and onto Beverly Boulevard and Burbank, near high-paid talent available for motion pictures.

In 1948 social area analysts, who were the successors to the surveys earlier in the century and to the Chicago sociologists, divided the county into 185 named places. They noted that the regional and city planners had defined their spatial categories; the county had designated sixteen major economic areas and 105 retail business areas (of

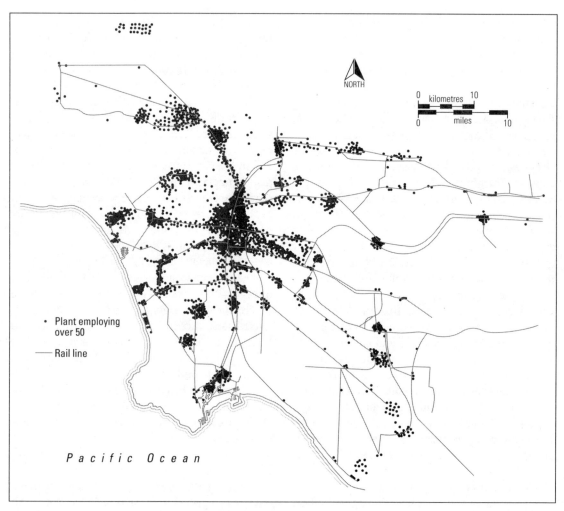

6.4   Manufacturing and service industries, 1960. The Los Angeles-built landscape was not a featureless sprawl. Many of the
industrial plants were still oriented along rail lines, even though many had given up using rail service. *Source:* Dudley F.
Pegrum, *Residential Population and Urban Transport Facilities in the Los Angeles Metropolitan Area* (Los Angeles: University of California Press, 1964):Map 11.

which seventeen were major). The city named
fifty-nine communities with which residents could
identify and receive certain decentralized city services. The Board of Education, the metropolitan
Welfare Council, and the Council of Churches created planning areas. All employed census tracts (a
small statistical unit introduced in 1940) to analyse
social characteristics. Through studies, government and social agencies would strive for greater
efficiency and equity. The Haynes Foundation subsidized non-governmental studies and other urban

research. John R. Haynes, as we have seen, was the
most prominent and tenacious of progressives. His
hope was that urban research would lead to action.
In the hands of experts, advocacy became less the
goal than analysis itself. In the two decades after
1950, urban analysis would pass through its
golden era of report after report.

The social analysts perceived residents as varying in three major dimensions: social rank, urbanization, and segregation. They then grouped census tracts into their named places. Social rank, the

first dimension, was based on occupation, rent, and education to yield high- middle- and low-income areas. The second dimension was a more difficult category: high urbanization meant low fertility, more women working, and the inverse of single-family dwelling units (The last seems somewhat odd, at least then, given that new construction was mostly single-family. Besides, older Philadelphia, for example, would have been relatively non-urbanized by this definition. No doubt this reflected the conclusions of the Chicago school.) The analysts put the 185 places within nine social area types, and imposed the degree of segregation on each (Figure 6.5).

Highlighted on the resulting map from 1940 census data were social rank sectors that were more or less as Homer Hoyt had described for many cities a few years before, based on special surveys. The most prominent were the high-income sector to the west through Beverly Hills and along the southern edge of the Santa Monica Mountains, and some places north and east, such as San Marino and a large part of Pasadena. The clearest low-income areas were to the south, adjacent to industry, and to the east of downtown.

Urbanization patterns, by contrast, showed up as zones rather than sectors, with higher levels closer to the centre, especially to the west of downtown, where the apartment district of the 1920s and Hollywood was located, attracting many single people, including the off-beat. In some of those areas in 1940, over four in five housing units were rented. The outer part of this zone was part of the higher-income sector. Areas with low urbanization (that is, with single-family houses) generally lay outward from the moderate zone. Blue-collar workers, not surprisingly from what we saw earlier in Chicago, were strongly represented in outlying single-family areas close to industrial districts.

As for high-degree segregation, only one high-income area (emerging Palos Verdes) showed up. The remainder of strongly segregated areas were mostly lower middle or lower income and mainly Black (South Central and Watts to the south) and Mexican (Chavez Ravine, Boyle Heights, and Belvedere to the east), all of them poorly serviced. Little Tokyo demonstrated low social rank, high

urbanization, and high segregation, though clusters of Japanese elsewhere were far less conspicuous. Obviously, this sophisticated map had defined class (though they avoided the term) in a statistical sense, as well as family status and ethnicity. This kind of study would be standard fare for social urbanists in the years ahead.

At the time, places beyond the built-up city (some of which were originally rural market centres) were still spatially separated one from another, so the basin had a patchy appearance from the air, in part because of the unplanned subdivisions laid out in the 1920s. Some remained as agricultural service centres; others (especially in the south) were producing oil; others were adjacent to large steel, auto, and aircraft plants; and some were old, marginal, and poor visible-minority communities, as in far northern San Fernando Valley. The lowlands had not yet become the 'plains of id', Subdivisions spread outward from urbanized Los Angeles, Long Beach, Burbank, and most other places. Municipal boundaries were sometimes clear, but often were not:

> Wonder who could laugh about this: mail, Torrance Post Office; gas, Compton; water from Dominguez; power, Long Beach; house in Keystone, listed in telephone books as Wilmington; pay property taxes in Los Angeles County, but half a block down the street is the school which is called Los Angeles [city] in which we pay a school tax.[139]

This was the peril to identity from living in a leftover piece in the county, yet it was stated with good humour; ordinary people might complain, but by 1950 most were living much better than they were earlier. To understand this a bit more, let us look at some communities for industrial workers.

South Gate

South Gate was a 1920s' suburb 7 mi. (11 km) from downtown in the central industrial belt that ran south down through Vernon and beyond. Tiny in 1920, it grew rapidly until it reached 50,000 in 1950. Industry attracted many people: Firestone, General Motors (GM), which in 1936 was the largest plant in the region (though, as was often the

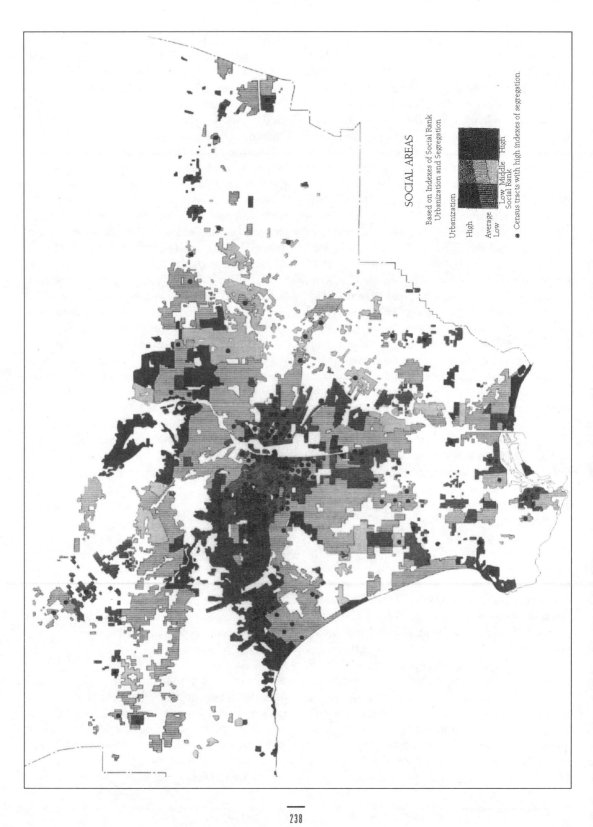

SOCIAL AREAS

Based on Indexes of Social Rank
Urbanization and Segregation

Urbanization

High

Average

Low

Social Rank

Low   Middle   High

● Census tracts with high indexes of segregation.

case, it was just outside the city boundaries for tax purposes), and about 200 more plants were located there, thus making South Gate the city's sixth largest industrial suburb. It incorporated in 1923, then added another residential area called Home Gardens a few years later.

In the 1920s Home Gardens was a shack town of poor White families, most of whom were from the southern United States. These migrants came to seek opportunity; what they found were low-paying jobs and large, half-acre unserviced lots that they could buy cheaply from a fly-by-night subdivider. Just like Mexicans in Chavez Ravine, they built tiny houses and garages (most with second-hand cars). With their chicken coops, rabbit hutches, goats, and large vegetable gardens, they were almost self-sufficient. They bartered goods with one another, such as home-made knitwear. Resistant to taxation, they put up with dusty roads, cesspools, and kerosene lighting and heating. They took in boarders if possible. Another nearby place, Bell Gardens, with its many southern Whites, was three-quarters self-built. Like some poor communities on the edge of Toronto and Chicago, Bell Gardens was a product of the era of Hoover's welfare capitalism.

The middle-class part of South Gate had higher aspirations. Downtown businessmen and the newspaper boosted shopping in the local stores, which included several chains like Woolworth. Competition from other shopping areas, including Los Angeles's downtown (reachable by the PE), generated countless buy-local campaigns. The élite also sought to improve services. They persuaded voters to take on debt for water, sewers, education, and a city hall. They annexed reluctant Home Gardens. As in many other marginal places of the 1920s in both Canada and the United States, South Gate was caught in the down-draft of the Great Depression: jobs were lost, houses were repossessed, and the city was burdened with unpayable infrastructure debts.

But with home-owner protection in 1933

through HOLC, FHA loan guarantees, the arrival of GM, and then military production to meet war demands, life improved well beyond 1920s' expectations. Although in 1940 South Gate's houses were the least plumbed in the whole region, HOLC loans helped some to improve. Others bought houses that were put up by builders on leftover subdivisions of the 1920s. The Wagner Act allowed workers to organize, and many did after overcoming initial resistance from GM and others.

During the early war years, wages shot up, women worked, and Home Gardens was spruced up even further, looking more like a respectable community. The median value of housing rose from near the bottom of the list of suburban places in 1930 to well above the middle in 1950. In fact, in all the region, the gap narrowed considerably between rich and poor places, a reflection of the strongest income redistribution ever on the continent. New services were added, partly at the behest of the unions in this city, half of whose workers were blue-collar. Although they tried, union members were not, however, able to overcome downtown power in running the city. Those who were not unionized resented the 'privileged' GM workers, who in 1946 struck for higher wages and shorter hours.[140] Then fear of communism gripped the unions; the press reviled its members as 'goons, featherbedders, and commies'.[141] Another difficulty was that South Gate was not like an industrial town of old: a majority of workers commuted to South Gate by car from elsewhere, and three-quarters of South Gate residents drove outside the city. Although a modest number rode the PE downtown to work in the 1920s, the number of commuters to and from elsewhere had increased sharply. The golden era for the working class had arrived under Washington's style of Keynesian demand management.

Watts

Watts is to the west of South Gate. Originally a village that housed workers who built railways, it was

6.5 *(Opposite)* Social areas of Los Angeles county, 1940. A sophisticated analysis correlating urbanization and class in nine dimensions overlain by a degree of segregation in built-up areas. This study was a logical technical progression from the Chicago School of Sociology and was followed by many similar ones of other cities over the next two decades. Source: Eshref Shevky and Marilyn Williams, The Social Areas of Los Angeles (Los Angeles: University of California Press, 1949).

a mixed community that was annexed to Los Angeles early on. In the 1920s, despite protests, the first Blacks moved in from South Central. During the war, encouraged by Roosevelt's 1942 order to desegregate war plants, 200,000 Blacks came to the region and settled in Watts. Tensions arose along the border that separated Watts from South Gate. By 1945 Watts was nearly all Black like South Central farther north. Whereas earlier Blacks and Whites had mixed to some extent in high schools and sports, and unions were sympathetic to integration in workplaces, now Alameda Street (the boundary) became a barrier. After 1945 'guarded tolerance broke down. As the economic security of South Gate residents rose, stringent segregation in all areas of life became affordable and by the 1950s, a reality.'[142] The outlawing of discrimination, which was previously maintained through restrictive covenants and zoning, did not prevent other means, such as the unwritten rules used by real estate salespeople. Blacks may have lived in single-family bungalows, and so were better housed than many in Chicago, but they were segregated. Discrimination remained 'severe'.[143]

Washington's Communities

Many new communities appeared in the basin from the late 1930s onward under Washington's auspices. Community builders finally had their day, mostly in creating housing projects near aircraft plants. In 1939 Fritz Burns and Fred Marlow built Westside Village (Mar Vista) near the Santa Monica Douglas plant, which was then expanding production for war. Setting up a staging site on National Boulevard for materials and milling, their carpenters, plumbers, and other construction workers mass-produced nearly 800 houses. They bought up whole nurseries for landscaping. Using a modest number of designs popularized by FHA, their houses were nearly identical in size, though some were larger for skilled workers and lower salaried workers. Most of the $3,000 units were sold to the semiskilled and unskilled, who were in demand for the mass production of airplanes. Burns and an associate built Toluca Wood in 1941 in Burbank near the Lockheed and Vega plants. By 1942 new housing was allowed only for war work-

ers. Westchester was an even more massive project, organized by Security Bank and situated near the main airport and the North American plant. Over 3,000 units on 3,000 acres (1,214 ha) required four developers—including Burns, of course. Clients were a diverse cross-section of occupations from labourers to managers, segregated only slightly by earlier standards through size of houses. A business centre, schools, churches, and local parks were included—another example of neighbourhood planning along rational lines. Although Westside Village, Toluca Wood, and Winchester were near places of employment, workers had to drive. Forty- and 50-ft (12- and 15-m) lots did not encourage walking.

These communities heralded many other similar home-ownership communities in the region for workers, who now had middle-class incomes. Henry Kaiser built Panorama in North Hollywood, a project near his steel plant at Fontana, and others. To guide developers in the still largely open San Fernando Valley, the city defined eighteen business nodes in its plan for that vast area. The most renowned community, though, was Lakewood in county space near Douglas's Long Beach plant that Louis Boyar laid out and started in 1950. Although most of those just noted were within the big city boundaries, Lakewood became, as we saw earlier, the first of the 'minimal-contract' cities, incorporated to keep out undesirable people and Long Beach tax levels.

The working class was thus enfolded into 'bourgeois utopias'.[144] Triply blessed by Washington with jobs, high pay, and cheap housing, they could buy cars to commute on the wide arterials and the freeways. They could, like the rich of Beverly Hills, drive to the beaches and places of pleasure, for example, Disneyland in Orange county. Loved ones could bury them in Forest Lawn. They could drive to shop at the increasing number of department stores in plazas. They could buy powerboats or fly small planes out of the numerous airports in the region. Their baby-boom children could attend public colleges in numbers as never before. For nearly a quarter of a century in this golden age, they would enjoy these benefits, forgetting their struggles during the Great Depression and the struggles of unionization and political action needed to achieve a larger share of national, continental, and global wealth. The cold war provided an outside enemy. Yet this now conspicuously democratic, open society continued to find enemies within, escalating protection and exclusion that rigidified the social fabric into armed class and racial camps. The grandchildren of the 1910 innocents had not shed their ancestors' contradiction: Los Angeles was officially open, but closed to those who questioned the dominant ethos. As someone said early in the century, there was no room for knockers.

CHAPTER 7

∎

Toronto, 1975:
The Alternative Future

The McLuhanesque CN Tower gleaming in the late evening sun greeted our British visitors, who arrived on a non-stop British Airways 747 flight from London. In their last look at some great North American cities, they came to what *Harper's* had just described as 'a city that works ... a model of the alternative future'. *National Geographic* piled on the praise, exclaiming that Toronto was 'worldly, wealthy, personable, and relatively problem free'. *Fortune* glowingly described this 'new great city'. A few years later in *Science*, the publicist for the 1981 American Association for the Advancement of Science would go all out: 'I have seen civilization and it works'.[1]

In fact, Toronto was in many ways like what it was earlier. In 1842 Charles Dickens found it 'full of life and motion, bustle, business and improvement', the streets in 'very good and clean repair', the populace 'orange', referring to Ulster Toryism, but he did not report that King Street glittered like Broadway.[2] In 1913 a less enthusiastic Rupert Brooke averred: 'It is all right. The only depressing thing it will always be what it is, only larger.' But it was not 'hellish' like New York.[3] It became much larger and took on a more cosmopolitan air. In a scant generation, Los Angeles became *passé*; Toronto's style, seemingly more European than American, was in.

The visitors had heard that Toronto's economy had been booming, that European, Caribbean, and Asian immigrants arrived in droves. It was the most important haven for American draft dodgers and deserters during the Vietnam War. As watchers of the urban scene, they were also aware that Jane Jacobs, New York City's prominent critic of public planning, had settled in Toronto in 1968. She arrived when Canada's world stature was rising and that of the United States was falling, mired in the quagmire of Southeast Asia. She moved north just as opposition to expressways, apartment redevelopment, and downtown expansion was vigorously on the upswing. By 1972 the opponents of development won over the voters and implemented reforms to protect neighbourhoods. In fact, the American Transit Association named the Premier of the Province of Ontario as Transit Man of the Year for decisively acting to stop an expressway in 1971. The urban growth ethic, so widely espoused in the wake of the Great Depression, was effectively challenged, ironically in a country that had been gaining dramatically in economic growth.

Our visitors took a limousine downtown on the expressways to the new Sheraton Hotel across from the spectacular city hall. They were not confused about where to go as they had been when arriving in Los Angeles. In Toronto, downtown was definitely still important, even though much office activity and certainly retailing had decentralized, as in many other cities. Besides, as Americans invariably commented, downtown Toronto was safe and astonishingly clean. Even more, Torontonians were extolling its multicultural cosmopolitanism. If most American cities were ailing, why had Toronto been doing so well? They would find the tale of Toronto easier, more straightforward, and more British than those of the American cities they visited. Toronto was less complicated, in large measure because ideological battles were less obvious across party lines. Few complained about government-run services, and almost everyone shared the belief that free enterprise had limits, unlike most Republicans, the extreme liberals south of the border. To Torontonians, the flag-waving American way was too utopian. Goldwater and Nixon could not have been elected in Canada. Not only that, the parliamentary system worked better than the checks and balances system that often seemed to be gridlocked.

To be Canadian (if English-speaking, that is) was now to be less consciously British than to be not American. Up against the huge and hegemonic United States (only one Canadian for every ten Americans), Canadians had opted for management and restraint built on British institutions. After all, the dominant myth of English-speaking Canadians had been that of Loyalists committed to the Crown. Those who fled the new United States after 1776 were obviously conservatives whose descendants would reject what they regarded as Yankee fads. The northern part of America was hardly hermetically sealed, yet the border filtered out some issues as a matter of 'survival', as Canadian writer Margaret Atwood, more strongly than most Canadians, put it.[4]

Much of the impulse to survive was found in the central city, in inner-city neighbourhoods where, in fact, Atwood and Jacobs lived. These residential districts represented the sharpest contrast to many big American cities. Between 1950 and 1975 American inner cities had deteriorated seemingly to the point of no return; Toronto and other Canadian cities had generally improved by mixing people of different incomes and ethnic groups. (Another story was found in Montreal where the two solitudes of French and English polarized.) The contrast between rich and poor was not nearly as great as it was south of the border. Population, economic growth, culture, and politics all played a part in making the differences between the two countries quite distinct. During this period, American excess in the face of slower growth was a contrast to Canada's restraint in the face of greater growth.

The Baby Boom, Immigrants, and Regional Shifts

For a generation after 1950, despite the baby boom, continental population growth increased modestly compared to that of earlier centuries. However, growth was faster than it was in the 1930s, considerably more so in Canada than in the United States. Canada's population rose by 70 per cent, and that of the United States by just over 40 per cent. As a proportion of the US, therefore, the Canadian population rose from 9 to 10.5 per cent, reflecting greater economic ebullience to the north.

The birth rate rose and then fell sharply after 1960. Both countries experienced the baby boom, which was already underway in the early 1940s and a rebound from the terrible early 1930s. In long-term perspective, the boom was nowhere near as dramatic as earlier population growth, yet the larger number of children increased the pace of school construction and obviously increased consumer demand for housing, furniture, and appliances. The birth rate peaked in 1957. The boom then fell off quickly, dropping every year until 1975 when the birth rate was at a historic low (only slightly more than half that of 1957). Perhaps this indicated less confidence in the future or an increasing desire to consume without the burden of the costs of raising children. More women joined the workforce. Average household size in cities dropped by one-quarter in the 1960s and 1970s, contributing to a decline in city populations. Combined with a dropping death rate (by 20 per cent), births contributed somewhat over half of the proportion of the overall increase.

Immigration to Canada (especially after the war) rose substantially compared to the years 1914 to 1945, which made the difference in population growth rates. Public policy encouraged immigration beginning in 1947 when Canadians adopted a credo, as policy makers had in 1896, that more people meant more economic growth. Building up Canada's population would also enhance its status in the world and give it more negotiating clout with the United States. Although fewer immigrants arrived in any one year after 1947 than between

1910 and 1913, the tide of immigrants remained strong and more varied. Annual migration levels were sensitive to slight rises and falls in the economy. A widely held consensus on the benefits of immigration was still strongly evident in 1975.

Initially, Europeans came to Canada: the British still did, and many Dutch, Germans, and 'displaced persons' came from eastern Europe. Italians arrived especially after 1950, followed by the Portuguese, Greeks and other southern Europeans, and Hungarian refugees. Canada opened the gates in the 1960s more quickly than the United States to what became known as visible minorities, that is, people from the Caribbean, Latin America, and south and east Asia. Emigration of migrants was generally low; fewer migrants returned to their homelands than in earlier times. For a few years, young American males (150,000 according to one guess) fled to Canada to escape military action during the Vietnam War. Perhaps about half of them stayed.

In contrast to earlier times, immigrants were almost all city bound, pushing urbanization of the population to its final limit of three-quarters or so of the population, or more or less the limit, though metropolitan growth continued. Although Montreal, Vancouver, and other metropolitan areas shared in absorbing the vast numbers, Toronto took in the most immigrants (over one-third in the early 1970s) when already housing one-eighth of Canada's population. What had been a very British and Protestant Toronto became ethnically and religiously diversified by 1960. Two in five Torontonians were foreign born by 1975 (more than twice the Canadian level), a pattern similar to that of Chicago and New York earlier. Their children added nearly as many. Unlike the earlier American metropolises, one-quarter of the immigrants were British. One in eight was Italian. Roman Catholics now made up one-third of the population, up from one-eighth early in the century. Sikhs and other groups made their appearance. Some schools in mixed neigh-

> Life became far more comfortable for the vast majority; this was the golden age of consumerism and leisure.

bourhoods claimed they taught children from more than 100 language backgrounds. Young people from rural Ontario, the depressed Maritimes, Quebec, and the West also moved to Toronto.

As a result, Toronto's population grew much more rapidly than that of the rest of the country, more than doubling the census metropolitan area to 2.8 million. In 1975 Toronto was about to surpass Montreal to become the tenth or even the eighth largest metropolitan area on the continent (the Canadian census definition was more restricted). The Toronto metropolitan area rose from one-quarter to one-third of Ontario's population. (This tendency to concentrate was true of all large metropolitan areas, especially if their vastly increased commutersheds defined metropolitanism. Urban boundaries were less clear, as many non-farmers occupied 'rural' areas.) The population of its central city, built up before 1930, rose very modestly until 1965, then fell markedly in the 1970s because household sizes dropped sharply and people moved to the oilfields of Alberta. The 1971 census was perhaps higher than it would have been otherwise had not most young Americans lived in the central city at least for a while after they arrived, thus raising the figure in the late 1960s. Even so, vacancy rates remained low as the affluent population took up more space. As we saw in Chapter 1, in the 1980s central-city population recovered considerably.

The population slump of the 1970s contrasted sharply with the continued declines in many American central cities, not least Toronto's near neighbours Detroit, Cleveland, and Buffalo. More than half of the top forty American central cities declined over the three decades after 1950. From 1970 to 1977 central cities as a whole suffered net losses of over 10 million people. More startlingly, populations of five large American *metropolitan* areas dropped too, while none did so in Canada. In the 1970s non-metropolitan populations increased through a modest wave of counterurbanization. To a degree this was a statistical illusion. Many urban people who were defined as non-metropolitan commuted to jobs in metropolises.

The regional shifts in the United States were remarkable and were part of a trend that began in the 1930s. The northeast and the Midwest lost many people to the West and some places in the South. However, northern urban regions still grew in the 1950s as many poor southern Blacks, poor Appalachians, and Puerto Ricans made the mistake of seeking opportunity in a weakening North; by the 1970s northern metropolitan areas were declining and many of the poor were stranded. More Blacks moved back to the South than came north. Not all of the South benefited, the poor cottonbelt states the least of all, though Atlanta was on the upswing. Florida and Texas increased dramatically. The West continued to expand, though California expanded at a slower rate. Florida's great expansion was now fuelled by retirees seeking the sun; many Canadians spent winters there, but most of the movement had to do with jobs.

Economic Growth, Stability, and Regional Shifts

The drive for speed, convenience, comfort, and safety continued as earlier, especially the latter three as the limits of speed were approached. Commercial enterprise and government sought to convince people of the need to spend. Speed and convenience of information flows picked up somewhat with the introduction of big computers, though in 1975 they were still an esoteric device used only by business, researchers, and governments. Long-distance phone systems were completed, though rates had not dropped as much as they would later. Jet-propelled aircraft increased speed in moving people, mail, and goods, but maximum practical levels seemed to be reached by 1975. Cars did not go faster, but the freeways reduced local travel time, even in rush-hour congestion. Freeways also made intercity travel faster, to the detriment of low-subsidized rail, yet again the limits were virtually reached by 1975. Certainly many machines, such as household appliances, provided greater convenience than earlier and were available almost everywhere and to more people down the income scale. Life became far more comfortable for the vast majority; this was the golden age of consumerism and leisure. There was great emphasis on safety; seat belts may have inconvenienced the many who resisted and increased discomfort a bit, but they reduced the number of deaths. Cold war initiatives maintained global or at least continental safety, though preoccupation with protection reached near paranoid levels before declining in the early 1970s. The economy slowed because limits of productivity per worker in manufacturing were reached by the mid-1960s. The number of workers stopped increasing. Besides, as machines displaced workers, more and more people were engaged in services, as we saw in Chapter 2, and their productivity was difficult to measure. As discussed in Chapter 1, a persistent 'silent depression' was beginning.

In 1950 the United States was by far the most powerful country in the world economically (with half the production), militarily, and hence politically. Its standard of living easily led, with Canada's just under three-quarters of its big neighbour's. The American position had weakened considerably by 1975, in large measure as western European countries (except Britain) and Japan recovered from the war and had higher increases of per capita productivity. On this score, several western European countries surpassed the United States in the 1970s. Between 1950 and 1975 the United States ranked fifteenth out of sixteen Western countries, and was behind Britain in total increase of growth. Canada was fifth, rather amazingly, since its infrastructure was not destroyed in the Second World War. Canada gained enormously over the US (if the data is believable) by twenty points to over 90 per cent (Figure 2.6).

In both countries, but especially in Canada, the economic fluctuations at the beginning of the century gave way to greater stability for a while, although by 1975 economic volatility and dramatically increasing inflation were again apparent. American per capita economic growth in conventional terms grew with only modest downturns until the oil crisis of 1973, but could not match Canada's. This was indeed a golden age for Canada, the equal of 1896 to 1913. Despite the cold war, which sustained productivity in the US, growth there was less. The recession around 1975 was more severe in the US than in Canada. Why Canada—which was a virtual economic satellite of the US in this era—did better is an interesting question. Economic historians have not provided clear answers. Here are some possible reasons.

Population growth probably contributed to economic growth, mostly before 1957, although the pace fell short of that of earlier centuries despite Canada's strong commitment to immigration. Immigrants responded to opportunities, and Canada was providing them. More likely sources of Canadian growth were mineral and forest resources, which were used for consumer durables that expanded greatly in both countries and even more for cold war military spending. The American Paley Report in 1950 showed that the US relied heavily on other countries, especially Canada, for raw materials. Replacing Minnesota's waning Mesabi iron ore ranges were rich veins in the Canadian Shield, northern Ontario, Quebec, and Labrador. The St Lawrence Seaway, which was opened in the late 1950s, expedited transport of ore to steel mills on or near the Great Lakes. In northern Ontario, miners dug up uranium for bombs and nuclear power.

Petroleum production also played a large part in Canada's growth after 1950. Alberta gushers supplied huge amounts of cheap oil, mostly for Canadian consumption when pipe lines were run to the industrial East in the 1950s and 1960s. Some oil moved south. As an indication of how much oil counted in the national accounts, Alberta's share of productivity jumped by 50 per cent between 1950 and 1975. The crisis inflicted by the Organization of Petroleum Exporting Countries in 1973 hurt the US in particular when prices quadrupled, adding to already rising inflation. Its oil surplus of the 1950s had evaporated; by then the United States was dependent on imports for over one-third of its consumption, and the amount was rising. (In 1979 the second oil crisis drove prices far higher, quadrupling again.) With its sub-

stantial energy supplies, including western natural gas and coal, Canada was protected. Besides, vigorous federal action kept down domestic oil prices.

Manufacturing probably helped Canada's economic growth, especially Toronto's region, in contrast to its American neighbours. A second wave of American branch plants were established in southern Ontario in the 1950s. Prior to 1965 Canadian auto plants produced small runs of several models for the Canadian market and for British commonwealth countries before 1940. The latter market dried up. Then in 1965 the Autopact agreement integrated Ontario and Quebec plants with American production (Figure 2.6). The 1965 deal allowed Detroit to exploit Canada's lower wages, its less cantankerous though unionized workforce, and (soon after) cheaper public health insurance. Yet even before 1965 Canadian gains in manufacturing were substantial. Canada produced plenty of goods for the cold war machine. In the 1960s Canadian exports of manufactured goods to the US military increased at a more rapid rate than US military spending, and increased again after 1975.

Political and social factors may have contributed to greater growth. Without having the US's burden of being the world's policeman nor its enormous Pentagon build-up, perhaps Canada's economy was more efficient since everyone was less stressed. Pentagon cost overruns were enormous, hammers billed at $300 and the like. Corporation taxes were cut more sharply in the US, just as spending for the Vietnam War rose, contrary to earlier war levies. Perhaps Canada's more complete social benefits system created greater security. Canada rounded out its social welfare system in the mid-1960s more effectively than its big ally. Certainly the euphoria of Expo '67 in Montreal, which was celebrated by all Canadians, just as the US was sagging under the Vietnam débâcle, suggests that Canadian workers were less demoralized.

While Canada's greater per capita economic growth could be partly attributed to the most rapid population growth, natural resources, and greater social security, the western European countries and Japan did well with slower population growth. Although they imported resources, collaboration of business, government, and often unions resulted in more effective and flexible productivity. Their social benefits, which were more complete than those in Canada, undoubtedly paid off. All English-speaking countries did not do as well as Japan, Germany, and a few others in the West, though Canada did the best.

In the postwar era, housing production took on a larger role in the economy as a major ingredient in encouraging demand and stability, but it is questionable as to how much it contributed to growth. As Keynes had argued during the Great Depression, maintaining steady consumption of housing and all the industries that supplied appliances and materials would smooth out the ups and downs of the economy. The pace of building partly reflected this countercyclical argument, but the bursts were patently political too. The American and Canadian federal governments were drawn in deeper as guarantors, even givers, of loans for builders and buyers. By the 1970s half of the new units in Canada had financial help. In the US the mortgage interest deduction was even more of a boon: the higher the income, the greater the benefit. To a modest degree, by European standards, Canadian and American governments financed low-income housing.

Over the generation after 1950, the rate of housing production in Canada lagged behind that in the US. The rate was initially slower in the 1950s, in part because large operators took longer to organize production. By 1970, though, the rate of Canadian production had increased. Canada's housing stock reached 7.6 million units in 1976, while America's was 79 million. To have been equal proportionately, given that Canada's population was 10.5 per cent that of the US, the Canadian total should have been 8.3 million units. By this measure there was a shortfall of nearly 10 per cent. (It is possible that second or more units in single-family housing were undercounted, though rapidly falling household sizes led to more space for everyone.)

Did that mean that Canadians were underhoused? Or did Americans overbuild? The latter was more likely; between 1963 and 1976 new American housing units totalled 27 million, but only some 17 million households were formed.

From 1960 to 1975 over one-third of new housing actually replaced older stock, probably most of it still usable, if the Canadian experience is a measure. Excess more clearly marked the American urban scene, as was so often the case in the past, for example, in building railways. New housing 'far exceeded household growth'.[5] Perhaps slower economic growth otherwise spurred a desperation for growth through housing, magnifying the Keynesian imperative. Ironically, the great American merchant builders, having slain the dragon of premature subdivision with federal aid, as discussed in the previous chapter, in turn overbuilt.

Building actually continued on the margins of slow-growing northern metropolises. By 1970 in Chicago and the surrounding area, housing production was 80 per cent over household formation, even though those suburbs added more people than other American metropolitan areas over ten years. This excess in housing production created a vacuum in the central city and 'produced faster abandonment of older housing', an issue we will examine later.[6] The massive regional shift of population from the north to the south also contributed to abandonment in the North. This was not at all the case in Canadian cities, even though the population thinned out in the 1970s. Perhaps new housing and all the accoutrements contributed less to economic growth than policy makers believed.

From 1950 to 1975 housing costs, at least in Canada, rose at the same rate as the cost of living, tripling (in nominal terms), but per capita and household income rose dramatically by seven times, so generally housing became more affordable. Home-ownership rose in the 1950s before decreasing. Since households were spending about the same proportion of income as earlier on shelter, they were buying more space, improved quality, and greater comfort. This was so even in older housing, as rooming-houses were converted to flats. Increasing affluence led many more city-dwellers than earlier to seek out cottages or weekend homes on run-down farms. Rapidly growing cities like Toronto would experience the highest costs and pressures. It never seemed to have enough houses, even though, unlike most American cities, it kept a large proportion of its older stock of houses.

The zero-sum game of one region winning at another's expense was played out, most obviously in the US. The economic and demographic shifts in the United States were the most startling regional changes. As we saw in the last chapter, Los Angeles's economy was already the great beneficiary of government largesse. This continued for aeronautics, space, and electronics. The so-called sunbelt emerged, or as wags have dubbed it, 'the rise of the gunbelt', reflecting cold war priorities, though certainly not all activity went directly into the war machine.[7] In truth, the 'gunbelt' was a series of clusters of high-technology research and production in Los Angeles, elsewhere in the West, a few in the South, and even some in the northeast. Silicon Valley became renowned, as did the North Carolina Research Triangle. In the north, Route 128 around Boston, the Connecticut Valley, and Nassau county shared in the boom. Southern New England and the West Coast benefited most from defence contracts in 1975, while the South and the West benefited in defence incomes. Space shots were organized in Houston and Huntsville, Alabama, though the other centres contributed expertise. Florida and California shared missile launches. They were also the locales of the new rapidly expanding leisure world, epitomized by Disney World and Disneyland. Low technology and low value-added manufacturing shifted to lower wage and weakly unionized areas in the South or to the Third World. At the same time, the economies of the Midwest frostbelt states declined as they received far less in government spending. However, they were sufficiently affluent to contribute more in taxes than they received.

Government spending and transfer payments resulted in regional equalization. Of four major regions in 1940, the South's per capita income was only 65 per cent of the national average, while the northeast's was 124 per cent. By 1977 the gap had been reduced 90 to 105 per cent. Military spending and social welfare transfers brought about near equalization. The lower cost of living in the South weakened the differential further. Even so, this shift masked sharp collective and individual differences; a composite of several measures of 'social well-being' and accentuating differences showed that Mississippi, the poorest state (minus 1,286) was still far behind Connecticut, the richest state (plus 545).[8] Mississippi's per capita income was only 70 per cent of the US average. More important, income inequality within these states remained greater than in western Europe, Japan, and Canada because social support in the US was less. Within cities, as discussed in Chapter 1, the differences among residential districts were becoming more acute.

Connecticut's continued high status was the result of military spending and a host of financiers living in New York's hinterland. Even so, manufacturing in the northeast and Midwest declined precipitously as measured by employment, if less so by value of production. Machines increased productivity per worker until the mid-1960s. The old manufacturing belt in 1950 employed 70 per cent of industrial workers, but only 56 per cent by 1965, after which absolute losses occurred. The downside of government induced regional action and a slower-growing economy in the manufacturing belt, dubbed the 'rustbelt'. Washington could not be persuaded to grant contracts to the Midwest to the same extent as it did to the South and West. Detroit, Buffalo, Dayton, and Chicago lost their aircraft manufacturing. Besides, their old mass-production industries fell behind world standards. The steel industry of Pittsburgh, Gary, Buffalo, and Cleveland rested on its oars, allowing more innovative Canadian and overseas companies to capture part of the American market. US Steel tried unsuccessfully to diversify into petroleum, while ignoring improvements in its main line of interest.

Auto production, the great mainstay of the Midwest, slowed after 1957 after catching up to pent-up demand following the depression and the war. Ralph Nader's widespread publicity on the poor quality of cars and the failure of the Edsel tarnished Detroit's image. Competition from overseas, such as the environmentally friendly Volkswagen 'Bug', began to undercut sales and, by the time of the 1973 oil crisis, small Japanese cars invaded America. Also, under the Autopact, Ontario's increased car production meant less in the adjacent

American Midwest. Detroit and Akron built more plants outside the frostbelt, which slid from 'chrome to corrosion'.[9] In other fields of production, the US lost market share to the revived economies of western Europe and Japan. By 1975 Japanese electronics were threatening even the American sunbelt.

Canadian regional personal income and public spending patterns became more equitable than earlier. Federal transfer payments to people and provinces, and then from provinces to municipalities helped greatly. The market economy of the Atlantic provinces, which sold lower-value fish and was industrially depressed since 1918, improved modestly, but was aided greatly by transfer payments for unemployment, pensions, and families. The Atlantic region's per capita income rose to 70 per cent of the Canadian average. Although the income gap between Ontario at the top to Newfoundland at the bottom was as great as that between Mississippi and Connecticut, the spread among individuals and cities was less. Infant mortality, perhaps the best indicator of social well-being, was lower in Canada by one-fifth. The gainer was the West, especially in Alberta, because oil production took off just before 1950. The dramatic price rise in the oil crisis in 1973, which was about the time that the cheap, accessible oil ran out, inflated that province's status even higher. However, to Albertans' chagrin, the federal government siphoned off many of the benefits. As a consequence, the industrial heartland of southern Ontario and Quebec contributed a smaller share to total national production, but were hardly depressed to the same degree as the adjacent rustbelt states. Thus, not all the frostbelt was cold economically. Quebec's nationalistic industrial strategy and stronger commitment than elsewhere to corporatist, European-like cooperation among government, business, and labour yielded somewhat more than its economy had earlier. With the Quiet Revolution after 1961, an aggressive manufacturing growth strategy based on cheap hydroelectricity raised Quebec's productivity by 5 per cent to four-fifths of Ontario's per capita. Quebec began to export electricity to the American northeast. Ontario, still the centre of economic activity, experienced substantial growth after 1950 in large part because of exports to America, so its standard of living exceeded that of most American states, and more equitably so.

Toronto's Economic Power

Torontonians were prime contributors and beneficiaries of this growth. Toronto's population growth exceeded that of Los Angeles in the 1960s. Like Los Angeles overtaking San Francisco, Toronto had been taking private-sector economic power from Montreal, though the latter retained some major Canadian headquarters. Like Los Angeles, Toronto's economy was more diverse, but it had more private-sector financial clout, similar to Chicago, which ranked second to New York. From a North American financial perspective, Toronto was the centre of the Canadian region, subsidiary to New York. Toronto's banks stood at the top of the Canadian financial pyramid. With one-eighth of the population in Canada, Toronto provided one-quarter of all the jobs in finance and insurance. Even Montreal's two large banks, the Royal Bank and the Bank of Montreal, moved their international operations to Toronto by 1965 because, as a Wall Street financier noted, it was a lot easier for top decision makers in New York to deal with their counterparts in only one place. The Toronto Stock Exchange was by far the largest in the country, with two-thirds of the trading. To maintain perspective, the TSE operated at only one-tenth the value of the New York exchange, though higher in volume because of its dominant position in mining stocks. Toronto was the number-one headquarters city in Canada, hosting three-fifths of

> From a North American financial perspective, Toronto was the centre of the Canadian region, subsidiary to New York.

resource companies, half of finance and manufacturing, and two-fifths of services. Most oil companies had their headquarters in Calgary, though the largest one, Imperial Oil (a subsidiary of Exxon), stayed in Toronto. After the 1976 victory of the Parti Québécois, Montreal's English élite lost further ground there to the rapidly rising French business élite. Sun Life, large by any standard, moved to Toronto in 1977, thus bolstering Toronto's already formidable share in insurance. Employment among the supporting cast of accountants and lawyers was high.

Manufacturing was diverse and continued to expand until about 1975. If employment growth gradually slowed, investment in plants and machinery grew by nearly five times (accounting for inflation) after 1950, raising value added per worker. Toronto held its share in the country, unlike its near Great Lakes neighbours or Montreal. Most Canadian steel production occurred just 40 mi. (64 km) to the west in Hamilton. Farm machinery manufacturer Massey-Ferguson had long been Toronto's most prominent firm, though it was experiencing troubled times by 1975. High-tech electrical equipment manufacturers, led by Northern Telecommunications, were turning adjacent Mississauga into a near Silicon Valley. Indeed, Northern Telecom and the CN Tower exemplified the assertion by Harold Innis and Marshall McLuhan that Canada's vast open space was linked by communications. Some earlier branch-plant General Electric high-tech plants still operated there. General Electric was only one of the many American corporations that arrived earlier in Toronto, which was the preferred market and labour-market location and hence had the largest concentration of foreign plants and offices in Canada. Several new auto plants in nearby places added to Toronto's importance in high-value added manufacturing. In fact, autos ranked first in Ontario manufacturing in revenues. Aircraft production, though nothing near the scale of Los Angeles or Seattle, was important nevertheless. In lower technological fields, upscale women's clothing was still prominent. Meat packing in Toronto had outlasted Chicago's, providing one justification for the epithet, Hogtown.

The Toronto media—and therefore the publishing and printing industries—dominated English-speaking Canada; its publishers and editors believed that they influenced national, provincial, and local politics and culture. The Toronto media filtered out powerful American influences. Historically conservatives were fervent nationalists. Now the left and unions took this position more frequently since conservatives gradually shifted from protectionism to free trade. Toronto led the way in entertainment and tourism in Canada, partly because of the proximity of Niagara Falls. Other attractions, such as the Shaw Festival and the Stratford Shakespeare Festival, attracted tourists in droves.

With so much high-value production and finance, Toronto's per capita income had long been the highest among large cities. The city's low unemployment rate was attractive to immigrants, though it rose in 1975. More than any American city, Toronto's economy benefited through social goods. Government had always been accorded greater respect in Canada than in the US. As Toronto was the capital of the largest province, the provincial government exerted a strong presence in the city. The federal government and the provinces worked out a refined system of transfers in health, education, and welfare that increased the share of provincial spending more than the federal share. This had meant more government jobs in Toronto. Ontario also subsidized municipalities. As a result, John Kenneth Galbraith's famous aphorism that American cities had public squalor amidst private opulence could not be applied as easily to Toronto in his native Ontario. Peter Ustinov regarded Toronto as 'a kind of New York operated by the Swiss'.[10]

Toronto Society and Culture

If, as Mark Twain remarked, 'In Boston they ask, How much does he know? in New York, How much is he worth? in Philadelphia, Who were his parents?' then in Toronto he might have asked, 'How carefully does he manage?' In the face of Yankee excesses, Torontonians stayed the course of order. In a 'vertical mosaic', Toronto's financial élite ran the country from their empire based at King and Bay streets and other nearby outposts.[11] As Peter Newman in *The Canadian Establishment* noted, 'Toronto is the only Canadian city where members of the local business establishment almost automatically qualify for cross-country status.'[12] The *Globe and Mail* was Canada's only English national newspaper; for the top civil servants in Ottawa, if the news 'isn't in the *Globe*, it hasn't happened'.[13] Almost three-quarters of Canada's élite lived in or around Toronto, and they were a pretty close-knit élite. One commentator believed that the financial institutions were under 'tight control'. The 'stock exchange and the banks have something of the character of a private club.'[14] Certainly scandals were few, or at least not brought to light. Old boys' networks formed in schools and universities continued over the generations. Still, in a growing economy, they were probably less exclusive than Boston's brahmins and Philadelphia's mainliners. Bright young Maritimers could still rise in banking ranks to the top positions. They were also augmented when many Anglo banking executives left Montreal as the French élite took over. The élite lived near one another in shady, stable neighbourhoods, mostly within the City of Toronto, not in separate enclaves as was usually the case in the US, an important point we will discuss later. For a long time, 'Toronto's aristocratic élite was unashamedly interventionist' in politics.[15]

The reality was that they had nowhere else to go. The élite were a stable crowd at the centre of Canada's economic arena. The border counted. To be sure, a few Canadians were successful in Britain; Toronto's Roy Thomson owned the *Times*. Some controlled companies in the United States, though they would not live in New York. A few lived in Bermuda to avoid paying taxes. Toronto's élite had been very successful in organizing business in the postwar era, though with a great deal of indirect help from Washington and Ottawa. Their immobility, nationally and internationally, had an important consequence locally: they had a stake in a workable city.

The élite were mostly of British ancestry,

which is not surprising because Toronto was a very British city before 1945, undergirding stability. Scottish names stood out in the ranks of bank executives. Roy Thomson's son, Kenneth, was then building a vast newspaper empire, as would Conrad Black later, like Roy Thomson from modest circumstances. Still, old family names were prominent in other fields: Eaton in retailing and Weston in food processing and retailing.

Few Jews had reached the commanding heights; in 1975 the Bronfman brothers, Edgar and Peter, were putting together one of Canada's largest conglomerates in an era when large disparate enterprises were concentrated within few large family firms. They had separated from the Montreal family of prohibition fame. On their inheritance from Seagram's, they built an impressive dynasty of a whole range of companies at home and overseas. Jews were prominent in real estate, though not exclusively so. The Reichmann development empire of Olympia and York was launched and about to take on New York City, investing in part the surplus that was built up in growing Canada.

Most of the élite were Tories. Destined, it had seemed from 1943 onward, to rule the province forever, they were the central organizing focus for political culture. In 1942 western progressives, who recognized the rising tide of social democracy, insisted that the Conservative Party be renamed the Progressive Conservative Party. Toronto's élite accepted the change, recognizing that same pressure in Ontario. The apparent oxymoron in the title was not actually the case. Toronto has been a relatively democratic place. In fact, even though some young Tories talked like Republicans, a strong streak of what has been called 'red Toryism' has run through the party, like the wets (to Mrs Thatcher) in British Toryism.[16] What this has meant is a concern for social welfare and the use of state enterprises, at least to a point, when it was necessary for general well-being. Since the turn of the century, Tories had not drawn a sharp line ideologically or often in practice between the state and business, as we will note later in the matter of public power. In contrast to social democrats, Progressive Conservatives showed a greater willingness to accept free enterprise (though within limits) to maintain a hierarchy of authority and to limit but not eliminate union power. In the postwar era, they accepted change, but sought to control its pace and thus dominated provincial and usually city politics in Toronto.

Representative of the red Tory view, at its apogee, was Bay Street lawyer, Frederick Gardiner, who was the dominant city politician in the 1950s and the dominant planner. Gardiner grew up in a working-class area and, although he moved to Forest Hill, which was a separate municipality between 1924 and 1967, he never lost a connection with those whom he considered less fortunate. While on the County Council as reeve of the Village of Forest Hill, he took on the thankless job of dealing with a reorganization of the Children's Aid Society. Hardly a Ben Franklin, he favoured public health insurance and broadening social programs, rights of collective bargaining, and equitable wages and working conditions. Not surprisingly, he was not exclusive on urban matters. He even allowed apartment structures for modest-income residents in posh Forest Hill. As chairman of the Municipality of Metropolitan Toronto (Metro), as we will see, he led the way in public housing in the suburbs.

Toryism can be recognized in other ways: Toronto remained in many ways a conservative, *petit bourgeois* society that was more inclined to the small-scale way of doing things and was less corporatized than America. In Chapter 2, we saw that in Canada private action was relatively free within the embrace of government. Certainly this was true in Toronto. Consider the following, which will be examined in more detail later. The élite did not hide away in gated retreats (except in their country estates); their neighbourhoods were not isolated from the city. Stable working-class districts persisted; 'invasion/succession' had been far less extensive than it was in Chicago. Rather, landscape

> Representative of the red Tory view, at its apogee, was Bay Street lawyer, Frederick Gardiner, who was the dominant city politician in the 1950s and the dominant planner.

changes were more ordered. Private and even vendor-take-back mortgages on residential properties rather than institutional financing persisted more strongly than they did in the US. Grand planning schemes were rejected early in the century, and they were not easily approved during the 1950s when planning was escalating. Even then, planning was subsumed under a management that was very concerned with making systems work and with equity rather than extravagance. House-style living, even if divided into apartment units, was preferred. Redevelopment for new apartment blocks was not popular with the populace. Public transit provided greater choice for citizens than in cities where cars were the only way to commute. Local politics remained largely non-partisan.

Then there was 'Toronto the Good'. Toronto may have become more exciting by 1975 with its multicultural celebrations, but it also had the image (in American and European circles) of being stuffy, exuding British evangelical moralism on Sunday closings and limiting liquor sales. Certainly Scottish Presbyterians kept Sunday low key; Sunday streetcars did not run until 1896, and even tobogganing was prohibited in parks. Most stores were closed on Sunday. In 1923 Ernest Hemingway, a reporter for the *Toronto Star*, complained that he could not buy peppermints at a drugstore on Sunday for a friend in hospital. In fact, as he rued in a letter, 'Christ, I hate to leave Paris for Toronto the City of Churches.'[17] Nearly everybody went to mostly Protestant churches. Sunday baseball and hockey were prohibited until 1950, and horseracing until the 1960s. Methodists had railed against booze, but Anglican Tory influence kept total prohibition at bay. The restrictions on Sunday and booze were gradually relaxed, though in 1975 they were still tighter in Ontario than they were in Quebec and most American cities. Ironically, though, by then Torontonians were more 'permissive' on sexual matters than

Americans, who were divided yet dominated by fundamentalism.[18] However, Torontonians still retained a dourness, as everything was taken so seriously.

Toronto was good in another respect. British-style Methodism in Toronto as elsewhere had been in the forefront of social reform, in sharp contrast to American individualistic Methodism. This kind of goodness paid off with a thicker federal safety net for all and the disappearance of its small slums after 1950. Toronto's goodness reflected the desire to manage public matters well. Violent crime stayed well below that of large American cities, where it was rapidly rising; Toronto's murder rates were not greater than those of the very rural Saskatchewan, about two per 100,000 annually compared to forty in some problem-beset metropolises south of the border. Canadians were far less antagonistic to cities than Americans were, so the lines between rural virtue and urban vice, and between city and suburb, were not so sharp.

Staid Toronto, however, experienced a social upheaval in the late 1960s. Unlike the United States, no race riots broke out in Toronto because there were far fewer Blacks (mostly West Indian), and they were not ghettoized and were less beleaguered. The central city was not a disaster; on the contrary, it was becoming more interesting. Toronto experienced a burgeoning of live theatre in old factories and warehouses. Because Canada was not directly involved in Vietnam, opposition to the war had a different quality: accusations of Canadian complicity were hardly as biting as the bitterness of Americans sent off to fight. In fact, Canada was feeling good about itself.

Even so, similarities to the US were apparent. Opposition to the Vietnam War, which was fuelled in part by the American draft dodgers and deserters, was concerted. The peak of opposition was reached in May 1970 after the Cambodian bombings and the Kent State shootings. A demonstration before the American consulate turned ugly as police on horseback hounded the protesters, but the affair was tame by American standards; no one was seriously hurt and few of those arrested were convicted. Also, democratization of universities spilled over from the Berkeley free-speech move-

ment, and from Paris and Prague and the 1968 civil rights movement in the US. Alternative educational experiments were more radical. The infamous Rochdale College, a brand new high-rise, was the most conspicuous. It took on the character of a vertical Haight-Ashbury, 'the biggest drug supermarket' in the east, 'the most Americanized place in Toronto ... a kind of tower of urban decay and social chaos'.[19] Even Satanists associated with Charles Manson camped out there. Rochdale College did not last long, though Yorkville Avenue remained the site of street counterculture for some time. More benignly, young people clamoured for action on environmental pollution through newly formed organizations like Pollution Probe.

Like the US, many of the same code words, such as 'citizen participation', came into use.[20] Citizen participation became a battle-cry, so much so that in the federal election of 1968, the new leader of the Liberal Party, Pierre Trudeau, successfully exploited that slogan. At the local level, a political reform movement arose, cutting across party lines, seeking to open up the governing of city councils and school boards, and having a marked influence on how the city worked in the longer term. It almost seemed that 'unsupervised gangs of self-taught city planners and land-use critics were permitted to roam the streets freely, writing essays and being aggressively earnest.'[21] By and large, though, people remained polite; the break with the past was hardly complete.

Governing: The Province, Metro, and the City

Stability and equity amidst growth were cardinal qualities in this era in Toronto, as well as in the province and the nation. With indirect federal financial help, the province contributed to municipal well-being while still allowing the cities some initiative. No one ever doubted that municipalities were creatures of the state. There was no need for a Dillon to espouse the rule nor cities to pursue charters for greater self-rule. Toronto and Ontario were not often seriously at odds, unlike New York or Chicago with their states, in part because for much of the past century Tories ran both, even though municipal government was officially non-partisan. But this was also the case in the nineteenth century when the Liberals directed the province for a quarter century. The province did not passively allow municipalities to exercise home rule (as in California), even though Ontario had a strong history of local policy making and management. Toronto as the capital city of Ontario probably reduced tension. The two levels worked together, usually with ease; each year the province passed legislation for municipalities, often without debate. Only rarely would the Premier and cabinet intervene directly in local disputes. We will consider four major ways the province dealt with cities: transfer payments, the resolution of local conflicts through an administrative tribunal, the restructuring of local governments (especially in the Toronto region), and supporting public enterprises. Situations in some American cities will also be discussed and compared with that of Toronto.

Transfer Payments

Since 1930, financial transfers had been the life blood of cities as never before. By the late 1960s the Canadian government, the ten provinces, and their municipalities worked out a finely tuned, complex system of transfers, including rules on national standards. Using its strong taxing power, the federal government supplied transfers directly to people and the provinces, but not directly to cities. A federal urban policy ministry of the 1970s was short-lived largely because the provinces saw it as an intrusion into their jurisdiction over municipal affairs. Although unlike the US, no Department of Housing and Urban Development was possible in Canada, the federal government guaranteed and financed housing mortgages, which influenced urban development. Its welfare state transfers to people helped cities too, if indirectly.

The Great Depression precipitated the need for funding from upper levels of government, though Canada's response was less vigorous than that of the American New Deal. Municipalities had carried the burden of relief until then. Few people had private insurance. The federal government

slowly increased the tiny, predepression old age pensions. Unemployment insurance was started in 1941. Beginning in 1945 the federal government sent family allowance cheques to every mother with children under the age of eighteen. Other than public education, it was Canada's first universal scheme. In concert with the provinces in 1966, old age security was expanded and universalized for everyone over sixty-five, a contributory Canada Pension Plan was introduced (Quebec's was separate), and, to the relief of cities or nearly so, the federal government financed half of general welfare to the able-bodied whose unemployment insurance benefits ran out. The provinces provided 30 per cent and local governments now only 20 per cent, though they retained administration. (The provinces provided benefits to the unemployable, who were formerly called the deserving poor.)

Also important were the transfers to equalize the financial abilities of provinces: the richer provinces subsidized the poorer ones on a scale of tax-raising capability. Likewise, federal transfers for higher education helped provinces to expand universities and colleges (all of which were public then) under the banner of 'human capital'. Most prominent was the financing of public health insurance. Rural Saskatchewan, which was the most devastated in the Great Depression, pioneered hospital insurance in 1948 and then health insurance in 1962 (though over doctors' fierce protests). Then suddenly everyone wanted it. The system was quickly put in place so that all were covered in 1968, except, unfortunately, for the large gaps in drugs and dental work. The provinces administered the plans as they had constitutional power from 1867. While slower off the mark in the 1930s than the US, Canada travelled much further along the road towards the welfare state. Still, it was not up to the standard of European countries.

Universal aid for children, seniors, and health care for everyone was widely accepted by Torontonians and others. Despite all of these measures, around 1970 poverty remained a problem for policy makers, though less so in Toronto than in poorer regions of the country. Yet health had improved everywhere: the infant mortality rate fell, people lived longer, and overcrowding in dwelling units had virtually disappeared. Compared to that of the United States, the Canadian system was more comprehensive and the outcome was more equitable (at least to the 1990s), as we saw in Chapter 1.

In the United States, Lyndon Johnson's Great Society programs actually reduced the level of urban and rural poverty, mostly through Medicaid and food stamps. The recipients were treated with a bit more attention, if still not much respect. Even so, life in the inner cities was deteriorating—rising unemployment, crime, and the drug traffic countered the gains. After 1975 progress would erode, as Aid to Families with Dependent Children fell behind rapidly rising inflation. Canada would also fall behind, if not as greatly, in the support of poor children. Meanwhile, government support for the elderly in both countries was indexed to inflation, thereby improving their financial status considerably.

The provinces substantially increased their own spending in financing communities after 1945. They contributed half of local revenues by 1975, compared to just over one-quarter in 1950, though amounts varied with need; large and affluent Toronto received far less per capita than rural townships and Indian reservations. Virtually all of the provincial transfers arrived in city coffers with strings attached, a sign of provincial policy limits on the local scene. These conditional grants allowed municipalities to reduce their reliance on property taxes. Early in the century property taxes financed 80 per cent of local operations; by 1975 property taxes financed just over one-third. The mayors of Canadian cities might complain that they were 'puppets on a shoestring', but earlier their predecessors had argued for relief from burdens such as support for the poor and education.[22] Although the provinces were more involved in local management, there was still a finely tuned dialogue between the two levels. Local initiatives, especially in Toronto, could have a powerful effect on provincial action at times.

> ## Equalizing local financial capabilities brought much greater equity in services than earlier.

Equalizing local financial capabilities brought much greater equity in services than earlier. Education was the largest of the grants, as over 40 per cent of all local expenditure was in this field. Thus school boards, which were elected on a ward basis like local councils in Toronto and other cities, could offer roughly equal programs; this was in sharp contrast to the US where rich suburbs often spent three times as much on schools as poor suburbs and central cities without adequate grants. Relative equality among Canadians and Torontonians largely laid to rest thoughts of the 'public choice' model of social life in urban areas, which was the model widely defended in the US, where those able to vote with their feet in favour of suburbs with the best services did so. Toronto real estate advertisements often mentioned particular schools, but few schools could claim superiority over others. Unlike the US, no one seriously entertained the idea of urban busing (except for disabled people) as a way to redress inequalities. Rather, extra funding was given to inner-city schools, which redressed inequities arising from home life and household incomes. Elementary schools provided day care and, for immigrant children, heritage language programs. Secondary schools ran English-as-a-second language courses. More egalitarian Canadians and Torontonians fostered few private schools. The province boosted local taxes to fund Roman Catholic schools; a decade later they were totally publically financed.

The mid-1970s' fiscal crisis of Canadian cities was a minor affair compared to what happened in many American cities, far less serious than that of the Great Depression when the province had to bail out some of Toronto's suburbs. However, it was bad enough for a few to warrant the label 'crisis': in the mid-1970s a gap of 6 per cent of expenditures over revenue led to cutbacks in service, but was followed by increased grants. It was hardly a disaster: Metro Toronto's bonds were still rated triple A by Moody's Rating Service. In contrast, in 1977 twenty-three American central cities had ratings of B or lower; most of them were large and in the northeast or Midwest. The best fiscal health was in smaller and newer central cities, primarily in Florida and southern California, though Los Angeles was not well. Among large cities, only Washington and Miami were among the top twenty. Between 1972 and 1982 the 'fiscal health' of seventy-one central cities deteriorated, though variations were considerable.[23]

New York is, of course, the best-known case of the 'fiscal crisis'. It experienced crises in 1914 and 1933, but 1975 was the worst. The big banks cut off their loans, even though hitherto they had benefited from tax-free municipal bonds. The city's finances were put in the hands of the Municipal Assistance Corporation (MAC), which was run by business leaders. The collapse was the result of mounting debt that tripled between 1969 and 1975. On the one hand, tax revenues fell when aggregate income fell, as many middle-class residents and businesses left New York, taking jobs away. Private sector employment fell by nearly 20 per cent between 1969 and 1977. On the other hand, the city had to spend more, as the inner-city poor relied on the already strained welfare system, which was the most expensive in the country in per capita terms. As private sector jobs evaporated, the city hired more workers. Blame was spread around: conservatives pointed the finger at waste, particularly at the high-paid and large unionized workforce, while the liberals blamed the free ride given to the banks and defended the need to create jobs. MAC prescribed financial austerity and the city and school system were forced to cut back severely on jobs. New York was in a league of its own when it came to per capita debt and interest rates, high wages, and taxes. Its per capita income was still higher than other cities, though only barely, yet the cost of living was higher. Manhattan residents were sharply divided into rich and poor with a weak middle class. Cleveland and Newark were also broke, and others teetered on the edge. But amazingly to non-Americans, it has been stated that: 'Fiscal crises should be regarded not as aberrations but as an integral part of urban politics.'[24] Or cities are in a 'state of permanent receivership'.[25] The

expansionists who drastically overcommit are inevitably replaced by those who curb expenditures sharply. This was partly a consequence of 'islands of functional power' in American big cities. 'The career bureaucrat' was 'the new Boss', hence the mayor stood 'impoverished'.[26] These assertions were hardly applicable to Toronto where the mayor was only the first among equals on city council, and the province made sure of stability.

The crises occurred despite the rapid increase in American federal support to cities in the late 1960s and early 1970s through conditional categorical grants, and then smaller unconditional block grants from 1974 to 1978. In contrast to Canada where federal support largely flowed through the provinces, city funding in the US was direct, reflecting the failure of states to keep up their share. 'Even more dependent on federal funding', central cities in 1978 received about one-fifth of their revenues from Washington, though some cities, generally the biggest and poorest ones, received considerably more.[27] In 1967 St Louis and Buffalo relied on the federal government for 1 to 2 per cent of their revenues; by 1978 they received 55 and 69 per cent respectively, which was a sign of crisis management. Many other cities received between one-quarter and half of their needs. There was a 'strong inverse relation between city needs and city capacity'.[28]

The fact that Washington had taken over and controlled the levers of action angered local and state politicians and bureaucrats. But under Nixon, when gifts shifted to unconditional block grants, local politicians put less into poor neighbourhoods than into more affluent ones. For a while in the late 1970s, all the measures for central cities were met with journalistic cheer. But when even further cuts came with the Reagan regime, the biggest and poorest cities lost even more economically and fiscally. The late 1970s tax revolts of affluent suburbanites in America only underlined the fragility of urban finances. Through the 1980s, in contrast, Canadian central cities showed far less stress.

Administrative Tribunal

The second way the province controlled cities and maintained stability was through the Ontario

Municipal Board (OMB), similar to bodies in some other provinces. A quasi-judicial body, it was first set up in 1906 to regulate and deal with conflicts between cities and the privately run street railways. With the first planning legislation in 1913, the OMB began to evaluate subdivisions, then reformed in 1932 to deal with the borrowing of municipalities in default. After 1945 it increasingly reviewed major planning and development schemes because these affected local infrastructure spending. Then in the 1960s the OMB took on the additional role of ombudsperson to ensure that local politicians were representing their constituents. Until then, most of the OMB's members, who were appointed by the government, tended to favour redevelopment projects. Provincial politicians originally set up the OMB to keep issues at arm's length; that is, so they could avoid making localized decisions. As in their dispute mechanisms, the OMB cost less than civil courts because lawyers did not have standing any more than ordinary citizens who have the right to appear before its panels. What better way of resolving disputes in Ontario, which was far less litigious than the United States? There were many disputes in rapidly developing Toronto; major ones involving the OMB will be considered later in this chapter.

Restructuring Local Governments

The province restructured local governments—most dramatically in creating the Municipality of Metropolitan Toronto (Metro) in 1953. It was the first metropolitan government on the continent since New York in 1898, and city officials elsewhere closely watched its progress. Metro is an upper-tier local government that covers 243 sq. mi. (630 km^2). Legislation retained the thirteen existing local urban and rural municipalities as a lower tier, though the central city (by far the largest) held half the seats on its council. New legislation in 1967 reduced these thirteen to six: the city and five boroughs (except one) were later all renamed cities for prestige. Startling by American standards, the richest enclave, the Village of Forest Hill, became part of the City of Toronto. Its mostly Tory residents protested little; after all, it was their party that governed, though they were given minor service perquisites. The process of creating and then

restructuring Metro and its local municipalities shows how the dominant power could exercise political will at a time of rapid growth and ebullient optimism. By 1975 Metro's usefulness as a regional body was beginning to· wear a bit thin, largely because growth was expanding beyond the 1953 boundary. As a result, the province created further metro-style governments around Metro.

Consider how Metro came to be. The system of local governments set up in 1849 worked well until 1913. Cities annexed urbanizing areas to provide services and to satisfy speculators. Stating that it could no longer afford the costs of servicing subdivisions that were often laid well in advance of building, the City of Toronto then stopped annexing adjacent small municipalities and parcels of York Township, despite requests from rich and poor districts. As early as 1914, progressives called for a metropolitan solution, pointing to the precedent of London's County Council. In the mid-1920s several of Toronto's suburbs split off from York Township, American-style, into rich and poor, as the province permitted new municipalities to form. Forest Hill incorporated as a village after Toronto rebuffed its annexation bid once again. In the mid-1920s and 1930s initiatives for metropolitan government withered, even though most of the suburban communities went broke in the Great Depression and were bailed out by the province. The élites of Cleveland, Pittsburgh, and St Louis also tried in vain to create metros in that era.

In the heady growth years of the 1940s, the City of Toronto was anxious to expand once again. Inspired by victory over the Nazis and worried about a return to depression, citizens were eager for new public projects, including subways and even public housing. Not since the era from 1905 to 1913 had they been so enthusiastic for action. The city appealed to the province in 1948 to permit amalgamation, which was actually the annexation of the urbanizing area and even farm land that was expected to sprout crops of houses and factories. By then, Frederick Gardiner, the well-connected Tory reeve of Forest Hill, was convinced of the need for a metropolitan solution to development.

The province turned the debate over to the OMB, whose chairman early in 1953 recommended a two-tier model rather than a unicity to satisfy suburban politicians who were fearful of losing their jobs and citizens of their traditional identities of place. Since hours and hours of public hearings were held previously, the province quickly legislated the necessary act. The Premier appointed Gardiner as chairman of Metro. Subsequently its council, composed of selected politicians elected at the local level, annually reappointed Gardiner as chairman until 1961.

Metro's major objective was to use City of Toronto property taxes (which were then still a large share of revenues) from its downtown offices, factories, and affluent neighbourhoods, as well as provincial grants to build suburban infrastructure to spur orderly growth. It soon charged subdividers upfront lot levies to replace slower-yielding local improvement taxes. Private companies supplied most of the development, as federal mortgages and mortgage guarantees helped greatly. Metro took over the administration of borrowing and wholesaling of material services to retailing local municipalities. It proceeded to build or upgrade water and sewer plants, and provided garbage dumps, arterial roads, and freeways. It took over the City of Toronto's transit commission and private suburban companies and extended transit lines throughout Metro. It provided low-income housing. The new Metro School Board acted as financier for local school boards. In 1957 the police (under a commission) became Metrowide. In 1966 Metro became the administrator of relief for the unemployed, by then called general welfare. Without question, as we will see in several cases, the new system worked well. Efficient, effective, and equitable delivery of services were the watchwords.

In the late 1960s and early 1970s Ontario (still growing economically) created several more regional governments similar to Metro in rapidly urbanizing areas of the province. In the Toronto

> **Metro's success raised envy among American advocates, who were caught in the gridlock of numerous local governments and state governments that were unable to act.**

area, four weak counties became regional governments, and the number of municipalities (now all with urban functions) were reduced to thirty from sixty-six. Compare this to the fragmentation that was so apparent in Los Angeles in 1950 and reinforced further there and elsewhere. Inequities and exclusion through 'snob' zoning in communities of very 'limited liability' with 'toy' governments were hardly as marked in Toronto's suburbs.[29] No unincorporated fragments existed, nor had they since 1849 for that matter (Figure 7.1). Elsewhere in Canada, Winnipeg was unified into one, though Montreal was less able to work out a Torontolike scheme and would remain much more fragmented.

Metro's success raised envy among American advocates, who were caught in the gridlock of numerous local governments and state governments that were unable to act. To many American politicians, however, the Toronto solution 'was more portent than promise, more the catastrophe to be avoided than the consummation devoutly to be wished.'[30] Indianapolis and Miami created new regional governments, and Columbus and Milwaukee annexed, as did Atlanta and several rapidly growing cities in the South and West, which probably improved equity and efficiency. However, most older cities found it impossible to act. Unlike Ontario, states 'displayed all the dynamism ... of the dinosaur', now hamstrung on urban issues by suburbanites.[31] Federal initiatives to open the suburbs to Blacks often ran into a figurative brick wall. Some upper-middle class places like Shaker Heights or Yeadon, outside Philadelphia, were more amenable, though affluent Blacks were still segregated.

Supporting Public Enterprises
Public bodies that provided revenue-producing services (some of them subsidized) were the fourth provincial influence on municipalities and an indi-

The Toronto Region

7.1   The Toronto region, 1975. Since 1988 it has been referred to as the Greater Toronto Area. It includes five regional governments and thirty local cities and towns. Beginning in 1953, the province combined many small places and rural townships
into larger units and created the second tier of urban government. Contrast with Figure 6.1. *Source:* Planning and Development, City of Toronto.

cator of stability. In the reform era early in the century in both countries, all levels of jurisdiction set
up boards and commissions to regulate private and
public action more or less at arm's length from government. English-speaking Canadians were more
aggressive and successful than Americans in running public bodies. Following the British Conservative Joseph Chamberlain's drive for civic gas and
water 'socialism', Tories were more willing to intervene. That made the task easier later for left-wing
initiators of public policy ideas. In some cases public enterprise occurred by default; the best-known
case was when the Canadian National Railways
pulled together a number of failed companies
in 1923. But it was the Conservatives who set

up the Canadian Broadcasting Corporation and
TransCanada Airlines (later Air Canada) in the
1930s. In Toronto, Tories were masters of power
and public transit because the province and city
cooperated.

In 1906 cities and the province established the
Ontario Hydro-Electric Power Commission (under
collective municipal ownership) to generate and
distribute Niagara power cheaply for industry,
commerce, and homes. Cities created their own
local bodies to retail power; these bodies were
responsible to elected councils. In the Ontario version of the progressive movement, which has
dubbed 'civic populism', a wide consensus, supported by the daily press, accused private power

producers as profit-hungry rings or trusts.[32] Unlike Chicago's Republicans, the red Tories who at the time captured the provincial Conservative party and dominated city council were not hobbled by free-enterprise notions or by localism, unlike California where Los Angeles went ahead with its own renowned Department of Water and Power. Local commissions elsewhere in Ontario at first competed with and then bought out the private retailing firms. Hydro was quickly transmitted throughout urban Ontario. In response to rapidly rising rural and industrial demand in the 1940s, Ontario Hydro expanded with hydro stations, conventional coal, then oil and gas plants, and finally nuclear operations. Most citizens in 1975 regarded publicly owned power as a rational and democratic system, though environmental groups raised worries about the costs and dangers of nuclear megaprojects.

Toronto's city council initiated other public enterprises, abetted by the province. In 1913 its council established a public abattoir to help small butchers to sell clean meat and to combat the meat 'trust', as they called it.[33] A short-lived Social Service Commission to assess welfare did not fare as well. The city and the federal government, which was responsible for ports, set up a harbour commission in 1911. The city often built its own public works. To avoid a take-over, the gas company agreed to have a city representative on its board of directors to watch prices. The city did not succeed in creating a telephone system because Bell Telephone was regulated federally and generally fairly, though in western Canada all telephone systems were publicly owned. Throughout rural Ontario, municipalities and cooperatives ran systems. In the twentieth-century, Canadians were positive about their publicly owned utilities.

Most relevant to city dwellers was their transit system on which they relied daily. From 1921 to 1975 and beyond, the Toronto system, which was praised by pro-transit advocates, consistently won prizes for efficiency and safety among large cities on the continent, but making it a publicly owned service took some doing. City council had never countenanced competition on the streets, as was the case in some American cities earlier. In 1849 it

franchised a private omnibus company with exclusive rights, then in 1861 a horse tram operation that ran for thirty years. City council could have run it as a division of its public works department in 1891, but despite electoral support, it had cold feet; the high cost of electrification deterred it too. Or perhaps, in one of the few cases of boodling recorded, private investors won a new franchise for thirty years. The Toronto Railway Co. (TRC), which was controlled by the private power magnates who also ran interurban radial lines in the region, expanded the earlier tram lines and quickly electrified the system.

Conflict began when the city expanded early in this century. The owners of the TRC, who were not into land speculation as was the case in at least Boston and Los Angeles, refused to extend transit lines beyond the 1891 boundaries of the city, so the city had to build its own streetcar routes into districts annexed after 1905. (The refusal yielded the important by-product of higher residential densities.) Like interurban riders, those transferring to the TRC to come downtown had to pay two fares. As elsewhere during this antimonopoly era, commuters complained about inadequate service, and everyone, influenced by the newspapers, thought the company was making too much money. Profits should go to the people, it was argued. By 1918 the voters were ready to take over the TRC. Overwhelming support did not let city council avoid responsibility this time. The province approved.

The Toronto Transportation Commission (TTC), appointed by city council (to which it was responsible and to which it had to turn to borrow), took over the lines in 1921 (and then the interurbans) extended routes into already urbanized areas, and bought new streetcars, but, like the TRC, it resisted expansion beyond city boundaries. Riders accepted an increase in fares, in the wake of inflation, which was something Americans resisted. The TTC was so aggressive, a reflection of its widespread legitimacy, that it bought out a private intercity bus company in 1927. Although ridership dropped off during the depression, high levels of ridership during the war put the system in a firm financial position.

The Toronto transit system was touted as the best. In the depths of the Great Depression, an American efficiency expert had concluded that 'there is no comparable system where the entire personnel appears to be working so harmoniously to the one end of furnishing the best possible transportation service in the most economical manner.'[34] He cited record keeping, frequency, speed of service, quality of the equipment, and safety. Accolades like these would pile up in the future. Metro took over the best system on the continent in 1953. Toronto's system remained the best, but, as we will see later, critics thought it could have done better.

The City of Toronto set up its own parking authority in 1952, building garages and lots and thus shamelessly and aggressively competing with private operators, but also controlling the flow of cars. Unlike electricity earlier, however, the city did not municipalize the private parking lots. Metro ran the ambulance service, cutting waste arising from competition. Torontonians still had few complaints about their public services, nor did they object that the city, Metro, and the province ran them. Toronto's public services were not quite as autonomous as the Los Angeles Department of Water and Power as they were responsible to the relevant government. Indeed, at the local level, city councillors sat on their boards. This did not mean that the quality of services was not debated, nor that some major decisions were not challenged. Much of the discussion later is on the specific issues that arose when critics, mostly from the central city, saw a shift in policy from untrammelled growth to their version of controlled growth when they won the council in 1972. Civic pride remained high; citizens applauded the spectacular new and modern city hall built a few years earlier, just across the street from an imposing 1890s structure.

City Politics, the Struggle to Control Growth

In the City of Toronto, strong leadership came and

went and then came again, but power was relatively dispersed. Unlike higher levels of government with their party machines and tight caucuses in the British tradition, local politics was largely non-partisan, though historically most council members were Tories of varied stripes from 'boosters' to 'cutters'.[35] During the Great Depression, some communists and Cooperative Commonwealth Federation members became aldermen and, in the heady late 1960s and early 1970s, New Democrats and Liberals were elected too. Non-partisanship meant that personality counted, though, unlike Los Angeles, relatively small wards meant that ordinary citizens could keep the politicians to the task of managing the city. Personality cults and machines never gained a foothold. Department heads and boards had considerable clout, but were not autonomous as they were in most American metropolises. Boards had no separate taxing power. From the late nineteenth century onward, Toronto's mayors were elected at large, and from 1904 to 1969 so were members of the executive Board of Control, which was originally set up to ensure financial integrity after one of the few cases of blatant boodling. Corruption was rare. In contrast to the US, a limited franchise curtailed patronage voting. Politicians could not hand out jobs to non-property owners for votes. Toronto agonized less than American progressives over structural changes in government; wards were retained and a civic service created with relatively little fuss. In the wake of the Great Depression, the post-1945 pro-development mood was hard to resist for most politicians and citizens. Ward politicians were cast in the role of protectors, often mediators of conflict inside their wards, though from the mid-1960s onward, many were criticized for being too pro-development. Compared to the US, election campaigns were not costly.

Voluntary associations of citizens were active periodically as watchdogs. The Board of Trade, as did the Bureau of Municipal Research and the Association of Women Electors, frequently made suggestions for change that were sometimes

> The Toronto transit system was touted as the best.

heeded. Ratepayer associations, British-style, had been active since early in the century, protecting affluent and middle-income neighbourhoods from excessive development pressures. Early in the century, the Central Council of Ratepayers tried to influence city hall on citywide issues. Ward politicians represented these organized groups to a point, even though the groups did not nominate them. Networking provided informal support for favoured candidates. From the mid-1960s to the mid-1970s neighbourhood residents' associations brought more clout to older ratepayer groups who opened their ranks to tenants (finally franchised in 1957). Working-class and poorer areas also organized with some help from community workers, who were funded by churches or even the federal government. The mood of the time, citizen participation, and slowing down change spurred them on. Members of residents' associations cut across party and class lines. Local groups combined into a new citywide body, the Confederation of Resident and Ratepayer Associations (CORRA). A metropolitanwide ratepayer group did not fare as well, in part because the suburban citizenry remained largely committed to growth. Home and school associations were also far more active in school reform during this period.

A loose coalition of reformers (mostly professionals) and members of all parties or of none, supported by CORRA activists, ran for city council in 1972 under the banner 'Community Organizing '72'. They took power in the city and repeated the feat in 1974. They were helped when the OMB chairman, J.A. Kennedy, a red Tory, redrew ward boundaries to give more electoral value to working-class districts. Likewise, the elimination of citywide Board of Control elections that had favoured those with large campaign funds made it easier for reform candidates. Although Toronto mayors had weak structural power, the mayor, David Crombie, red Tory and erstwhile New Democrat, was in a strong position.

Crombie had a clear mandate to 'preserve neighbourhoods', which was his campaign slogan. The residents' groups and these politicians represented a reform movement, labelled by one academic activist as 'radical conservative', advocating a

break with the past but also committed to conservation of the city. They sought a return to stability in the city. Ironically, advocates of growth and progress, who believed they had been doing good, were dubbed the 'old guard'. One could add that this was hardly a revolutionary movement, since the reformers sought to conserve most of the governing apparatus. Some of the more radical called unsuccessfully for a greater decentralization of power. In this era certain city and Metro department heads were verbally attacked, but in most cases their positions were not. The issue had more to do with managing the city, preserving a humane scale and proactively reshaping planning and housing programs. Previously, those who worried about doing too much too fast were overwhelmed by pro-growth 'boomtown' enthusiasm. Many young people joined in the fray, alongside older stalwarts of political battles. At a city committee meeting in 1972 Marshall McLuhan lent his voice, lambasting 'more, more, more'. Suburban neighbourhoods by and large felt less threatened, so their politicians remained more committed to continuing development and to a more closed style of decision making.

The 1972 victory for reformers altered the style of politics sufficiently to protect neighbourhoods in the centre by controlling development. The new regime rejected the notion that planning issues were beyond politics, so city council subsequently abolished the planning board (which had statutory powers and was composed mostly of non-elected people) and set up a citizen advisory board when the province agreed with the city's view. Planning and development were combined into one department. The deputation process became more open, and some city hall employees were decentralized to site offices. Other jurisdictions followed. As it turned out, many of Crombie's strongest allies were somewhat dismayed by his subsequent compromises. Nonetheless, much was accomplished over his three two-year terms, recalling earlier spurts of rapid change.

Why did this outburst of central-city political action happen? At bottom, the shift was from growth to slower controlled growth. A new generation—the baby boomers—was growing up without having experienced privation during the Great Depression as many of their parents had, so their urge to promote growth through urban development and redevelopment was less. They expected growth, and with it higher incomes, to continue in other ways. American draft dodgers, who were mostly from suburban communities, liked the lively action in Toronto's relatively dense downtown and living in rooms and flats in old houses. More professionals who dominated reform settled in the central city. In the 1960s they found houses were cheaper in the city than in the suburbs. For many, private and public-sector service jobs, which were increasing rapidly in the city centre, were more accessible by transit, bicycles, or on foot. But even those who had to commute outward found a greater richness in life in the old city. Redevelopment of the inner city threatened stability, and property values rose relative to suburban prices by the late 1960s.

Ideological justification for action to control growth for a livable city arose from images of American freeways, suburban sprawl, and the White flight from American central cities. Ideological support came from Jane Jacobs's fight against the Lower Manhattan Expressway, but more basically from British democratic tradition. All this action was based on the work of others before them who had fought long battles over change; this was an intensification of the protective instinct of inner-city dwellers, many of whom had grown up in the city. To those who are tempted to think that this was a case of NIMBYism that favoured affluent gentrifiers, Chapter 1 has made it clear that the differential in incomes between Toronto's central city and its suburbs has hardly changed since 1950.

Urban Planning: Management First

The reformers sought to control private development and reshape the planning goals of the postwar era. They succeeded to a remarkable degree when compared to other cities on the continent and especially to Toronto's woeful neighbours on the Great Lakes. The costs were minimal, the gains considerable. The following is a consideration of

the slow rise of comprehensive planning, its golden era of the 1950s, then the reaction against changes that the reformers regarded as ill-advised, such as inner-city expressways, the car favoured over transit, excessive private apartment redevelopment, public housing ghettoization through urban renewal, too much office and institutional pressure on central neighbourhoods, and low-density postwar suburbs. This is compared to the American urban experience of excessive suburbanization and the dereliction of many inner cities, as already suggested in the earlier discussion on fiscal ill-health.

Comprehensive planning gained legitimacy in Toronto only in the early 1940s, then lost it by the late 1960s in the city, though the province persisted in pushing grand regional planning until 1975 when it gave up. Torontonians had never been very eager to plan. Early twentieth-century grand planning failed to grab public attention. A group of architects, who were exercising new ideological power and following American precedents, presented a substantial plan in 1905, primarily touting a grand entrance to the city from the bay, a parkway around the city, and diagonal streets. The mayor responded that it was nice but completing water and sewer systems had a higher priority. Then, following Burnham's Chicago initiative, the Guild of Civic Art enticed city council to set up a commission to review and expand ideas. A regional map of diagonals was added, as was a civic square, but when the plan's chief political advocate was defeated in 1912 after the plan was tabled, interest petered out. Most politicians were not interested. In fact, unlike Chicago, too few businessmen could be persuaded to support the scheme in the first place. Piecemeal practical management was more important than visions.

Toronto did, however, get one piece of planning legislation at the time, paralleling American states: it established the right of the city surveyor to comment on suburban subdivision plans at the OMB in order to forestall premature subdivision and integrate streets and highways. Toronto was notorious in not requiring this earlier; within the original 1793 township survey, the urban pattern became a hodgepodge of disconnected streets. The 1912 legislation came too late to have much effect;

urban expansion would not revive to the extent it had previously until after 1945.

Another opportunity for planning seemed to open up in 1914. Thomas Adams was hired away from the British planning establishment to head the federal government's Commission on Conservation. Adams drew up a model planning act, a decade before Herbert Hoover's, and then relentlessly pushed provinces to pass legislation, but with little success. Ontario passed a weak act in 1917 that did not require master plans. Finally, Adams gave up on small-minded Canadians and went to New York, where he became the chief architect of the regional plan of 1929.

In the late 1920s Toronto's growth advocates had a brief whirl once more with a central area plan that again promoted a civic square and diagonals, but the voters said no. The influential public works commissioner, R.C. Harris, expressed the dominant view: since Toronto was not 'aiming at aesthetic preeminence', the city should avoid 'unnecessary extravagance'.[36]

Early in the Great Depression, city planners virtually disappeared, though there were some modest housing initiatives as we will see later. Higher incomes, full employment, and anticipation of victory in the war further raised expectations. Servicemen's reports from Britain on the rising demand for social welfare measures led to enthusiasm among workers and the middle class to vote for the social democrats (then known as the Cooperative Commonwealth Federation). Their near victory in the 1943 provincial election helped to force action in urban and social planning, as the Tories moved strongly to the left. The Board of Trade, the Bureau of Municipal Research, the Trades and Labour Council, and the Ontario Association of Architects wanted action. These groups persuaded a still reluctant city council to establish an advisory Planning Board.

A regional plan, published at the end of 1943 with some fanfare, was replete with expressways like the Los Angeles scheme. In the suburbs, the plan designated industrial areas and adjacent residential districts. It also designated some private and public redevelopment areas, including a model public housing district. The planners estimated the population and urbanized area thirty years ahead. (These were underestimated by a considerable margin, the latter more so, as it turned out, but then all projections at the time were low following the depression, not surprisingly.) Toronto thus began planning seriously.

In 1944 the board staff published a map of seventy-eight neighbourhoods within the city. This map would help define census tracts in 1951 and then local planning districts soon after. Following the Chicago sociologist, the staff designated most districts with older pre-1914 housing as declining or vulnerable, and only 18 per cent as sound. Neglect during the depression and war took its toll on the quality of housing. However, only 2 per cent of the districts, totalling far less than a square mile, were defined as blighted or as slums. They included Regent Park and the Ward. The latter was Toronto's most conspicuous and long-standing slum, just across the street from city hall. These slums were far smaller than those in Chicago or Los Angeles. Toronto's problems were clearly not as severe. More equitable incomes compared to those in the United States was the crucial factor; few people were destitute. City rules and inspections in the past had also helped to maintain housing. Little wonder then that most of Toronto's politicians were not very interested in housing redevelopment; the problems were isolated, indeed scattered, as the planning commissioner in the 1930s had averred.

Besides, a comprehensive zoning scheme was stalled in the 1920s. Following a series of *ad hoc* district designations, city council resisted comprehensive American zoning action. It finally agreed to a proposal on administrative grounds in 1937. The planning commissioner and his tiny staff took four years to develop the scheme to the point of public hearings. Then they shelved it during the war years. The comprehensive scheme, which covered every square inch of private property, did not have legal force until 1952. The reluctance to move swiftly was another sign of a society that showed little interest in a heavy bureaucratic hand.

Still, a head of steam was building up for planning. As in American cities, planning was believed to promote growth in an orderly fashion. During the war, the Toronto Reconstruction Council promoted postwar employment and public works improvements. The council, which was composed mostly of progressive businessmen, supported the city plan. After the war and renamed the Civic Advisory Council, it recommended a metropolitan government among other reforms as positive steps towards comprehensive planning.

Metro Planning

More important, the province had to be seen as taking action to thwart the rise of the social democrats and to get on with housing for a city bursting at the seams. The province created a department of planning, and passed a planning act in 1946 that encouraged urban municipalities to prepare master plans. To cities, this meant drawing up a plan for public works (mainly roads and water and sewer provision) to allow housing to be built, as well as subdivision controls. Toronto and other places prepared their wish lists with cost estimates. At the behest of Frederick Gardiner, the province also created a regional planning body that included all of York County and therefore a lot of agricultural land. Local plans, if they existed, were required to fit the regional plan.

The creation of Metro in 1953 expedited plans already defined for transit, roads, schools, water, and sewer mains, and added more. Primarily this meant that Metro Chairman Gardiner organized action, covering a vast region of 720 sq. mi. (1,865 km^2), three times the size of Metro itself. The first Metro planning commissioner, who had worked at the city for nearly five decades, died in 1955. Gardiner did not want to appoint another. He was impatient with planning: 'plans, plans, we've got millions of plans. Let's get the godamn shovels into the ground.'[37] Gardiner eventually gave in, but he was in charge; it is not going too far to say that he was in fact the planner, often out on the terrain checking his fiefdom. He controlled the development agenda throughout the 1950s.

The new commissioner and his staff tabled the most massive planning document ever in 1959. The plan incorporated many earlier ideas, refined the expressway pattern, and designated industrial areas in the suburbs (too many, as it later turned out). It also promoted redevelopment of the down-

town in the expectation of raising property taxes to pay for further infrastructural improvements. Although modified over the next few years, the plan was never officially adopted by Metro. Neither Gardiner nor many of the council members or non-politicians on the planning board wanted to commit Metro to a long-term plan. The province never forced Metro to do so. Even so, the plan remained the guide for action.

Gardiner, unlike New York's dominant developer Robert Moses, was upfront, always explained what he was doing, and was careful to build coalitions for his planned schemes. Unlike Mayor Richard Daley of Chicago, Gardiner did not have to run a patronage machine. The province, as a large subsidizer of the roads and other infrastructure, could always check the books and limit Metro's borrowing. Luckily for Toronto, and unlike Moses and Daley, Gardiner was not very interested in tearing down vast areas of the city for redevelopment. Yet in the 1950s he represented a wide consensus for suburban growth. No one questioned the provincial-federal initiative of acquiring nearly 10,000 acres (4,050 ha) of farm land for the subsequent development of public housing, subsidized ownership housing, and York University. All this development seemed so rational, at least until the 1960s. Then challenges from the province and the local municipalities undercut Metro's power.

Provincial regional planning came to the fore in the mid-1960s. Then subdivisions had to be tied to trunk sewer systems; septic tanks were forbidden near the city. As the chief financier, the province could thus control the pace of building. Under the new Tory premier, John Robarts, the government sought to tighten the spread of development. Gardiner had been too anxious to expand. In the 1960s the province slowed subdivision approvals to half the 1950s' level. Besides, Metro's boundaries were being reached as people continued to move in.

As a mechanism for growth, the province now questioned Metro's value. Provincial politicians were reluctant to draw Metro's borders farther out because that might give Metro politicians too much power compared to the rest of the province. It cut back Metro's planning area to Metro itself. As a result, the province became its own agent for planning the region. To define the rapidly growing area, the province created the Toronto Centred Region (TCR) a vast area of 8,700 sq. mi. (22,531 km²). Based on many studies of commuting and industrial indicators, and a wide public interest in urban affairs, John Robarts and Darcy McKeough (who was the cabinet minister in charge of finance, economics, and intergovernmental affairs) put forth a radical vision of controlling the extent and direction of growth in the name of infrastructural and energy efficiency. This was unveiled in 1970 as the 'Design for Development'.

The scheme proposed limiting Toronto's immediate growth to an east-west lakeshore corridor (where it had been strongest), severely *restricting* urban growth to the north to maintain agriculture and allow for recreation, and designating 'growth poles', an arc of smaller cities roughly 75–100 mi. (120–60 km) from Toronto. The scheme also defined a wide 'parkway' belt within the lakeshore corridor for already-planned freeways and megapylons transmitting power. Closely connected to this plan was land just to the east of Toronto for a large new town, south of the second major federally owned airport for the region.

The Toronto region was thus to be brought under tight control to foster compactness. The Ontario government's control was extended elsewhere in the province, if briefly. It restructured local governments along the lines of Metro, around Metro itself and in other growing urban areas, and, despite protests from local politicians, unified small municipalities. To spur growth in economically weaker regions, the province assembled land for towns and industry. (This was the heyday of regional economic programs in the country.) The only successful venture, as it turned out, was a new town near a heavy industry development on Lake Erie. Despite a great deal of talk of new towns at a distance from big cities, to allay commuting, following Britain and Europe, little else was accomplished.

The impulse to control growth died by 1975, just as the era of the 'silent depression' (noted in Chapter 1) began. McKeough quit the government over local political discontent with some regional

governments. Soon after, big developers cashed in on their land holdings in the protected zone to the north of Metro when a new large trunk sewer permitted development in Markham, Vaughan, and other municipalities. Because of public fiscal constraint, the province delayed the already-planned new town to the east. The airport plans gathered dust too. Retrenchment in direct government action in planning, as in social welfare, began; private developers were given more leeway. Even so, in the 1960s, the province had slowed development to the north, which helped to increase the price of property in Metro and the old built-up city. This encouraged redevelopment and indirectly contributed to the fierce political action generated in the central city.

Central City Planning

Following the 1946 planning act, City of Toronto planners drew up a detailed plan. By law, it was to be in accordance with the wider regional plan regarding expressways, major arterials, and industrial areas. Yet when Metro was formed in 1953, direct control over development and zoning remained with the local municipality. Hence, by and large, the controversies between public and private rights were fought on the municipal level. Local politics were usually more heated than Metro conflicts, especially when big projects, and sometimes small ones, were on the agenda. The major exception was the expressway fight with Metro. Because big developers vigorously pursued redevelopment in the already built-up City of Toronto, it was the scene of the most intense action.

In 1954 the city hired a new chief planner and a staff, who responded to federal legislation on urban renewal with an interim plan in 1956, then gradually drew up a new, more comprehensive official plan. The planners prepared detailed neighbourhood plans that elicited critical views from residents. The 1956 urban renewal map suggested large areas for redevelopment well beyond the blighted 2 per cent designated in 1944 for commercial, industrial, institutional, and high-rise residential. It urged for commercial and apartment intensification redevelopment around subway stops. City politicians were so anxious to promote new schemes that they even created the post of development commissioner. The development commissioner from England and pro-development politicians lobbied upper levels of governments for funds. By the early 1960s, the chief planner (also British) was cast more in the role of defending neighbourhoods from a city council bent on more development. Finally, after much debate and argument, the city passed an official plan in 1969. It specified large parts of the city for residential and institutional redevelopment. In this plan, the Spadina Expressway came down from the north into the heart of the city. Together with what critics assayed was too much redevelopment, the expressway became the focus of public reaction.

The Great Expressway Debate

A striking difference between the city's 1969 plan and the 1959 Metro plan (and refined version of 1965) was the lack of agreement on expressways. As in the United States, intercity expressways (first called parkways) had been on the agenda since the 1930s. In fact, in 1939 (the same year as some in the US) Ontario opened its first parkway, the Queen Elizabeth Way (QEW) from Toronto to Hamilton, and then later to Niagara Falls and Buffalo. Plans were underway for more. In 1943 the first regional plan for Toronto incorporated projected provincial routes and boldly marked out urban expressways. Some of these would have been destructive to housing and others to wildlife in ravines. After the war, the province started to build a four-lane, limited-access Toronto bypass, a link in the east-west route from Detroit to Montreal, and an intercity route to the north.

With the formation of Metro in 1953 and its direct responsibility for main highways within the area, serious building of the inner-city routes began. Prominent planner, Hans Blumenfeld, brought the latest ideas from the United States, not least inner and outer rings or boxes, with radial links from one to the other, which was a rational revision in transportation engineering thinking of 1943 ideas. In the US, 90 per cent of federal funding greatly expedited construction after 1956. This encouraged Metro planners. In Ontario, though, the pace of building would be slower because the province supplied most of the funds without federal aid. Every stretch had to be assessed for provincial funding support, adding a note of piecemeal decision making that inevitably roused political passions.

The plan was pretty well in place by 1959 (Figure 7.2). Work had already begun on the waterfront route (later named after Gardiner), which was connected with the upgraded QEW on the west to downtown and then to the recently started Don Valley Parkway to the northeast. These road developments did not proceed easily. Soon after in 1961, at the insistence of Gardiner, who had long espoused the route, the Spadina radial link between the outer and inner boxes was approved. Metro councillors, who usually agreed with Gardiner about 80 per cent of time, gave him only 70 per cent support on this. The Spadina interchange's link with the now upgraded northern bypass was promoted by Eaton's, Canada's largest department store chain, which had initiated a huge shopping centre called Yorkdale where the two would meet. Metro assembled land along the proposed route, but it was not long before resistance arose to the Metro expressway plan and to the Spadina Expressway specifically. Ratepayer groups spoke against it at hearings.

The City of Toronto ruined the plan. Resolve was supplied by the élite of Rosedale, an affluent central-city neighbourhood. In 1961 they argued successfully with politicians that they did not want a freeway running down a ravine adjacent to their Victorian houses. The effect of this was to knock out the key inner box link—the Crosstown—from the city's official plan of 1969, and ultimately Metro's. The Spadina radial route would spew traffic out onto arterial streets at whatever point it stopped.

Construction on Spadina from the northern end was undertaken slowly after 1959, the financing of each piece debated at the Ontario Municipal Board. An irresolvable argument continued over the southern end-point. Metro appeased resistant neighbourhoods through which the Spadina Expressway would run; politicians persuaded lead-

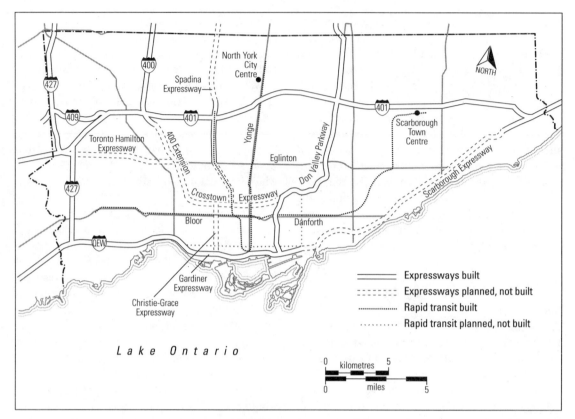

7.2   Plan for expressways and subways, 1959, and those built by 1985. The first expressway was the Queen Elizabeth Way (QEW), opened in 1939. Toronto planners followed American precedents in defining inner and outer rings or boxes. Central city opposition stopped several. Toronto's subway system was the first built on the continent since early in the century. Additions have been few since 1978. *Source:* Metro Planning Board brochure, 1961.

ers in Cedarvale that the route would pass through their district without interchanges, thus limiting local disruption. Farther south, in the Annex, it would run below grade.

Beginning in the summer of 1969, central-city professionals mounted a massive campaign against the Spadina Expressway. Several organizations were formed. Resistance to the Embarcadero in San Francisco and the inner box in London heartened opponents of the expressway. Hours of strategizing led to various protests. The *Globe and Mail* supported the opponents, which was important because influential businessmen and Tories read it.

After acrimonious debates at Metro and the OMB, in 1971 the newly selected leader of the governing Progressive Conservative Party, William Davis, cancelled the Spadina Expressway beyond

the first mile or so that was already constructed. This was one of those rare, direct provincial interventions into local affairs. The immediate result was that the successful opponents celebrated on the Yonge Street mall.

Subsequently, the two other planned radial links, the Scarborough to the east and the 400 extension from the northwest, were put on hold because coalitions of resident and ratepayer groups held firm. The western side of the inner box, the Christie-Grace, still only a line on Metro's plan, dropped from sight. That 2.5-mi. (4-km) link would have been the most destructive of all to housing, which was mostly that of the ethnic working class. Toronto was therefore largely spared the experience of Chicago or Los Angeles; no more inner-city routes have been built. Metro planners fell far short of their goal. In fact, the siting of routes that have been built resulted in the destruction of few houses. Canada's metropolitan freeway lane mileage per capita in 1975 was only a quarter of that of the United States. Canadians were less 'hooked on the freeway drug', to use Jane Jacobs's phrase.[38] That was in large measure the result of phased building: without federal subsidies (as in the US), construction could not proceed as rapidly. Even Chairman Fred Gardiner, who pushed expressways resolutely, told Metro council in 1956 that 'It is the experience of every large city in America that a succession of new expressways is not the answer to efficient and economical movement of traffic. Each successive one is filled the day it is opened ... Additional rapid transit is the only answer.'[39]

The Transit City: The Struggle for Balance

Gardiner, following his planners, argued for a balanced 'modal split', as planning jargon put it. Despite his particular pet freeways, he was actually more enthusiastic about transit than most suburban politicians or, for that matter, the provincial premiers. The latter were willing to pay half the costs of Metro's freeways, but were reluctant to do so for subways. But, argued the reformers, Gardiner's notion of balance was too tilted towards the car. In the end, they achieved more than Metro's planners had targeted. As we saw earlier, citizens had long been committed to the publicly run transit system.

Clearly the most important public transport initiative was building a rapid transit system, which was opened in 1954. In 1946, the electors of the City of Toronto voted for it. It was a smart decision to replace the heaviest streetcar route in the city; running to the northern high-income sector meant that business people could ride the trains downtown very quickly to the financial district. Equally important was the bright idea of connecting streetcar and bus lines with the subway stations, so that transfers would not be needed in many cases. The TTC added more lines, first on the west side of downtown, followed by an east-west route that replaced another heavily used streetcar line, the Bloor-Danforth. Opened in 1966, it served

a more diverse population of riders than had the Yonge line. By 1975 additions were made to each end. Unfortunately, due to a legacy of Gardiner's attempt to balance transit with cars, the Spadina subway line, when completed in 1978, unlike the others, did not directly replace a heavily travelled surface route. Rather, at its north end it ran in the median of the partially completed Spadina Expressway, so buses still had to run on what had been the potential routes. High costs, coupled with a failure of planning, stopped construction after 1978 until the early 1990s.

Streetcars remained on many streets. In fact, in the 1950s Toronto bought some streamlined US streetcars. As the streetcars were more efficient and less polluting than buses, as well as distinctive symbols of Toronto, a vigorous citizens' protest stopped the commission's proposal in 1962 to phase out the streetcars. Keeping streetcars, opening subways, and spreading out the buses raised total fares, but did not prevent a sharp drop in per capita ridership when the system expanded into the lower-density suburbs in 1953; rides per capita fell sharply from 256 to 170 per year. Low density (thus encouraging car use for commuting), not the size of the city, was the enemy of transit. Yet that level of ridership persisted until 1975 in contrast to absolute declines south of the border. The good news was that per capita ridership actually rose after 1975, at least until the depression of the early 1990s. Now more plentiful in numbers, 70 per cent of central area office workers continued to commute by transit, which was more than New York. Around 1975 in Canada's large cities, one-quarter of commuters used public transportation compared to only one in eight in the US because the number of transit revenue miles per capita ran at 2.5 times the figure for American metropolises, 'a statistically significant difference'.[40]

Even so, the need, for financial support rose. By 1960 the province began to subsidize capital construction; after all it had been subsidizing the car, argued transit protagonists. Even though fares rose because inflation increased, the province in 1971 also began to subsidize operations. In part the subsidy covered a decline in revenue when Metro abolished the two-zone fare system to appease suburbanites who were denied more expressways. So transit was added to the list of provincial grants, though these would remain lower than in any other city on the continent. The province even more heavily subsidized commuter rail and bus lines of the government-run GO system, which eased the pressure on roads in transporting more distant residents, especially those who worked in the centre.

By North American standards, Toronto's transit system was successful for the following reasons. When Metro took it over in 1953, the legacy of three decades of city ownership was very positive. At the end of the war, it had a large surplus, enough to encourage citizens to vote for a subway. The TTC developed its own competent engineering and other staff to build more lines at reasonable cost. Despite a further shift to auto commuting, Metro's new development was denser than transit systems south of the border. One system served the whole area, so that integration of buses with rapid transit was rationally planned. Despite a reluctance to subsidize, the province ensured orderly settlement. Despite their enthusiasm for expressways, the suburbs were supportive of both transit and fairly high densities, so the city-suburban split was not pronounced.[41] But this was all relative: European cities did better, as critics pointed out. The only clear answer to transit health was obvious: restrict the use of the car directly. That was unthinkable.

The Curse of Corbusier Confronted

Private redevelopment of residential areas for high-rise apartment buildings became a bane to radical conservatives, a third issue on their plate. Although apartments raised population densities, the reformers argued that medium density was enough to sustain transit. Toronto had always been somewhat uneasy about apartment buildings. A rash of building between 1909 and 1912 had led city council to ban them in the belief that they would become like New York tenements—slums— though in fact most were occupied by middle-class single people or childless couples. The politicians occasionally broke their own rule, not least in the late 1920s, when the number of units increased

dramatically, similar to Chicago and Los Angeles. Most apartment buildings were built on previously undeveloped land.

After the Great Depression and war, during which many single-family houses were subdivided, new apartment blocks began to sprout up around 1955 in place of old single-family houses in the inner city. These early structures rose no more than fifteen storeys. Then in the 1960s developers put up even higher ones that were twenty, even thirty storeys. City council became enamoured with the idea. More areas were considered ripe for redevelopment, including several low-income areas, as shown clearly in the 1969 plan. The plan sensibly designated sites around subway stops, but the largest swath would have covered an enormous rectangular area of about a square mile east of downtown. Private developers cleared and redeveloped many lots before 1969.

The largest, St Jamestown, was a cluster of eighteen high towers that accommodated 11,000 people in 6,000 units, over 300 people per acre. It was billed as a city within a city, like Stuyvesant Town earlier in New York. The project replaced over 400 houses on 35 acres (14 ha) assembled by speculators, with government approval for rezoning and generally without public subsidies. By 1966, in part because of these tall towers, nearly half of the City of Toronto's dwellings were apartment units, up from under one-third in 1951. Private redevelopment housed primarily young, lower-middle class households, which were then shrinking markedly in size.

Reformers were offended; they favoured low-rise, walk-up blocks, many of which had been put up earlier. They continued to support conversions of houses to flats, or partially to flats, as in basements, which had been common too. Both would maintain medium-density and guarantee viable transit. Private redevelopment was debated in city council's chambers before capacity audiences, which were sometimes divided between construction workers and opponents. From 1969 to 1972 skirmishes between reformers and the pro-development old guard became increasingly intense. On the eve of the 1972 election, the old guard won approval for several large new projects.

When the Reform Council was elected in 1972, the radical conservatives sought repeal of the bylaws, but Mayor Crombie sought compromise through negotiation. He was successful in one case; the Windlass project was scaled down to mixed use and called Village by the Grange. The reformers were successful in one, Sherbourne-Dundas. There Torontonians witnessed aldermen tearing down hoardings, and Jane Jacobs and the mayor of the Borough of East York, True Davidson, holding up placards to persuade passing motorists to the join the cause. The outcome was the preservation of early and mid-Victorian houses and the building of infill housing behind in the deep lots. West St Jamestown was allowed to proceed by an OMB panel that was annoyed by the resistance. The OMB became a frequent battleground over developments. Unresolved South St Jamestown remained contentious for years afterward.

One project was dropped because the market slowed down for several years after 1972. The City of Toronto's population declined temporarily as the migrant flow could no longer compensate for the outflow and fall in household size. After 1975 condominium tenure became very popular for baby-boom singles and childless couples, who were now much more affluent, but the scale of production and the allowable sites for redevelopment were more constrained than they were in the late 1960s. Low-rise, inner-city neighbourhoods were largely protected.

Neighbourhood resistence to the many high-rise developments in the suburbs was minimal because they occupied greenfield sites. Compared to most cities in North America, Toronto's high-rise clusters were quite remarkable; some of them were at the very northern edge of Metro in North York (Figure 7.3). These stood out in contrast to farms across Steeles Avenue, where the government at least until then had drawn a sharp boundary. Between 1965 and 1973 in the whole of Metro the number of apartment blocks that were over twenty storeys increased from eight to 142, most of them in the suburbs. In Canada generally, apartments accounted for two-thirds of all units started at the peak in 1969; Toronto had the most.

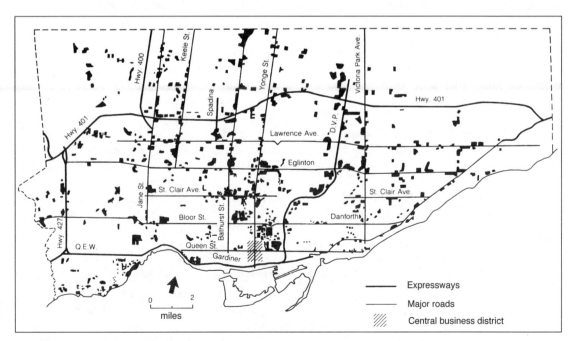

7.3 Apartments in Toronto. Restrictions on the spatial expansion of development contributed to the push skyward on green-fields in the suburbs and redeveloped land in the older city. The building process slowed down soon after 1970, in part owing to neighbourhood opposition. *Source:* Metropolitan Toronto Planning Board.

Public Housing Transformed to Social Housing

The fourth issue that bothered reformers was the quality of public housing in the city and, to a degree, in the suburbs. They were not at all opposed to public subsidies for housing. Public housing had not come easily to Toronto, given the city's *petit bourgeois* disposition, though it finally led the way in Canada. The city had guaranteed two limited-dividend philanthropic projects in 1912, but was only lukewarm to post-1918 federal funding of 'homes for heroes', even though Britain was already providing council housing. The Lieutenant-Governor of the province (the King's representative) spoke vigorously in 1934 for public housing in the suburbs to replace Toronto's slums. The policy group that followed, and then groups such as the predecessor of the Community Planning Association, instead advocated public housing on redeveloped land in a slum area called Cabbagetown. City officials rejected this in 1934, so Toronto did not get New Deal-style garden apartments. Rather, it undertook a modestly ambitious renovation program in 1936. Before-and-after photos by the city (now in the City Archives) made the point that old units could be revamped. Certainly the 9,000 units that were inspected and half repaired sat better with most citizens than redevelopment. The finance commissioner claimed that this action 'undoubtedly obviated' the need for new low-cost housing.[42] Making do remained Toronto's watchword. The programs helped to reduce urban blight, and were a harbinger of later stronger interest in renovation in the 1960s.

By 1946 Toronto's housing advocates had won over the activist mayor, Robert Saunders (another red Tory), to the cause of building new housing to replace slums. In one of the last of Toronto's money bylaw referendums, ratepayers actually voted for a renewal project for low-income working men and their families on the same proposed site of 1934, Regent Park in Cabbagetown. Not until 1949 were reluctant federal and provincial governments persuaded to finance part of the scheme. That year, as in the US, the federal government supported new suburban housing as never before through a new housing act. Not until 1956 was the federal government serious enough about urban renewal to supply public housing. As a result, in the inner city, four other projects followed Regent Park by the mid-1960s: one was a mixture of townhouses and high-rise apartments, another only of high-rises, and the last two had largely low-rises.

Reformers reacted by the mid-1960s. They had a number of concerns. First, they recognized that public projects, as in the United States, were becoming ghettos of the poor instead of housing for working men and their families. These projects were increasingly for single mothers and their children, who would stay on welfare for long periods. Second, reformers were angry about redevelopment itself, as in the case of private rental, because low-income people were displaced. Third, expropriation or compulsory purchases of properties for developments did not give the mostly low-income owners fair market value for their homes. By then inflation was afflicting the housing market, so many of the displaced were unable to buy elsewhere in the city. Remedial legislation calling for 'a home for a home' was finally passed in 1968. Design was also a concern. Advocates for low-income people were complaining of the incompatability of high-rise living, especially for families with children because of the inadequate play space. Poor people, as always, had a hard time managing, though, owing to better management, Toronto's projects did not deteriorate nearly as much as those in the US. Advocates of change also attacked the design of largely garden apartment and row housing projects because they looked different from the surrounding areas. The garden city notion with its open space and pedestrian walkways, which cut off through traffic and street parking, set off such projects as ghettos of the poor, even if most of the residents were White.

The inner city actually hosted few projects. In sharp contrast to the US, where public housing was built mostly on redeveloped land in central cities by the 1950s, Metro put up low-income public housing on greenfields in the suburbs, beginning in 1957. Gardiner steamrolled the first project, Lawrence Heights, over neighbours' protests.

He argued for green grass for children, as had earlier advocates. Why should the poor not enjoy middle-class environments? Of course, the design with its spartan walk-up apartments, row housing, and a 'sea of grass' made the 'project' (a pejorative term in itself) stand out among the adjacent conventional, single-family neighbourhoods.[43] The federal-provincial partnership went on to assemble land for more such projects.

Because production was slow, the Ontario government took over ownership of both the city and Metro projects in 1965 and built vigorously. The Ontario Housing Corporation (OHC) built several projects, mostly in the suburbs, but concentrated the poor in them. Unfortunately, many of the projects were high-rises. Blacks and other minorities were not segregated from Whites. As housing advocates' protests had attracted the attention of the chief financier of mortgages, the federal government, the province sharply reduced the number of such projects in 1973. It had built nearly 30,000 heavily subsidized units in Metro. In 1978 management of over 100 sites of diverse sizes returned to Metro's Housing Authority. Another type of quasi-public housing was 'limited dividend' operated by private owners, but financed by low-interest government loans. About 15,000 of these units served a somewhat higher-income population than the strictly defined public housing.

Far less controversial than OHC's public housing, the Metro Housing Company, supported by Metro, began building small projects for senior citizens in the 1950s, mostly on open land with some tucked into built-up areas. By the 1970s its projects averaged 300 units, more of them in high-rises than earlier. The MHC built over 9,000 units by 1975, doubling that over the next decade and a half, and bought up some existing apartments. It also began to serve a more diversified population of disabled people, even families, though these were a minority. A major social question rises from these actions: why was family public housing so hard to justify, and even harder for single mothers and their children, while public housing for seniors was easy? The obvious answer is that poor children who became teenagers with little hope for personal improvement meant trouble. More fundamentally, Canadians were reluctant to give up the notion of the undeserving poor.

The city reformers sought to overcome ghettoization of the poor. One of their agreed-on major policies was to revive public housing in 1973, but with a radically different approach and more in keeping with some European models. Now the schemes would include home-owners, non-equity cooperative members, market-rent tenants, and means-tested (rent geared to income, as in public housing) tenants in the same projects. Besides mixing incomes, reformers would build houses and small apartments at medium densities. Retreating from garden apartment-style landscaping and culs-de-sac in public housing projects, they would follow the adjacent grid pattern of streets; later this design would be called 'neotraditional'. Several projects were underway by 1976; the most prominent was the St Lawrence on derelict rail and industrial land near downtown. The location fitted another goal of the reform council: to mix land uses and increase core-area housing.

Federal and provincial mortgage funding and subsidies financed these projects, though at first aggressive persuasion was needed. Vancouver used this approach in building the spectacular False Creek scheme. Through its City Home Corporation, Toronto's city council continued to build over the next two decades. It encouraged non-equity cooperatives, that is, cooperatives without investment from resident members. These were federally financed at low interest rates, and served those who could afford to pay the full rent. Even though the reform fever decreased considerably in the 1980s, the idea stuck. Central-city dwellers, both rich and poor, were almost unanimous on the need for mixed housing and provision for low-income people. Most of the latter, however, would still live in private accommodations throughout Metro rather than in council housing. While most people enjoyed more space, many new immigrants doubled up in apartment units to make ends meet. Many people lived in basement flats, which remained illegal outside the City of Toronto until 1994. Only about 12 per cent of households lived in social housing—welfare, limited dividend, elderly, and mixed income. But then as a developer put it around 1970, if NHA mortgage-supported suburban single-family housing were included, the total of social housing in Canada would be over 50 per cent! South of the border, deducting mortgage interest from income taxes was an even more handsome subsidy.

Scaling Down the Core and Protecting the Central Area

Downtown redevelopment for offices and institutions was the reformers' fifth concern. As in American cities, the flurry of office and hotel construction in the late 1920s created a glut until the mid-1950s. Then employment in offices began to increase. Downtown business people began to worry that Toronto was not in the same league as Baltimore, Pittsburgh, or even Montreal, where big developments were occurring. Toronto built a new city hall with a grand plaza, and the Toronto-Dominion Bank led other big Canadian banks onto the path of redevelopment, hiring big development firms and prominent architects. By 1965 the first of several TD towers was opened. It seemed the process would be endless for banks: the Commerce, the Royal Bank, the Bank of Montreal, and eventually the Bank of Nova Scotia put up new main office buildings too.

A massive Metro Centre proposal by the real estate divisions of the two national railways, Canadian National and Canadian Pacific, and by office-builder Cadillac-Fairview would have redeveloped the rail yards south of the central business district. Citizen groups strongly resisted the project, in part because it would have meant the destruction of the elegant, classical 1919 Union Station. Only the CN Tower, the tallest free-standing structure in the world, was the result after all the debate. Another scheme by the Eaton retail empire on its superblock would have destroyed the huge 1890s' old city hall, except the tower. The reformers, including the mayor, were also concerned that developers were allowed not only excessive densities but also added bonuses if they allowed more open space. These plazas were usually windswept, which did not make sense either. By 1973 the downtown core had doubled its office space from what it was in 1962.

After considerable debate and controversy, the reform council dramatically redrew the 1969 plan for the core and for the whole central area. The council scaled down densities and tightened the rules for negotiations over bonuses, which now had to include social goods like day care and the provision of land for social housing. It cut back the 1969 plan's generous land-use designations for institutions to the west of downtown and for high-rise apartments just to the east. The OMB and the province finally agreed. Perhaps the weakened interest in development during the recession after 1973 muted the developers' protests and allowed the new rules to be put in place. One of the most successful developments approved was a scaled-back Eaton Centre. Besides offices and eventually a hotel, the developers preserved the old city hall and an old church. A grand arcade of shops on four levels was spectacular. The mall was anchored at each end by two department stores, which was the pattern of suburban malls. The reformers' insistence that the street facade be opened for access helped to maintain business interest on Yonge Street.

To take pressure off redevelopment on the margins of the central business district and its effects on adjacent, still largely house-form neighbourhoods, reformers promoted outlying office centres, though city council had only persuasive power to bring this about. In Toronto substantial office, retail, and apartment redevelopment around a few major subway stations had been going on for two decades, though Eglinton and Yonge, which had the most office space, did not match the scale of Century City in Los Angeles, but it was on the subway route. By 1975 the decentralization of offices and the creation of focused reconcentration on new greenfield sites was underway around many American cities. The nascent edge town centres had stores, hotels, and even housing. In Metro Toronto about then, the Borough (later the City) of North York, or at least its development-minded mayor, decided to build its own new downtown with a new city hall, other municipal offices, private offices, shopping, hotels, and high-density housing. Soon after 1975 Metro planners too were ready to accept this idea of suburban downtowns and higher densities. In the 1978 Metro plan, two major and two secondary centres were proposed for the three largest outlying boroughs, North York, Scarborough, and Etobicoke, all of which were on rapid transit lines, far fewer than the surfeit of centres in, say, Houston.

Suburban politicians thought that having a downtown with high density would symbolize their cities' coming of age. As a consequence, North York, which was most in favour of the Spadina Expressway, gradually lost interest in its construction as it sought to attract business for itself. North York was one of the largest redevelopment ventures in the urban area. Scarborough, the only part of Metro that was still growing, also built a town centre, only this was on open land, like Mississauga to the west. With the revival of office construction, much office space was, in fact, decentralized. By the late 1980s half of the floor space in offices was still in the central area, half of that in the financial core. Of the rest, about one-quarter was near rapid transit stations, probably the best record outside of Manhattan. Unfortunately, as the depression of the 1990s revealed, office and hotel overbuilding (repeating the pattern in the 1880s and 1920s) created a glut. Toronto too was subject to forces beyond political control.

Low-Density Suburbs

If the Corbusier-style apartment towers offended reformers, the low-density suburbs, which are often associated with Frank Lloyd Wright's Broadacre City, also bothered them. Most of the reformers lived in medium-density neighbourhoods largely created by the 1930s. In suburbs, they saw sprawl, that evocative, overused term. Low densities encouraged people to commute by car rather than by transit, and by 1975 the issues of energy consumption and loss of agricultural land were added to the list of problems.

For scholars and policy makers without a historical perspective, the postwar era was the golden age of suburbanization. However, as we have seen in earlier chapters, suburbanization—expansion outward—was part and parcel of city growth since the beginning. Only during depressions, especially

in the early 1930s, did the process of building virtually stop. The *pace* of suburbanization was slower after 1950 than a century earlier simply because the *pace* of urbanization of the population could not be as fast. More and more people were needed to increase the percentage growth of already large cities. What was different in the generation after 1950 was that the *number* of suburbanites was larger and, even more significantly, they appropriated *far more land* than suburbanites had earlier. Open spaces around houses took up more space than the floor space of buildings themselves. An American study calculated that between 1950 and 1970 each new single-family home added an astonishing 'six-tenths of an acre' to the urbanized area, just over half on the house lot alone; that is 3.5 to an acre! The remainder was for streets, schools, churches, stores and parks.[44]

In Toronto's suburbs single-family densities were higher than they were in the US, at perhaps 6.5 per net acre, but reformers were hardly placated by the comparison. Images of sprawling American suburbs fuelled the protest. In the reformers' view, Metro planners and politicians were pulling the wool over citizens' eyes with their advocacy of a 'balanced' transportation system favouring the car. Reformers were offended by housing on 40- to 60-ft (13- to 18-m) wide lots instead of 17- to 30-ft (6- to 10-m) lots as earlier; the open space around high-rise apartments that cancelled out part of the high density; the vast open spaces and parking lots around low-slung industrial plants and office parks; and a higher proportion of land for roads. Metropolitan parks, created in the 1950s mostly in ravine lands, were largely exempt from criticism, though they added to gross densities and thus increased the pressure to use cars. On its then 240 sq. mi. (622 km^2) Metropolitan Toronto housed 2.1 million people by 1975, nearly double the population in 1950. Even though there was substantial apartment construction in the suburbs, with half of all units built from 1965 to 1975, the gross density was only 8,750 per square mile compared to more than twice that (18,000 or so) in the old pre-1940 city. One-quarter of Torontonians in the census metropolitan area beyond Metro's borders at that time lived at low density too, indeed lower. The introduction of commuter rail, which subsequently allowed for distant exurban developments, worked against overall compactness.

Redesigning low-density suburbs was much more difficult than altering the shape of redevelopment in the city. How could planners, entrepreneurs, or anyone increase the density of the post-1945 suburbs? It was hardly a likely possibility, though perhaps new suburbs could be at higher densities. Even that would prove difficult when it seemed most suburbanites and their politicians seemed to enjoy their large lots and cars.

In the 1940s, one of the local prominent advocates of the corporate suburb, Humphrey Carver, called for neighbourhoods of 7,500 people living in 2,000 units at densities of thirty people per acre, roughly seven units. Some projects were considerably below that. In those days, land was cheap, incomes had increased, so many people bought homes in the suburbs, some even beyond Metro boundaries. By 1960 some planners who were worried about the low density level proposed smaller lots to increase the numbers of units to ten per acre. As we have seen, the province limited subdivision approvals. By 1976 the province's Ministry of Housing suggested twelve and a half single-family units per *net* acre instead of the usual six and a half to seven, and narrower local streets, though this was obviously stretching the bounds of possibility. By 1978 Metro planners were arguing to increase the number of units in low-rise developments to an astonishing fifteen, a pre-1914 level, though politicians could not accept this. Still, a lot of 1970s' single-family housing was to be built at higher densities than earlier, and rental and later condominium apartments went up in the suburbs. By one international comparison, Toronto equalled the density of Hamburg and Copenhagen, if not other European cities. Building office parks with vast open spaces and parking lots, however, con-

> **By one international comparison, Toronto equalled the density of Hamburg and Copenhagen, if not other European cities.**

tinued, which was a larger problem for infrastructure costs, but was too little controlled. Nonetheless, compared to American metropolises, Toronto lost far fewer office jobs to the suburbs.

The pull to the suburbs was thus more constrained than it was in the United States. Compactness was served by rising new housing costs: between 1961 and 1976, the cost of new homes nearly quadrupled, whereas incomes lagged behind. Land prices became a larger portion of the price—one-third of the total. The number of potential purchasers dropped. Not surprisingly during this period, apartment construction was strong.

The reason for higher housing prices was hotly debated in the 1970s: some said corporate control of land was the key factor. One study said nine firms owned or had options on over 36,000 acres (14,570 ha) around Toronto. This was certainly enough to supply lots for many years. Even if developers had oligopolistic power by then, why would they want lot prices to rise so much that sales would fall behind demand? Certainly this was not the case in the US. Some conservative voices blamed the demand of the baby-boom generation and the demand for more space per person as household size decreased. Everyone agreed that government limitations on infrastructural expansion and restrictions against scattered subdivisions in the 1960s to the north were undoubtedly a factor. Whether developers were overcharging could possibly have been proven had the governments banked suburban land outside Metro Toronto as they had within it during the 1950s. Clearly that action had kept prices down for a while. Basically, Toronto was growing rapidly: the pressure on prices of both new and old houses resulted in very low vacancy levels. That led to rising rents, so much so that in 1975 the province imposed rent review, which was later strengthened. Reform action in Toronto was a far cry from what was happening south of the border.

The Sad Tale of American Central Cities

By 1975 vast tracts of abandoned houses, many torched by their owners, and vacant lots had turned inner residential areas of most large American cities into wastelands, so land values evaporated. As presaged as early as the 1910s, central-city populations fell dramatically, especially in many midwestern and northeastern old, slow-growing cities. As abandonment expanded, the pace of decline actually accelerated from 1950 to 1980. St Louis lost nearly half of its inner-city residents, and Cleveland, Buffalo, Pittsburgh, and Detroit were not far behind. Even the populations of Minneapolis and St Paul fell more than could be accounted for by dropping household size. The fall accelerated in the 1970s, and began even in Los Angeles. Although New York City's population dropped less than others, it lost 100,000 units a year in the early 1970s. The South Bronx became a symbol of urban malaise, sometimes compared to fire-bombed Dresden or Tokyo. While inspecting the South Bronx, President Jimmy Carter, as shown in a 1977 photo, looked almost as forlorn as the burned-out buildings themselves.

In big cities in the past, fires provided opportunities to rebuild. This was not the case now; the thinned-out cities were increasingly occupied by low-income people who paid few taxes. New York City's average income in 1929 was 45 per cent above the national average; by 1980 it only equalled it. One-fifth of New Yorkers were below the poverty line, despite Great Society efforts. Jobs declined as department stores closed, factories locked their gates, and rail companies shut down their yards and moved elsewhere. As we saw earlier, the fiscal health of most big cities deteriorated in the 1970s. Those that annexed higher-income areas were somewhat protected from fiscal woes, but not from abandonment and malaise. Less dense cities with an increasing number of poor people meant providing more services with fewer revenues. Central-city incomes continued to fall compared to the suburbs, as we saw in Chapter 1. Writers referred to cities in 'distress', not only in fiscal crises.[45] The great northeastern and midwestern central cities—and not a few others—were becoming unsustainable. Many urban scholars were dismayed; titles like *Cities in Trouble*, *Dark Ghettos*, *The Urban Wilderness*, *America's Ailing Cities*, or *Newsweek*'s 'The Sick, Sick Cities' all told of the worry. A few commentators were slightly optimistic, while most were very pessimistic that anything could be done.

Residents abandoned not only old housing but new public housing as well. Public projects served mostly poor families; an increasing number of them were single mothers and their children on welfare. Many of the working slum dwellers, displaced by redevelopment, refused to move into these presumed crime- and drug-ridden projects. The welfare poor were stuffed into high-rise apartments, the worst possible environments for young children. If in Toronto the few high-rise projects were barely liveable, many of the American ones became intolerable. Despite the hope that new buildings would 'wipe away all the shame and degradation', the result was even worse than the slums. As a horrified writer noted of one New York project, 'the shoddy shiftlessness, the broken windows ... the cold draughty corridors ... the plaintive women ... the gigantic masses of brick, concrete, or asphalt, the inhuman genius with which our know-how has been perverted to create human cesspools worse than those of yesterday.'[46] One solution was to knock them down (Figure 7.4).

The riots of the mid-1960s in Watts, Newark, Detroit, Cleveland, and elsewhere (over eighty of them in 1967 alone) added to environmental mayhem. Unlike riots early in the century when Whites attacked Blacks, now Blacks attacked Whites' property. Surely the destruction was demoralizing to those who believed that communities had to be based on a degree of stability.

These disasters were the result of excessive suburbanization (which drew people outward), the condition of older housing, the perception that Blacks in the central city were pushing White people out (probably the predominant view), combined with misguided public redevelopment and renewal policies. The White 'flight' was encouraged by the massive building of housing and shopping

7.4   Creative destruction? The dynamiting of virtually new public housing in St Louis, the Pruitt-Igoe apartments, in 1972. The design had won an architectural award. The poor lost a lot more older housing to urban renewal schemes to build civic centres and the like. In 1990 the site was still vacant. *Source:* St Louis Post-Dispatch.

Escape from the perceived problems of the city was as important as the pull. It almost seemed that many people could not wait to get out, at least that was the view of those who wrote that White Americans were antiurban. The flight from increasing poverty in the city was far more pronounced in the United States than it was in Canada. The flight to good housing coincided with escape from the rising influx of poor minorities, Blacks, and, to a lesser degree, Latinos. Postwar immigration to cities from the rural South added to those who came during the war. White flight took off in the 1950s: one-fifth of Atlanta families dispersed over a decade after 1956. The share of Whites fell in all central cities, dramatically so in many of them, and was still falling in 1975. At the same time, the share of Blacks rose in every large city, to 70 per cent in Washington (the best place for Blacks to find jobs) and over 50 per cent in Detroit, and was still increasing though slowly by 1975 in nearly all large cities. Among older big cities, only Minneapolis–St Paul and a few western cities had fewer than the average ratio of Blacks in America, which was about one in eight.

The flight to the suburbs was aided by 'block busting'. Block busting happened in Harlem around 1905. Covenants (if now unwritten because they were illegal after 1948) and zoning (which was put in place to stem the tide in the 1920s) now evaporated. On a street towards which Blacks were inexorably advancing, one uneasy White family would sell to a Black family through a realtor who knocked on the door to exploit their fear. The other White families would express anger and then panic, so that the destabilizing realtors made a killing in buying cheaply and selling dear to the new settlers, thus creating a dual housing market. Blacks paid more, at least for some years. That was the case when my cousins in Detroit were the first on their block to sell their house on a per-

centres on the outskirts, followed by the factories and offices that provided jobs, and suburban expressways. Many would say it was the car that created these suburbs, but that by itself is too simple, too technologically deterministic. Perhaps there was an urge to own, after all, on this new frontier after opportunities to own on the old frontier had evaporated for most people. Certainly home-ownership rose from one-half to two-thirds of urban units, if only for a decade after 1950. However, the will to possess needed a kick-start, as we noted earlier; it was not automatic. The land industry and government promoted suburban home-ownership and occupation of space more than ever. The great or large merchant tract builders accommodated White people with lower-middle class incomes who were moving upward socially into their communities and, by the late 1960s, the Black middle class as well, and thus drew people to the suburbs. Institutional lending increased, home-ownership rose, and the slow-growing cities were emptied. Not enough people migrated to fill the vacant houses.

fectly nice, middle-class street. The index of racial dissimilarity between the central city and the suburbs rose, most markedly in the 1970s. The flight (or the rout by realtors) in America seemed to start suddenly in the early 1950s.

But more than racism was at work. Flight occurred even in cities with few Blacks, such as the Twin Cities. Working-class Whites left to improve the quality of their housing. We must remember that earlier in the century, writers referred to the grey areas of dilapidated housing occupied by working-class Whites surrounding CBDS; those grey areas were zones of transition that were pushing outward. The 1960 census reported 2.3 million units of substandard housing in cities. Inner-city housing for the working class had not been as well serviced as middle-income areas. In 1950 in the City of St Louis, over one-quarter of units had only cold water. That was a lot. If residents of those units, two-thirds of whom were tenants, were now earning more money, they were likely to seek greener pastures in the suburbs. Although not all northern cities were as underserviced, most southern cities were far worse. For the many who were poor during the Great Depression but now had money, a new house with a mortgage was far superior to renting an old house or a flat with poor plumbing, for which it was hard to borrow money for renovations. Smoke and grime were also push factors; they were still a problem in 1950 in most cities, though as trains switched from coal to diesel fuel and factories to electricity, the air became cleaner in many cities, except, ironically, when smog struck. Possibly adding to urban insecurity was the threat of atomic war; the bombs were surely focused on the urban core. But then Canadian cities had problems with smoke and fear of attack too. In sum, the expanding Black presence in cities was an important push factor, but the condition of inner-city housing and environment in creating demand for better living conditions in the suburbs was, it would appear, more fundamental to the rout. The reality of decay preceded the wartime rush of Blacks to the North.

Public policies for urban renewal, downtown redevelopment, and inner-city expressways added to the urban decay, despite all the high hopes, even

while suburbanization policies improved the material lot of the majority. A key concern to Toronto reformers, as we have just seen, was what they perceived as excessive redevelopment, both private and public. Urban renewal, the 1950s' catch phrase, carried an almost religious imperative. After all, America was about newness; its religious movements had espoused new covenants. Revitalization, another term used by boosters, also generated emotion. In the US, saving the city by tearing it down and redeveloping it became a mania in the 1950s and 1960s. In 1958 it was innocently asserted that 'some 17 million Americans live in dwellings that are beyond rehabilitation', which was yet another assertion justifying destruction and encouraging the rout.[47] Some Canadians came close to emulating this desire to get rid of the old; one advocate blustered that Canadians should accept 'the idea that a spent and outmoded neighbourhood can be profitably amputated through a single operation'.[48] Why was an amputation needed?

Utopian visions played a part, but so did power. Early in the century, Burnham's plan for Chicago and those for many other cities carried this same imperative, though they largely failed. Piecemeal, private commercial downtown redevelopment obviously took place back then, but large swaths of old housing remained. The ravages of the Great Depression made living conditions worse. Shabby, neglected houses were conspicuous. Something had to be done. The ideology of growth came on the agenda as never before. After the war, the mayors heading 'pro-growth coalitions' in large cities added their messianic thrust; to be considered world-class, their downtowns and central run-down residential areas had to be rebuilt.[49] The desire was widely shared; no one wanted the depression again. Even conservative Toronto was bitten by the bug for a while, though fortunately, leaders like Fred Gardiner were not convinced about wholesale redevelopment.

So with this consensus of good intentions, redevelopment became the path towards recovery, growth, and even greatness. In the United States, the federal government aided the process much more so than in Canada by financing two-thirds of the demolition of old central residential and industrial areas. In Canada cities had to provide half the funding for this. Little wonder that urban renewal earned the epithet of 'federal bulldozer'.[50] With this funding, between 1949 and 1965 local redevelopment authorities expropriated property from low-income home-owners, absentee landlords, and speculators (27,000 acres/10,926 ha by one estimate), an area equalling the size of the City of Boston or of the City of Toronto and a bit more. According to another estimate, which is probably conservative given the high figures cited by others, 425,000 housing units were demolished up to 1968 to make way for the new. Development authorities in the US sold, or tried to sell, the land at a very deep discount to entice private buyers or municipalities themselves. By 1965, of the 27,000 acres (10,930 ha), only about 10,000 acres (4,047 ha) had been sold. Of the remainder, 7,500 acres (3,035 ha) were still uncleared, but the buildings were obviously becoming increasingly derelict. Much of the land would never be sold. Urban renewal schemes and expressways (especially after 1956 when the federal government flooded states and cities with money for freeway construction) were also a great slum clearance program, cutting great gashes through lower-income, inner-city neighbourhoods.

Most cleared areas did not rehouse the poor. The original goal of the 1949 act was to provide decent new housing for the working poor, though, as earlier, it was not certain that this should be temporary or permanent, and was thus a matter of great confusion. All the discussion did not result in an adequate number of units. In fact, the 1949 act promised over 800,000 units in six years. That level was not reached until 1968, partly because redevelopment authorities did not help housing authorities to provide housing, so housing administrators could not rehouse many, and in effect turfed out many poor people. In fact, only a few of the displaced actually moved into new public housing, and many resisted the idea. Production of new housing slowed down until the Johnson programs in the late 1960s increased the levels. Then the new programs built increasing numbers of housing units for seniors rather than for families,

so that by 1978 seniors lived in over one-quarter of the 1.3 million American public housing units. Of course, too, those that were erected were mostly the wrong kind—the high-rises—that had failed families. Low-rise suburban public housing for poor Blacks was not in the cards: central-city mayors wanted all of the public housing projects; the suburbs resisted any thought of public housing.

More upper-income market housing (financed by insurance companies and heavily subsidized by government at deep discounts) than low-income housing was provided. 'Heavily tilted toward the land interest', cities built monumental structures or organized 'centers' for new office construction.[51] Boston's Prudential Center and the Government Center, as well as projects on Bunker Hill in Los Angeles and elsewhere, were widely praised by all but the downtrodden. Cities built new convention centres, sports stadiums, and universities, but they left plenty of excess space. If the loss of one-quarter of St Louis's population can be attributed to the decline in household size (which is a generous assumption), then one-quarter of its citizens left because of urban renewal combined with declining living conditions and more attractive prospects in the suburbs.

By the early 1960s many Americans, inspired by Jane Jacobs, reacted strongly to the waste and extravagance, but by that time, it was too late. The image of the federal bulldozer grabbed attention; tearing down low-income housing for monumental public properties or building expensive apartments for those who could pay market rents, or even low-income vertical ghettos prompted disgust. American liberals and government officials defended urban renewal, but it was obvious that they were wrong. Too many put redevelopment ahead of decent housing for the poor to raise the tax base, create employment, and promote grandeur. In a real sense, the federal, state, and local authorities bullied the poor (who numbered 750,000 by 1964, two-thirds of them Black) out of their homes in the name of saving the city through higher and better use of land. Thousands of marginal but vital businesses were lost, and many did not begin again in the poor areas where they settled. Only in 1964 did small relocation grants begin to help those who were dislocated. When housing authorities did build for the poor, they hired architects who had no sensitivity to people's needs.

Yet conservatives (actually classical liberals) among the critics were responsible for emasculating the idea of public housing for low-income people while saying that the market would take care of people. In contrast, conservatives in Europe had long recognized the need for the state to act for up to 20 per cent or more of the population, putting the social imperative ahead of the economic. Besides, in Europe suburban land was not out of bounds for social housing. American conservatives were not averse to taking bargain-priced lands for profitable ventures while preaching the virtues of free enterprise. Hypocrisy abounded on all sides.

Public policies in the late 1960s and early 1970s did not help inner cities any more than they had over the previous two decades. All the huff about the Great Society and Model Cities programs did little to improve quality, though policy advocates tried to engage residents in what was called maximum feasible participation and, as we saw earlier, raised the incomes of the poor a bit. Participation turned out to be largely unfeasible. What could go wrong did go wrong. Good intentions had perverse results in a country that could not come to grips with social well-being for all.

The threat of urban renewal programs undoubtedly destabilized neighbourhoods, but they probably just accelerated the inevitable decline. Landlords, tenants, and home-owners would hardly make improvements while watching the advancing federal bulldozer. Many landlords cashed in if they sold their properties in time and were well located. Apparently appraisers agreed with them that central-business uses would expand, so the land was worth a lot. These prices drove up the need for high densities. Of course, central business spatial expansion had slowed by 1913 and stopped by 1930. Poor home-owners on

less desirable land were paid far too little to buy decent housing.

Large areas of American central cities were abandoned by owners simply because too much new housing was built outside the city. If federal action through redevelopment and highway building contributed to creating half the vacant areas bereft of housing, abandonment created the other half of 'boarded up' units.[52] Rents fell, so absentee slum landlords gave up as one block after another became socially unstable, even if the buildings could have been made habitable. Also, government, finance companies, and realtors red-lined run-down neighbourhoods, starting as early as the 1930s, so loans were not available for rehabilitation. This was the final result of the trickling down of housing (and of the theory justifying it), but not quite. Until 1968 middle-class Blacks occupied the better housing. Block busting dwindled by the 1960s after so many Whites left, thus prices fell, providing cheaper housing for Blacks. But after 1968, middle-class Blacks began to move when open-housing deals (if not laws) allowed them to move to some generally older suburbs. The underclass that was left behind was victimized by crime and drugs, so by 1975 the central cities declined by yet another notch. Mayors, an increasing number of whom were Black, presided over less and less wealth, so abandonment of the inner city was hastened. These processes of excess on the edge and decline in the centre added up to a national policy for urban development, despite the proclamations to save the old areas of cities through weak rehabilitation schemes.

Rather than too many Blacks, as many Whites believed, there were not enough people to fill up houses. Few foreign immigrants were arriving in any case, and the influx of the American poor from the South slowed to a trickle and others began to return. The filtering down of housing to the poor had reached the bottom; land values evaporated in many inner cities. The enormous 'rent gap' of vir-

> **Public policies in the late 1960s and early 1970s did not help inner cities any more than they had over the previous two decades.**

tually free land in the centre made little difference.[53] Urban homesteading did not take off. Nostrums like *The City Is the Frontier* by an advocate of redevelopment drew few back to inner cities.[54]

Of course, optimists pointed to Greenwich Village, Boston's South End, Rittenhouse Square, Georgetown, and Capitol Hill, which had survived or were upgraded and now housed mostly upper-middle class Whites who valued centrality. A modest number of upper-middle class people lived in apartments, as on Society Hill, and stable upper-income neighbourhoods persisted, as in Indian Village in Detroit, on the Gold Coast in Chicago, and next to Forest Park in St Louis. The optimists could also say that St Louis was still half full, not half empty, and many working-class residents still remained, but these were conspicuous exceptions to the rule. Although the housing act of 1974 encouraged more gentrification, the enormous decline of inner cities could not be halted, as we saw in Chapter 1. Most industrial workers and the lower-middle class could not be enticed back. They blamed Blacks for making them move out to the suburbs and away from the problems. Most suburbanites thought that what went on in cities was not their problem. Why should they be concerned when a factory worker in the suburbs had two cars and television sets, and could 'clothe his wife in excellent copies of Paris fashions the same year they are designed'?[55] The excess of new building laid waste to large parts of perfectly useful inner cities.

Toronto's Inner City: No Empty Houses

Torontonians and Canadian urbanites were spared the worst excesses, in large part because public policies were less desperately put together. Policy makers estimated that, compared to the US, far fewer houses in Canada needed to be replaced. Besides, cities had to provide half the money for renewal, thus discouraging redevelopment. Urban renewal in Canadian cities resulted in clearing just over 700 acres (283 ha) up to 1964, and most of that in a handful of old industrial areas, far less

proportionately than in US cities. In Toronto, authorities cleared 111 acres (45 ha) for four public housing projects, one-sixth of a square mile, subsequently little more. Besides, every bit of slum clearance was replaced. In fact, the first public housing project was phased so that not one family was forced to move, though later on, clearance occurred first, lending credence to the reformers' criticism of a bulldozer approach. Certainly private and some other public action redeveloped nearly 1,900 acres (770 ha) in the City of Toronto between 1952 and 1962, but it was done incrementally at 1 per cent a year—and was *site specific* like the public housing. In some cases, when plans went awry, demolition resulted in low-tax yielding parking lots, showing that local government had limited power over developers. Fortunately, growth always picked up again. Toronto's one great superblock development, private St Jamestown, still stands today with few vacancies, though living there is not the choice of many residents. None of the public units are boarded up, and few private houses are. Without extensive red-lining and with a strong tradition of private vendor-take-back mortgages, sellers had confidence not only in buyers but also in the central city and in neighbourhood stability. Thus the central city did not collapse.

As we saw earlier, the City of Toronto's population, which was constant from 1941 to 1971, declined in the early 1970s, but the number of households did not decline in Toronto or in other Canadian cities in general. In the 1980s Toronto and other large cities added people. Central-city average household incomes as a share of those in the metropolitan area fell somewhat: of the three largest cities, Montreal's fell from 89 per cent in 1960 to 81 per cent in 1980, Vancouver's from only 94 to 90 per cent, and Toronto's from 93 to 89 per cent. In fact, it was Metro's suburbs, not the central city, that lost more ground to some of the new suburban communities beyond Metro. The City of Toronto actually improved its ratio in the 1980s, as we saw in Chapter 1. Most American central-city average household incomes fell precipitously, some down to only half the average of suburban household incomes.

Although Canadian central cities still housed more of the poor than the suburbs, they retained their high-income and middle-income professional populations. In fact, in inner-city Toronto, professionals increased in numbers, resulting in a larger pool of reformers bent on protecting their inner neighbourhoods. Federal rehabilitation and neighbourhood improvement grants modestly helped some Canadian inner-city areas, though obviously these were in less dire need than American cities. To protect tenants in a relatively high-priced city and when prices were rising dramatically in the early 1970s because of inflation, Ontario introduced rent control across the province, though its chief impact was on Toronto. That helped to maintain the social mix. If professionals took over many older houses, those with low incomes were more than likely to move within the city than seek public housing in the suburbs. Continuity was more obvious than discontinuity; the reformers made sure of that.

As a result, Toronto and other Canadian inner cities generally retained their densities more so than American cities. Even in 1950 the City of Cleveland's density was 50 per cent less than that of the City of Toronto, which was of similar size; by 1975 the gap had widened to 100 per cent even though Toronto's density dropped from 20,000 per square mile to 18,000. Toronto had a legacy to draw on. Before 1921 this medium-density city (with ten to twenty units of often semidetached, two- and three-storey houses per net acre) was in part sustained by the tight street railway network. Pre-1930 medium-density housing in the city was adaptable for a wide range of people and the maintenance of services. After the war, the earlier housing stock remained largely intact despite renewal. Apartment construction raised densities in a few areas, but more were built in the suburbs. Reform reaction after 1965 showed that few houses had to be torn down.

By 1975 the racial differences between the inner cities of the two countries were startling. Toronto's Black population was small (only 1 or 2 per cent) and scattered, with an index of segregation lower than that of European immigrants because Blacks spoke English. In the nineteenth century, Blacks in Canada formed ghettos only in Halifax's Africville and a few small places in southwestern Ontario, as the modest number who arrived by the underground railroad dispersed. To be sure, Blacks were often discriminated against in employment, but less obviously so than in the US.

More fundamentally, despite suburbanization in many projects around growing Canadian cities, there was no surplus of houses. Public policies were more restrained because cheap housing was much less of a goal than it was across the border. The movement to suburbs was less pronounced in the 1950s in Canada than it was in US. Although some movers from the city left because of congestion and deterioration, fear drove few outward. The word 'flight' could hardly be applied in Canada. The inconvenience and boredom of suburban life encouraged some to return to the city.

One could say that Canadian central cities were saved partly by luck because of the faster-growing economy in Canada. The zero-sum regional game was not played out in Canada nearly as strongly as it was in the US. Although the economically stronger western, Texas, and Florida urban areas were similar in character to those in Canada, even their central cities in growing metropolitan regions lost population and all of them had slums. American planning officials who visited Toronto found no American-style slums. In the last analysis, less income disparity in Canada was the fundamental reason for the difference. However, over the next two decades the gulf between the rich and poor would widen even in Toronto.

Toronto's Mixed and Safe Spaces

The City of Toronto's reform politicians sought to mix land uses, in sharp contrast to the earlier twentieth-century drive to separate uses, though the urge to separate and purify was never as strong as it was in the US. A wag later dubbed Toronto as Vienna on the inside and Phoenix on the outside, though that overstates the difference between the centre and the edge. Nonetheless, it was clear that 1950s' low-density suburban development was at variance with the central city, which, despite great

Downtown

Consider Toronto's central business district, still powerful by North American standards, so powerful that, as we have seen, reformers sought to limit its expansion to prevent it from becoming like Manhattan. Many office towers and hotels were built, many acres were redeveloped, and Toronto's CBD remained viable. Redevelopment was not, however, carried out with the same degree of urgency as in, say, Boston or Pittsburgh. The projects were carried out one by one, and blended into the existing fabric. Toronto's financial core, the strongest in the country, remained compact while

changes, was recognizably still the same environment in 1975 as earlier.

expanding employment sharply, at least until 1972. The great railway hotel of the late 1920s, the Royal York, now scrubbed of soot, was still a splendid sight. The grand railway station, having been saved from a proposed excessive redevelopment, still hummed with commuter traffic, though with fewer intercity trains. Older warehouses were being renovated, some for live theatre, which was a sign of affluence.

Downtown retail trade dropped off as suburban malls sprouted up, but the Eaton Centre under construction confirmed that retailing in the centre was still strong and in a league with the best in the United States, such as the Loop (Figure 7.5). The extensive underground walkways with shops attracted shoppers throughout the winter. The

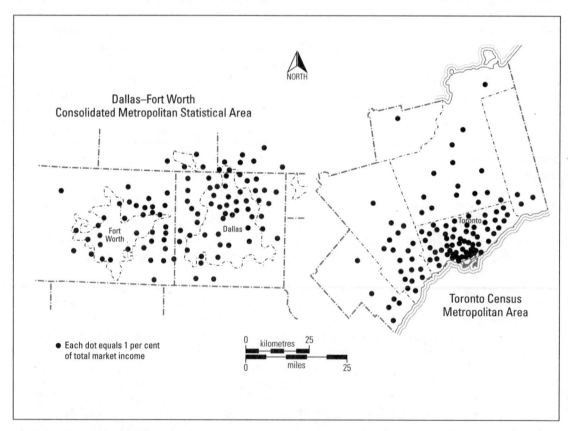

7.5 Retail market areas, Dallas–Fort Worth and Toronto in 1990–1. Some American cities still have relatively strong retail activity downtown, but in many of them residents are so dispersed that strong central centres are impossible. Canadian consumers are not as dispersed. *Source:* J.W. Simmons, *International Comparisons of Commercial Structure* (Toronto: Centre for the Study of Commercial Activity, Ryerson Polytechnic University, 1995):Figure 1.

plaza in front of the new city hall was a place of action. To counter after-five emptiness in the downtown canyons, the Reform Council began to build housing on the margins of the district, complementing places of entertainment. By then, most of the old downtown slum, the Ward, was redeveloped in a piecemeal fashion. On the northern margins of the area, hospitals expanded. Just to the north, the provincial government increased its employment by 50 per cent over the generation after 1950, so it built new office structures, and to its west the University of Toronto spread as it also increased employment at a similar rate when enrolments climbed.

As American visitors invariably noted, the downtown area was clean and safe. One of the few parts of the central area in which women would not walk late at night was a small district to the east, but even that was being refurbished. There was already a great deal of mixing of uses. Redevelopment continued after 1975, if under new rules. Even when relatively more offices were built in the suburbs, downtown office space again doubled before the collapse of the early 1990s.

Waterfront

If Toronto's CBD showed great stability while growing, its waterfront changed dramatically. As in many other cities, beginning in the 1850s and through several subsequent stages, landfill had created accessible and valuable land for trains, ships, and factories in the central area on the southern margin of the CBD. The Toronto Harbour Commission filled in the bay further on the central waterfront in the 1910s and improved the western and eastern beaches for recreation. It drained the great marshy Ashbridge's Bay to the east for industry, storage, and shipping. Across the bay, the islands, which were accessible only by ferry, became covered with temporary tents, then permanent cottages and parkland as city dwellers escaped the summer heat. Unlike Chicago, where industry pulled away from the centre, leaving the lakefront

> **As American visitors invariably noted, the downtown area was clean and safe.**

available for parkland, Toronto's immediate doorstep was industrialized, but residents of the city had the islands across the harbour and other beaches nearby along Lake Ontario. Ducks and geese still had spots of wilderness for breeding. In the 1930s the federal government built the small island airport, but decided that passenger traffic should flow primarily through an airport far from downtown to avoid fog. Lakegoing ships entered the harbour by the western and eastern gaps.

The central waterfront became less and less important after 1945 as industry continued to move outward along rail lines or later along highways, and shipping was concentrated in the eastern end. The great promise of the St Lawrence Seaway, opening the Great Lakes to ocean-plying vessels, was never fulfilled, however. Containerization did not help Toronto, as most shipping companies sought harbours open year round on the East Coast. Even Ashbridge's Bay, the locale for shipping, never lived up to the expectation that it would be a great industrial district. Low-value bulk storage predominated along the central waterfront, though the oil companies built refineries to the west along the Lake Ontario shore.

By the 1960s the central waterfront was ripe for new activities. Several proposals for recreation and living roused interest, including a daring, Venicelike project called Harbour City, which had Jane Jacobs's support, but it failed, along with the others. As an election ploy in 1972, the federal government offered part of the land it owned or controlled to the city for parkland. Several proposals sought to make it a lively precinct. To raise revenue to mount cultural programs, the Harbourfront Corporation (a non-profit federal public corporation) leased land for high-rise condominiums. Few citizens were pleased with the towers.

At the very centre of the central waterfront, very tall condominiums were built next to a large hotel on Harbour Square. The 1920s' Terminal building, used earlier for transfers from ship to rail, was transformed to shopping, a dance theatre, upscale offices, and condominium apartments. In the process, Harbourfront Corporation demolished one of three sets of grain silos. Knocking down one of these silos signalled the decline of industrial Toronto. Toronto's waterfront was only one of several in North America that were being transformed from production and transfer of goods to the production of pleasure.

Meanwhile, the portlands district, as the drained Ashbridge's Bay came to be known, languished. The environmentally minded mused that it should have remained a marsh in the first place, but they were largely happy with another development: the headland or Leslie Street Spit. In the expectation of ocean traffic in the 1950s, the Harbour Commission began another project that would have created an outer harbour. To achieve this, they constructed a long headland. Trucks carted and dumped landfill progressively out at an angle into the lake. This artificial wilderness attracted plants and birds. Ring-billed gulls made their niche there and established their largest nesting ground in the world. The wilderness advocates, however, had to fight off the sailboat crowd, who made concerted attempts to occupy much of the space. In the end, the environmentalists and bird-watchers won most of the space.

The fight over the islands was the most contentious debate over open space. It was a vignette in the struggle between those who advocated the separation of land uses and those who wanted to keep them mixed. The purists had long considered the islands an eyesore and a preserve of privilege for those who lived in their little cottages year round on very low-priced leases. With the formation of Metro in 1953, which took over the management of large parks, the parks commissioner and politicians were determined to tear down the cottages to create more park space for the working poor, such as the vast number of Italians flooding into the city. It was a worthy notion, most thought at the time, so Metro demolished many cottages. Residents at the far east end of the islands, however, strengthened their resolve to stay. Some politicians supported them, and then their battle became a case in the repertoire of the reformers. The residents won in 1973, though the often tense

negotiations would go on for two decades over their leasing rights. A few hundred people had survived in this odd neighbourhood, watching the clock to see when their only access to the city, the Ward's Island ferry, came and went. Today some new housing is planned to enhance the commercial and public service capabilities. Some semblance of the notion that parks, marinas, and houses could coexist was established. Political tugs-of-war were normal in making open space. Like many cities after 1945, Toronto sought to create more and larger areas of parkland. The conservation authority converted many of the valleys of Toronto's two small rivers, the Don and the Humber.

Manufacturing Suburbanized

After 1950 manufacturing of great diversity continued to expand, if gradually more slowly, in the Toronto area. Until the 1970s, investment per worker rose resulting in higher value-added productivity. Spatial shifts were obvious, first to the suburbs within Metro where trucks were displacing rail, then increasingly outside Metro's boundaries because of cheaper land and lower taxes. Nearby the auto assembly plants in Oakville, Oshawa, and Brampton turned out cars for North America. Parts suppliers expanded on suburban sites. Industrial districts within the original city (on the waterfront and along rail lines) lost over half of their firms and half of their employees, a phenomenon not unique to Toronto. As elsewhere, but not as extensively, these areas took on a shabby appearance. However, the garment district on Spadina, which shifted to manufacturing upscale women's clothing, remained viable, and Lever Brothers still made soap and Gooderham and Worts still made whiskey near the mouth of the Don River.

Compared to Chicago, Toronto hardly measured up as a rail hub. As in Chicago, though, rail freight sorting yards had long been on the margins. In the 1960s the companies erected new high-tech yards in the far outskirts. The downtown and three older inner suburban rail yards were largely abandoned. Only the passenger train yards remained downtown, but the companies proposed to move these when they tried unsuccessfully to renew their downtown property for offices. Even now, after a second round of proposals that resulted in the Sky-dome Stadium, at extravagant cost to provincial taxpayers, many of the lands remain empty. Toronto may have lost industry less rapidly than Buffalo, but the old city's role was now in decline. During the recession in the early 1980s regional industrial employment fell, and then in the early 1990s industrial production itself declined. Toronto would not be immune from wider forces at work. Meanwhile, financial and business services grew strongly.

A Medley of Neighbourhoods with a Mélange of Incomes

Before looking at particular neighbourhoods, let us consider income differentials throughout the area. What was striking in the central city was the *mélange* of incomes within many neighbourhoods; the suburbs tended to be more homogeneous, especially if they were new. Mixing can be seen through comparison of the median as a percentage of the average household income within tracts. (When median equals average, the neighbourhood is very homogeneous.) What was most striking was that the richest but oldest tracts in Forest Hill and Rosedale had the greatest discrepancies between median and average incomes. In Forest Hill and Rosedale, there were modest-income households (defined as one or more people living in one unit). Many affluent families rented out units in their large houses, and apartment buildings housed those of lesser means. The Regent Park public housing tracts showed sharply increasing differences too, as if the Ontario Housing Corporation, like the city, was fostering income mixes in its inner-city projects. Many other old inner neighbourhoods became more varied by income after 1950. By contrast, lower-middle incomes of residents of St Jamestown's high-rise apartment cluster were remarkably equal. By and large also, the suburbs showed greater internal equity as did public housing in the Jane-Finch area. As the baby-boom generation grew up, many of the postwar suburbs became more heterogeneous; empty nesters rented out basement apartments in their bungalows.

Greater income differentiation through the 1970s resulted in a greater income gap between the richest and poorest neighbourhoods, yet the pattern was not consistent. Within Metro in 1970, average household income in the top census tract in the new and very upscale Bayview area was six times that of the lowest census tract in one of two Regent Park's public housing projects; the income gap grew to sixteen times in 1980. But these were the extreme cases; the next highest income census tract in rich Forest Hill exceeded the next lowest, also in Regent Park, by only five times in both years. That tract in Regent Park had an astonishingly 50 per cent higher average household income than just across the street, where apparently authorities placed relatively more single mothers, but that was deceptive too. As Parkdale, west of downtown, began to absorb those released from the nearby mental hospital, it lost ground, so spots in Parkdale seemed like 'landscapes of despair'.[56] Even so, Parkdale's three tracts retained many higher-income people.

Most older residential districts in the central city were improved largely through private and public investment. Except for Parkdale, talk of vulnerable, declining, and even blighted neighbourhoods, which were so ubiquitous in the 1940s, fell by the wayside by 1975. The number of conversions in older housing was high: the number of units was altered in over one-third of properties over the decade before 1971. Owners added nearly one-quarter more new units, and about one in ten removed units. The processes of conversion and deconversion of versatile older houses continued, though deconversion back to single-family housing picked up somewhat in the 1970s. Governments were now supporting renovations, if only modestly; rising affluence was the basic reason for better-quality housing. Older streetcar shopping strips were improved too, though augmented by a few small malls. Within the central city, as we have already seen, the reform movement sought to protect and strengthen neighbourhoods.

In the postwar era, home-ownership in Metro rose to over 60 per cent before falling back to 50 per cent by 1970. In the central city, owning house-holders fell by one-third to only 40 per cent, partly because many who earlier would not have been classified as householders (such as roomers and boarders) were now living in apartments or flats. Households were smaller. Single people and childless couples, who now had more money to spend, could afford to upgrade.

Let us look at three kinds of neighbourhoods as defined by the people who improved their properties: southern European immigrants who took over working-class, even middle-class housing that was vacated by suburbanizing Anglos, Jews, and other earlier ethnic groups; so-called gentrifiers; and incumbent upgraders. The last is hard to distinguish from the other two: anyone fixing a property *after* occupation could be classed as an incumbent upgrader, and this was widespread. In this

sense, filtering up replaced filtering down. Yet Toronto had not experienced the process earlier to the same degree as American cities, so the filtering-up process was not really new, only stronger with rising incomes.

Ethnic Neighbourhoods

Economic opportunity and government encouragement attracted immigrants to Toronto in the largest numbers in Canada, even for some years on the continent. The major area of occupation was to the west of downtown, extending earlier immigrant reception areas (Figure 7.6). Many of the west-side working-class areas, often within walking distance of factories, turned from British to Italian, whose presence in the old city quadrupled

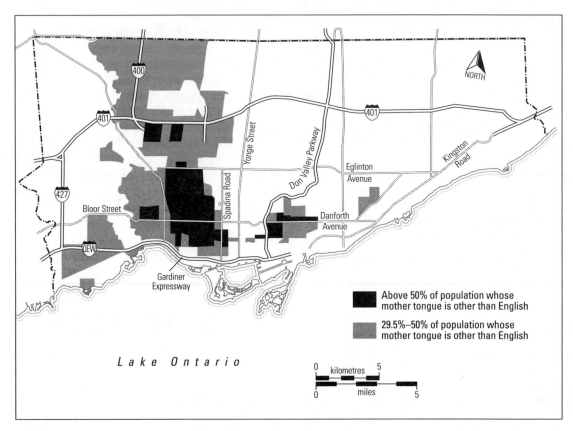

7.6   Immigrant neighbourhoods in Toronto, 1971. Jews were the most segregated, but Italians, the largest ethnic group, were still strongly defined in a corridor to the northwest. By 1970 many immigrants immediately moved into suburban housing, bypassing the old reception area west of downtown. *Source:* George Nader, *Cities of Canada* (Toronto: Macmillan, 1975–6):221.

from 1950 to 1960. Mostly refugees from rural poverty in southern Italy, they often lived on the same streets with fellow villagers. They partially replaced Jews in the Spadina district, who bolted from the city for North York in a corridor along Bathurst, where Orthodox Jews remained tightly clustered. The Italians contributed over one-third of the construction workers in building Toronto. They focused their commerce on 'Little Italy', which was established at the turn of the century. They were also conspicuous in other old, near west-side shopping strips.

Like their predecessors in Chicago, many Italians moved to the suburbs by the 1960s, mostly north into York, then North York west of the Jews, and in the next decade to the northwest in and around the old village of Woodbridge beyond Metro's boundary. They were dispersing widely too. A second more prominent Little Italy emerged at Dufferin and St Clair. As many Italians left, others took their place, some of them Anglo professionals with modest incomes. Italians had upgraded and Mediterraneanized many Victorian working-class houses; multiple earners in households made that possible. Unlike the Italian community at the turn of the century in New York, Toronto's were far less crowded and had an obviously higher standard of living; they owned their houses and kept them well. The neighbourhoods of 'such hard working people' could not be called slums.[57] They mixed with others, so they were not ghettoized.

Later arriving Portuguese immigrants also occupied houses on the west side and upgraded them, painting the old red brick houses with bright blues and purples. Some took over shops in Kensington Market (a cluster of stores, not a farmers' market), which Jews had originally created around 1910, so it became known as the Portuguese market. They would stay in the area, and were somewhat more clustered than the Italians. Ukrainians and Poles, who originally settled earlier in the century just to the west of downtown, moved westward *en masse* to the neighbourhoods of Parkdale and High Park. Most of them lived in pre-1930 housing. In contrast, the Greeks settled just east of the Don River, their shops concentrated on a strip

of Danforth Avenue. Most of the east side would, however, remain British. In 1956 Hungarians tripled their numbers, many settling in middle-income housing in the Annex, with their restaurants and shops nearby on what has been called the 'goulash strip'. Germans were scattered. Without question, the immigrant communities added vitality and stability. By 1960 the British stock recognized this, and even began celebrating the diverse neighbourhoods and shopping districts.

Visible non-European minorities were far more obvious by 1975 than they were earlier. Toronto's multicultural composition became more complex. Caribbean Blacks were scattered, though some gradually concentrated to a degree in public housing projects in the suburbs. However, families of British origin, many from the chronically depressed Atlantic provinces, predominated in public projects. Black businesses clustered on Bathurst near Bloor and on St Clair, though the adjacent residential areas could not by any stretch of the imagination be called ghettos. South Asians created a shopping cluster in the east end.

The Chinese were increasing rapidly and many settled in the turn-of-the-century immigrant Spadina neighbourhood, just to the west of downtown. The new Chinatown, a few blocks west of the 1920s' Chinatown, arose partly because of the redevelopment of part of the Ward (Toronto's long-standing slum) for the new city hall and plaza. Here was the most obvious postwar case of invasion/succession and urban renewal of mixed residential and commercial put to public use. To the west, the Chinese took over the shops hitherto run by earlier immigrants, including Jews, and converted some synagogues to churches. They mixed residentially with Whites around this cluster of shops in the still-useable 1870s' housing. It was even less of a ghetto than the Ward had been. Vietnamese shops began to appear in Chinatown, and Korean ones elsewhere.

Newly arrived Chinese were clustered in many locales by 1975: in one Chinatown east of the Don River, and in another in Scarborough at Agincourt, far from downtown. The Chinese, most of whom were from Hong Kong, became the largest immigrant group after 1975, and so were conspicuous in business and universities. Adding to the incredible mix were South Asians, Southeast Asians, Arabs, and many other European groups, who demonstrated microsegregation only here and there. A South Asian business cluster developed in the east end. Business concentrations were more obvious than residential concentrations.

To generalize, the city and increasingly the suburbs showed ethnic mixing with few signs of ghettoization. Many groups showed degrees of concentration, especially Orthodox Jews, but members of all groups were found everywhere. After learning English, the propensity of those who moved up in the world was to move, some even to staunchly British, affluent north Toronto. Institutions like churches and in some cases Catholic separate schools, food shops, and restaurants provided a degree of institutional completeness. The city too had available a wide range of community services built up over the decades; the central city was rich institutionally. Toronto was about as diverse and tolerant a place as one could find in the world; it could hardly be otherwise as two in five people in the metropolitan area were foreign born and, of those, less than one in ten were British or American in origin. Largely gone, or driven far underground, was the racism of the 1920s and early 1930s that was directed at Jews. Blacks complained (rightly sometimes) about police harassment and job discrimination, and Whites complained about Black crime, but the racial tension was nowhere near the levels in American cities.

A large part of the reason was affluence; since 1945 Canada and Toronto in particular had experienced substantial economic growth, hence most ethnic neighbourhoods improved. Although families of Italian construction workers had to live in tight quarters earlier on, often renting out rooms as did many others in high-priced Toronto, their late nineteenth-century houses and communities were well serviced, and they carefully manicured them,

adding grape arbours and vegetable plots. The working class became comfortably competent. The concepts of slums and blighted areas were only memories by 1975, as the reform impulse put them to rest. If there were concentrations of the poor (those on welfare), they were now more obvious in the new public housing projects in the suburbs.

So-Called Gentrification

Gentrification in a narrow sense referred to suburban returnees deconverting houses for multiple use back to single-family use, in some cases by gutting the house to its structure and totally refurbishing it. But 'white painting', to use a Toronto term, could apply to any new owners who grew up in the city or elsewhere, not just suburban returnees who had a higher income than the previous owner or tenants, and who might still rent out space. Of course, with rising affluence and the increase in professional white-collar jobs, gentrification could hardly have been otherwise, given that a large number of people, now in shrinking families, preferred accessible properties in still safe, mixed, and relatively sound inner-city neighbourhoods.

The professionalization of a few areas that were hitherto largely working class began after 1960, and then dramatically in only one instance, in Don Vale, a district of very diverse house sizes and home to many low-income people. Renovation companies bought up old houses in the early 1960s when they were relatively cheap. Prices began to rise markedly after 1964 largely because of limited new building and partly as a result of renovators. These companies bought up rows of century-old bay and gable houses, and gutted and sandblasted them for the professionals to buy. But just around the corner, there were still the rooming-houses for the marginal and houses occupied by working-class families who had not moved into the new public housing in Regent Park just to the

> ... since 1945 Canada and Toronto in particular had experienced substantial economic growth, hence most ethnic neighbourhoods improved.

south. The professionals were confident that their properties would appreciate.

A few old Don Vale houses became cooperatives by 1975, and some new houses were built, but it was still a pretty mixed area, sufficiently so that social service agencies still operated for the poor, and the Winchester House still served up beer by the glass. The middle class and their shopping strip gloried in the name Cabbagetown, though actually the original Cabbagetown (named after a vegetable grown in the front yards of Irish immigrants a century before) had been farther south. Professionals gradually filtered into other working-class areas, such as north Riverdale and west of the university, most of them doing some degree of renovation. In the 1980s some less-affluent residents worried that gentrification was 'spreading like a blight'—how ironic![58]

Incumbent Upgraders

Another central-city neighbourhood, the Annex, a district of a half square mile, a mile or so north and west of city hall, became attractive to new professionals. Here there was more incumbent upgrading than gentrification. It too was very mixed by income, and by 1975 it was polarized somewhat. Only one in five households were owners, though those one in five owned over half the houses, as many rented out space.

The neighbourhood had experienced substantial changes. By 1910 builders filled the district. On the margins were some earlier working-class cottages, but most of the rest were substantial three-storey, semidetached houses for the upper quarter of the income scale. In the late 1880s and early 1890s Richardsonian Romanesque houses for the rich graced major streets. By the 1920s husbands were dying off, and mature affluent children moved to the new upper-income districts like Forest Hill and Lawrence Park to the north. Many of the bigger houses were converted to apartments, rooming- and boarding-houses, (the latter persisting in Toronto longer than in US cities) and for institutional use, many for national non-profit organizations. This filtering down intensified during the Great Depression as many owners needed to take in tenants to augment incomes. The Ratepayers Association, composed of prominent industrialists and downtown lawyers, tried unsuccessfully to stem the tide. The Annex seemed headed the way of the rooming-house slum on the near north side of Chicago.

Yet a countertrend arose in the late 1930s when builders built small detached houses here and there in the large yards of big houses. This infill surely signalled an underlying confidence in older areas, despite the planning board's warning in 1944 that the district was vulnerable to decline. Wartime brought more doubling up, as war plants in Toronto enticed many to leave small towns and farms for the city. By 1945 the neighbourhood housed twice the number of people it had in 1910. Although many houses were absentee-owned, professionals and lesser businessmen (mostly Anglo and Protestant) still owned many of others. Many of them even rented out third-floor rooms, often acting in loco parentis for students and others.

In the postwar era, transformations were extensive, yet the neighbourhood remained stable. Its income was now a bit below the average of the city, but with a very wide range because there were singles as well as families. When properties turned over, private vendor-take-back mortgages predominated, a continued sign of confidence in the future of the neighbourhood, in contrast to similar districts in Detroit. Huron Street School remained a respectable place to send children, if now much more diverse ethnically as better-off ethnic residents moved in. Stability was enhanced by new professional families in the 1950s and after, sustained by the lively commercial strips on the edges, and intensified by neighbourhood organizations.

After the depression and war, not a few properties were shabby and streetscapes neglected. Like Toronto generally, some Annex residents turned out in tidy-up campaigns in the spring. In the Annex, groups started by professional married women moved vigorously to clean it up. They badgered their aldermen to step up inspection of rundown properties and tighten rooming-house rules. To ensure neighbourhood stability, they worked endlessly on the creation of the zoning system, which was passed in 1954 to make sure that the politicians did not give up too much of the Annex for redevelopment. Most of the area was indeed designated medium-density residential to fit the house styles and to confirm the social mix, permitting apartments and rooms in homes and a limited number of professional offices.

On the other hand, the residents had to allow some apartment blocks. The city blessed these, partly to increase pedestrian traffic through adjacent subway stations. Developers found the large lots with houses of the erstwhile rich easier to assemble than the usual 25-ft (8-m) lots in the area. Besides, what were the owners going to do with them? Only so many institutional uses, rooming-houses, student co-ops, and fraternity and sorority houses could take over them. Many residents were glad to see the properties put to better use than have the homes become run-down rooming-houses; after all, these apartments would attract a higher-income clientele than did many rooming-houses, and none of the activists themselves would have wanted to manage those late-Victorian monsters. A *modus vivendi* was reached in the late 1950s. The activists persuaded the city planners to undertake one of the first detailed local neighbourhood plans in the city. In 1962 they could with certainty advertise the area with the slogan, 'It's Smart to Live in the Annex.'[59]

But then developers built the apartments higher, to over twenty floors, so opposition mounted. The Annex Ratepayers Association, which was resigned earlier to the Spadina and Crosstown expressways, rejected them outright in 1969. With the help of neighbourhood planners, residents drew up a new plan after 1972, though the changes, after interminable discussion, would be relatively modest in the end. All the while, more professionals arrived and fixed up houses. Speculators altered rooming-houses to bachelorettes; that is, each room now had a little bath and kitchen; roomers no longer had to share, but they paid more. Those were affluent times. People became more tolerant too: group homes for disabled people, shelters for battered women, Natives, and even halfway houses for released criminals appeared. The neighbourhood held a fall fair, and the school held a spring fun fair. Property crimes were infrequent, murders very rare. Even if people showed a

preference for privacy by sitting in their backyards rather than on verandahs, all seemed serene in 1975, in sharp contrast to what similar neighbourhoods in nearby American cities went through.

The Annex may have lost many of its wealthy residents, but the neighbourhoods of the élite—Rosedale, Forest Hill (Crestwood Heights in a well-known 1950s study), and Lawrence Park—remained stable. All were more densely built up than similar American districts like Beverly Hills, Grosse Pointe, Shaker Heights, or obviously Scarsdale. Few of the old élite sought new postwar suburban housing, though some of the newly rich did. Before 1930 the Kingsway to the west in Etobicoke and the upper Beaches in the east end of the city attracted a modest number of the middle élite. What is most striking is that they were not separated from the rest of their cities; they paid their taxes to the wider community, not to some exclusive municipality. Forest Hill actually became part of the city in 1966.

Suburban Communities

Toronto's postwar suburbs, which were increasingly corporate-built, housed two-thirds of the metropolitan population by 1975. Many of the suburbs in the 1950s were similar to American designs using the neighbourhood unit, refined by Clarence Perry in the 1920s, which included schools, commercial plazas, and a hierarchy of open space along with a hierarchy of arterial and curvilinear, looping and cul-de-sac streets for residential areas. The houses, which were often ranch-style with picture windows and recreation rooms as in America (split-level for the more upscale), were oriented parallel to the streets on wider and shallower lots in contrast to those of prewar districts. Backyards were used for barbeques (though not many pools), but not for vegetable production, in contrast to houses in immigrant areas. These suburbs were of low density, though not as low as those of some American suburbs, as we saw earlier. In contrast to many American tracts, apartment buildings often punctuated the landscape.

The most prominent Toronto project was Don Mills, 7 mi. (11 km) from downtown. Industrialist E.P. Taylor assembled over 2,000 acres (810 ha) of rural land in the late 1940s to house 29,000 people (somewhat more than had been planned for Radburn, New Jersey, two decades before), and provided a lot of open space and land for manufacturing plants. Architect and planner, Macklin Hancock, experimented with the current principles of neighbourhood design. Focused on a major intersection that neatly allowed four precincts, each precinct had two elementary schools surrounded by lots of green space. In the central area, which was about one-eighth of the area and encircled by a ring road, Hancock set aside a shopping centre (with a department store), offices, a community centre with a skating and hockey arena and a curling rink, high schools, churches, and a series of low-rise rental apartments. In the valley he built a private golf course. Of the over 8,000 units on 800 residential acres (324 ha), more than half were to be in apartments, and one-quarter were to be in row and semidetached houses. Thus, only one-quarter would have been ranch-style bungalows. A major reason for this mix was the goal of housing many working-class people who would work in factories on the margin, similar to a European new town where people would not need to commute.

But politicians in North York (which was then still a municipal township) did not like the idea of tiny working-class row houses, and Taylor wanted too much rent from the federal government's Central Mortgage and Housing Corporation, which would have subsidized the smaller units, so the smallest houses were dropped from the plan. In the end, despite a degree of mixed styles and incomes, Don Mills was a decidedly Anglo middle-class community, which attracted people seeking to improve their status. It was a community of the 'organization man', of activists in churches, lodges, and service clubs, in common with many new suburban communities across the continent.[60] Eventually, though, some moved back to the central city, which they perceived as more lively and accessible. More than 90 per cent commuted to work outside Don Mills. The 4,500 employees of the manufacturing plants, such as IBM and later office parks, had to commute to the area. The new town concept just could not work in Toronto or North

America generally when most people had to drive. In income, though, Don Mills was more diverse than the two American new towns of Reston and Columbia. The *Architectural Forum* asserted that this 'new town ... is a planner's dream coming true'.[61]

Erin Mills was E.P. Taylor's next big venture. In 1955 he bought 8,700 acres (3,520 ha) of farm land in the Township of Toronto (later part of the City of Mississauga). Over four times the size of Don Mills, Erin Mills was built at greater densities. Others developed Bramalea on 5,200 acres (2,104 ha) northwest of Toronto. To the east on 1,700 acres (688 ha) of land assembled by the federal-provincial partnership, the provincial government created Malvern in the late 1960s for lower-middle income owners. Although private builders did the actual construction, it was a government project. Despite some public housing projects on assembled land, a majority of houses there were single-family dwellings. All of these were large assemblies. A host of other subdivisions were more limited in scale. By and large, municipalities allowed the developers and planners what they wanted, but within the tougher constraints of provincial government rules.

The Contrast

Although many of Toronto's suburbs resembled the design and spread of those south of the border, to the bane of central-city reformers, the differences were marked: Toronto's suburbs had higher residential densities generally, public housing, private apartments, many walk-up and high-rise apartments, greater use of public transit, and indeed fewer shopping centres. But differences between the central cities and Toronto's neighbouring cities on the Great Lakes were even more marked and became more so over the next twenty years.

Reformers confronted with some success the curse of Corbusier (high-rises) and that of Wright (suburbs), to extend the medium-density city of the past. They restrained the urge for skyscrapers that began with Chicago architects. They curbed the push for freeways; unlike the strident advocacy of the Southern California Automobile Association for freeways, the Ontario Motor League hardly beeped. Probably more strongly than reformers in English-speaking metropolises anywhere in the world, they endeavoured to overcome the segregation of the poor through mixed housing. Even so, their predecessors had created a far more inclusive urban environment than any south of the border. A wise province reduced the municipal fragmentation—and with it the enormous inequities and exclusiveness—that were plaguing big American metropolises, even though it moved too late to limit the expansion of low-density suburbs. A wise federal government had created universality in health care.

Even so, Torontonians would experience tougher days ahead than had the generation after the war. Renewed growth in the 1980s was, like the 1920s, largely illusory, as it was based on the deceptive hope of riches through telecommunications, luxury offices, the lifestyles of the vulgar rich, expectations of growth through the stock market, and, of course, supplying Star Wars. Less clearly so than in the 1920s, the economic decline came in stages after 1987, like thumping down the stairs step by step. By the 1990s, the gains of the twentieth century that had blessed Canada and especially Toronto were fading. The urbanization of society was completed; Toronto, like the others, just seemed to expand without purpose.

CHAPTER 8

■

Great Cities in North America: Their Past, Their Future

The weight on mankind of time and space, of physical surroundings and history—in short, of geography—is bigger than any earthbound technology is ever likely to lift.[1]

In this study, great American cities have represented their ages—Philadelphia about 1760, New York 1860, Chicago 1910, Los Angeles 1950, and Toronto 1975. No candidate on the continent seems clearly appropriate to represent the turn of the millennia, though some might argue otherwise for would-be contenders, such as Atlanta, Vancouver, Seattle, Orlando, or Las Vegas. But, following Andy Warhol, urban fame today seems to come in fifteen-minute segments, which is hardly enough time to achieve greatness. These new prominent cities, especially the last two, bespeak entertainment rather than production, ephemera rather than permanence, dissipation rather than hard work, and people looking for something to do in a world of stagnation.

This is not to deny change, but changes today seem to be on a treadmill. Technologies, even cybernetics, are not delivering speed, convenience, and comfort as they once did. The pursuit of protection is aggressive, but not productive. Many will object, nonetheless, that technology will fulfil the promises of the gurus of the information society. Before debating the future, let us briefly review how North American society and its big cities have exhibited change, but also persistence in the midst of change. It has been said that the only certainty is change, that all is in flux, yet *plus ça change, plus*

c'est la même chose. Both of these long-standing assertions have been true.

Social Changes and Persistence over 400 Years

Social change obviously occurs at different historical scales, from personal histories through the long term. Here I focus on some long-term salient social transformations. In modern times—that is, since the waning of the Middle Ages (from the Black Death onward), and especially after 1750—westerners have expected growth. Progress as an expectation increasingly picked up steam. Most people defined human betterment as material well-being, though many never entirely lost sight of religious and social ends. Immigrants flocked to the continent in waves, and the economy and cities grew. Then after 1860 population growth began to slow and consumption gradually overtook production as the chief goal of North Americans. Instead of production running at full tilt, almost autonomously, consumption was harnessed to maintain production and became the central public policy in the 1930s. One senses that the 400 years of growth have run their course. Consumption has, as it were, caught up to the capabilities of production, and the limits have been reached.

The dominant thrust of a commitment to growth has been to speed up movement, make tasks more convenient, create comfort, and foster safety. As we saw in Chapter 2, technologies through business organization and government support brought speedier and more convenient flows of information, goods, and people. For information, the telegraph was the greatest leap forward, though certainly long before that, people had tried various means to increase velocity. The telephone, a different way of sending messages, did not displace the telegraph; the computer has done that. Because information (in our society the most important of which has to do with money) has been moving at the speed of light for some time, and transactions are now automated, further gains seem unlikely. Businesses have discovered that adding more power brings diminishing returns. Besides, human beings can handle only so much information. The internets are clogged with material, most of which no one wants. Unlike the telephone, to the majority of people in their private lives, cyberspace is not relevant.

Steamships, steam trains, automobiles, trucks driven by internal combustion engines, and then aircraft dramatically raised the thresholds for speed in transporting goods and people. Refinements of convenience gradually added to their efficiency. In the 1960s a balance in use among various modes of transporting goods was virtually worked out when jet airplanes, freeways, bulk loaders, and containerization came into play. Over the long run, ships and trains were the losers, but in recent times their share of traffic has become more stabilized. Many imagine dramatic new means, but note that supersonic air travel has had only a marginal impact. Breakthrough products of high magnitude seem less likely. Companies fight over market share. The huge mergers are detrimental to workers rendered redundant in downsizing, to consumers who have subsidized oligopolies through higher prices, and to citizens who have bolstered corporations through various incentives.

The limits can be seen in other fields. Construction techniques contributed to the building of cities. The greatest gains occurred in the nineteenth century with balloon framing using two-by-four lumber for building houses, and the steel

frame that, along with the elevator, allowed office buildings to rise to new heights after 1880 (in the process overshadowing church spires, an earlier symbol of power). Prestressed concrete was the final major addition. The process of urbanization, as a great contributor to growth, was essentially complete by 1970. The addition of edge cities and further residential suburbs are only marginal. Now too there is an excess of suburban malls and offices.

How nature restricts the expansion of human capital is difficult to measure. On the one hand, knowledge seems like a bottomless pit. Creativity is quite apparent in biotechnology and writing software. Genetic analysis will occupy some people for some time, though the early pioneers probably made the largest gains. Computer chip size has reached its miniscule limit. The calls for more investment in human capital continue. On the other hand, for the running of society (that is, for most people), formal education had its greatest impact a century ago when compulsory attendance impelled most children to learn how to read and write. Obviously, along with other goals of schooling, basic skill training has to continue. The failure to raise the compulsory age beyond sixteen suggests that a limit of impact was reached by about 1920. The great expansion of universities and colleges in the 1960s probably had as much to do with holding back unemployment as human capital, its chief justification. In today's zero-sum employment situation, parents are very anxious about their children's future. Policy makers are anxious about growth. As a consequence, far too much is expected of the education system. Education was a key ingredient in economic development theory, which was so popular in the 1950s. The trouble with that theory is that it never anticipated the limits to education and stagnation in the developed world.

Nature has drawn boundaries around progress. It still provides the raw materials on which we all depend, though today North Americans who have been the most affluent on earth in the past

> **Nature has drawn boundaries around progress.**

rely more and more on other world regions for fossil fuels and metallic minerals. The continent is no longer self-sufficient. Other countries have caught up to North America, just as Canadians caught up to Americans in material wealth. But they reach the plateau of growth imposed by nature much faster than the originators of the Industrial Revolution.

As a result of these massive changes over the past few centuries, business organization and workings of the state have changed greatly. Whereas in earlier times most people worked on the land, now most work in cities, many in financial and other services, whose products are largely insubstantial. Economic growth and the drive to growth induced specialization. Specialization of tasks brought larger-scale enterprises and bureaucratization. Owners and managers had to integrate the work of many. Not surprisingly, government later became bureaucratized and large scale too because politicians had to respond to demands for subsidies and to calls for more intensive and widespread regulation over the large corporations, which, while making goods, brought about nasty side effects like pollution. The Great Depression, expressing the failure of the great corporations to deliver growth, resulted in even more massive government, not only to regulate but to spur on spending among consumers. The federal governments themselves, more so in the United States, became major consumers of military hardware. The Pentagon in Washington was the symbol and pivotal decision-making centre for America's part in the cold war.

The great changes in technology and organization gradually (if erratically) improved the material well-being of North Americans. In early colonial cities, including Philadelphia, the rich and poor were strongly divided by income, but in the countryside, farmers—along with many craftsmen, the first middle class—shared the wealth from the land more equitably, though the differences grew over time in older settled areas. As the population became more urbanized, fewer people managed their own work and more became employees. Income differences sharpened, as evident in the new rich of New York, though as development delivered more growth and more goods. With

bureaucratization, as in Chicago, a new middle class emerged to run and operate the different layers of organization. Skilled manual workers lost out to the waves of new technologies, but these new technologies created new skills. Industrial workers had a harder time keeping their incomes up, though many achieved shorter working hours in spurts. Henry Ford helped to improve their condition, although at the cost of collective rights. Still, by 1930, too few could buy from the rising tide of consumer goods. Then government began to subsidize them, so working-class people joined the mainstream of middle-class lifestyles.

The postwar era of plenty—plenty of guns and butter—began to unravel after 1970, more quickly in the United States than in Canada. Wages and welfare support began to fall behind inflation. In the early 1980s and even more so in the 1990s, workers lost jobs and security, weakening the economy even further. Companies downsized, their managers said, to remain competitive. This impulse was partly the result of mechanization of work in factories and then offices. It was also the result of production of lower-value goods and services elsewhere in the world, helped along by an enthusiasm for free trade. It seems also to have been the consequence of a drive towards custom-made goods among the now many affluent who, by the 1960s, were tired of sharing the benefits of mass production with ordinary mortals; they wanted to distinguish themselves from others, just like the élite in early colonial cities who set themselves off from the leather-aproned workers with their trappings of conspicuous finery. Productivity per worker began to fall and affluence declined. Thus social competition for status became fiercer. Finally, the top managers (mostly those born into or who married into wealth) pulled far ahead in the reward game. Prominent lawyers and accountants, the acolyte courtiers who attended them, did well. Rippling down the income scale, the status game was played out. Near the bottom, those with little accuse those with nothing of free loading. The postwar egalitarian world was, it seems, fleeting. North America seems to have become like the Europe that immigrants sought to escape—violent and divided.

Despite the rise of material progress and then its sudden recent reversal, cultural continuity remains powerful. The liberal dream still resonates through the souls of dwellers on the continent. That dream of opportunity, of owning property and achieving independence, inspired Europeans to escape the constraints of court, custom, and crop failures. Land provided the first frontier. Franklin and Jefferson spurred it on. Despite massive urbanization, the mass system of production, severe depressions, and the obvious power of corporations and government, the liberal dream has remained deeply entrenched. The phenomenon of Orange county engineers voting Republican while Washington paid their salaries through corporate aerospace employers underlines the depth of this culture. The flagrantly high salaries paid to professional sports heroes and entertainment celebrities are justified in the minds of most people as the results of hard work and commitment. The rags-to-riches stories still inspire the media. Individual success remains the cardinal virtue.

Obviously, though, no American is an island any more than anyone else. Thus communitarian urges have countered the naked and blatant side of the pursuit of success. Not everyone could take kindly to the excesses of the Horatio Algers of stock manipulators, the rotten apples in the barrel. Certainly too people have sought to participate in voluntary bodies. Neighbourliness is praised. Churches are the most obvious examples, in the United States especially, of people banding together. United Way campaigns foster a sense that business is concerned about local communities, and therefore everyone else should be too. Without doubt, these do provide a counterweight to the individualistic drive.

Still, an enormous gulf separates the belief and the reality, the reality that America is old and no longer young. Contradictions cleave the minds and hearts. First, voluntary bodies are seen as substitutes for government. Politicians tell mothers to seek day care from friends rather than government subsidies when their friends also need help. They tell those on welfare to rely on neighbours when neighbours are too busy chasing other neighbours for help. They tell those who cannot find jobs to go back where there is no employment. Ironically, those neoliberals who argue for voluntarism also argue for the commercialization of everything. The constant refrain of individualism and voluntary action drowns out the voices of those who say wait: did not the government bring the economy out of the Great Depression? Politicians say they will cut entitlements to the middle class and subsidies to business. Business people applaud. Yet because their competitors are 'grubbing with two hands' in the public trough, they do too.[2]

Then too we see the contradiction of certain churches preaching love, but excluding those who cannot stand their other, often hidden, message of pursuing success. We see too a long line of charismatic preachers cajoling the wayward into their approved means of salvation and comforting their flock while picking their pockets. Then too we see people's deep suspicion of unions because unions are the agents that destroy individual initiative. Then we note too that many unions accede to the laying off of their members rather than share work. And not least in this riven culture, the leaders are insecure in their positions in society. 'Plutocrats' complain that they are 'victims of a conspiracy raised against them by the institutions they themselves' control, like the media.[3] According to polls, encouraged by the media, most people fervently believe that America is the land of opportunity, even though most fail to make the grade. The result is a 'Jeffersonianism—the poisonous amalgam of white supremacy, state's rights, and anti-government rhetoric, which has come in both agrarian and pro-industrial forms—that has legitimated or promoted the grossest forms of racial and class inequality for centuries in the United States.'[4] Ironically, the very people who suffer the most from class inequality are often also the most boldly racist, yet they are often the most fatalistic. The bumper sticker of a few years ago proclaiming that 'God, guns and guts made America' was pure bravado, a cry from the outsider, the dispossessed.

In Canada these contradictions have not been as sharp. Unlike Americans who want the future now, Canadians give the impression of being willing to wait for a while. Those who cannot wait have migrated to the United States. (Some Americans have gone the other way.) The luxury of free land did not give Canadians the same sense of expansiveness. They are not quite, like many Americans, 'suckers for utopian promises'.[5] Governments were so conspicuously involved at every turn in nineteenth-century development that citizens could hardly not notice. Churches have been more socially conscious by preaching what has been called the 'social gospel'. Even so, Canadians have been very reluctant to accept the legitimacy of unions and of the need for equity. Beneath the collective surface is an underlying ethos that people should not rely on the state, as evidenced in recent strong shifts to the right in some elections. So they too are still drawn to the liberal dream.

Change and Continuity in Cities

An array of cities covers the continent, thinner in sparsely populated areas and thicker where agricultural settlement was originally strong. Hierarchies of urban places are obvious on maps, in our imaginations, and in our actions. Despite regional shifts in economic power, the great cities in the East are still conspicuous. Inertia has remained powerful. At this scale of cities as points on a map, North American cities are like those in the Old World with similar economies. All expedite the economic, cultural, and political life of their regions, even their country and the continent. Cities are not likely to fade away, nor are nation states. Mixed in with the popular notion that cities are the engines of growth is the idea of a new world order, of city states, running only on the fuel of transnational corporations' financial power, as if cities can be useful without hinterlands. In the past, reciprocal relations with peripheries have always been necessary. No city has survived without a hinterland to harness. The best-known city state, Singapore, is more vulnerable than cities with national hinterlands.

> **Beneath the collective surface is an underlying ethos that people should not rely on the state, as evidenced in recent strong shifts to the right in some elections.**

At the local scale, the array of activities and patterns in North American cities are both similar and different from those in Europe. Similar to Europe are the reality of central business districts, manufacturing areas, and class-divided neighbourhoods. Yet many American cities have financially weak and residentially poor central cities. They are the result of the exceptional dynamics within American society that we have associated with America's special brand of capitalism's 'creative destruction'. Less creative Canadians have not been as inclined to destroy their cities.

The original cores remain in all cities, largely on or near where the founders first established them. City halls, financial firms, and major cultural institutions have remained, even in Detroit, which has the most threadbare central district. Central areas have sloughed off many functions from earlier times, partly because these could not pay the costs of remaining and (notably in this century) because suburban locations became more attractive. In the small cities of early times, residences were often close by, even mixed with central activities. But from late eighteenth-century Philadelphia onward, business spread to adjacent zones, though with varying intensity. Then wholesalers had to occupy the periphery of the district as retailing expanded. As the cities grew larger, some retailers started up stores in suburban residential areas. In this century most retailing activity largely left the centre, more obviously so in the United States than in Canada, but not all: midtown Manhattan is still the premier shopping area on the continent. If the Detroit CBD has lost Hudson's, most other cities have still kept downtown department stores.

In recent years much has been made of edge cities as new downtowns. These clusters sport shiny, postmodern office towers, huge shopping malls, and culture emporiums. They have drawn off a great deal of the office activity from the downtown—head offices of corporation not engaged in day-to-day finance, back offices of the banks, and the like. Many believe these are the cities of the future. But these commentators need to note that Wall Street is still there, and that the Capitol has not moved to Tyson's Corner. Unlike the United States where corporations govern edge cities, in Canada the fewer edge cities are also municipal centres.

Manufacturing has almost entirely abandoned the centre. In the beginning, manufacturing was not so much in the centre as in rural areas, in the edge cities of the time, in nearby urban villages (like Germantown near Philadelphia), and in mill towns that the city later engulfed. Even within cities, early large-scale shipbuilding and sugar refining was on the periphery waterfront development because higher-value wholesaling occupied the central sites adjacent to wharves. The railways gradually drew new manufacturing along their lines outward from their central yards and stations. More powerful was the rise of the large industrial complexes of mass production, such as steel making, in creating manufacturing districts on the margins. These satellite towns, the edge cities of their day, also housed workers and local services. In more recent times, the margins are very attenuated: new auto plants have sprung up in seemingly distant greenfields. Still, they are connected by freeways and rail, the former to haul in parts, the latter to ship out vehicles. Urbanization has thus overwhelmed places in the countryside, bringing manufacturing back to rural areas in a sense as it was in the beginning. In the meantime, of course, rural dwellers had gradually taken on urban ways. Within a few years from its outset, the telegraph reached every small town, even villages, through lines strung on poles.

Residential districts in the city are more homogeneous than earlier. Differentiation on the microscale became macroscale. Because wealth was unevenly held in early cities, houses were of different size and quality. Even then, some spatial differentiation was apparent: if overstating the separation, the rich lived on the main streets, the middling on the side streets, and the poor in alleys. But the rich had country estates as they do now, and shack towns of the poor sprang up on the edge

around garbage dumps and clay pits that supplied brick makers. Poor suburbs persist today, more conspicuously so in a still very wealth-divided America.

As the cities grew, developers and builders created residential sectors based on class. Wedges of different classes extended outward, the new young households (at least over the last century) were more likely on the margins. In the era of rapid urbanization in the nineteenth century, the rich often abandoned their houses after one generation to move out farther. By 1900 some of these wedges of the rich became fixed; in Chicago the rich did not leave the inner reaches of the Gold Coast. At the other end of the scale, the immigrant poor usually occupied central areas, many occupying the old mansions of the rich. In the nineteenth century the slums expanded enormously. The middle classes had to move too as central functions expanded and the poor encroached on their residential areas. As the nineteenth century progressed, the pace of suburban subdividing opened up many more lots for the builders.

Over the past century, suburbanization has gobbled up even more land, though the pace of population growth was slower than in the previous century. Affluence and advertising begat the desire for more space per person. Following precedents early in the century, after 1945 the great community builders built neighbourhoods replete with services. Industrial workers, who were vilified earlier by employers, could now have new cars in every garage. Today many of the low-quality tracts have become suburban slums. The well-off now wall themselves off in suburban enclaves from the unlucky.

Although Canadian cities had gone through a transformation paralleling that across the border, the changes were less marked. Scholars finally rejected the notion of 'the North American city'.[6] They recognized that Canadian cities were not the same. Because governments restrained excessive expansion, suburbanization was less explosive and public transit was sustained. They allowed fewer edge cities, and the most prominent became municipal centres too. More sharply, inner-city neighbourhoods, which had never gone through nearly as strong an invasion/succession process, remained liveable. Even affluent areas were mixed by income and were not exclusive. With post-1945 affluence, residents improved nearly all of them. Unlike American cities, urban renewal programs were far more modest and few houses were abandoned. In America, urban decline and the flight of the American White middle class has often been blamed on Blacks. But long before, landlords did not keep up the quality of housing and bought off city inspectors. Besides, the case is strong for saying that financiers, developers, and government subsidies drew the middle class out from central cities. Not enough Blacks came to fill the housing that was left.

The major difference between American and Canadian cities is revealed in the inner city. This can be explained partly by stronger economic growth in Canada in this century, particularly in the thirty years before 1990. Yet Canadian urban dwellers have long expected stronger urban management. Besides, the collective impulse has been stronger as indicated by the widespread support of public health insurance. The fundamental basis for the difference, however, lies in the greater gap between rich and poor in the United States.

Now, in the face of slow growth and high public debts, Canadians have slipped into the dark side of the liberal dream: success can only be achieved in a zero-sum economy by putting others down. Even so, Canadian inner cities are not likely to go through the changes experienced in the United States, but the future is less bright.

The Future

Predicting the long or short term is dangerous, but often indulged in. Experts frequently give contradictory advice. 'Prophecies tell us more about the present than what is going to come.'[7] The information highway advocates have been predicting a grand future. On the other hand, some demographers and environmentalists who take an overall view of the world predict apocalyptic doom. As humanity has skimmed off the best nature has to offer, bleak conditions are portended. Like many others, I am more circumspect about foreseeing a wondrous time ahead, but, like most people, I would rather not think apocalyptically. Yet it is almost certain that North Americans will not experience the economic growth of the past because it seems that 'the sources of material progress have finally and irreversibly dried up'.[8] The most benign scenario is that of twentieth-century slow-growth Britain, at least until the vain attempts of the Conservatives to revive the economy, only to benefit the privileged. But the future could be worse; continued stagnation could lead to a collapse worse than the Great Depression. On hitting the bottom, we might not recover. As is often stated, it was war and cold war that prevented a return to the depressed 1930s. If extraordinary social and environmental measures are implemented, then it is possible to conceive of a steady-state society and sustainable cities with a lower level of material waste.

The optimistic visionaries bank on human ingenuity and the technology of the information society. Solutions will run ahead of impending danger, so that, for example, new substitutes will be found for material shortages. But much of the hucksterism on interactive television that will somehow foster ingenuity is surely bunk. The boss of Bell Atlantic exclaims: 'We stand on the verge of a great flowering of intellectual property, a true Renaissance that will unleash the creative energies of inventors, entrepreneurs, hackers, artists and dreamers.' Bill Gates of Microsoft says, 'It'll change everything.' Futurist George Gilder is firmly fixed on the notion that 'the human spirit—emancipated and thus allowed to reach its rarest talents and aspirations—will continue to amaze the world with heroic surprises.'[9] Symbolic analysts buy into this notion; toys for the boys are irresistible. The Toronto visionary Marshall McLuhan was right, it seems—the medium is the message in the global village. The messages do not really matter. Quality does not matter because quality cannot be measured. Indeed, does anyone know what quality is when we deal with such insubstantial symbols as Mickey Mouse?

What could the information revolution do for beleaguered cities? Telecommuting—working at home—has excited many. Look at the savings in gasoline, safety, and nerves. Everyone will be hooked into modems with screens. No centres will be needed; the people handling money will no longer have to cluster together. Manufacturing will be virtually out of sight, most of it in the sweatshops of the Third World. Through virtual reality, the proponents seem to say, one would not have to go out to restaurants. No one would have to meet with anyone else, not even sex partners. Farmers using remote-controlled robots to measure fertility, plant and harvest crops, and tend livestock could then retire to the pleasures of the information highway. This will not happen, of course, because people are gregarious; 'cyberspace is a pretense at circumventing true space, not a genuine replacement of it'.[10]

But what is true space? The glorious future will, it seems, only extend the current trend: even higher incomes for media moguls, more machinery replacing people to increase presumed productivity per worker, overtime for those with the best full-time jobs, part-time jobs for most, higher unemployment, and reduced earnings and lowered social benefit incomes for the vast majority. Future generations will not earn enough to maintain pension plans at their current levels. Incomes will become as distorted as those in Mexico or Zaire. Homelessness will be endemic. Then the manufacturing industries and trade will supply only the fortunate upper fifth or even tenth of the income scale, while the others will not be able to buy goods other than necessities, if that. Then the masses will sift through garbage dumps when even food stamps, food banks, and school breakfasts become relics of another age.

The trend could turn the United States and Canada into Third World countries, even without masses of Third World refugees. Hitherto middle-class consumers will become peasants tugging their forelocks in deference to the wealthy. The poor are coming from the South in ever larger numbers until the US and Canada so closely resemble their deteriorating homelands that they will no longer try to migrate. The apocalypse of 'the coming plague' may be nigh; it is already engulfing Africa and gradually the rest of the world.[11] For most people, the glorious future scenario will be anguished.

But this may be going too far. What other possibilities are there? For Canada and the US and other First World countries, an economic collapse like that of 1929 could be in the making. This could actually bring some kind of redress with those at the top losing, and with increased (but hopefully only temporary) misery at the bottom. The collapse could alter attitude. The Great Depression resulted in a host of (now frayed) stabilizers from social welfare to insurance to prevent financial failures. But can the corporate world go on for long before capsizing under the weight of merger deals that no longer work in a stagnant economy where the pursuit is market share? Can the city—that is, those who control the money in finance and government—sustain society any longer? If the city is to sustain society, and society is to sustain the city, then the change of consciousness among leaders and citizens will have to be extraordinary.

> If the city is to sustain society, and society is to sustain the city, then the change of consciousness among leaders and citizens will have to be extraordinary.

Ponder this scenario. Suppose a few leaders at the top of the financial pyramid suddenly realize that they can no longer run the world as they have. They have come to see that, like most people, they are prisoners of a dilemma, that of a zero-sum game where they win and everyone else loses. They are now aware that most people in America and many in Canada 'loathe government in the abstract but love it in the particular', the particular benefits and subsidies that governments bestow particularly on them.[12] They have to agree with the devastating notion that 'to allow the market mechanism to be sole director of the fate of human beings and their natural environment … would result in the demolition of society'.[13] They are now aware that Pecksniffian moralizing—so popular among the privileged, who in their zeal to reform society, hector the poor about their habits and sloth like inquisitors promoting salvation by condemning sinners and heretics—hardly works.

Taking their cue from Adam Smith, who asserted that 'People of the same trade seldom meet together, even for merriment and diversion, but the conversation ends in a conspiracy against the public, or in some contrivance to raise prices,' they decide to meet not to fleece the public but because they have the common life, with children and the environment, in mind.[14] In short, they see that North America is no longer young, and that they should follow another of Smith's concerns, empathy with a sympathy for others.

Suppose these hitherto capitalists and conservative defenders of the liberal dream put forward a framework for managing a balanced, sustainable steady-state society. Their agenda could probably include the following:

1. *Income distribution*. They would enjoin the politicians to redistribute income, not to total equality, but sufficiently so that every person has enough for the basic needs of food, shelter, and clothing, plus some modest pleasures. These leaders would start the process by cutting their own million-plus salaries to $100,000 to ensure that the after-tax differential is not more than five times between the top and bottom 1 per cent of incomes. Incomes would be high enough to let mothers stay at home if they wish.

2. *Full employment*. Instead of cutting jobs to increase productivity per worker and therefore profit, they would cut the knot tying labour to productivity. Together with union leaders, they would push for fewer working hours to reach full employment. Employers would commit their operations to hiring and training. Employment would keep people occupied and give them a sense of worth. Together with the income proposal, this would ensure that welfare was a thing of the past. Even disabled peo-

ple who could not work would have high enough incomes to hire helpers.

3. *Non-profits*. They would push governments to support small, non-profit enterprises to complement the large corporations. Those with creative instincts could be either self-employed or organized cooperatively.

4. *Election of politicians*. Ensure proportional representation so that a wide variety of people have a voice in governing. The German model of both constituency and party-slate members would work better than the 'first-past-the post' system now in place in both Canada and the United States.

5. *Debts and taxes*. They would heed Oliver Wendell Holmes's admonition in 1904 that taxes are what we pay for a civilized society. Reduce current government deficits and debts over the long term by raising capital gains and inheritance taxes, obviously more so for those at the top. Corporation taxes, which have fallen since 1950, would rise substantially. Currency exchange transaction taxes would be sensible. Unemployment insurance would be used as originally intended, to help those when readjustments are made. On private debts, they could follow the Biblical regulation of the Jubilee Year, recorded in Leviticus 25:28. Although not likely ever followed, every fifty years (shades of the long wave), creditors were to forgive debtors.

6. *Technology as servant*. Rather than chasing new technologies to keep ahead of the Joneses and to reduce employment, leaders would change technology only in increments, not in a vain attempt to increase growth or raise market share, but to optimize comfort for everyone.

7. *Reused resources*. They would lead the way in reducing reliance on non-renewable resources and, before the eradication of fish stocks, what used to be called renewable resources. Resources would be reused as much as possible. The derelict buildings in cities would be renovated. Although this would not help to boost productivity because the work is so labour intensive, it would provide employment.

8. *Living together*. Leaders would lead by moving to modest-income areas in central cities to

encourage others to follow. Affluent suburbs would include people of all incomes and races. Mixed-income areas with local leadership would be pleasant neighbourhoods to live in and open to everyone. Imagine leaders and other citizens holding street dances, together tending community gardens, and riding bicycles. One would find identity through participation. Other programmatic notions could be included to move society from the maximum of 'creative destruction' to managed 'creative construction and conservation'.

This program is fanciful, yet because one can dream of, it is not altogether impossible. It is not radical because it does not attempt to rid the world of money or even of status distinctions. Not everyone is endowed with the ability to lead. It accepts large-scale organizations. Actually, something like this could be possible. In November 1994 William J. McDonough, president of the Federal Reserve Bank of New York, brought together a group of fellow citizens—academics, business people, and journalists—to discuss the 'growing disparity in wages' that raise profound issues ... of equity and social cohesion', and whether Americans 'will be able to go forward together as a unified society ...'[15] Early in 1996 the organizers of the Davos, Switzerland, conference that brought together global business leaders and politicians also issued a warning about the consequences of social divisions. Unless the leaders, those with power, set aside their personal drive for status and slow the desperate pursuit to control resources, a permanent collapse into an age of savagery could be our lot. No one would, to borrow lines from Dylan Thomas, '... go gentle into that good night ...', but all would 'Rage, rage against the dying of the light'.

NOTES

■

Preface

1. Hochschild 1995:15–23
2. Warner 1968, [1972] 1995
3. Schumpeter 1950:83
4. Susman 1984:84

Chapter 1

1. Saporito 1994:112
2. *Globe and Mail*, 18 January 1995
3. Lowi 1969:194
4. Booth 1993:17
5. *Globe and Mail*, 15 February 1995
6. Anderson 1994:38
7. Shefter 1993
8. Paul DiMaggio, 'On Metropolitan Dominance: New York in the Urban Network' in ibid., 212
9. Harvey 1985:254
10. Rieff 1991
11. *The Economist*, 3 July, 28 August 1993
12. Pierce F. Lewis, 'The Galactic Metropolis' in Platt and Macinko 1983:23–50
13. Knox 1994:137
14. Garreau 1991
15. Phillips 1994:36
16. Useem 1984:150
17. Gibson 1993:11
18. Frey 1993:760
19. Ley 1995
20. Epstein 1992:45
21. *New York Times*, 30 April 1987
22. Beauregard 1993
23. *The Economist*, 9 July 1994
24. Anderson 1994:42
25. *Globe and Mail*, 11 October 1993
26. Guterson 1992:285
27. Ibid.
28. Danielson 1976
29. Wilson 1978: 987
30. Bell 1992:13
31. Massey and Denton 1993
32. *The Economist*, 10 September 1994
33. *Globe and Mail*, 4 March 1995
34. Frey 1995:333
35. Caraley 1992:12
36. Ibid., 1
37. Smith 1994:171–5
38. *The Economist*, 8 May 1993
39. *The Economist*, 6 November 1993, 16 April 1994
40. Toronto data is from Wendy Sweazy of the Metropolitan Toronto School Board.
41. *The Economist*, 16 April 1994
42. Rusk 1993
43. Rosenblatt 1994:501
44. Garreau 1991:185
45. Monkkonen 1988:217, italics added
46. Savitch and Thomas 1991:249
47. Caraley 1992:1
48. *The Economist*, 2 December 1995
49. *The Economist*, 11 December 1993
50. *The Economist*, 5 November 1994
51. *Business Week*, cover story, 15 August 1994
52. *The Economist*, 5 November 1994
53. Agnew 1987:130
54. *New York Times*, 20 November 1994
55. Howe and Strauss 1993:40
56. *The Economist*, 17 July 1993
57. *Globe and Mail*, 31 January 1996
58. *The Economist*, 23 April 1994
59. *Globe and Mail*, 28 June 1994
60. Madrick 1995:102
61. *Toronto Star*, 11 September 1993
62. *Toronto Star*, 3 July 1994
63. Phillips 1994:33
64. Mann 1994:15
65. Howe and Strauss 1993:45
66. *New York Times*, 20 November 1994
67. Jacobs 1961:270
68. Ibid., 291
69. Jacobs 1992:24
70. Quoted in Beauregard 1993:208
71. Lowi 1969:201
72. Jacobs 1961:300
73. Ibid., 310, italics original
74. Harrison 1994:18
75. Tocqueville [1835, 1840] 1954, vol. 1:271–2
76. Ibid., 272, 273
77. Ibid., 273
78. Jacobs 1961:316
79. Garreau 1991

Chapter 2

1. Chandler 1977
2. Bruchey 1990:418
3. Madrick 1995:172
4. Earle 1992:174
5. Tomlins 1985:326
6. Roe 1994; *The Economist*, 8 October 1994
7. *Toronto Star*, 13 June 1993
8. Card and Freeman 1993:184
9. North 1966:176
10. Hughes 1991:xiii
11. Ibid., 7
12. Aramony 1987:58
13. Beniger 1986
14. Quoted in Kasson 1976:22
15. Boswell [1791] n.d.:808
16. *The Economist*, 18 June 1994
17. Quoted in Clark 1975:61
18. Marx 1967
19. Noble 1977:8
20. Paepke 1993:99
21. Quoted in Lemon 1972:92
22. Smith 1950

23. Hughes and Allen 1988
24. Halpern 1995:227; Moore 1994
25. Horowitz 1968:22
26. Quoted in Meinig 1979:183
27. Lustig 1982:19, 21
28. Lipset 1990:150
29. *Globe and Mail*, 20 February 1995
30. Perloff et al. 1960:295
31. *Toronto Star*, 24 December 1987
32. *The Economist*, 11 February 1995
33. Quoted in Daly and Cobb 1989:109
34. *Globe and Mail*, 1 June 1993
35. The Brundtland Report, World Commission on Environment and Development 1987
36. Cobb and Cobb 1994
37. Hirsch 1976
38. Madrick 1995
39. Kennedy 1987
40. Boswell [1791] n.d.:383
41. Berry 1991; see Earle 1992:446–540
42. Hall and Preston 1988
43. Berry 1995
44. Pred 1966:15
45. Wills 1995
46. Walker 1950:329
47. Hoselitz 1960

Chapter 3

1. Burnaby [1775] 1960:54
2. Quoted in Bridenbaugh and Bridenbaugh 1942:11
3. Lemon 1972
4. Rutman 1965:22
5. Ibid., 45, 87
6. Vickers 1994:19
7. McCusker and Menard 1985
8. Doerflinger 1986
9. Rothenberg 1992
10. Meinig 1986:200
11. Scharf and Westcott 1884,

3:1806–7
12. Brown 1989:111
13. Warner 1968
14. Jones 1980:209
15. Nash 1979:257
16. Salinger and Wetherell 1985
17. Hodges 1986:3, 40
18. Nash 1979:399
19. Alexander 1980:17
20. Quoted in Nash 1979:328
21. Egnal 1975:161
22. Franklin 1964:147
23. Illick 1976:105
24. Judith Diamondstone, 'The Government of Eighteenth-Century Philadelphia' in Daniels 1978:238–63.
25. Schultz 1993:26
26. Alexander 1980:14
27. Ibid., 9
28. Klepp 1989a
29. Michael McMahon, '"Publick Service" Versus "Mans Properties": Dock Creek and the Origins of Urban Technology in Eighteenth-Century Philadelphia' in McGaw 1994:114–47.
30. Bridenbaugh and Bridenbaugh 1942:225
31. Ibid., 199
32. Bowden 1994; Schweitzer 1993
33. Warner 1968:11
34. Thompson 1989:287
35. Quoted in Bridenbaugh and Bridenbaugh 1942:21
36. Quoted in Nash 1979:268–9
37. Harris 1989; Warner 1968:15
38. Alexander 1980:21; Smith 1990
39. Lord Adam Gordon, 'Journal … 1764–1765' in Mereness 1961:367–456
40. Ibid., 412

Chapter 4

1. Lankevich and Furer

1964:90
2. Spann 1981:281
3. Quoted in ibid., 16–17
4. Soltow 1975
5. Bruchey 1990:195–6
6. Quoted in Still 1956
7. Blondheim 1994:75
8. Marvin 1988:199
9. Albion 1939:12–13
10. Hunter 1979–91, 2:411
11. Quoted in Bliss 1987:173
12. Marx 1967
13. Pred 1980:151
14. DuBoff 1983:257
15. Quoted in Pred 1980:155
16. Chandler 1977:197; Thompson 1947:98
17. Blondheim 1994:67, 100
18. Werner and Smith 1991:42
19. Myers 1931–2:201
20. David R. Meyer, 'Integration of Regional Economies, 1860–1920' in Mitchell and Groves 1987:321
21. Spann 1981:96
22. Bridges 1984:46
23. Quoted in Still 1956:133
24. Blumin 1989:1
25. Taylor 1951:296
26. Scobey 1994:218
27. Ackroyd 1990:373
28. Gilfoyle 1992:88, 103
29. Quoted in Stansell 1986:199
30. Spann 1981:384
31. Ibid., 352
32. Bridges 1984:158
33. Spann 1981:53
34. Ibid., 333
35. Ibid., 297
36. Albion 1939:213
37. Spann 1981:351
38. Mikkelson 1898:33–4; Buttenweiser 1987
39. Moehring 1981:17
40. Spann 1981:284
41. Still 1956:197
42. Spann 1981:117
43. Quoted in Tebbel and Zuckerman 1991:19
44. Eric Monkkonen, 'Nineteenth-Century Institutions:

Dealing with the Urban "Underclass'" in Katz 1993:334–65
45. Katz et al. 1982:14
46. Spann 1981:75
47. Quoted in Stansell 1986:209
48. Quoted in Mikkelson 1898:34–5
49. Porter 1931:940
50. Blackmar 1989:222
51. Still 1956:57
52. Spann 1981:351
53. Quoted in ibid., 3
54. Ibid., 4, 10
55. Quoted in Still 1956:137
56. Boyer 1985:52
57. Spann 1981:98
58. Martyn Bowden, 'Growth of Central Districts in Large Cities' in Schnore 1975:83
59. Spann 1981:97
60. Albion 1939:266
61. Boyer 1985:89, 92
62. Ibid.,43
63. Quoted in ibid.,90
64. Chandler 1977:225–6
65. Boyer 1985:67
66. Gilfoyle 1992:31, 33
67. Halttunen 1982:198
68. Wilentz 1984:30
69. Quoted in Blumin 1989: 111
70. Spann 1981:347
71. Stott 1990:29
72. Spann 1981:403
73. Stott 1990:52–3
74. Pessen 1973:179
75. Quoted in Boyer 1985:45
76. Marsh 1990:14
77. Scherzer 1992:286
78. Spann 1981:108
79. Lubitz 1970:96
80. Boyer 1985:50
81. Spann 1981:16
82. Ibid., 17
83. Ibid., 420
84. Still 1956:78
85. Spann 1981:314
86. Plunz 1989:58
87. Quoted in Spann 1981:42
88. Ibid., 177

Chapter 5

1. Quoted in Pierce 1993:430
2. Hines 1974:336
3. Briggs 1968:56
4. Quoted in Condit 1973:73
5. Steffens [1904] 1957:163
6. Pierce 1933:251
7. Flanagan 1996:163
8. Bruchey 1990:310
9. Eric Lampard, 'Introductory Essay' in Taylor 1991:373
10. Chandler 1977
11. Horowitz 1976:14, 90
12. Leach 1993
13. Jaher 1982:477
14. Ronald Abler, 'The Telephone and the Evolution of the American Metropolitan System' in Pool 1977:334
15. Jaher 1982:473, 489
16. Ibid., 477, 490
17. Pierce 1933:438
18. Quoted in Hogan 1985:2
19. Zueblin 1916:v
20. Ibid., v
21. Ibid., 36
22. Ibid., 62
23. Ibid., v, vi, 36, 62
24. Quoted in Jaher 1982:474
25. Klein and Kantor 1976:74
26. Pierce 1933:401
27. Quoted in Hogan 1985:21
28. Ibid., 20, 31
29. Einhorn 1991:203
30. Quoted in Pierce 1933:412
31. Hogan 1985:16
32. Ibid.
33. Ibid., 95
34. Boyer 1978:190
35. Jaher 1988:504
36. Hogan 1985:14
37. Teaford 1984
38. Keating 1988:114
39. Quoted in Gilbert 1991:209
40. Platt 1991
41. Jaher 1982:481
42. Quoted in Keating 1988:76
43. Harris 1996
44. Weiss 1987
45. Hines 1974:331

46. Rosen 1986:146
47. Condit 1964:79
48. Quoted in Pierce 1933:236
49. Mayer and Wade 1969:218
50. Wendt and Kogan 1952
51. Quoted in Pierce 1933:448
52. Ibid.
53. Quoted in ibid., 421
54. Condit 1864:165
55. Barth 1980:146
56. Pierce 1933:251
57. Condit 1973:58
58. Quoted in Pierce 1933:431
59. Quoted in Mayer and Wade 1969:215–16
60. Pierce 1933:289
61. Quoted in ibid., 436
62. Mayer and Wade 1969:216
63. Barrett 1983:18
64. Taylor [1915] 1970:165
65. McKenzie 1933:218
66. Pierce 1933:426
67. Doucet and Weaver 1991:103
68. Baumann 1980:212
69. If in 1930 one-third of Chicagoans lived in two-unit buildings, then let us assume that half of these (16.5 per cent) were owning families, while the other half rented from them. If the ownership level was 31.4 per cent as reported in the census, then 14.9 would have lived in single-family structures. Conversely, of the 68.6 who rented, 52.1 would have lived in buildings with more than one rental unit. Needless to say, this is speculative arithmetic, but will we ever know for certain?
70. Abbott 1936:368
71. Hogan 1985:114
72. Quoted in Pierce 1933:398
73. Quoted in Keating 1988:54
74. Quoted in Pierce 1933:399
75. Ibid., 409
76. Ibid., 430

77. Quoted in Slayton 1986:31
78. Hogan 1985:28
79. Philpott 1978:141
80. Gilbert 1991:217
81. Schiesl 1977:189
82. Teaford 1993:198
83. Jaher 1982:508
84. Ibid., 538
85. Ibid.
86. Quoted in Cronon 1991:356
87. Teaford 1993:149

Chapter 6

1. Scott 1949:1
2. Nadeau 1960:279
3. Quoted in Fogelson 1967:3
4. David Fine, 'Introduction' in Fine 1984:11
5. Quoted in Werner Hirsch, 'Los Angeles Leading City? A Leading Question' in Hirsch 1971:237
6. Quoted in Cleland 1947:212
7. Quoted in Weiss 1987:82
8. Robbins and Tilton 1941:6
9. Scott 1949:49
10. Leuchtenburg 1958:245
11. R. Marchand 1986:239
12. Thurow 1976:110
13. Leuchtenburg 1958:248
14. Bernstein 1987:212
15. Jaher 1982:729
16. Nadeau 1960:158–9
17. Ibid.
18. Viehe 1991
19. Davis 1990:117
20. Cleland 1947:271
21. Ibid., 273
22. Scott 1991:441
23. Nash 1992:144
24. Davis 1990:392
25. Minsky 1965:89–91
26. Ibid., 117, 129, 155
27. Ibid.
28. Jaher 1982:649–50
29. Ibid., 675
30. Fogelson 1967:191
31. Starr 1990:62

32. Jaher 1982:719
33. Ibid., 615
34. Davis 1990:101
35. Gottlieb and Wolt 1977:236
36. Hopkins 1972:153
37. Ibid.
38. Stimson 1955:427, 429
39. Gottlieb and Wolt 1977:224
40. Cleland 1947:277
41. Carr [1935] 1955:264–5
42. Jaher 1987:655
43. Ibid., 675
44. Fogelson 1967:192
45. Hirsch 1971:28
46. Davis 1990:331
47. Nadeau 1960:237
48. Finney 1945:99
49. Jaher 1982:685
50. Ibid., 683
51. Ibid., 668
52. Ibid.
53. Gottlieb and Wolt 1977:278
54. Davis 1990:251
55. Ibid., 295
56. Cleland 1947:312
57. Ibid., 260
58. Liahna K. Babaner, 'Raymond Chandler's City of Lies' in Fine 1984:110
59. Ibid., 123
60. Quoted in Reid 1992:xx
61. Basso 1950:26, 45
62. Hirsch 1971:101
63. Hundley 1992:168
64. Cleland 1947:183
65. Crouch and Dinerman 1963:367
66. Ibid., 220
67. Miller 1981
68. Crouch and Dinerman 1963:409
69. Weiss 1987:26
70. Fogelson 1967:249
71. Ibid., 249
72. Weiss 1987:145
73. Quoted in Robbins and Tilton 1941:9
74. Ibid., 254
75. Scott 1949:165
76. Ibid.
77. Ibid., 181

78. John C. Ries and John J. Kirlin, 'Government in the Los Angeles Area: The Issue of Centralization and Decentralization' in Hirsch 1971:92
79. Theodore Lowi quoted in ibid., 93
80. Nadeau 1960:193
81. Ostrom 1953:26
82. Davis 1990:196
83. Nelson 1983:97
84. David R. Reynolds, 'Progress toward Achieving Efficient and Responsive Spatial-Political Systems in Urban America' in Adams 1976:514–15
85. Scott 1949:133, 131
86. Ibid., 132
87. Banham 1971:216
88. Bottles 1987:172
89. Ibid., 109
90. Ibid., 129
91. Ibid., 141
92. Ibid., 215
93. Ibid., 213
94. Scott 1949:95
95. Mumford 1961:510
96. Scott 1949:252
97. Banham 1971:175
98. Weiss 1987
99. Ibid., 66
100. Nadeau 1960:155
101. Fogelson 1967:257
102. Quoted in Toll 1969:207
103. Scott 1949:89, 110
104. Nadeau 1960:150
105. Ibid., 145
106. Eichler 1982:55
107. Ibid., 152
108. Ibid., 152–3
109. Ibid., 47
110. Ibid., 9
111. Zierer 1956:267
112. Eichler 1982:75
113. Colean 1975:75–6
114. Bacher 1993
115. Robbins and Tilton 1941:173
116. Scott 1971:192
117. Weiss 1987:104

118. Davis 1990:165
119. Ibid., 164
120. B. Marchand 1986:31
121. Ward 1971:158, 162 (Burgess map)
122. Bartholomew 1932:43; McKenzie 1933:218
123. Baar 1992:39; Dennis 1994:305
124. Marsh 1990:131
125. Jackson 1985:196
126. Ibid., 200
127. Scott 1971:322
128. Jackson 1985:198
129. Bauer 1934:129
130. Goulden 1976:237
131. Funigiello 1978:205
132. Robbins and Tilton 1941:192
133. Scott 1942:98
134. Klein and Schiesl 1990:18
135. Alexander and Bryant 1951:vii
136. Fishman 1987:156; Banham 1971:161
137. Steiner 1981
138. Hancock 1949:175
139. Quoted in Cottrell and Jones 1952:47
140. Nicolaides 1993:365
141. Goulden 1976:112
142. Nicolaides 1993:368
143. Collins 1980:24
144. Fishman 1987

Chapter 7

1. Lemon 1985:11
2. Dickens [1842] 1908:202–3
3. Lemon 1985:17
4. Atwood 1972
5. Brian J.L. Berry, 'Inner City Futures: An American Dilemma' in Bourne 1982:562
6. Ibid., 582
7. Markusen et al. 1991
8. Smith 1973:100
9. Teaford 1993:220
10. *Globe and Mail*, 1 August 1987
11. Porter 1965
12. Newman 1975:218
13. Quoted in Hayes 1992:5
14. Merrijoy Kelner, 'Changes in Toronto's Élite Structure' in Mann 1970:197
15. Kaplan 1982:608
16. Horowitz 1968:22
17. Lemon 1985:57
18. Lipset 1990:114
19. Fetherling 1994:130, 135
20. Arnstein 1960
21. Fetherling 1994:128
22. Higgins 1977:64
23. Ladd and Yinger 1989:121–2
24. Quoted in Brecher and Horton 1993:9
25. Elkin 1987:131
26. Lowi 1969:201
27. Chudacoff and Smith 1988:299
28. Lowi 1969:196
29. Toll 1969:296; Gottdeiner 1977:26
30. Scott 1971:516
31. Fried 1971:133
32. Armstrong and Nelles 1986:141
33. Lemon 1985:17
34. Quoted in Lemon 1985:77
35. Warren Magnusson, 'Toronto' in Magnusson and Sancton 1983:98
36. Lemon 1989
37. Colton 1980:82
38. *Globe and Mail*, 1 November 1969.
39. Colton 1980:166
40. Goldberg and Mercer 1986:152
41. Frisken 1991:268
42. Lemon 1985:68
43. Sewell 1993:105
44. Berry 1982:562
45. Bradbury, Downs, and Small 1982:42
46. Quoted in Mayer 1978:184
47. David Seligman, 'The Enduring Slums' in Editors of Fortune 1958:92
48. Quoted in Bacher 1993:220
49. Mollenkopf 1983:3
50. Martin Anderson, 'The Federal Bulldozer' in Wilson 1966:491
51. Elkin 1987:90
52. Bourne 1967:182
53. Smith and Williams 1986:22
54. Abrams 1965
55. Berger 1960:93
56. Dear and Wolch 1987
57. Iacovetta 1992
58. Caulfield 1994:202
59. Johnson 1962
60. Clark 1966:92
61. Quoted in Sewell 1993:93

Chapter 8

1. *The Economist*, 30 July 1994
2. Savoie 1990:290
3. Lapham 1995:33
4. Lind 1995:373
5. Hughes 1995:76
6. Goldberg and Mercer 1986
7. Hughes 1995:76
8. Paepke 1993:viii
9. *The Economist*, 12 February 1994
10. *The Economist*, 30 July 1994
11. Garrett 1995
12. *The Economist*, 27 January 1996
13. Polanyi 1957:72–3
14. Smith [1776] 1970:232
15. Cassidy 1995:113

SELECTED BIBLIOGRAPHY

Abbott, Carl. 1974. 'The Neighborhoods of New York, 1760–1775'. *New York History* 55:35–54.

———. 1987. *The New Urban America: Growth and Politics in Sunbelt Cities.* Chapel Hill: University of North Carolina Press.

Abbott, Edith. 1936. *The Tenements of Chicago 1908–1935.* Chicago: University of Chicago Press.

Abrams, Charles. 1965. *The City Is the Frontier.* New York: Harper and Row.

Ackroyd, Peter. 1990. *Dickens.* London: Sinclair-Stevenson.

Adams, John S., ed. 1976. *Urban Policy Making and Metropolitan Dynamics: A Comparative Geographical Analysis.* Cambridge: Ballinger.

———. 1987. *Housing Americans in the 1980s.* New York: Sage.

Adams, Thomas, Harold M. Lewis, and Theodore T. McCrosky. 1929. *Population, Land Values and Government, Volume 2, Regional Survey.* New York: Regional Plan of New York and Its Environs.

Addams, Jane. 1910. *Twenty Years at Hull House.* New York: Macmillan.

Adler, Gerald M. 1971. *Land Planning by Administrative Regulation: The Policies of the Ontario Municipal Board.* Toronto: University of Toronto Press.

Agnew, John. 1987. *The United States in the World Economy: A Regional Geography.* Cambridge: Cambridge University Press.

Albion, Robert G. 1939. *The Rise of New York Port, 1850–1860.* New York: Charles Scribners Sons.

Alexander, John K. 1980. *Render Them Submissive: Responses to Poverty in Philadelphia, 1760–1800.* Amherst: University of Massachusetts Press.

Alexander, Robert G., and Drayton Bryant. 1951. *Rebuilding a City: A Story of the Redevelopment Patterns in Los Angeles.* Los Angeles: Haynes Foundation.

Allinson, Edward P., and Boies Penrose. 1887. *Philadelphia 1681–1887: A History of Municipal Development.* Philadelphia: Allen, Lane and Scott.

Altman, Morris. 1988. 'Economic Growth in Canada, 1645–1739: Estimates and Analysis'. *William and Mary Quarterly* 44:684–711.

Anderson, Kurt. 1994. 'Las Vegas, USA.' *Time* 142 (10 June):36–43.

Aramony, William. 1987. *The United Way: The Next Hundred Years.* New York: Donald I. Fine.

Arkes, H. 1981. *The Philosopher and the City: The Moral Dimensions of Urban Politics.* Princeton: Princeton University Press.

Armstrong, Christopher, and H.V. Nelles. 1986. *Monopoly's Moment: The Organization and Regulation of Canadian Utilities, 1830–1930.* Philadelphia: Temple University Press.

Arnold, Robert K., et al. 1961. *The California Economy 1947–1980.* Menlo Park, CA: Stanford Research Institute.

Arnstein, Sherry R. 1960. 'Eight Rungs on the Ladder of Citizen Participation'. *Journal of the American Institute of Planners* 35:216–24.

Artibise, Alan F.J., and Gilbert A. Stelter, eds. 1979. *The Usable Urban Past: Planning and the Politics in the Modern Canadian City.* Toronto: Macmillan.

Atwood, Margaret. 1972. *Survival: A Thematic Guide to Canadian Literature.* Toronto: Anansi.

Baar, Kenneth. 1992. 'The National Movement to Halt the Spread of Multifamily Housing, 1890–1926'. *Journal of the American Planning Association* 58:39–48.

Bacher, John C. 1993. *Keeping to the Marketplace: The Evolution of Canadian Housing Policy.* Montreal and Kingston: McGill-Queens University Press.

Bailyn, Bernard, and Barbara DeWolfe. 1986. *Voyages to the West: A Passage in the Peopling of America on the Eve of the Revolution.* New York: Knopf.

Bairoch, Paul. 1988. *Cities and Economic Development: From the Dawn of History to the Present.* Chicago: University of Chicago Press.

Balakrishnan, T.R. 1982. 'Changing Patterns of Ethnic Residential Segregation in the Metropolitan Areas of Canada'. *Canadian Review of Sociology and Anthropology* 19:92–110.

Ball, Norman R., ed. 1991. *Building Canada.* Toronto: University of Toronto Press.

Baltzell, E. Digby. 1958. *Philadelphia Gentlemen: The Making of a National Upper Class.* Glencoe: Free Press.

Banfield, Edward C. 1968. *The Unheavenly City: The Nature and Future of Our Urban Crisis.* Boston: Little, Brown.

Banham, Reyner. 1971. *Los Angeles: The Architecture of Four Ecologies.* Harmondsworth: Penguin.

Barnet, Richard J., and John Cavanaugh. 1994. *Global Dreams: Imperial Corporations and the New World Order.* New York: Simon & Schuster.

Barrett, Paul. 1983. *The Automobile and Urban Transit: The Formation of Public Policy in Chicago 1900–1930.* Philadelphia: Temple University Press.

Barrows, Robert G. 1983. 'Beyond the Tenement: Patterns of American Urban Housing 1870–1930'. *Journal of Urban History* 9:395–407.

Barth, Gunther. 1980. *City People: The Rise of Modern City Culture in Nineteenth-Century America.* New York: Oxford University Press.

Bartholemew, Harland. 1932. *Urban Land Uses: An Aid to Scientific Zoning Practice.* Cambridge: Harvard University Press.

Basso, Hamilton. 1950. 'Los Angeles'. *Holiday* (January):26–45.

Bauer, Catherine. 1934. *Modern Housing.* Boston: Houghton Mifflin.

Bauman, John F. 1980. 'Housing the Urban Poor'. *Journal of Urban History* 6:211–20.

Beauregard, Robert A. 1993. *Voices of Decline: The Postwar Fate of US Cities.* Oxford: Blackwell.

Beito, David T. 1989. *Taxpayers in Revolt: Tax Resistance During the Great Depression.* Chapel Hill: University of North Carolina Press.

Bell, Derrick. 1992. *Faces at the Bottom of the Barrel: The Permanence of Racism.* New York: Basic Books.

Beniger, James R. 1986. *The Control Revolution: Technological and Economic Origins of the Information Society.* Cambridge: Harvard University Press.

Berger, Bennett. 1960. *Working-Class Suburb: A Study of Auto Workers in Suburbia.* Berkeley: University of California Press.

Berman, Marshall. 1988. *All That Is Solid Melts in Air: The Experience of Modernity.* New York: Viking Penguin.

Bernstein, Iver. 1990. *The New York City Draft Riots: Their Significance for American Society and Politics in the Age of the Civil War.* New York: Oxford University Press.

Bernstein, Michael A. 1987. *The Great Depression: Delayed Recovery and Economic Change in America, 1929–1939.* Cambridge: Harvard University Press.

Berry, Brian J.L. 1991. *Long-Wave Rhythms in Economic Development and Political Behavior.* Baltimore: Johns Hopkins University Press.

_____. 1995. 'Long Swings in American Inequality: The Kuznets Conjecture Revisited'. *Papers in Regional Science* 74:153–74.

Biggar, Richard, and James D. Kitchen. 1952. *How the Cities Grew: A Century of Municipal Independence and Expansionism in Metropolitan Los Angeles.* Los Angeles: Bureau of Governmental Research, Haynes Foundation.

Binford, Henry. 1985. *The First Suburbs.* Chicago: University of Chicago Press.

Bird, Richard M., and N. Enid Slack. 1993. *Urban Public Finance in Canada,* 2nd ed. Toronto: Wiley.

Bish, Robert. 1971. *The Public Economy of Metropolitan Areas.* Chicago: Markham.

Blackmar, Elizabeth. 1989. *Manhattan for Rent, 1785–1850.* Ithaca: Cornell University Press.

Blake, Nelson Manfred. 1956. *Water for the Cities: A History of the Urban Water Supply Problem in the United States.* Syracuse: Syracuse University Press.

Bliss, Michael. 1987. *Northern Enterprise: Five Centuries of Canadian Business.* Toronto: McClelland and Stewart.

Blondheim, M. 1994. *News Over the Wires: The Telegraph and the Flow of Public Information in America, 1844–1897.* Cambridge: Harvard University Press.

Blumenfeld, Hans. 1967. *The Modern Metropolis: Its Origins, Growth, Characteristics, and Planning.* Cambridge: MIT Press.

Blumin, Stuart M. 1989. *The Emergence of the Middle Class Social Experience in the American City, 1760–1900.* New York: Cambridge University Press.

Booth, Cathy. 1993. 'Miami!' *Time* 142 (6 September):14–21.

Borchert, John. 1967. 'American Metropolitan Evolution'. *Geographical Review* 57:301–32.

Boswell, James. [1791] (n.d.) *The Life of Johnson.* New York: Modern Library.

Bothwell, Robert. 1992. *Canada and the United States: The Politics of Partnership.* Toronto: University of Toronto Press.

Bottles, Scott L. 1987. *Los Angeles and the Automobile: The Making of the Modern City.* Berkeley: University of California Press.

Bourne, Larry S. 1967. *Private Redevelopment of the Central City: Spatial Processes of Structural Changes in the City of Toronto.* Chicago: Department of Geography, University of Chicago.

_____. 1981. *The Geography of Housing.* London: Arnold.

_____, ed. 1982. *Internal Structure of the City: Readings on Urban Form, Growth and Policy*, 2nd ed. New York: Oxford University Press.

_____. 1993. 'Close Together and Worlds Apart: An Analysis of the Changes in the Ecology of Income in Canadian Cities'. *Urban Studies* 30:1293–1317.

_____, and David Ley, eds. 1993. *The Changing Social Geography of Canadian Cities.* Montreal and Kingston: McGill-Queens University Press.

_____, R.D. MacKinnon, and J.W. Simmons, eds. 1973. *The Form of Cities in Central Canada: Selected Papers.* Toronto: Department of Geography, University of Toronto.

Bowden, Martyn J. 1994. 'The Mercantile City in North America: Theory and Reality'. Paper presented at the Annual Meeting of the Association of American Geographers.

Bowman, Lynn. 1974. *Los Angeles: Epic of a City.* Berkeley: Howell North.

Boyer, Christine. 1983. *Dreaming the Rational City: The Myth of American City Planning.* Cambridge: MIT Press.

_____. 1985. *Manhattan Manners: Architecture and Style, 1850–1900.* New York: Rizzoli.

Boyer, Paul. 1978. *Urban Masses and Moral Order in America, 1820–1920.* Cambridge: Harvard University Press.

Bradbury, Katherine L., Anthony Downs, and Kenneth A. Small. 1982. *Urban Decline and the Future of American Cities.* Washington: Brookings Institution.

Brecher, Charles, and Raymond D. Horton. 1993. *Power Failure: New York City Politics and Policy Since 1960.* New York: Oxford University Press.

Breton, Raymond, and Jeffrey Reitz. 1994. *The Illusion of Difference: Realities of Ethnicity in Canada and the United States.* Montreal: C.D. Howe Institute.

Bridenbaugh, Carl. 1950. *The Colonial Craftsman.* New York: New York University Press.

_____. 1955. *Cities in Revolt: Urban Life in America, 1743–1776.* New York: Knopf.

_____, and Jessica Bridenbaugh. 1942. *Rebels and Gentlemen: Philadelphia in the Age of Franklin.* New York: Oxford University Press.

Bridges, Amy. 1984. *A City in the Republic: Antebellum New York and the Origins of Machine Politics.* Cambridge: Cambridge University Press.

Briggs, Asa. 1968. *Victorian Cities.* Harmondsworth: Penguin.

Brilliant, Eleanor L. 1990. *The United Way: Dilemmas of Organized Charity.* New York: Columbia.

Britton, John N.H., ed. 1996. *Canada and the Global Economy.* Montreal/Kingston: McGill-Queen's University Press.

Broadway, Michael J. 1989. 'A Comparison of Patterns of Urban Deprivation between Canadian and US Cities'. *Social Indicators Research* 21:531–51.

Bromley, N.K. 1994. *Law, Space and the Geographics of Power.* New York: Guildford Press.

Brown, Ralph H. 1948. *Historical Geography of the United States.* New York: Harcourt Brace.

Brown, Richard A. 1989. *Knowledge Is Power.* New York: Oxford University Press.

Bruchey, Stuart. 1990. *Enterprise: The Dynamic Economy of a Free People.* Cambridge: Harvard University Press.

Brunn, Stanley D., and Thomas R. Leinbach. 1991. *Collapsing Space and Time: Geographic Aspects of Communication and Information.* London: HarperCollins Australia.

Buder, Stanley. 1969. *Pullman: An Experiment in Industrial Order and Community Planning 1880–1930.* New York: Oxford University Press.

Bunting, Trudi, and Pierre Filion, eds. 1991. *Canadian Cities in Transition.* Toronto: Oxford University Press.

Burnaby, Andrew. [1775] 1960. *Travels through the Middle Settlements in North America in the Years 1759 and 1760*, 2nd ed. Ithaca: Cornell University Press.

Burnham, Daniel H., and Edward H. Bennett. [1909] 1970. *Plan of Chicago*, edited by Charles Moore. New York: Da Capo.

Burt, Nathaniel. 1963. *The Perennial Philadelphians: The Anatomy of an American Aristocracy.* Boston: Little, Brown.

Burton, Lydia, and David Morley. 1979. 'Neighborhood Survival in Toronto'. *Landscape* 23, 3:33–40.

Business Week. 1994. 'Inequality: How the Gap Between Rich and Poor Hurts the Economy' (15 August).

Buttenweiser, Ann L. 1987. *Manhattan Water-Bound: Planning and Developing Manhattan's Waterfront from the Seventeenth Century to the Present.* New York: New York University Press.

California. [1939] 1984. *The WPA Guide to California.* New York: Pantheon Books.

Canada Bureau of Census. 1876. *Census of Canada 1870–1871*, vol. 4. Ottawa: Canada Bureau of Census.

Canada Mortgage and Housing Corporation. 1946. *Greater Toronto Housing Atlas.* Ottawa: Canada Mortgage and Housing Corporation.

Canadian Institute of Planners. 1994. *1919–1994: Special Edition of Plan Canada.* Ottawa: Canadian Institute of Planners.

Caplow, Theodore, et al. *Recent Social Trends in the United States 1960–1990.* Montreal and Kingston: McGill-Queens University Press.

Cappon, Lester J., et al., eds. 1976. *Atlas of Early American History: The Revolutionary Era.* Princeton: Princeton University Press for The Newberry Library and Institute of Early American History and Culture.

Caraley, Demetrius. 1992. 'Washington Abandons the Cities'. *Political Science Quarterly* 107:1–30.

Card, David, and Richard B. Freeman, eds. 1993. *Small Differences That Matter: Labor Markets and Income Maintenance in Canada and the United States.* Chicago: University of Chicago Press.

Careless, J.M.S. 1984. *Toronto to 1918: An Illustrated History.* Toronto: Lorimer; Ottawa: Museum of Civilization.

_____. 1989. *Frontier and Metropolis: Regions, Cities, and Identities in Canada before 1914.* Toronto: University of

Toronto Press.

Carr, Harry. [1935] 1955. *Los Angeles: City of Dreams.* New York: Grosset & Dunlap.

Carver, Humphrey. 1962. *Cities in the Suburbs.* Toronto: University of Toronto Press.

Cassidy, John. 1995. 'Who Killed the Middle Class?' *New Yorker* (16 October):113–24.

Castells, Manuel. 1983. *The City and the Grassroots: A Cross-cultural Theory of Urban Social Movements.* Berkeley: University of California Press.

Caulfield, John. 1994. *City Form and Everyday Life: Toronto's Gentrification and Critical Social Practice.* Toronto: University of Toronto Press.

Chandler, Alfred D. 1977. *The Visible Hand: The Managerial Revolution in American Business.* Cambridge: Harvard University Press.

Choko, Marc. 1994. 'La "Boom" des immeubles d'appartements à Montreal de 1921 à 1951'. *Urban History Review* 23:3–18.

Chudacoff, Howard P., and Judith E. Smith. 1988. *The Evolution of American Urban Society*, 3rd ed. Englewood Cliffs: Prentice-Hall.

City of Chicago. 1975. *Population of the City of Chicago Per Square Mile 1900–1970.* Chicago: Department of Development and Planning.

Ciucci, Giorgio, et al. 1979. *The American City: From the Civil War to the New Deal.* Cambridge: MIT Press.

Clark, Dennis, ed. 1975. *Philadelphia: 1776–2076, a Three Hundred Year View.* Port Washington: Kennikat.

Clark, Gordon L. 1985. *Judges and the Cities: Interpreting Local Autonomy.* Chicago: University of Chicago Press.

Clark, Kenneth B. 1965. *Dark Ghetto: Dilemmas of Social Power.* New York: Harper & Row.

Clark, S.D. 1966. *The Suburban Society.* Toronto: University of Toronto Press.

Cleland, Robert G. 1947. *California in Our Time.* New York: Knopf.

Cobb, Clifford W., and John B. Cobb, Jr. 1994. *Green National Product: A Proposed Index of Sustainable Economic Welfare.* Lanham: University Press of America.

Cochran, Thomas C., and William Miller. 1961. *The Age of Enterprise: A Social History of Industrial America*, rev. ed. New York:

Coclanis, Peter A. 1990. 'The Wealth of British America on the Eve of the Revolution'. *Journal of Interdisciplinary History* 21:245–60.

Colean, Miles L. 1953. *Renewing Our Cities.* New York: Twentieth Century Foundation.

_____. 1975. *A Backward Glance in Oral History: The Growth of Government Housing Policy in the United States 1934–1975.* Washington: Mortgage Bankers Association of America

Collins, Keith. 1980. *Black Los Angeles: The Maturing of the Ghetto 1940–1950.* Saratoga, CA: Century 21 Publishing.

Colton, Joel, and Stuart Bruchey, eds. 1987. *Technology and Society: The American Experience.* New York: Columbia University Press.

Colton, Timothy. 1980. *Big Daddy: Frederick C. Gardiner and the Building of Metropolitan Toronto.* Toronto: University of Toronto Press.

Condit, Carl W. 1952. *The Rise of the Skyscraper.* Chicago: University of Chicago Press.

_____. 1964. *The Chicago School of Architecture: A History of Commercial and Public Buildings in the Chicago Area.* Chicago: University of Chicago Press.

_____. 1973. *Chicago 1910–1929: Building, Planning and Urban Technology.* Chicago: University of Chicago Press.

_____. 1980. *The Port of New York, Volume 1, A History of the Rail and Terminal System from the Beginning to Pennsylvania Station.* Chicago: University of Chicago Press.

Conzen, Michael P. 1977. 'The Maturing Urban Systems in the United States, 1840–1910'. *Annals of the Association of American Geographers* 67:88–108.

_____. 1981. 'The American Urban System in the Nineteenth Century'. *In Geography and the Urban Environment: Progress in Research and Applications*, vol. 4, edited by D.T. Herbert and R.J. Johnston, 295–347. New York: Wiley.

Cose, Ellis. 1993. *The Rage of the Privileged Class.* New York: Harper Collins.

Cottrell, Edwin A., and Helen Jones. 1952. *Metropolitan Los Angeles: Characteristics of the Population.* Los Angeles: Haynes Foundation.

Cray, Robert E. 1988. *Paupers and Poor Relief in New York City and Its Rural Environs.* Philadelphia: Temple University Press.

Cronon, William. 1991. *Nature's Metropolis: Chicago and the Great West.* New York: Norton.

Crook, Harold. 1993. *Giants of Garbage: The Rise of the Global Waste Industry and the Politics of Pollution Control.* Toronto: Lorimer.

Crouch, Winston W., and Beatrice Dinerman. 1963. *Southern California Metropolis: A Study in Development of Government for a Metropolitan Area.* Berkeley: University of California Press.

Cuff, Robert D., and J.L. Granatstein. 1978. *American Dollars and Canadian Prosperity: Canadian-American Economic Relations, 1945–1950.* Toronto: Samuel-Stevens.

Cullingworth, J. Barry. 1993. *The Political Culture of Planning: American Land Use Planning in Comparative Perspective.* New York: Routledge.

Cunningham, William Glenn. 1951. *The Aircraft Industry: A Study in Industrial Location.* Los Angeles: L.L. Morrison.

Daly, Herman E., and John B. Cobb, Jr. 1994. *For the Common Good: Redirecting the Economy toward Community, the Environment, and a Sustainable Future*, 2nd ed. Boston: Beacon Press.

Dana, Julian. 1947. *J.P. Giannini: Giant in the West.* New York: Prentice-Hall.

Daniels, Bruce, ed. 1978. *Towns and Country: Essays on the Structure of Local Government in the American Colonies.* Middleton, Connecticut: Wesleyan University Press.

Danielson, Michael N. 1976. *The Politics of Exclusion.* New York: Columbia University Press.

_____. 1982. *New York: The Politics of Urban Regional Development.* Berkeley: University of California Press.

Daunton, Martin J. 1988. 'Cities of Homes and Cities of Tenements: British and American Comparisons, 1870–1914'. *Journal of Urban History* 14:283–319.

Davidson, James Dale, and William Rees-Mogg. 1993. *The Great Reckoning: Protect Yourself in the Coming Depression*, rev. ed. New York: Simon & Schuster.

Davies, Pearl J. 1958. *Real Estate in American History.* Washington: Public Affairs Press.

Davis, Donald R. 1988. *Conspicuous Production: Automobiles and Elites in Detroit, 1899–1933.* Philadelphia: Temple University Press.

Davis, Mike. 1990. *City of Quartz: Excavating the Future in Los Angeles.* London: Verso.

Dear, Michael, and Jennifer Wolch. 1987. *Landscapes of Despair: From Deinstitutionalization to Homelessness.* Princeton: Princeton University Press.

Dennis, Richard. 1994. 'Interpreting the Apartment House: Modernity and Metropolitanism in Toronto, 1900–1930'. *Journal of Historical Geography* 20, no. 3:305–22.

de Vries, Jan. 1984. *European Urbanization 1500–1800.* Cambridge: Harvard University Press.

Diamondstone, Judith. 1966. 'Philadelphia's Municipal Corporation, 1701–1776'. *Pennsylvania Magazine of History and Biography* 90:183–201.

Dickens, Charles. [1842] 1908. *American Notes and Pictures from Italy.* London: Dent.

Doerflinger, Thomas M. 1986. *A Vigorous Spirit of Enterprise: Merchants and Economic Development in Revolutionary Philadelphia.* Chapel Hill: University of North Carolina Press.

Dolan, Jay P. 1975. *The Immigrant Church: New York's Irish and German Catholics, 1815–1865.* Baltimore: Johns Hopkins University Press.

Domosh, Mona. 1990. 'Shaping the Commercial City: Retail Districts in Nineteenth-Century New York and Boston'. *Annals of the American Association of Geographers* 80:268–84.

Doucet, Michael J. 1982. 'Urban Land Development in Nineteenth-Century North America: Themes in the Literature'. *Journal of Urban History* 8:299–342.

_____, and John Weaver. 1991. *Housing the North American City.* Montreal and Kingston: McGill-Queens University Press.

Douthwaite, Richard. 1992. *The Growth Illusion: How Economic Growth Has Enriched the Few, Impoverished the Many, and Endangered the Planet.* Bideford: Green Books; Dublin: Lilliput Press.

Downs, Anthony. 1973. *Opening Up the Suburbs: An Urban Strategy for America.* New Haven: Yale University Press.

Drache, Daniel, and Meric Gertler. 1991. *The New Era of Global Competition: State Policy and Market Power.* Montreal and Kingston: McGill-Queen's University Press.

Du Boff, Richard B. 1983. 'Telegraph and the Structure of Markets in the United States, 1845–1890'. *Research in Economic History* 8:253–77.

Duffy, John. 1968. *A History of Public Health in New York City, 1625–1866.* New York: Sage.

Earle, Carville. 1992. *Geographical Inquiry and American Historical Problems.* Stanford: Stanford University Press.

Ebner, Michael H. 1988. *Creating Chicago's North Shore: A Suburban History.* Chicago: University of Chicago Press.

Economic Council of Canada. 1991. *A Joint Venture: The Economics of Constitutional Options, Twentieth Annual Review.* Ottawa: Minister of Supply and Services Canada.

Editors of *Fortune.* 1958. *The Exploding Metropolis.* Garden City: Doubleday.

Egnal, Marc. 1975. 'The Politics of Ambition: A New Look at Benjamin Franklin's Career'. *Canadian Review of American Studies* 6:151–64.

_____. 1988. *A Mighty Empire: The Origins of the American Revolution.* Ithaca: Cornell University Press.

Eichler, Ned. 1982. *The Merchant Builders.* Cambridge: MIT Press.

Einhorn, Robin. 1991. *Property Rules: Political Economy in Chicago, 1833–1872.* Chicago: University of Chicago Press.

Eldridge, Hope T., and Dorothy S. Thomas. 1964. *Population Redistribution and Economic Growth: United States, 1870–1950: Demographic Analysis and Interrelations*, vol. III. Philadelphia: American Philosophical Society.

Elkin, S.L. 1987. *City and Regime in the American Republic*. Chicago: University of Chicago Press.

Engerman, Stanley L., and Robert E. Gallman, eds. 1986. *Long-Term Factors in American Economic Growth*. Chicago: University of Chicago Press.

Epstein, Jason. 1992. 'The Tragical History of New York'. *New York Review of Books* (9 April):45–52.

Ernst, Robert. 1949. *Immigrant Life in New York City, 1825–1863*. New York: King's Crown Press.

Ewen, Stuart. 1976. *Captains of Consciousness: Advertising and the Social Roots of the Consumer Society*. New York: McGraw-Hill.

Ewers, Hans-Jugen, J.R. Goddard, and Horst Matarath. 1986. *The Future of the Metropolis: Berlin, London, Paris, New York*. Berlin: Walter de Gueyter.

Fellman, Jerome. 1957. 'Pre-building Growth Patterns of Chicago'. *Annals of the Association of American Geographers* 47:49–82.

Fetherling, Douglas. 1994. *Travels by Night: A Memoir of the Sixties*. Toronto: Lester Publishing.

Field, Alexander. 1992. 'Uncontrolled Land Development and the Duration of the Depression in the United States'. *Journal of Economic History* 32:785–805.

Fine, David, ed. 1984. *Los Angeles in Fiction: A Collection of Original Essays*. Albuquerque: University of New Mexico Press.

Finney, Guy W. 1945. *Angel City in Turmoil: A Story of the Minute Men of Los Angeles in Their War on Civic Corruption, Graft, and Privilege*. Los Angeles: American Press.

Fishman, Robert. 1987. *Bourgeois Utopias: The Rise and Fall of Suburbia*. New York: Basic Books.

_____. 1980. 'The Anti-Planners: The Contemporary Revolt Against Planning and Its Significance for Planning History'. In *Shaping An Urban World*, edited by G. Cherry, 243–52. London: Mansell.

Flanagan, Maureen A. 1987. *Charter Reform in Chicago*. Carbondale: Southern Illinois University Press.

_____. 1996. 'The City Profitable, the City Livable: Environmental Policy, Gender, and Power in Chicago in the 1910s'. *Journal of Urban History* 22:163–90.

Flink, James J. 1975. *The Car Culture*. Cambridge: MIT Press.

Fogelson, Robert M. 1967. *The Fragmented Metropolis: Los Angeles 1850–1930*. Cambridge: Harvard University Press.

Foley, Donald L., et al. 1965. *Characteristics of Metropolitan Growth in California*. Berkeley: Institute of Urban and Regional Development, University of California.

Ford, James. 1936. *Slums and Housing, with Special Reference to New York City: History, Conditions, Policy*, 2 vols. Cambridge: Harvard

Ford, John Anson. 1961. *Thirty Explosive Years: Los Angeles County*. San Marino: Huntington Library.

Ford, Larry R. 1994. *Cities and Buildings: Skyscrapers, Skid Rows and Suburbs*. Baltimore: Johns Hopkins University Press.

Foster, Mark. 1989. *Henry J. Kaiser: Builder in the Modern American West*. Austin: University of Texas Press.

Franklin, Benjamin. 1964. *The Autobiography of Benjamin Franklin*, edited by Leonard W. Labaree et al. New Haven: Yale University Press.

Frey, William. 1993. 'The New Urban Revival in the US'. *Urban Studies* 30:741–74.

_____. 1995. 'The New Geography of Population Shifts: Trends toward Balkanization'. *In State of the Union—America in the 1990s, Volume 2, Social Trends*, edited by Farley Reynolds, 271–334. New York: Sage.

_____, and Alden Speare, Jr. 1988. *Regional and Metropolitan Growth and Decline in the United States*. New York: Sage.

Fried, Joseph P. 1971. *Housing Crisis USA*. New York: Praeger.

Friedrich, Otto. 1987. *City of Nets: A Portrait of Hollywood in the 1940s*. New York: Harper & Row.

Fries, Sylvia D. 1977. *The Urban Idea in Colonial America*. Philadelphia: Temple University Press.

Friis, Herman R. 1968. *A Series of Population Maps of the Colonies and the United States, 1625–1790*. New York: American Geographical Society.

Frisken, Frances. 1988. *City Policy-Making in Theory and Practice: The Case of Toronto's Downtown Plan*. London, ON: Department of Political Science, University of Western Ontario.

_____. 1991. 'The Contributors of Metropolitan Government and the Success of Toronto's Public Transit System: An Empirical Dissent from the Public Choice Paradigm'. *Urban Affairs Quarterly* 27:268–92.

_____, ed.. 1994. *The Changing Canadian Metropolis: A Public Policy Perspective*, 2 vols. Berkeley: Institute of Governmental Studies Press; Toronto: Canadian Urban Institute.

Frost, Lionel. 1991. *The New Urban Frontier: Urbanisation and City Building in Australasia and the American West*. Sydney: New South Wales University Press.

Frug, Gerald E. 1980. 'The City as a Legal Concept'. *Harvard Law Review* 93:1059–154.

Fukuyama, Francis. 1992. *The End of History and the Last Man.* New York: Free Press.

Funigiello, Philip J. 1978. *The Challenge of Urban Liberalism: Federal-City Relations During World War II.* Knoxville: University of Tennessee Press.

Furer, Howard B., comp. and ed. 1974. *Chicago: A Chronological and Documentary History.* Dobbs Ferry, NY: Oceana.

Gabler Neal. 1988. *An Empire of Their Own: How the Jews Invented Hollywood.* New York: Crown.

Gad, Gunter, and Deryck Holdsworth. 1987. 'Corporate Capitalism and the Emergence of the High Rise Office Building'. *Urban Geography* 8:212–31.

Galbraith, John Kenneth. 1994. *A Journey Through Economic Time: A First-Hand View.* Boston and New York: Houghton-Mifflin.

Gallman, Robert F., and John Joseph Wallis, eds. 1992. *American Economic Growth and Standards of Living before the Civil War.* Chicago: University of Chicago Press.

Garreau, Joel. 1991. *Edge City: Life on the New Frontier.* New York: Doubleday.

Garrett, Laurie. 1995. *The Coming Plague: Newly Emerging Diseases in a World Out of Balance.* New York: Farrar, Straus and Giroux.

Garvan, Anthony N.B. 1966. 'Proprietary Philadelphia in Artifact'. In *The Historian and the City,* edited by Oscar Handlin and John Burchard. Cambridge: MIT Press.

Gelfand, Mark I. 1975. *A Nation of Cities: The Federal Government and Urban America 1933–1965.* New York: Oxford University Press.

Gere, Edwin A., Jr. 1982. 'Dillon's Rule and the Cooley Doctrine: Reflections of the Political Cultures'. *Journal of Urban History* 8:271–98.

Gertler, Leonard O., and Ronald W. Crowley. 1977. *Changing Canadian Cities: The Next 25 Years.* Toronto: McClelland and Stewart.

Gibson, Paul. 1993. *The Bear Trap: Why Wall Street Doesn't Work.* New York: Atlantic Monthly.

Giedion, Siegfried. 1948. *Mechanization Takes Command: A Contribution to Anonymous History.* New York: Oxford University Press.

Gilbert, James. 1991. *Perfect Cities: Chicago's Utopias of 1893.* Chicago: University of Chicago Press.

Gilchrist, David T., ed. 1967. *The Growth of Seaport Cities, 1790–1825.* Charlottesville: University Press of Virginia.

Gilfoyle, Timothy J. 1992. *City of Eros: New York City: Prostitution and the Commercialization of Sex 1790–1920.* New York: Norton.

Gilje, Paul A. 1987. *The Road to Mobocracy: Popular Disorder in New York City, 1765–1834.* Chapel Hill: University of North Carolina Press.

Glaab, Charles N., and A. Theodore Brown. 1967. *A History of Urban America.* New York: Macmillan.

Gluck, Shuna Berger. 1987. *Rosie the Riveter Revisited: Women, the War, and Social Change.* Boston: Twayne.

Goheen, P.G. 1990. 'The Changing Bias of Inter-urban Communication in Nineteenth-Century Canada'. *Journal of Historical Geography* 16:177–96.

Goldberg, Michael A., and John Mercer. 1986. *The Myth of the North American City: Continentalism Challenged.* Vancouver: UBC Press.

Goldfield, David R., and Blaine A. Brownell. 1990. *Urban America; A History,* 2nd ed. Boston: Houghton Mifflin

Goldsmith, William W., and Edward J. Blakely. 1993. *Separate Societies: Poverty and Inequality in US Cities.* Philadelphia: Temple University Press.

Goldstein, Joshua S. 1988. *Long Cycles: Prosperity and War in the Modern Age.* New Haven: Yale University Press.

Goodrich, Carter. 1960. *Government Promotion of American Canals and Railroads.* New York: Columbia University Press.

Gottdiener, Mark. 1977. *Planned Sprawl: Private and Public Interests in Suburbia.* Beverly Hills: Sage.

Gottlieb, Manuel. 1976. *Long Swings in Urban Development.* New York: National Bureau of Economic Research.

Gottlieb, Robert, and Irene Wolt. 1977. *Thinking Big: The Story of the Los Angeles Times.* New York: Putnam.

Gottman, Jean. 1961. *Megalopolis: The Urbanized Northeastern Seaboard of the United States.* Cambridge: MIT Press.

Goulden, Joseph C. 1976. *The Best Years 1945–1950.* New York: Atheneum.

Grebler, Leo, ed. 1963. *Metropolitan Contrasts.* Los Angeles: UCLA.

Green, Paul M., and Holli Melvin, eds. 1987. *The Mayors: The Chicago Political Tradition.* Carbondale: Southern Illinois University Press.

Greene, Jack P. 1988. *Pursuits of Happiness: The Social Development of Early Modern British Colonies and the Formation of American Culture.* Chapel Hill: University of North Carolina Press.

———, and J.R. Pole, eds. 1984. *Colonial British America: Essays in the New History of the Early Modern Era.* Baltimore: Johns Hopkins University Press.

Gregory, James Noble. 1989. *American Exodus: The Dust*

Bowl Migration and Okie Culture in California. New York: Oxford University Press.

Griffith, Ernest S. 1958. *History of American City Government in the Colonial Period*. New York: Oxford University Press.

Guest, Dennis. 1980. *The Emergence of Social Security in Canada*. Vancouver: UBC Press.

Gunn, L. Ray. 1988. *The Decline of Authority: Public Economic Policy and Political Development in New York, 1800–1860*. Ithaca: Cornell University Press.

Guterson, David. 1992. 'No Place Life Home: On the Manicured Streets of a Master-Planned Community'. *Harper's* 285 (November):55–64.

Gutman, Herbert G. 1966. *Work, Culture and Society in Industrializing America: Essays in American Working-Class and Social History*. New York: Random House.

Haar, Charles M., and Jerold S. Kayden, eds. 1989. *Zoning and the American Dream: Promises Still to Keep*. Chicago and Washington: Planners Press.

Hacker, Andrew. 1992. *Two Nations: Black, White, Separate, Hostile, Unequal*. New York: Scribners.

Hall, Peter, and Paschal Preston. 1988. *The Carrier Wave: New Information Technology and the Geography of Innovation, 1846–2003*. London: Unwin Hyman.

Halpern, Robert. 1995. *Rebuilding the Inner City: A History of Neighborhood Initiatives to Address Poverty in the United States*. New York: Columbia University Press.

Halttunen, Karen. 1982. *Confidence Men and Painted Women: A Study of Middle-Class Culture in America, 1830–1870*. New Haven: Yale University Press.

Hamel, P. 1993. 'Crisis, Modernity and Postmodernity: The Crisis in Urban Planning'. *Canadian Journal of Urban Research* 2:16–29.

Hamer, David. 1990. *New Towns in the New World: Images and Perceptions of the Nineteenth-Century Urban Frontier*. New York: Columbia University Press.

Hammack, David C. 1983. *Power and Society: Greater New York at the Turn of the Century*. New York: Sage Foundation.

Hancock, Ralph. 1949. *Fabulous Boulevard*. New York: Funk & Wagnalls.

Hanson, Earl, and Paul Beckett. 1944. *Los Angeles: Its People and Its Homes*. Los Angeles: Haynes Foundation.

Harrington, Michael. 1962. *The Other America: Poverty in the United States*. New York: Macmillan.

Harris, P.M.G. 1989. 'The Demographic Development of Colonial Philadelphia in Comparative Perspective'. *Proceedings of the American Philosophical Society* 133:262–304.

Harris, Richard. 1996. *Unplanned Suburbs: Toronto's American Tragedy*. Baltimore: Johns Hopkins University Press.

Harrison, Bennett. 1994. *Lean and Mean: The Changing Landscape of Corporate Power in an Age of Flexibility*. New York: Basic Books.

Hartmann, Susan M. 1982. *The Home Front and Beyond: American Women in the 1940s*. Boston: Twayne.

Hartog, Hendrik. 1978. *Public Property and Private Power: The Corporation of the City of New York in American Law, 1730–1870*. Chapel Hill: University of North Carolina Press.

Hartz, Louis. 1955. *The Liberal Tradition in America*. New York: Harcourt Brace Jovanovich.

———, et al. 1964. *The Founding of New Societies*. New York: Harcourt, Brace and World.

Harvey, David. 1989. *The Condition of Postmodernity*. Oxford: Blackwell.

Hayden, Delores. 1995. *The Power of Place: Urban Landscapes as Public History*. Cambridge: MIT Press.

Hayes, David. 1992. *Power and Influence: The Globe and Mail and the News Revolution*. Toronto: Key Porter Books.

Hayes, Samuel P. 1974. 'The Changing Political Structures of the City in Industrial America'. *Journal of Urban History* 1:6–25.

Henstall, Bruce. 1980. *Los Angeles: An Illustrated History*. New York: Knopf.

Herrtje, Arnold, and Mark Perlman, eds. 1991. *Evolving Technology and Market Structure*. Ann Arbor: University of Michigan Press.

Hershberg, Theodore, ed. 1981. *Philadelphia: Work, Space, Family and Group Experience in the Nineteenth Century*. Oxford: Oxford University Press.

Higgins, Donald J.H. 1977. *Urban Canada: Its Government and Politics*. Toronto: Macmillan.

Higonnet, Patrice, David S. Pandes, and Henry Rosovsky, eds. 1991. *Favorites of Fortune: Technology, Growth, and Economic Development Since the Industrial Revolution*. Cambridge: Harvard University Press.

Hilliard, Celia. 1979. '"Rent Reasonable to Right Parties": Gold Coast Apartment Buildings, 1906–1929'. *Chicago History* 8:66–77.

Hines, Thomas S. 1974. *Burnham of Chicago: Architect and Planner*. Chicago: University of Chicago Press.

———. 1982. 'Housing, Baseball and Creeping Socialism: The Battle of Chavez Ravine, Los Angeles 1949–1959'. *Journal of Urban History* 8:123–44.

Hirsch, Fred. 1976. *The Social Limits of Growth*. Cambridge: Harvard University Press.

Hirsch, Werner Z., ed. 1971. *Los Angeles: Viability and Prospects for Metropolitan Leadership*. New York:

Praeger Books.

Hise, Greg. 1993. 'Home Building and Industrial and Decentralization in Los Angeles: The Roots of the Postwar Urban Region'. *Journal of Urban History* 19:95–125.

Hochschild, Jennifer L. 1995. *Facing Up to the American Dream: Race, Class, and the Soul of the Nation*. Princeton: Princeton University Press.

Hodge, Gerald. 1986. *Planning Canadian Communities: An Introduction to the Principles, Practice and Participants*, 1st ed. Toronto: Methuen.

_____. 1991. *Planning Canadian Communities: An Introduction to the Principles, Practice and Participants, 2nd ed.* Toronto: Nelson.

Hodges, Graham. 1986. *New York City Cartmen 1667–1850*. New York: New York University Press.

Hoffman, Charles. 1970. *The Depression of the Nineties: An Economic History*. Westport: Greenwood Press.

Hofstadter, Richard. 1948. *The American Political Tradition and the Man Who Made It*. New York: Knopf.

Hogan, David J. 1985. *Class and Reform: School and Society in Chicago 1880–1930*. Philadelphia: University of Pennsylvania Press.

Hohenberg, Paul M., and Lynn H. Lees. 1985. *The Making of Urban Europe, 1000–1950*. Cambridge: Harvard University Press.

Hood, Adrienne D. 1994. 'The Gender Division of Labor in the Production of Textiles in Eighteenth-Century Rural Pennsylvania: Rethinking the New England Model'. *Journal of Social History* 27:537–61.

Hopkins, Ernest J. 1972. *Our Lawless Police: A Study of the Unlawful Enforcement of the Law*. New York: De Capo Press.

Horowitz, Daniel. 1985. *The Morality of Spending: Attitudes towards the Consumer Society in America 1875–1940*. Baltimore: Johns Hopkins University Press.

Horowitz, Gad. 1968. *Canadian Labour in Politics*. Toronto: University of Toronto Press.

Horowitz, Helen L. 1976. *Culture and the City: Cultural Philanthropy in Chicago from the 1880s to 1917*. Lexington: University of Kentucky Press.

Hoselitz, Bert F. 1960. *Sociological Aspects of Economic Growth*. London: Collier-Macmillan.

Hounshell, David. 1984. *From the American System to Mass Production*. Baltimore: Johns Hopkins University Press.

Howe, Neil, and Paul Strauss. 1993. *13th Gen: Abort, Retry, Ignore, Fall?* New York: Random House.

Hoyt, Homer. 1933. *One Hundred Years of Land Values in Chicago*. Chicago: University of Chicago Press.

_____. 1939. *The Structure and Growth of Residential Neighbourhoods in American Cities*. Washington: Federal Housing Administration.

Huang, Nian-Sheng. 1994. *Benjamin Franklin in American Thought and Culture 1790–1990*. Philadelphia: American Philosophical Society.

Hughes, Everett C. 1931. *The Chicago Real Estate Board: The Growth of an Institution*. Chicago: Society for Social Research of the University of Chicago.

Hughes, Jonathan R.T. 1991. *The Governmental Habit Redux: Economic Controls of Colonial Times to the Present*. Princeton: Princeton University Press.

Hughes, Richard T., and C. Leonard Allen. 1988. *Illusions of Innocence: Protestant Primitivism in America, 1630–1875*. Chicago: University of Chicago Press.

Hughes, Robert. 1995. 'Take the Revolution'. *Time*, special issue (Spring).

Hughes, Thomas P. 1983. *Networks of Power: Electrification in Western Society 1880–1930*. Baltimore: Johns Hopkins University Press.

Hugill, Peter. 1993. *World Trade Since 1431: Geography, Technology and Capitalism*. Baltimore: Johns Hopkins University Press.

Hundley, Norris. 1992. *The Great Thirst: California and Water 1770s–1990s*. Berkeley: University of California Press.

Hunter, Louis C. 1979–91. *A History of Industrial Power in the United States, 1780–1930*, 3 vols. Charlottesville: University Press of Virginia for the Eleutherian Mills-Hagley Foundation.

Hutson, James H. 1971. 'An Investigation of the Inarticulate: Philadelphia's White Oaks'. *William and Mary Quarterly* 3rd ser., 23:3–25.

Hylen, Arnold. 1981. *Los Angeles Before the Freeways 1850–1950: Images of an Era*. Los Angeles: Dawson Publishing.

Iacovetta, Franca. 1992. *Such Hardworking People: Italian Immigrants in Postwar Toronto*. Montreal and Kingston: McGill-Queens University Press.

Illick, Joseph. 1976. *Colonial Pennsylvania: A History*. New York: Scribners.

Ingham, John N. 1978. *The Iron Barons: A Social Analysis of an American Urban Elite, 1874–1965*. Westport: Greenwood Press.

Innis, Harold. 1951. *The Bias of Communication*. Toronto: University of Toronto Press.

Irving, John. 1989. *A Prayer for Owen Meany*. Toronto: Lester & Orpen Dennys.

Isin, Engin. 1992. *Cities without Citizens: The Modernity of the City as a Corporation*. Montreal: Black Rose.

Jackson, Kenneth T. 1985. *Crabgrass Frontier: The Subur-*

banization of the US. New York: Oxford University Press.

———, and Stanley K. Schultz, eds. 1972. *Cities in American History*. New York: Knopf.

Jacobs, Jane. 1961. *The Death and Life of Great American Cities*. New York: Random House.

———. 1992. *Systems of Survival*. New York: Random House.

———. 1993. 'Are Planning Departments Useful?' *Ontario Planning Journal* 8 (July-August):4–5.

Jaher, Frederic C. 1982. *The Urban Establishment: Upper Strata in Boston, New York, Charleston, Chicago and Los Angeles*. Urbana: University of Illinois Press.

Jakle, John A., and David Wilson. 1992. *Derelict Landscapes: The Wasting of America's Built Environment*. Baltimore: Johns Hopkins University Press.

Johnson, Norah. 1962. 'It's Smart to Live in the Annex'. *Habitat* (Jan.-Feb.):107.

Jones, Alice Hanson. 1980. *Wealth of a Nation to Be: The American Colonies on the Eve of the Revolution*. New York: Columbia University Press.

Jones, Barclay G. 'Applications of Catastrophic Techniques to the Study of Urban Phenomenon: Atlanta, Georgia, 1940–1975'. *Economic Geography* 56:201–22.

Jones, David W., Jr. 1985. *Urban Transit Policy: An Economic and Political History*. Englewood Cliffs: Prentice-Hall.

Kalbach, Warren E. 1980. *Historical and Generational Perspectives of Ethnic Residential Segregation in Toronto, 1851–1871*. Toronto: Centre for Urban and Community Studies, University of Toronto.

Kaplan, Harold. 1982. *Reform, Planning, and City Politics: Montreal, Winnipeg, and Toronto*. Toronto: University of Toronto Press.

Kasson, John F. 1976. *Civilizing the Machine: Technology and Republican Values in America, 1776–1900*. New York: Grossman.

———. 1990. *Rudeness and Civility: Manners in Nineteenth-Century America*. New York: Hill and Wang.

Katz, Michael. 1989. *Undeserving Poor: From the War on Poverty to the War on Welfare*. New York: Pantheon.

———, et al. 1982. *The Social Organization of Early Industrial Capitalism*. Cambridge: Harvard University Press.

———, ed. 1993. *The Underclass Debate: Views from History*. Princeton: Princeton University Press.

Katznelson, Ira. 1981. *City Trenches: Urban Politics and the Patterning of Class in the United States*. New York: Pantheon.

Keating, Ann Durkin. 1988. *Building Chicago: Suburban Developers and the Creation of a Divided Metropolis*. Columbus: Ohio State University Press.

Keating, Michael. 1991. *Comparative Urban Politics: Power of the City in the USA, Canada, Britain and France*. Aldershot: Edward Elgar.

Kennedy, Paul. 1987. *The Rise and Fall of the Great Powers: Economic Change and Military Conflict from 1500 to 2000*. New York: Random House.

Kern, Stephen. 1983. *The Culture of Time and Space 1880–1918*. Cambridge: Harvard University Press.

Kidner, Frank L., and Philip Neff. 1945. *An Economic Survey of the Los Angeles Area*. Los Angeles: Haynes Foundation.

Kindleberger, Charles P. 1989. *Manias, Panics, and Crashes: A History of Financial Crises*, rev. ed. New York: Basic Books.

Kirk, Carolyn, and Gordon W. Kirk, Jr. 1981. 'The Impact of the City on Home Ownership: A Comparison of Immigrants and Native Whites at the Turn of the Century'. *Journal of Urban History* 7:474–5.

Klein, Maury, and Harvey A. Kantor. 1976. *Prisoners of Progress: American Industrial Cities 1850–1920*. New York: Macmillan.

Klein, Norman M., and Martin J. Schiesl, eds. 1990. *20th Century Los Angeles: Power, Priorities and Social Conflict*. Claremont: Regina Books.

Klepp, Susan E. 1989a. 'Demography of Early Philadelphians'. *Proceedings of the American Philosophical Society* 133:85–111.

———. 1989b. *Philadelphia in Transition: A Demographic History of the City and Its Occupational Groups, 1720–1830*. New York: Garland.

Kneier, Charles M. 1957. *City Government in the United States*, 3rd ed. New York: Harper Bros.

Knox, Paul L. 1994. *Urbanization: An Introduction to Urban Geography*. Englewood Cliffs: Prentice-Hall.

Kouwenhoven, John A. 1953. *The Columbia Historical Portrait of New York: An Essay in Graphic History*. New York: Doubleday.

Krueckeberg, Donald A., ed. 1983. *Introduction to Planning History in the United States*. New Brunswick: Rutgers University Center for Urban Policy Research.

Krugman, Paul. 1990. *The Age of Diminished Expectations: US Economic Policies in the 1990s*. Cambridge: MIT Press.

Kulikoff, Allan. 1992. *The Agrarian Origins of American Capitalism*. Charlottesville: University Press of Virginia.

Ladd, Helen F., and John Yinger. 1989. *America's Ailing Cities: Fiscal Health and the Design of Urban Policy*. Baltimore: Johns Hopkins University Press.

Lamoreaux, Naomi. 1988. *The Great Merger Movement in American Business, 1895–1904*. New York: Cambridge University Press.

Langdon, Philip. 1994. *A Better Place to Live: Reshaping the American Suburb*. Amherst: University of Massachusetts.

Lankevich, George J., and Howard B. Furer. 1964. *A Brief History of New York City*. Port Washington, NY: Associated Faculty Press.

Lapham, Lewis. 1995. 'Reactionary Chic'. *Harper's* 290 (March):31–42.

Lasch, Christopher. 1991. *The True and Only Heaven: Progress and Its Critics*. New York: Norton.

Lazonick, William. 1991. *Business Organization and the Myth of the Market Economy*. New York: Columbia University Press.

Leach, William. 1993. *Land of Desire: Merchants, Power and the Rise of a New American Culture*. New York: Pantheon.

Leacy, F.H., ed. 1983. *Historical Statistics of Canada*. Ottawa: Statistics Canada.

Lees, Andrew. 1985. *Cities Perceived: Urban Society in European and American Thought, 1820–1940*. Manchester: Manchester University Press.

Lemann, Nicholas. 1991. *The Promised Land: The Great Black Migration and How It Changed America*. New York: Knopf.

Lemon, James T. 1972. *The Best Poor Man's Country: A Geographical Study of Early Southern Pennsylvania*. Baltimore: Johns Hopkins University Press.

_____. 1985. *Toronto Since 1918: An Illustrated History*. Toronto: Lorimer; Ottawa: Museum of Civilization.

_____. 1989. 'Plans for Early 20th Century Toronto: Lost in Management'. *Urban Historical Review* 18:11–31.

_____. 1990. *The Toronto Harbour Plan of 1912: Manufacturing Goals and Economic Realities*. Toronto: Canadian Waterfront Resource Centre.

Leuchtenburg, William. 1958. *The Perils of Prosperity 1914–1932*. Chicago: University of Chicago Press.

Levy, Frank. 1988. *Dollars and Dreams: The Changing American Income Distribution*. New York: Norton.

Ley, David. 1995. *The New Middle Class and the Remaking of the Central City*. Oxford: Oxford University Press.

Lichtenstein, Nelson. 1982. *Labor's War at Home: The CIO in World War II*. Cambridge: Cambridge University Press.

Lind, Loren J. 1974. *The Learning Machine: A Hard Look at Toronto's Schools*. Toronto: Anansi.

Lind, Michael. 1995. *The Next American Nation: The New Nationalism and the Fourth American Republic*. New York: Free Press.

Lindstrom, Diane. 1978. *Economic Development in the Philadelphia Region, 1810–1850*. New York: Columbia University Press.

_____, and John Sharpless. 1978. 'Urban Growth and Economic Structure in Antebellum America'. *Research in Economic History* 3:161–216.

Lipset, Seymour M. 1990. *Continental Divide: The Values and Institutions of the United States and Canada*. New York: Routledge.

Lo, Clarence Y.H. 1990. *Small Property versus Big Government: Social Origins of the Property Tax Revolt*. Berkeley: University of California Press.

Lockwood, Charles. 1976. *Manhattan Moves Uptown: An Illustrated History*. Boston: Houghton Mifflin.

Logan, John R., and Harvey L. Molotch. 1987. *Urban Fortunes: The Political Economy of Place*. Berkeley: University of California Press.

Logan, John R., and Todd Swanstrom. 1990. *Beyond the City Limits: Urban Policy and Economic Restructuring in Comparative Perspective*. Philadelphia: Temple University Press.

Long, Clarence D. 1940. *Building Cycles and the Theory of Investment*. Princeton: Princeton University Press.

Lorimer, James. 1978. *The Developers*. Toronto: Lorimer.

Los Angeles Regional Planning Commission. 1941. *Business Districts*. Los Angeles: Los Angeles Regional Planning Commission.

Lowi, Theodore. 1969. *The End of Liberalism: Ideology, Policy and the Crisis of Public Authority*. New York: Norton.

Lubitz, Edward. 1970. 'The Tenement Problem in New York City and the Movement for Its Reform, 1856–1867'. PhD thesis, New York University.

Lubove, R. 1963. *Community Planning in the 1920s: The Contribution of the Regional Planning Association*. Pittsburgh: University of Pittsburgh Press.

Lupo, Alan, Frank Colcord, and Edmund P. Fowler. 1971. *Rites of Way: The Politics of Transportation in Boston and the US City*. Boston: Little, Brown.

Lustig, R. Jeffrey. 1982. *Corporate Liberalism: The Origins of Modern Political Theory, 1890–1920*. Berkeley: University of California Press.

Lynch, Kevin. 1960. *The Image of the City*. Cambridge: MIT Press.

_____. 1971. *The Visual Environment of Los Angeles*. Los Angeles: City Planning Commission.

Maddison, Angus. 1991. *Dynamic Forces in Capitalist Development: A Long-Run Comparative View*. New York: Oxford University Press.

Magnusson, W., and Andrew Sancton, eds. 1983. *City*

Politics in Canada. Toronto: University of Toronto Press.

Mahoney, Timothy R. 1990. *River Towns in the Great West: The Structures of Provincial Urbanization in the American Midwest, 1820–1870.* Cambridge: Cambridge University Press.

Maier, Pauline. 1993. 'The Revolutionary Origins of the American Corporation'. *William and Mary Quarterly,* 3rd ser., 50:51–84.

Mandelbaum, Seymour J. 1965. *Boss Tweed's New York.* New York: J. Wiley and Sons.

Mann, Christopher B. 1994. 'Our Generation's Nihilism Is Inevitable'. *Pomona College Today* (Summer):15.

Mann, W.E., ed. 1970. *The Underside of Toronto.* Toronto: McClelland and Stewart.

Marchand, Bernard. 1986. *The Emergence of Los Angeles: Population and Housing in the City of Dreams 1940–1970.* London: Pion.

Marchand, Roland. 1986. *Advertising the American Dream: Making Way for Modernity 1920–1940.* Berkeley: University of California Press.

Markusen, Ann, et al. 1991. *The Rise of the Gunbelt: The Military Remapping of Industrial America.* New York: Oxford University Press.

Marmor, Theodore R., J.L. Mashaw, and P.L. Harvey. 1990. *America's Misunderstood Welfare State: Persisting Myths, Enduring Realities.* New York: Basic Books.

Marsden, George M. 1980. *Fundamentalism and American Culture 1870–1925.* New York: Oxford University Press.

Marsh, Margaret. 1990. *Suburban Lives.* New Brunswick, NJ: Rutgers University Press.

Martin, John F. 1991. *Profits in the Wilderness: Entrepreneurship and the Founding of New England Towns in the Seventeenth Century.* Chapel Hill: University of North Carolina Press.

Marvin, Carolyn. 1988. *When Old Technologies Were New: Thinking About Electric Communication in the Late Nineteenth Century.* New York: Oxford University Press.

Marx, Leo. 1967. *The Machine in the Garden.* New York: Oxford University Press.

Mason, Joseph B. 1982. *History of Housing in the United States 1930–1980.* Houston: Gulf Publishing.

Massey, Douglas S., and Nancy Denton. 1993. *American Apartheid: Segregation and the Making of the Underclass.* Cambridge: Harvard University Press.

Mayer, Harold M. 1943. *The Railroad Pattern of Metropolitan Chicago.* Chicago: Department of Geography, University of Chicago.

_____, and Clyde F. Kohn, eds. 1959. *Readings in Urban Geography.* Chicago: University of Chicago Press.

_____, and Richard C. Wade, Jr. 1969. *Chicago: The Growth of a Metropolis.* Chicago: University of Chicago Press.

Mayer, Martin. 1978. *The Builders: Houses, People, Neighbourhoods, Government, Money.* New York: Norton.

McBain, Howard L. 1925. 'The Legal Status of the American Colonial City'. *Political Science Quarterly* 40:177–200.

McCann, L.D., ed. 1987. *Heartland and Hinterland: A Geography of Canada.* Toronto: Prentice-Hall.

McCarthy, Kathleen D. 1982. *Noblesse Oblige: Charity and Cultured Philanthropy in Chicago 1849–1929.* Chicago: University of Chicago Press.

McCusker, John J., and Russell R. Menard. 1985. *The Economy of British America, 1607–1789.* Chapel Hill: University of North Carolina Press.

McGaw, Judith A., ed. 1994. *Early American Technology: Making and Doing Things from the Colonial Era to 1850.* Chapel Hill: University of North Carolina Press for the Institute of Early American History and Culture.

McGerr, Michael E. 1986. *The Decline of Popular Politics: The American North, 1865–1928.* New York: Oxford University Press.

McGrane, Reginald C. [1924] 1965. *The Panic of 1837: Some Financial Problems of the Jacksonian Era.* New York: Russell and Russell.

McKenzie, Roderick D. 1933. *The Metropolitan Community.* New York: McGraw Hill.

McMahon, Michael. 1990. *Metro's Housing Company: The First 35 Years.* Toronto: The Company.

McWilliams, Carey. 1949. 'Look What's Happened to California'. *Harper's Magazine* 199 (October):21–9.

_____. [1946] 1973. *Southern California: An Island on the Land.* Santa Barbara: Peregrine Smith.

Meinig, Donald W., ed. 1979. *The Interpretation of Ordinary Landscapes.* New York: McGraw-Hill.

_____. 1986. *Atlantic America, 1492–1800, Volume 1, The Shaping of America: A Geographical Perspective on 500 Years of History.* New Haven: Yale University Press.

_____. 1993. *Continental America, 1800–1867, Volume 2, The Shaping of America: A Geographical Perspective on 500 Years of History.* New Haven: Yale University Press.

Melosi, Martin V., ed. 1980. *Pollution and Reform in American Cities 1870–1930.* Austin: University of Texas Press.

_____. 1981. *Garbage in the Cities: Refuse Reform and the Environment 1880–1980.* College Station: Texas A & M Press.

Mereness, N.D., ed. 1961. *Travels in the American Colonies*. New York: Antiquarian Press.

Meyer, Gladys Eleanor. 1941. *Free Trade in Ideas: Ideas in American Liberalism Illustrated in Franklin's Philadelphia*. New York: Kings Cross Press.

Mikkelson, M.A. 1898. *A History of Real Estate, Building, Architecture in New York City*. New York: New York Public Library.

Miller, Gary. 1981. *Cities by Contract: The Politics of Municipal Incorporation*. Cambridge: MIT Press.

Miller, Howard S. 1989. 'The Politics of Public Bathing in Progressive St Louis'. *Gateway Heritage* 10 (Fall):2–21.

Miller, Richard G. 1976. *Philadelphia—the Federalist City: A Study of Urban Politics, 1789–1801*. Port Washington, NY: Kennikat Press.

Mills, C. Wright. 1951. *White Collar: The American Middle Classes*. New York: Oxford University Press.

Minsky, Hyman, ed. 1965. *California Banking in a Growing Economy: 1946–1975*. Berkeley: University of California Press.

Miron, John. 1988. *Housing in Postwar Canada: Demographic Change, Household Formation and Housing Demand*. Montreal and Kingston: McGill-Queens University Press.

Mitchell, Robert, and Paul Groves, eds. 1987. *North America: The Historical Geography of a Changing Continent*. Totowa, NJ: Rowman & Littlefield.

Moehring, Eugene P. 1981. *Public Works as the Patterns of Urban Real Estate Growth in Manhattan, 1835–1894*. New York: Arno Press.

Mohl, Raymond. 1985. *The New City: Urban America in the Industrial Age, 1860–1920*. Arlington Heights, Ill.: Harlan Davidson.

Mollenkopf, John Hall. 1983. *The Contested City*. Princeton: Princeton University Press.

_____. 1988. *Power, Culture and Place: Essays on New York City*. New York: Russell Sage Foundation.

Monchow, Helen C. 1939. *Seventy Years of Real Estate: Subdividing in the Region of Chicago*. Evanston: Northwestern University Press.

Monkkonen, Eric H. 1988. *America Becomes Urban: The Development of US Cities and Towns 1780–1980*. Berkeley: University of California Press.

Moore, Lawrence. 1994. *Selling God: American Religion in the Marketplace of Culture*. New York: Oxford University Press.

Morgan, D. 1993. *Riding in the West: The True Story of an 'Okie' Family from the Great Depression through the Reagan Years*. New York: Knopf.

Morrill, Richard L., and O. Fred Donaldson. 1972. 'Geographic Perspectives on the History of Black America'. *Economic Geography* 48:1–23.

Morton, Desmond, and Terry Copp. 1980. *Working People: An Illustrated History of Canadian Labour*. Toronto: Deneau and Greenberg.

Mowery, David, and Nathan Roseberg. 1989. *Technology and the Pursuit of Economic Growth*. Cambridge: Cambridge University Press.

Mowry, George Edwin. 1951. *The California Progressives*. Chicago: Quadrangle Press.

Muller, Edward. 1977. 'Regional Urbanization and the Selective Growth of Towns in North American Regions'. *Journal of Historical Geography* 3:21–39.

Muller, Peter O. 1981. *Contemporary Suburban America*. Englewood Cliffs: Prentice-Hall.

Mumford, Lewis. 1961. *The City in History: Its Origins, Its Transformation, and the Prospects*. New York: Harcourt, Brace & World.

Myers, Margaret. 1931–2. *The New York Money Market, Volume 1, Origins and Development*, 4 vols. New York: Columbia University Press.

Nadeau, Remi A. 1960. *Los Angeles: From Mission to Modern City*. New York: Longman Green & Co.

Nadel, Stanley. 1990. *Little Germany: Ethnicity, Religion, and Class in New York City, 1845–80*. Urbana: University of Illinois Press.

Nader, George A. 1975–6. *Cities of Canada*, 2 vols. Toronto: Macmillan.

Nash, Gary B. 1978. 'City Planning and Political Tension in the Seventeenth Century: The Case of Philadelphia'. *Proceedings of the American Philosophical Society* 112:54–78.

_____. 1979. *The Urban Crucible: Social Change, Political Consciousness and the Origins of the American Revolution*. Cambridge: Harvard University Press.

Nash, Gerald D. 1972. 'Stages in California's Economic Growth 1870–1970: An Interpretation'. *California Historical Quarterly* 51:315–30.

_____. 1985. *The American West Transformed: The Impact of the Second World War*. Bloomington: University of Indiana Press.

_____. 1992. *A.P. Giannini and the Bank of America*. Norman: University of Oklahoma Press.

Nelson, Howard J. 1983. *The Los Angeles Metropolis*. Dubuque: Kendall/Hunt.

Newbury, Darryl. 1989. *Stop Spadina: Citizens Against an Expressway*. Toronto: Commonact Press.

Newman, Katherine S. 1993. *Declining Fortune: The Withering of the American Dream*. New York: Basic Books.

Newman, Peter. 1975. *The Canadian Establishment*, vol.

1. Toronto: McClelland and Stewart.

Newman, Peter W., and Jeffrey R. Kenworthy. 1988. 'Gasoline Consumption and Cities: A Comparison of US Cities with a Global Survey'. *Journal of the American Planning Association* 55:24–37.

Nicolaides, Becky Marianna. 1993. 'In Search of the Good Life in Community and Politics in Working-Class Los Angeles'. PhD thesis, Columbia University.

Noble, David. 1977. *America by Design.* New York: Oxford University Press.

Norrie, Kenneth, and Douglas Owram. 1991. *A History of the Canadian Economy.* Toronto: Harcourt Brace Jovanovich.

North, Douglas C. 1966a. *The Economic Growth of the United States, 1790–1860.* New York: Norton.

_____. 1966b. *Growth and Welfare in the American Past: A New Economic History.* Englewood Cliffs: Prentice-Hall.

Novick, Marvyn. 1979–80. *Metro's Suburbs in Transition,* 2 vols. Toronto: Metropolitan Toronto Social Planning Council.

Olton, Charles S. 1974. 'Philadelphia's First Environmental Crisis'. *Pennsylvania Magazine of History and Biography* 97:90–100.

Ostrom, Vincent. 1953. *Water & Politics: A Story of Western Policies and Administration in the Development of Los Angeles.* Los Angeles: Haynes Foundation.

Pacyga, Dominic A., and Ellen Skerrett. 1986. *Chicago: City of Neighborhoods.* Chicago: Loyola University Press.

Paepke, C. Owen. 1993. *The Evolution of Progress: The End of Economic Growth and the Beginning of Human Transformation.* New York: Random House.

Parsons, James J. 1949. 'California Manufacturing'. *Geographical Review* 39:229–41.

Patterson, James T. 1981. *America's Struggle against Poverty 1900–1980.* Cambridge: Harvard University Press.

Paullin, Charles O. 1932. *Atlas of the Historical Geography of the United States.* Washington: Carnegie Institution.

Pease, Otis, ed. 1962. *The Progressive Years: The Spirit and Achievement of American Reform.* New York: Braziller.

Pegrum, Dudley F. 1964. *Residential Population and Urban Transport Facilities in the Los Angeles Metropolitan Area.* Los Angeles: University of California Press.

Peirce, Neal R., Curtis W. Johnson, and John S. Hall. 1993. *Citistates: How Urban American Can Prosper in a Competitive World.* Washington: Seven Locks Press.

Pendergrast, Eudora S. 1981. *Suburbanizing the Central City: Analysis of the Shift in Transportation Policies Governing the Development of Metropolitan Toronto 1959–1978.* Toronto: Department of Urban and Regional Planning, University of Toronto.

Perloff, Harvey S., et al. 1960. *Regions, Resources and Economic Growth.* Baltimore: Johns Hopkins University Press.

Pessen, Edward. 1973. *Riches, Class and Power before the Civil War.* Lexington: Heath.

Peterson, Paul E. 1981. *City Limits.* Chicago: University of Chicago Press.

Peterson, Wallace. 1993. *Silent Depression: The Fate of the American Dream.* New York: Norton.

Phillips, Kevin. 1994. *Arrogant Capital: Washington, Wall Street, and the Frustration of American Politics.* Boston: Little, Brown.

Philpott, Thomas Lee. 1978. *The Slum and the Ghetto: Neighbourhood Deterioration and Middle Class Reform, Chicago 1880–1930.* New York: Oxford University Press.

Pierce, Bessie, ed. 1933. *As Others See Chicago: Impressions of Visitors, 1673–1933.* Chicago: University of Chicago Press.

Platt, Harold L. 1991. *The Electric City: Energy and the Growth of the Chicago Area, 1880–1930.* Chicago: University of Chicago Press.

Platt, R.H., and R. Macinko. 1983. *Beyond the Urban Fringe.* Minneapolis: University of Minnesota Press.

Plunz, Richard A. 1989. *A History of Housing in New York City: Dwelling Type and Social Change in the American Metropolis.* New York: Columbia University Press.

Polanyi, Karl. 1957. *The Great Transformation: The Political and Economic Origins of Our Time.* Boston: Beacon Press.

Pomerantz, Sidney Irving. 1938. *New York, an American City, 1783–1803: A Study of Urban Life.* New York: Columbia University Press.

Pool, Ithiel de Sola, ed. 1977. *The Social Impact of the Telephone.* Cambridge: MIT Press.

Porter, John A. 1965. *The Vertical Mosaic: An Analysis of Social Class and Power in Canada.* Toronto: University of Toronto Press.

Porter, Kenneth. 1931. *John Jacob Astor: Business Man,* 2 vols. Cambridge: Harvard University Press.

Posados, Barbara. 1983. 'Suburb into City—the Transformation of Urban Identity on Chicago's Periphery: Irving Park as a Case Study 1870–1920'. *Journal of the Illinois State Historical Society* 76:162–76.

Pred, A.R. 1966. *The Spatial Dynamics of US Urban Industrial Growth 1800–1914: Interpretive and Theoretical Essays.* Cambridge: MIT Press.

_____. 1973. *Urban Growth and the Circulation of Information: The United States System of Cities, 1790–1840.*

Cambridge: Harvard University Press.

———. 1980. *Urban Growth and City Systems in the United States, 1840–1860*. Cambridge: Harvard University Press.

Price, Jacob M. 1974. 'Economic Function and the Growth of American Port Towns in the Eighteenth Century'. *Perspectives in American History* 8:123–86.

Radford, John P. 1981. 'The Social Geography of the Nineteenth-Century US City'. *In Geography and the Urban Environment: Progress in Research and Application*, vol. 4, edited by D.T. Herbert and R.J. Johnson, 257–93. New York: Wiley.

Rae, John B. 1968. *Climb to Greatness: The American Aircraft Industry 1920–1960*. Cambridge: MIT Press.

Rainwater, Lee. 1970. *Behind Ghetto Walls: Black Families in a Federal Slum*. Chicago: Aldine.

Rand, Christopher. 1967. *Los Angeles: The Ultimate City*. New York: Oxford University Press.

Randall, Frank A. 1949. *History of the Development of Building Construction in Chicago*. Urbana: University of Illinois Press.

Ray, D. Michael. 1965. *Market Potential and Economic Shadow: A Quantitative Analysis of Industrial Location in Southern Ontario*. Chicago: Department of Geography, University of Chicago.

Rea, K.J. 1985. *The Prosperous Years: The Economic History of Ontario*. Toronto: University of Toronto Press.

Regehr, Ernie. 1975. *Making a Killing: Canada's Arms Industry*. Toronto: McClelland and Stewart.

———. 1987. *Arms Canada: The Deadly Business of Military Exports*. Toronto: Lorimer.

Reid, David, ed. 1992. *Sex, Death and Good in LA*. New York: Pantheon.

Reitz, Jeffrey G. 1980. *The Survival of Ethnic Groups*. Toronto: McGraw-Hill.

Relph, Edward. 1987. *The Modern Urban Landscape*. Baltimore: Johns Hopkins University Press.

Reps, John W. 1980. *Town Planning in Frontier America*. Columbia: University of Missouri Press.

Rieff, David. 1991. *Los Angeles: Capital of the Third World*. New York: Simon & Schuster.

Riesman, David, Nathan Glazer, and Reuel Denney. 1950. *The Lonely Crowd: A Study of Changing American Characteristics*. New Haven: Yale University Press.

Robbins, George W., and Leon D. Tilton. 1941. *A Preface to a Master Plan*. Los Angeles: Pacific Southwest Academy.

Robinson, William W. 1968. *Los Angeles: A Profile*. Norman: University of Oklahoma Press.

Roe, Mark. 1994. *Strong Managers, Weak Owners: The Political Roots of American Corporate Finance*. Princeton: Princeton University Press.

Romo, Ricardo. 1983. *East Los Angeles: History of a Barrio*. Austin: University of Texas Press.

Rose, Albert. 1972. *Governing Metropolitan Toronto: A Social and Political Analysis 1953–1971*. Berkeley: University of California Press.

Rose, Mark H. 1979. *Interstate: Expressway Highway Politics, 1941–1956*. Lawrence: University of Kansas Press.

Rosen, Christine M. 1986. *The Limits of Power: Great Fires and the Process of City Growth in America*. New York: Cambridge University Press.

Rosenberg, Nathan. 1982. *Inside the Black Box: Technology and Economies*. Cambridge: Cambridge University Press.

Rosenwaike, Ira. 1972. *Population History of New York*. Syracuse: Syracuse University Press.

Rosenzweig, Roy, and Elizabeth Blackmar. 1992. *Central Park: The Contradictions of a Democratic Public Space*. Ithaca: Cornell University Press.

Ross, David P., Richard Shillington, and Clare Lochhead. 1994. *The Canadian Fact Book on Poverty—1994*. Ottawa: Canadian Council on Social Development.

Rothblatt, Donald N. 1994. 'North American Metropolitan Planning: Canadian and US Perspectives'. *Journal of the American Planning Association* 60:501–20.

———, and Andrew Sancton. 1993. *Metropolitan Governance: American/Canadian Intergovernmental Perspectives*. Berkeley: Institute of Governmental Studies, University of California Press; Kingston: Institute of Intergovernmental Relations, Queens University.

Rothenberg, Winifred B. 1992. *From Market-Places to a Market Economy: The Transformation of Rural Massachusetts, 1750–1850*. Chicago: University of Chicago Press.

Rubenstein, James M. 1992. *The Changing US Auto Industry: A Geographical Analysis*. London: Routledge.

Rusk, David. 1993. *Cities Without Suburbs*. Washington: Woodrow Wilson Center Press.

Russell, Victor L., ed. 1984. *Forging a Consensus: Historical Essays on Toronto*. Toronto: University of Toronto Press.

Rutherford, Paul. 1994. *The New Icons: The Art of Television Advertising*. Toronto: University of Toronto Press.

Rutman, Darrett. 1965. *Winthrop's Boston: Portrait of a Puritan Town, 1630–1649*. Chapel Hill: University of North Carolina Press.

Ryan, Mary. 1990. *Women in Public: Between Banners and Ballots, 1825–1880*. Baltimore: Johns Hopkins University Press.

Salinger, Sharon. 1987. *'To Serve Well and Faithfully'*:

Labor and Indentured Servants in Pennsylvania, 1682–1800. Cambridge: Harvard University Press.

_____, and Charles Wetherell. 1985. 'Wealth and Poverty in Prevolutionary Philadelphia'. *Journal of American History* 71:826–40.

Saporito, Bill. 1994. 'The World's Best Cities for Business'. *Fortune* (14 November):112–42.

Sassen, Saskia. 1991. *The Global City: New York, London, Tokyo.* Princeton: Princeton University Press.

Savitch, H.V., and J.C. Thomas, eds. *Big City Politics in Transition.* Newbury Park, CA: Sage.

Savoie, Donald. 1990. *The Politics of Public Spending.* Toronto: University of Toronto Press.

Schafer, Robert. 1974. *The Suburbanization of Multi-Family Housing.* Lexington: Lexington Books.

Scharf, J. Thomas, and Thompson Westcott. 1884. *History of Philadelphia, 1609–1884,* 3 vols. Philadelphia: L.H. Everts.

Scherzer, Kenneth A. 1992. *Unbounded Communities: Neighborhood Life and Social Structure in New York City, 1830–1875.* Durham: Duke University Press.

Schiesl, Martin J. 1977. *The Politics of Efficiency: Municipal Administration and Reform in America 1880–1920.* Berkeley: University of California Press.

_____. 1980. 'City Planning in World War II'. *California History* 59:126–43.

Schlereth, Thomas J. 1991. *Victorian America: Transformations in Everyday Life 1876–1915.* New York: Harper Collins.

Schnore, Leo F., ed. 1975. *The New Urban History: Quantitative Explorations by American Historians.* Princeton: Princeton University Press.

Schor, Juliet. 1991. *The Overworked American: The Unexpected Decline of Leisure.* New York: Basic Books.

Schultz, Ronald. 1993. *The Republic of Labor: Philadelphia Artisans and the Politics of Class, 1720–1830.* New York: Oxford University Press.

Schultz, Stanley K. 1989. *American Cities and City Planning, 1800–1920.* Philadelphia: Temple University Press.

_____, and Clay McShane. 1978. 'To Engineer the Metropolis: Sewers, Sanitation and City Planning in Late Nineteenth-Century America'. *Journal of American History* 65:389–411.

Schumpeter, J.A. 1950. *Capitalism, Socialism, and Democracy,* 3rd ed. New York: Harper & Row.

Schurr, Sam, and Bruce C. Netschert. 1960. *Energy in the American Economy 1850–1975.* Baltimore: Johns Hopkins University Press.

Schuyler, David. 1986. *The New Urban Landscape: The Redefinition of City Form in Nineteenth-Century America.* Baltimore: Johns Hopkins University Press.

Schwartz, Barry, ed. 1970. *The Changing Face of the Suburbs.* Chicago: University of Chicago Press.

Schweitzer, Mary M. 1987. *Custom and Contract: Household, Government and the Economy of Colonial Pennsylvania.* New York: Columbia University Press.

_____. 1993. 'The Spatial Organization of Federalist Philadelphia, 1790'. *Journal of Interdisciplinary History* 24:31–57.

Scobey, David. 1994. 'Anatomy of the Promenade: The Politics of Bourgeois Sensibility in Nineteenth-Century New York'. *Social History* 17:203–27.

Scott, A.J. 1988. *Metropolis: From the Division of Labor to Urban Form.* Berkeley: University of California Press.

_____. 1991. 'The Aerospace-Electronics Industrial Complex of Southern California: The Formative Years 1940–1960'. *Research Policy* 20:439–56.

Scott, Mellier G. 1942. *Cities Are for People: The Los Angeles Region Plans for Living.* Los Angeles: Pacific Southwest Academy.

_____. 1949. *Metropolitan Los Angeles: One Community.* Los Angeles: Haynes Foundation.

_____. 1971. *American City Planning Since 1890.* Berkeley: University of California Press.

Scranton, Philip. 1983. *Proprietary Capitalism: The Textile Manufacture at Philadelphia, 1800–1885.* Cambridge: Cambridge University Press.

Seeley, John R., R. Alexander Sim, and B.W. Loosley. 1956. *Crestwood Heights.* New York: Basic Books.

Segal, Howard P. 1985. *Technological Utopianism in American Culture.* Chicago: University of Chicago Press.

Sellers, Charles. 1991. *The Market Revolution: Jacksonian America, 1815–1846.* New York: Oxford University Press.

Semple, R. Keith. 1973. 'Recent Trends in the Spatial Concentration of Corporate Headquarters'. *Economic Geography* 49:318.

Sennett, Richard. 1970. *Families Against the City: Middle-Class Homes of Industrial Chicago 1872–1890.* Cambridge: Harvard University Press.

Sewell, John. 1993. *The Shape of the City: Toronto Struggles with Modern Planning.* Toronto: University of Toronto Press.

_____. 1994. *Houses and Homes: Housing for Canadians.* Toronto: Lorimer.

Shammas, Carole. 1990. *The Pre-Industrial Consumer in England and America.* Oxford: Clarendon Press.

Shefter, Martin, ed. 1993. *Capital of the American Century: The National and International Influences of New York City.* New York: Sage.

Shevky, Eshref, and Marilyn Williams. 1949. *The Social*

Areas of Los Angeles. Los Angeles: University of California Press.

Siegel, Adrienne, comp. and ed. 1975. *Philadelphia: A Chronological and Documentary History, 1615–1970.* Dobbs Ferry, NY: Oceana Publications.

Simler, Lucy. 1990. 'The Landless Worker: An Index of Economic and Social Change in Chester County, Pennsylvania, 1750–1820'. *Pennsylvania Magazine of History and Biography* 114:163–99.

Simmons, James W. 1974. *The Growth of the Canadian Urban System.* Toronto: Centre for Urban and Community Studies.

_____. 1995. *International Comparisons of Commercial Structure,* Research Paper 2. Toronto: Centre for the Study of Commercial Activity, Ryerson Polytechnic University.

Sitton, Tom. 1992. *John Randolph Haynes: California Progressive.* Palo Alto: Stanford University Press.

Sjoberg, Gideon. 1960. *The Pre-industrial City.* Glencoe: Free Press.

Sklar, Martin J. 1988. *The Corporate Reconstruction of American Capitalism 1890–1916: The Market, Law and Politics.* New York: Cambridge University Press.

Skogan, Wesley. 1977. 'The Changing Distribution of Big City Crime'. *Urban Affairs Quarterly* 13:33–45.

Slayton, Robert A. 1986. *Back of the Yards: The Making of a Local Democracy.* Chicago: University of Chicago Press.

Smith, Adam. [1776] 1970. *The Wealth of Nations,* edited by Andrew Skinner. Harmondsworth: Pelican.

Smith, Billy G. 1990. *'The Lower Sort': Philadelphia's Laboring People, 1750–1800.* Ithaca: Cornell University Press.

Smith, David M. 1973. *The Geography of Social Well-Being in the United States: An Introduction to Territorial Social Indicators.* New York: McGraw-Hill.

_____. 1994. *Geography and Social Justice.* Oxford: Blackwell.

Smith, Henry Nash. 1950. *Virgin Land: The American West as Symbol and Myth.* New York: Vintage.

Smith, Neil, and Peter Williams, eds. 1986. *Gentrification of the City.* Boston: Allen & Unwin.

Smith, Wilson, ed. 1964. *Cities of Our Past and Present: A Descriptive Reader.* New York: Wiley.

So, F.S., and Judith Getzelo, eds. 1988. *The Practice of Local Government Planning,* 2nd ed. Washington: International City Management Association.

Soltow, Lee. 1975. *Men and Wealth in the United States, 1850–1870.* New Haven: Yale University Press.

Spann, Edward K. 1981. *The New Metropolis: New York City, 1840–1857.* New York: Columbia University Press.

Spear, Allan H. 1967. *Black Chicago: The Making of a Chicago Ghetto.* Chicago: University of Chicago Press.

Spera, Elizabeth Gray K. 1980. 'Building for Business: The Impact of Commerce on the City Plan and Architecture of the City of Philadelphia, 1750–1800'. PhD thesis, University of Pennsylvania.

Stansell, Christine. 1986. *City of Women: Sex and Class in New York, 1789–1860.* New York: Knopf.

Starr, Kevin. 1990. *Material Dreams: Southern California through the 1920s.* New York: Oxford University Press.

Statistics Canada. (various years) *Canada Yearbook.* Ottawa: Statistics Canada.

Steffens, Lincoln. [1904] 1957. *The Shame of the Cities.* New York: Hill and Wang.

Steiner, Rodney. 1981. *Los Angeles: The Centrifugal City.* Dubuque: Kendall/Hunt.

Stelter, Gilbert A., and Alan F.J. Artibise, eds. 1977. *The Canadian City: Essays in Urban History.* Toronto: McClelland and Stewart.

_____, eds. 1982. *Shaping the Urban Landscape: Aspects of the City-Building Process.* Ottawa: Carleton University Press.

_____, eds. 1986. *Power and Place: Canadian Urban Development in the North American Context.* Vancouver: UBC Press.

Sternlieb, George, and Robert W. Burchall. 1973. *Residential Abandonment: The Tenement Landlord Revisited.* New Brunswick, NJ: Rutgers University Press.

Sternlieb, George, et al. 1979. 'Housing Abandonment in the Urban Core'. *Journal of the American Institute of Planners* 40:321–32.

Stilgoe, John R. 1988. *Borderland: Origins of the American Suburb 1820–1939.* New Haven: Yale University Press.

Still, Bayrd. 1956. *Mirror for Gotham: New York as Seen by Contemporaries from Dutch Days to the Present.* New York: New York University Press.

Stinson, Grace H. 1955. *Rise of the Labor Movement in Los Angeles.* Berkeley: University of California Press.

Stokes, I.M. Phelps. 1915–28. *The Iconography of Manhattan Island, 1498–1909.* New York: R.H. Dodd.

Stone, Leroy O. 1967. *Urban Development in Canada: An Introduction to the Demographic Aspects.* Ottawa: Dominion Bureau of Statistics.

Stott, Richard B. 1990. *Workers in the Metropolis: Class, Ethnicity, and Youth in Antebellum New York.* Ithaca: Cornell University Press.

Struthers, James. 1995. *The Limits of Affluence: Welfare in Ontario, 1920–1970.* Toronto: University of Toronto Press.

Susman, Warren I. 1984. *Culture as History: The Trans-*

formation of American Society in the Twentieth Century. New York: Pantheon.

Tarr, Joel. 1971. *A Study of Boss Politics: William A. Lorimer of Chicago.* Urbana: University of Illinois Press.

————, and Gabriel Dupuy, eds. 1988. *Technology and the Rise of the Networked City in Europe and America.* Philadelphia: Temple University Press.

Taylor, George Rogers. 1951. *The Transportation Revolution, 1815–1860.* New York: Holt Rinehart and Winston.

Taylor, Graham R. [1915] 1970. *Satellite Cities: A Study of Industrial Suburbs.* New York: Arno.

Taylor, William R., ed. 1991. *Inventing Times Square.* New York: Sage Foundation.

Teaford, Jon C. 1975. *The Municipal Revolution in America: Origins of Modern Urban Government, 1650–1825.* Chicago: University of Chicago Press.

————. 1979. *City and Suburb: The Political Fragmentation of Metropolitan America.* Baltimore: Johns Hopkins University Press.

————. 1984. *The Unheralded Triumph: City Government in America, 1870–1900.* Baltimore: Johns Hopkins University Press.

————. 1986. *The Twentieth-Century American City: Problem, Promise and Reality.* Baltimore: Johns Hopkins University Press.

————. 1990. *The Rough Road to Renaissance: Urban Revitalization in America 1940–1985.* Baltimore: Johns Hopkins University Press.

————. 1993. *Cities of the Heartland: The Rise and Fall of the Industrial Midwest.* Bloomington: Indiana University Press.

Tebbel, John, and Mary Ellen Zuckerman. 1991. *The Magazine in America, 1741–1990.* New York: Oxford University Press.

Temin, Peter. 1969. *The Jacksonian Economy.* New York: Norton.

Thernstrom, Stephen. 1973. *The Other Bostonians: Poverty and Progress in the American Metropolis, 1880–1970.* Cambridge: Harvard University Press.

Thomas, Dorothy S., and K.C. Zachariah. 1961. 'Some Temporal Variations in Internal Migration and Economic Activity, United States 1880–1950'. *International Population Conference, 1961*, vol. 1, 525–32. London: International Council for Scientific Theory of Population.

Thomas, William L., et al. 1956. *Man's Role in Changing the Face of the Earth.* Chicago: University of Chicago Press.

Thompson, Peter John. 1989. 'A Social History of Philadelphia Taverns, 1603–1800'. PhD thesis, University of Pennsylvania.

Thompson, Richard H. 1989. *Toronto's Chinatown: The Changing Social Organization of an Ethnic Community.* New York: AMS Press.

Thompson, Robert Luther. 1947. *Wiring a Continent: The History of the Telegraph Industry in the United States 1832–1866.* Princeton: Princeton University Press.

Thompson, Warren S. 1955. *Growth and Change in California's Population.* Los Angeles: Haynes Foundation.

Thurow, Lester C. 1976. *Generating Inequality: Mechanics of Distribution in the US Economy.* New York: Basic Books.

Tiebout, Charles. 1956. 'A Pure Theory of Local Expenditures'. *Journal of Political Economy* 64:416–24.

Tocqueville, Alexis de. [1835, 1840] 1954. *Democracy in America*, 2 vols., edited by P. Bradley. New York: Vintage.

Toll, Seymour. 1969. *Zoned American.* New York: Grossman.

Tolles, Frederick B. 1948. *Meeting House and Counting House.* New York: Norton.

Tomlins, Christopher L. 1985. *The State and the Unions: Labor Relations, Law and the Organized Labor Movement in America, 1880–1960.* Cambridge: Cambridge University Press.

————. 1993. *Law, Labor, and Ideology in the Early American Republic.* Cambridge: Cambridge University Press.

Turner, Jonathan, and Charles Staines. 1976. *Inequality: Privilege and Poverty in America.* Pacific Palisades: Goodyear.

Urbanomics Research Associates. 1969. *Southern California Economy: Trends 1940–1970, Current Issues and Recommended Goals for the Region.* Prepared for the Southern California Association of Government. Claremont: Urbanomics Research Associates.

US Bureau of Census. 1975. *Historical Statistics of the United States: From Colonial Times to 1970*, 2nd ed. Washington: US Government Printing Office.

————. (various years) *Statistical Abstracts.* Washington: US Department of Commerce.

Useem, Michael. 1984. *The Inner Circle: Large Corporations and the Rise of Business Political Activity in the United States and United Kingdom.* New York: Oxford University Press.

US Housing & Home Finance Agency, Office of Housing Economics. 1948. *Housing Statistics Handbook.* Washington: US Housing & Home Finance Agency.

Vance, James E., Jr. 1990. *The Continuing City: Urban Morphology in Western Civilization.* Baltimore: Johns

Hopkins University Press.

Vickers, Daniel. 1994. *Farmers and Fishermen: Two Centuries of Work in Essex County, Massachusetts, 1630–1850*. Chapel Hill: University of North Carolina Press for the Institute of Early American History and Culture.

Viehe, Fred. 1991. 'The Social-Spatial Distribution in the Black Gold Suburbs of Los Angeles, 1900–1930'. *Southern California Quarterly* 73:35–54.

Vietor, Richard H.K. 1984. *Energy Policy in America Since 1945*. New York: Cambridge University Press.

Wachs, Martin. 1984. 'Automobiles, Transport, and the Sprawl of Los Angeles'. *Journal of the American Planning Association* 50:297–310.

Wade, Louise Carroll. 1987. *Chicago's Pride: The Stockyards, Packingtown and Environs in the Nineteenth Century*. Urbana: University of Illinois Press.

Wade, Richard. 1967. *The Urban Frontier: The Rise of Western Cities, 1790–1830*. Cambridge: Harvard University Press.

Wakstein, Allen M. 1970. *The Urbanization of America: An Historical Anthology*. Boston: Houghton Mifflin.

Walker, Robert A. 1950. *The Planning Function in Urban Government*, 2nd ed. Chicago: University of Chicago Press.

Wallach, Leonard. 1991. 'The Myth of the Master Builder: Robert Moses, New York and the Dynamics of Metropolitan Development Since World War II'. *Journal of Urban History* 17:339–62.

Wallerstein, Immanuel. 1974–80. *The Modern World in Systems,* 3 vols. New York: Academic Press.

Walton, Gary M., and James F. Shepherd. 1979. *The Economic Rise of Early America*. Cambridge: Cambridge University Press.

Ward, David. 1971. *Cities and Immigrants: A Geography of Change in Nineteenth-Century America*. New York: Oxford University Press.

_____. 1989. *Poverty, Ethnicity, and the American City 1840–1925*. Cambridge: Cambridge University Press.

_____. 1990. 'Social Reform, Social Surveys, and the Discovery of the Modern City'. *Annals of the Association of American Geographers* 80:491–503.

Ward, Sally K. 1994. 'Trends in the Location of Corporate Headquarters, 1969–1989'. *Urban Affairs Quarterly* 29:468–78.

Warner, Sam Bass, Jr. 1962. *Streetcar Suburbs: The Process of Growth in Boston 1870–1900*. Cambridge: Harvard University Press/MIT Press.

_____. 1968. *The Private City: Philadelphia in Three Periods of Its Growth*. Philadelphia: University of Pennsylvania Press.

_____. [1972] 1995. *The Urban Wilderness: A History of the American City*. Berkeley: University of California Press.

Weber, Adna. [1899] 1963. *The Growth of Cities in the Nineteenth Century: A Study in Statistics*. Ithaca: Cornell University Press.

Weigley, Russell F., ed. 1982. *Philadelphia: A 300 Year History*. New York: Norton.

Weiler, N. Sue. 1979–80. 'Walkout: The Chicago Men's Garment Workers Strike 1910–1911'. *Chicago History* 8:238–49.

Weimer, David R. 1966. *The City as Metaphor.* New York: Random House.

Weiss, Marc A. 1987. *The Rise of the Community Builders: The American Real Estate Industry and Urban Land Planning*. New York: Columbia University Press.

Wendt, Lloyd, and Herman Kogan. 1952. *Give the Lady What She Wants: The Story of Marshall Field and Company*. Chicago: Rand & McNally.

_____. 1967. *Bosses in Lusty Chicago: The Story of Bathhouse John and Hinky Dink*. Bloomington: Indiana University Press.

Werner, Walter, and Steven T. Smith. 1991. *Wall Street.* New York: Columbia University Press.

Wilentz, Sean. 1984. *Chants Democratic: New York City and the Rise of the American Working Class, 1788–1850*. New York: Oxford University Press.

Wilkins, Mira. 1970. *The Emergence of Multinational Enterprise: American Business Abroad from the Colonial Era to 1914*. Cambridge: Harvard University Press.

Williamson, Jeffrey G., and Peter H. Lindert. 1980. *American Inequality: A Macroeconomic History*. New York: Academic Press.

Wills. Garry. 1995. 'The Visonary Struggle in the House'. *New York Review of Books* 42 (23 March):6–8.

Wilson, James Q., ed. 1966. *Urban Renewal: The Record and the Controversy*. Cambridge: MIT Press.

Wilson, William H. 1989. *The City Beautiful Movement.* Baltimore: Johns Hopkins University Press.

Wilson, William Julius. 1978. *The Declining Significance of Race: Blacks and Changing American Institutions*. Chicago: University of Chicago Press.

_____. 1987. *The Truly Disadvantaged: The Inner City, the Underclass, and Public Policy*. Chicago: University of Chicago Press.

Wolf, Edwin, and Kenneth Finkel. 1990. *Philadelphia: Portrait of an American City*, 2nd ed. Philadelphia: Canio Books/Library Company of Philadelphia.

Wolf, Stephanie G. 1976. *Urban Village: Population, Community, and Family Structure in Germantown, Pennsylvania, 1683–1800*. Princeton: Princeton University

Press.

Woodbury, Coleman. 1931. *Apartment House Increases and Attitudes toward Home Ownership*. Chicago: Chicago Institute for Economic Renewal.

World Commission on Environment and Development. 1987. *Our Common Future* (Brundtland Report). Oxford: Oxford University Press.

Wright, Gwendolyn. 1980. *Moralism and the Modern Home: Domestic Architecture and Cultural Conflict in Chicago 1873–1913*. Chicago: University of Chicago Press.

Wrightson, Keith. 1982. *English Society, 1580–1680*. New Brunswick, NJ: Rutgers University Press.

Yates, JoAnne. 1989. *Control through Communication: The Rise of Systems in American Management*. Baltimore: Johns Hopkins University Press.

Zorbaugh, Harvey W. 1929. *The Gold Coast and the Slum*. Chicago: University of Chicago Press.

Zueblin, Charles. 1916. *American Municipal Progress*, 2nd ed. New York: Macmillan.

Zunz, Olivier. 1990. *Making America Corporate 1870–1920*. Chicago: University of Chicago Press.

INDEX